VOLUME SIX HUNDRED AND THIRTY TWO

METHODS IN ENZYMOLOGY

Tumor Immunology and Immunotherapy – Cellular Methods Part B

METHODS IN ENZYMOLOGY

Editors-in-Chief

ANNA MARIE PYLE
Departments of Molecular, Cellular and Developmental Biology and Department of Chemistry
Investigator, Howard Hughes Medical Institute
Yale University

DAVID W. CHRISTIANSON
Roy and Diana Vagelos Laboratories
Department of Chemistry
University of Pennsylvania
Philadelphia, PA

Founding Editors

SIDNEY P. COLOWICK and NATHAN O. KAPLAN

VOLUME SIX HUNDRED AND THIRTY TWO

METHODS IN ENZYMOLOGY

Tumor Immunology and Immunotherapy – Cellular Methods Part B

Edited by

LORENZO GALLUZZI
Department of Radiation Oncology, Weill Cornell Medical College, Sandra and Edward Meyer Cancer Center, Caryl and Israel Englander Institute for Precision Medicine New York, NY; Department of Dermatology, Yale University School of Medicine, New Haven, CT, United States; Université de Paris, Paris, France

NILS-PETTER RUDQVIST
Department of Radiation Oncology, Weill Cornell Medical College, New York, NY, United States

ELSEVIER

ACADEMIC PRESS
An imprint of Elsevier

Academic Press is an imprint of Elsevier
50 Hampshire Street, 5th Floor, Cambridge, MA 02139, United States
525 B Street, Suite 1650, San Diego, CA 92101, United States
The Boulevard, Langford Lane, Kidlington, Oxford OX5 1GB, United Kingdom
125 London Wall, London, EC2Y 5AS, United Kingdom

First edition 2020

ISBN: 978-0-12-818675-6
ISSN: 0076-6879

For information on all Academic Press publications
visit our website at https://www.elsevier.com/books-and-journals

Publisher: Zoe Kruze
Acquisition Editor: Ashlie M. Jackman
Editorial Project Manager: Shellie Bryant
Production Project Manager: James Selvam
Cover Designer: Greg Harris

Typeset by SPi Global, India

Contents

Contributors

Mahmoud S. Alghamri
Department of Neurosurgery; Department of Cell and Developmental Biology, University of Michigan Medical School, MSRB II, Ann Arbor, MI, United States

David Altshuler
Department of Neurosurgery; Department of Cell and Developmental Biology, University of Michigan Medical School, MSRB II, Ann Arbor, MI, United States

Giada Amodio
San Raffaele Telethon Institute for Gene Therapy (SR-Tiget), San Raffaele Scientific Institute IRCCS, Milan, Italy

Niroshana Anandasabapathy
Department of Dermatology; Meyer Cancer Center, Weill Cornell Medical College, New York, NY, United States

Romina E. Araya
Inflammatory Cell Dynamics Section, Cancer and Inflammation Program, Center for Cancer Research, National Cancer Institute, NIH, Bethesda, MD, United States

Jirina Bartunkova
Department of Immunology, Charles University, 2nd Faculty of Medicine and University Hospital Motol; Sotio, Prague, Czech Republic

Ina Becker
Department of Radiation Oncology, Universitätsklinikum Erlangen, Friedrich-Alexander-Universität Erlangen-Nürnberg, Erlangen, Germany

Martina Borghi
Department of Infectious Diseases, Istituto Superiore di Sanità, Rome, Italy

Natalia K. Botelho
Department of Fundamental Oncology, Faculty of Biology and Medicine, University of Lausanne, Epalinges, Switzerland

Laura Bracci
Department of Oncology and Molecular Medicine, Istituto Superiore di Sanità, Rome, Italy

Sven Brandau
Research Division, Department of Otorhinolaryngology, University Hospital Essen, West German Cancer Center, Essen, Germany

Vincenzo Bronte
Department of Medicine, Section of Immunology, University of Verona, Verona, Italy

Demi Brownlie
MRC Centre for Reproductive Health, Queen's Medical Research Institute, The University of Edinburgh, Edinburgh, United Kingdom

Kirsten Bruderek
Research Division, Department of Otorhinolaryngology, University Hospital Essen, West German Cancer Center, Essen, Germany

Annika M. Bruger
de Duve Institute, Université catholique de Louvain, Brussels, Belgium

Kyle Burrows
Department of Immunology, University of Toronto, Toronto, ON, Canada

Luca Businaro
Institute for Photonics and Nanotechnology, Italian National Research Council, Rome, Italy

Yonina Bykov
Ludwig Collaborative Laboratory, Memorial Sloan Kettering Cancer Center, New York, NY, United States

Stefania Canè
Department of Medicine, Section of Immunology, University of Verona, Verona, Italy

Luca Cassetta
MRC Centre for Reproductive Health, Queen's Medical Research Institute, The University of Edinburgh, Edinburgh, United Kingdom

Maria Graciela Castro
Department of Neurosurgery; Department of Cell and Developmental Biology, University of Michigan Medical School, MSRB II, Ann Arbor, MI, United States

Giulia Cerrato
Metabolomics and Cell Biology Platforms, Gustave Roussy Cancer Campus, Villejuif; INSERM, U1138; Equipe 11 labellisée par la Ligue Nationale contre le Cancer, Centre de Recherche des Cordeliers; Université Paris Descartes/Paris V, Sorbonne Paris Cité, Paris, France

Anne-Marie Chalupova
Sotio, Prague, Czech Republic

Pailin Chiaranunt
Department of Immunology, University of Toronto, Toronto, ON, Canada

Maria Ciccolella
Department of Oncology and Molecular Medicine, Istituto Superiore di Sanità, Rome, Italy

Andrea Comba
Department of Neurosurgery; Department of Cell and Developmental Biology, University of Michigan Medical School, MSRB II, Ann Arbor, MI, United States

Stéphanie Corgnac
Department of Fundamental Oncology, Faculty of Biology and Medicine, University of Lausanne, Epalinges, Switzerland; INSERM UMR 1186, Integrative Tumor Immunology and Genetic Oncology, Gustave Roussy, EPHE, PSL, Fac. de Médecine—Univ. Paris-Sud, Université Paris-Saclay, Villejuif, France

Aurélien Corneau
Sorbonne-Université, UMS037 PASS, Plateforme de Cytometrie CyPS, Faculté des Sciences Sorbonne-Université, Paris, France

Isabelle Cremer
Team Inflammation, Complement and Cancer, INSERM, Centre de Recherche des Cordeliers; Sorbonne Université, Paris, France

Adele De Ninno
Institute for Photonics and Nanotechnology, Italian National Research Council; Department of Civil Engineering and Computer Science, University of Rome Tor Vergata, Rome, Italy

Karin E. de Visser
Division of Tumor Biology & Immunology, Oncode Institute, Netherlands Cancer Institute, Amsterdam, The Netherlands

Anna-Jasmina Donaubauer
Department of Radiation Oncology, Universitätsklinikum Erlangen, Friedrich-Alexander-Universität Erlangen-Nürnberg, Erlangen, Germany

Alena Donda
Department of Fundamental Oncology, Faculty of Biology and Medicine, University of Lausanne, Epalinges, Switzerland

Danique E.M. Duits
Division of Tumor Biology & Immunology, Oncode Institute, Netherlands Cancer Institute, Amsterdam, The Netherlands

Günes Esendagli
Department of Basic Oncology, Hacettepe University Cancer Institute, Ankara, Turkey

Rainer Fietkau
Department of Radiation Oncology, Universitätsklinikum Erlangen, Friedrich-Alexander-Universität Erlangen-Nürnberg, Erlangen, Germany

Olivera J. Finn
Department of Immunology, University of Pittsburgh School of Medicine, Pittsburgh, PA, United States

Lesley M. Forrester
MRC Centre for Regenerative Medicine, University of Edinburgh, Edinburgh, United Kingdom

Benjamin Frey
Department of Radiation Oncology, Universitätsklinikum Erlangen, Friedrich-Alexander-Universität Erlangen-Nürnberg, Erlangen, Germany

Jitka Fucikova
Department of Immunology, Charles University, 2nd Faculty of Medicine and University Hospital Motol; Sotio, Prague, Czech Republic

Udo S. Gaipl
Department of Radiation Oncology, Universitätsklinikum Erlangen, Friedrich-Alexander-Universität Erlangen-Nürnberg, Erlangen, Germany

Claudia Galassi
Istituto di Patologia Generale, Università Cattolica del Sacro Cuore, Rome, Italy

Jérôme Galon
INSERM, Laboratory of Integrative Cancer Immunology, Équipe Labellisée Ligue Contre le Cancer, Sorbonne Université, Sorbonne Paris Cité, Université Paris Descartes, Université Paris Diderot, Centre de Recherche des Cordeliers, Paris, France

Maria Belen Garcia-Fabiani
Department of Neurosurgery; Department of Cell and Developmental Biology, University of Michigan Medical School, MSRB II, Ann Arbor, MI, United States

Romina S. Goldszmid
Inflammatory Cell Dynamics Section, Cancer and Inflammation Program, Center for Cancer Research, National Cancer Institute, NIH, Bethesda, MD, United States

Silvia Gregori
San Raffaele Telethon Institute for Gene Therapy (SR-Tiget), San Raffaele Scientific Institute IRCCS, Milan, Italy

Alfonso Grimaldi
Center for Life Nano Science @ Sapienza, Istituto Italiano di Tecnologia, Rome, Italy

Esra Güç
MRC Centre for Reproductive Health, Queen's Medical Research Institute, The University of Edinburgh, Edinburgh, United Kingdom

Santiago Haase
Department of Neurosurgery; Department of Cell and Developmental Biology, University of Michigan Medical School, MSRB II, Ann Arbor, MI, United States

Esther Heideveld
MRC Centre for Reproductive Health, Queen's Medical Research Institute, The University of Edinburgh, Edinburgh, United Kingdom

Marta Horcas-Lopez
MRC Centre for Reproductive Health, Queen's Medical Research Institute, The University of Edinburgh, Edinburgh, United Kingdom

Qingrong Huang
Department of Dermatology; Meyer Cancer Center, Weill Cornell Medical College, New York, NY, United States

Padma Kadiyala
Department of Neurosurgery; Department of Cell and Developmental Biology, University of Michigan Medical School, MSRB II, Ann Arbor, MI, United States

Neha Kamran
Department of Neurosurgery; Department of Cell and Developmental Biology, University of Michigan Medical School, MSRB II, Ann Arbor, MI, United States

Lenka Kasikova
Department of Immunology, Charles University, 2nd Faculty of Medicine and University Hospital Motol; Sotio, Prague, Czech Republic

Oliver Kepp
Faculty of Medicine, University of Paris Sud, Kremlin-Bicêtre; Cell Biology and Metabolomics Platforms, Gustave Roussy Cancer Campus, Villejuif; Equipe 11 labellisée par la Ligue Nationale contre le Cancer, Centre de Recherche des Cordeliers; Institut National de la Santé et de la Recherche Médicale (INSERM), UMR1138, Equipe labellisée Ligue Nationale Contre le Cancer; Centre de Recherche des Cordeliers, Equipe labellisée par la Ligue contre le cancer, Université de Paris Descartes, Université Sorbonne Paris Cité, Université Paris Diderot, Sorbonne Université, INSERM; Pôle de Biologie, Hôpital Européen Georges Pompidou, AP-HP; Université Paris Descartes/Paris V, Sorbonne Paris Cité; Université Pierre et Marie Curie/Paris VI, Paris, France

Sarah H. Kim
Department of Surgery, Memorial Sloan Kettering Cancer Center, New York, NY, United States

Takanori Kitamura
MRC Centre for Reproductive Health, Queen's Medical Research Institute, The University of Edinburgh, Edinburgh, United Kingdom

Kevin Kos
Division of Tumor Biology & Immunology, Oncode Institute, Netherlands Cancer Institute, Amsterdam, The Netherlands

Guido Kroemer
Metabolomics and Cell Biology Platforms, Gustave Roussy Cancer Campus, Villejuif; Faculté de Medicine, Université Paris Saclay/Paris XI, Le Kremlin-Bicêtre; Equipe 11 labellisée par la Ligue Nationale contre le Cancer, Centre de Recherche des Cordeliers; Institut National de la Santé et de la Recherche Médicale (INSERM), UMR1138, Equipe labellisée Ligue Nationale Contre le Cancer; Pôle de Biologie, Hôpital Européen Georges Pompidou, AP-HP; Centre de Recherche des Cordeliers, Equipe labellisée par la Ligue contre le cancer, Université de Paris Descartes, Université Sorbonne Paris Cité, Université Paris Diderot, Sorbonne Université, INSERM; Université Paris Descartes/Paris V, Sorbonne Paris Cité; Université Pierre et Marie Curie/Paris VI, Paris, France; Suzhou Institute for Systems Biology, Chinese Academy of Sciences; Suzhou Institute for Systems Medicine, Chinese Academy of Medical Sciences, Suzhou, China; Department of Women's and Children's Health, Karolinska Institute, Karolinska University Hospital, Stockholm, Sweden

Peng Liu
Faculty of Medicine, University of Paris Sud, Kremlin-Bicêtre; Cell Biology and Metabolomics Platforms, Gustave Roussy Cancer Campus, Villejuif; Equipe 11 labellisée par la Ligue Nationale contre le Cancer, Centre de Recherche des Cordeliers; Institut National de la Santé et de la Recherche Médicale (INSERM), UMR1138, Equipe labellisée Ligue Nationale Contre le Cancer; Université Pierre et Marie Curie; Université Paris Descartes/Paris V, Sorbonne Paris Cité, Paris, France

Martha Lopez-Yrigoyen
MRC Centre for Regenerative Medicine, University of Edinburgh, Edinburgh, United Kingdom

Pedro Ricardo Lowenstein
Department of Neurosurgery; Department of Cell and Developmental Biology, University of Michigan Medical School, MSRB II, Ann Arbor, MI, United States

Pauline Maby
INSERM, Laboratory of Integrative Cancer Immunology, Équipe Labellisée Ligue Contre le Cancer, Sorbonne Université, Sorbonne Paris Cité, Université Paris Descartes, Université Paris Diderot, Centre de Recherche des Cordeliers, Paris, France

Jacopo Mancini
Department of Oncology and Molecular Medicine, Tumor Immunology Unit, Istituto Superiore di Sanità, Rome, Italy

Gwenola Manic
Unit of Tumor Immunology and Immunotherapy, Department of Research, Advanced Diagnostics, and Technological Innovation, IRCCS Regina Elena National Cancer Institute, Rome, Italy

Eugenio Martinelli
Department of Electronic Engineering, University of Rome Tor Vergata, Rome, Italy

Amaia Martinez-Usatorre
Department of Oncology UNIL-CHUV, Faculty of Biology and Medicine, University of Lausanne, Epalinges, Switzerland

Isabelle Martins
Metabolomics and Cell Biology Platforms, Gustave Roussy Cancer Campus, Villejuif; INSERM, U1138; Equipe 11 labellisée par la Ligue Nationale contre le Cancer, Centre de Recherche des Cordeliers; Université Paris Descartes/Paris V, Sorbonne Paris Cité; Université Pierre et Marie Curie/Paris VI, Paris, France

Fabrizio Mattei
Department of Oncology and Molecular Medicine, Tumor Immunology Unit, Istituto Superiore di Sanità, Rome, Italy

Denize A. Meijer
Division of Tumor Biology & Immunology, Oncode Institute, Netherlands Cancer Institute, Amsterdam, The Netherlands

Arianna Mencattini
Department of Electronic Engineering, University of Rome Tor Vergata, Rome, Italy

Michelle L. Miller
Department of Immunology, University of Pittsburgh School of Medicine, Pittsburgh, PA, United States

Arthur Mortha
Department of Immunology, University of Toronto, Toronto, ON, Canada

Martina Musella
Unit of Tumor Immunology and Immunotherapy, Department of Research, Advanced Diagnostics, and Technological Innovation, IRCCS Regina Elena National Cancer Institute; Department of Molecular Medicine, University "La Sapienza", Rome, Italy

Donatella Negri
Department of Infectious Diseases, Istituto Superiore di Sanità, Rome, Italy

Louis Ngai
Department of Immunology, University of Toronto, Toronto, ON, Canada

Felipe Javier Núñez
Department of Neurosurgery; Department of Cell and Developmental Biology, University of Michigan Medical School, MSRB II, Ann Arbor, MI, United States

Ilenia Pacella
Dipartimento di Medicina Interna e Specialità Mediche, Sapienza Università di Roma, Rome, Italy

Lenka Palova-Jelinkova
Department of Immunology, Charles University, 2nd Faculty of Medicine and University Hospital Motol; Sotio, Prague, Czech Republic

Silvia Piconese
Dipartimento di Medicina Interna e Specialità Mediche, Sapienza Università di Roma; Istituto Pasteur Italia—Fondazione Cenci Bolognetti, Rome, Italy

Jeffrey W. Pollard
MRC Centre for Reproductive Health, Queen's Medical Research Institute, The University of Edinburgh, Edinburgh, United Kingdom; Department of Developmental and Molecular Biology, Albert Einstein College of Medicine, New York, NY, United States

Carolina Rodriguez-Tirado
Department of Developmental and Molecular Biology, Albert Einstein College of Medicine, New York, NY, United States

Pedro Romero
Department of Oncology UNIL-CHUV; Department of Fundamental Oncology, Faculty of Biology and Medicine, University of Lausanne, Epalinges, Switzerland

Daniela Rozkova
Sotio, Prague, Czech Republic

Paul F. Rühle
Department of Radiation Oncology, Universitätsklinikum Erlangen, Friedrich-Alexander-Universität Erlangen-Nürnberg, Erlangen, Germany

Catherine Sautes-Fridman
Team Inflammation, Complement and Cancer, INSERM, Centre de Recherche des Cordeliers; Sorbonne Université, Paris, France

Giovanna Schiavoni
Department of Oncology and Molecular Medicine, Tumor Immunology Unit, Istituto Superiore di Sanità, Rome, Italy

Paola Sestili
National Centre for the Control and the Evaluation of Medicines, Istituto Superiore di Sanità, Rome, Italy

Antonella Sistigu
Unit of Tumor Immunology and Immunotherapy, Department of Research, Advanced Diagnostics, and Technological Innovation, IRCCS Regina Elena National Cancer Institute; Istituto di Patologia Generale, Università Cattolica del Sacro Cuore, Rome, Italy

Klara Sochorova
Sotio, Prague, Czech Republic

Radek Spisek
Department of Immunology, Charles University, 2nd Faculty of Medicine and University Hospital Motol; Sotio, Prague, Czech Republic

Agnieszka Swierczak
MRC Centre for Reproductive Health, Queen's Medical Research Institute, The University of Edinburgh, Edinburgh, United Kingdom

Ece Tavukçuoğlu
Department of Basic Oncology, Hacettepe University Cancer Institute, Ankara, Turkey

Iva Truxova
Department of Immunology, Charles University, 2nd Faculty of Medicine and University Hospital Motol; Sotio, Prague, Czech Republic

Martijn van Baalen
Flow Cytometry Facility, Netherlands Cancer Institute, Amsterdam, The Netherlands

Pierre van der Bruggen
de Duve Institute, Université catholique de Louvain; WELBIO, Brussels, Belgium

Christophe Vanhaver
de Duve Institute, Université catholique de Louvain, Brussels, Belgium

Ilio Vitale
Italian Institute for Genomic Medicine (IIGM), Turin; Candiolo Cancer Institute—FPO, IRCCS, Candiolo, Italy

Sarka Vosahlikova
Sotio, Prague, Czech Republic

Max D. Wellenstein
Division of Tumor Biology & Immunology, Oncode Institute, Netherlands Cancer Institute, Amsterdam, The Netherlands

Dmitriy Zamarin
Ludwig Collaborative Laboratory; Department of Medicine; Parker Institute for Cancer Immunotherapy, Memorial Sloan Kettering Cancer Center; Department of Medicine, Weill Cornell Medical College, New York, NY, United States

Liwei Zhao
Faculty of Medicine, University of Paris Sud, Kremlin-Bicêtre; Cell Biology and Metabolomics Platforms, Gustave Roussy Cancer Campus, Villejuif; Equipe 11 labellisée Ligue Nationale contre le Cancer, Centre de Recherche des Cordeliers; Institut National de la Santé et de la Recherche Médicale (INSERM), UMR1138, Equipe labellisée Ligue Nationale Contre le Cancer; Université Paris Descartes, Sorbonne Paris Cité; Université Pierre et Marie Curie, Paris, France

Preface: More than two decades of modern tumor immunology

Lorenzo Galluzzi[a,b,c,d,e,*], Nils-Petter Rudqvist[a,*]
[a]Department of Radiation Oncology, Weill Cornell Medical College, New York, NY, United States
[b]Sandra and Edward Meyer Cancer Center, New York, NY, United States
[c]Caryl and Israel Englander Institute for Precision Medicine, New York, NY, United States
[d]Department of Dermatology, Yale University School of Medicine, New Haven, CT, United States
[e]Université de Paris, Paris, France
*Corresponding authors: e-mail address: deadoc80@gmail.com; nils.rudqvist@gmail.com

The roots of modern tumor immunology extend back to the pioneering work of William B. Coley, an American surgeon who—at the end of the 19th century—was the first to successfully use a partially inactivated bacterial preparation (the so-called "Coley toxin") for treating soft tissue sarcoma (Coley, 1893), and the revolutionary proposals of Paul Ehlrich, a German-Jewish physician who—at the beginning of the 20th century—introduced for the first time in history the concept of cancer immunosurveillance (Ehrlich, 1909; Rao, Gharib, & Han, 2019). However, with the exception of punctual episodes including the establishment of the Cancer Research Institute of New York (in 1957), whose specific purpose was the development of immunological approaches to treat cancer (Decker et al., 2017), and the publication of the first cancer vaccine study (in 1959) (Graham & Graham, 1959), the field remained virtually unborn until the 1990s. Indeed, the 1990s witnessed not only the discovery that cytotoxic T lymphocyte associated protein 4 (CTLA4) mediates robust immunosuppressive functions, hence constituting a potential target for cancer therapy (Krummel & Allison, 1995; Leach, Krummel, & Allison, 1996), but also the realization that oncogenesis occurs in the context of a bidirectional crosstalk between malignant cells and the immune system of the host (Dighe, Richards, Old, & Schreiber, 1994; Kaplan et al., 1998), laying the foundation for the modern cancer immunosurveillance and immunoediting theories (Dunn, Bruce, Ikeda, Old, & Schreiber, 2002; Dunn, Old, & Schreiber, 2004). In the 2000s, the field gathered additional momentum from preclinical and clinical findings elucidating multiple mechanisms that underlie the ability of the immune system to control developing tumors and the capacity of progressing neoplasms to evade immunosurveillance (Bargou et al., 2008; Dudley et al., 2002; Galon et al., 2006;

Hildner et al., 2008; Jaiswal et al., 2009; Koebel et al., 2007; Majeti et al., 2009; Morgan et al., 2006; Savage et al., 2008; Shankaran et al., 2001), as well as from the discovery that the efficacy of multiple conventional anticancer agents relies (at least in part) on the (re-)activation of tumor-targeting immune responses as a consequence of immunogenic cell death (ICD) induction (Bezu et al., 2018; Casares et al., 2005; Galluzzi, Bravo-San Pedro, Demaria, Formenti, & Kroemer, 2017; Galluzzi, Buque, Kepp, Zitvogel, & Kroemer, 2015, 2017; Galluzzi, Kepp, Chan, & Kroemer, 2017; Galluzzi, Vitale, et al., 2018; Galluzzi, Zitvogel, & Kroemer, 2016; Obeid et al., 2007; Pol et al., 2015). These data paved the way to the cancer immunotherapy revolution of the 2010s (Kelly, 2018).

The 2010s represent indeed the decade in which modern tumor immunology revolutionized the clinical management of multiple tumor types upon the approval of no less than 12 different immunotherapeutic agents for use in cancer patients by the US Food and Drug Administration (FDA) and various equivalent regulatory agencies worldwide (Galluzzi, Chan, Kroemer, Wolchok, & Lopez-Soto, 2018). In January 2019, when this Preface was being redacted, these agents included: (1) sipuleucel T, a myeloid cell-derived product originally approved for the treatment of asymptomatic or minimally symptomatic metastatic castration resistant (hormone refractory) prostate cancer in April 2010 (Hagihara et al., 2019; Kantoff et al., 2010); (2) ipilimumab, a CTLA4-targeting immune checkpoint inhibitor (ICI) (Aranda et al., 2014) initially licensed for use in patients with unresectable or metastatic melanoma who received at least one prior systemic treatment in March 2011 (Hodi et al., 2010; McDermott, Haanen, Chen, Lorigan, & O'Day, 2013; Robert, Schadendorf, Messina, Hodi, & O'Day, 2013), and currently approved for no less than two additional indications (Coens et al., 2017; Eggermont et al., 2015; Eggermont et al., 2016; Geoerger et al., 2017; Merchant et al., 2016); (3) nivolumab, an ICI targeting programmed cell death 1 (PDCD1, best known as PD-1) first licensed for the management of patients with unresectable or metastatic melanoma and disease progression following ipilimumab and, if $BRAF^{V600}$ mutation positive, a BRAF inhibitor in December 2014 (Weber et al., 2015), and currently available for no less than 10 additional indications (Ansell et al., 2015; Borghaei et al., 2015; Cella et al., 2016; El-Khoueiry et al., 2017; Escudier, Motzer, et al., 2017; Escudier, Sharma, et al., 2017; Ferris et al., 2016; Harrington et al., 2017; Horn et al., 2017; Kazandjian et al., 2016; Kiyota et al., 2017; Motzer et al., 2015; Overman et al., 2017; Sharma et al., 2017; Vokes et al., 2018; Weber et al., 2017; Xu et al., 2017; Younes et al., 2016);

(4) pembrolizumab, a PD-1-targeting ICI originally licensed for the treatment of advanced or unresectable melanoma patients who are no longer responding to ipilimumab and BRAF inhibitors (when indicated) in September 2014 (Hodi et al., 2016; Joseph et al., 2018; Patnaik et al., 2015; Ribas et al., 2016; Robert et al., 2018, 2014), and currently approved for no less than 10 additional indications (Alley et al., 2017; Balar, Castellano, et al., 2017; Brahmer et al., 2017; Brogden et al., 2018; Chatterjee et al., 2016; Chen et al., 2017; Chow et al., 2016; Doi et al., 2018; Frenel et al., 2017; Fuchs et al., 2018; Gandhi et al., 2018; Garon et al., 2015; Hsu et al., 2017; Hui et al., 2017; Le et al., 2017, 2015; Muro et al., 2016; Nanda et al., 2016; O'Neil et al., 2017; Ott, Elez, et al., 2017; Ott, Piha-Paul, et al., 2017; Patnaik et al., 2015; Petrella et al., 2017; Plimack et al., 2017; Reck et al., 2016; Robert et al., 2015; Schachter et al., 2017; Seiwert et al., 2016; Shaverdian et al., 2017; Tahara et al., 2018; Zinzani et al., 2017); (5) cemiplimab, yet another ICI specific for PD-1 originally approved for the treatment of metastatic or locally advanced cutaneous squamous cell carcinoma in September 2018 (Migden et al., 2018); (6) durvalumab, an ICI specific for CD274 (best known as PD-L1) originally licensed for the treatment of locally advanced or metastatic urothelial carcinoma progressing during or following platinum-containing chemotherapy in May 2017 (Levy, Massard, Soria, & Deutsch, 2016; Powles et al., 2017), and currently also available for the management of patients with unresectable nonsmall cell lung carcinoma (Antonia et al., 2017); (7) avelumab, a PD-L1-targeting ICI originally approved for the treatment of adult and pediatric patients with metastatic Merkel cell carcinoma in March 2017 (D'Angelo et al., 2018; Kaufman et al., 2016, 2018), and currently licensed also for use in patients with locally advanced or metastatic urothelial carcinoma progressing during or following platinum-containing chemotherapy (Apolo et al., 2017; Heery et al., 2017); (8) atezolizumab, yet another PD-L1 inhibitor originally approved for the treatment of locally advanced or metastatic urothelial carcinoma progressing during or following platinum-containing chemotherapy in May 2016 (Balar, Galsky, et al., 2017; Necchi et al., 2017; Perez-Gracia et al., 2018; Rosenberg et al., 2016), and currently approved for no less than two additional indications (Fehrenbacher et al., 2016; Hida et al., 2018; Rittmeyer et al., 2017); (9) PEGylated interferon alpha 2B, a recombinant cytokine (Garcia-Martinez et al., 2018) originally approved for the treatment of stage III melanoma patients after surgery in March 2011 (Bottomley et al., 2009; Bouwhuis et al., 2010; Di Trolio et al., 2012; Eggermont et al., 2008; Eggermont et al., 2010; Fusi et al., 2009);

(10) talimogene laherparepvec, an oncolytic virus originally approved for the treatment of metastatic, unresectable melanoma patients in October 2015 (Andtbacka et al., 2015; Pol et al., 2018; Pol, Kroemer, & Galluzzi, 2016; Vacchelli et al., 2013), (11) tisagenlecleucel, a preparation of autologous T cells engineered to express a chimeric antigen receptor (CAR) (Borrie & Maleki Vareki, 2018; Raikar et al., 2018) that recognizes CD19, originally approved for the treatment of pediatric and young adults with acute lymphocytic leukemia who relapsed or are refractory to recommended prior therapy in August 2017 (Maude et al., 2018), and now also licensed for the therapy of subjects with various B-cell lymphomas relapsing on or refractory to recommended prior therapy (Schuster et al., 2019); and (12) axicabtagene ciloleucel, another variant of CAR-expressing T cells specific for CD19 (Kellner, Peipp, Gramatzki, Schrappe, & Schewe, 2018), originally approved for the treatment of several variants of refractory or relapsing B-cell lymphoma in October 2017 (Neelapu et al., 2017).

Alongside, considerable translational advances have been made in the 2010s, including (but not limited to): (1) the realization of the major impact of the immunological tumor microenvironment on disease outcome among cancer patients treated with a variety of therapeutic regimens beyond immunotherapy (Atherton et al., 2018; Bantug, Galluzzi, Kroemer, & Hess, 2018; Chevrier et al., 2017; De Henau et al., 2016; Herbst et al., 2014; Humbert & Hugues, 2019; Lavin et al., 2017; Li et al., 2019; Mariathasan et al., 2018; Riaz et al., 2017; Spranger, Bao, & Gajewski, 2015; Straussman et al., 2012; Tauriello et al., 2018; Zheng et al., 2017); (2) the identification of the gut microbiota as a major determinant of the sensitivity of cancer patients to multiple treatments including ICIs (Bhutiani et al., 2018; Gopalakrishnan et al., 2018; Iida et al., 2013; Routy et al., 2018; Vetizou et al., 2015; Viaud et al., 2013); (3) the discovery that novel antigens expressed by cancer cells as a consequence of nonsynonymous mutations play a key role in tumor progression and response to treatment (Gubin et al., 2014; Janikovits et al., 2018; Ozcan, Janikovits, von Knebel Doeberitz, & Kloor, 2018; Rizvi et al., 2015; Schumacher & Schreiber, 2015; Vitale et al., 2019); (4) the elucidation of innate immune mechanisms that provide a major contribution to both natural and therapy-elicited immunosurveillance (Andre et al., 2018; Dou et al., 2017; Ferrari de Andrade et al., 2018; Galluzzi, Vanpouille-Box, Bakhoum, & Demaria, 2018; Harding et al., 2017; Lopez-Soto, Gonzalez, Smyth, & Galluzzi, 2017; Mackenzie et al., 2017; Oyer, Gitto, Altomare, & Copik, 2018; Ruscetti et al., 2018;

van Montfoort et al., 2018; Vanpouille-Box et al., 2017; Vanpouille-Box, Demaria, Formenti, & Galluzzi, 2018); and (5) the identification of exosomes as a prominent vehicle for malignant cells to condition distant anatomical sites and interface with the host immune system (Chen et al., 2018; Diamond et al., 2018; Gabrusiewicz et al., 2018; Hoshino et al., 2015; Mirzaei et al., 2018; Yaddanapudi et al., 2019; Zhang et al., 2015). Finally, major technological advances have been achieved in the 2010s, including (but not limited to): (1) the development of reliable and cost-effective platforms for single cell DNA or RNA sequencing (Alemany, Florescu, Baron, Peterson-Maduro, & van Oudenaarden, 2018; Mulqueen et al., 2018; Zhang et al., 2018); (2) the design of robust algorithms enabling not only the analysis of large single cell omic datasets (Becht et al., 2019; Grobner et al., 2018; Wu et al., 2014), but also various other *in silico* functions that are widely employed in modern tumor immunology, such as neoantigen and TCR motif prediction (Balachandran et al., 2017; Brown & Holt, 2019; Dash et al., 2017; Formenti et al., 2018; Glanville et al., 2017; Hundal et al., 2019; Karasaki et al., 2017); (3) the refinement of the CRISPR/Cas system for gene editing (Halperin et al., 2018; Hille et al., 2018; Kalhor et al., 2018; Nishimasu et al., 2018; Shen et al., 2018; Wang et al., 2018; Yan et al., 2019); and (4) the introduction of mass cytometry (Chevrier et al., 2017; Galluzzi, Yamazaki, & Demaria, 2017; Lavin et al., 2017; Spitzer & Nolan, 2016; Wierz, Janji, Berchem, Moussay, & Paggetti, 2018). Beyond any reasonable doubt, the preclinical, clinical and technological progress we witnessed in the 2010s will pave the way to an exciting decade of achievements in the fields of tumor immunology and immunotherapy as the 2020s are on their way.

In *Tumor Immunology and Immunotherapy* (a group of volumes of the *Methods in Enzymology* series), world leaders in the field gathered to provide the first comprehensive compendium of methods for the study of cancer immunology in both its fundamental and translational components. Each of the chapters of *Tumor Immunology and Immunotherapy* provides a detailed description of one or a few protocols that are suitable for the characterization of (components of) the tumor–host interaction, *in vitro*, *in vivo* or *in silico*. The book consists of five volumes: "Molecular Methods," "Cellular Methods—Part A," "Cellular Methods—Part B," "Integrated Methods—Part A," and "Integrated Methods—Part B," and is equally addressed to expert researchers who may want to extend their technical portfolio, and to beginners in this exciting and rapidly growing field of investigation.

Acknowledgments

L.G. is supported by a Breakthrough Level 2 grant from the US Department of Defense (DoD), Breast Cancer Research Program (BRCP) [#BC180476P1], by a startup grant from the Department of Radiation Oncology at Weill Cornell Medicine (New York, USA), by industrial collaborations with Lytix (Oslo, Norway) and Phosplatin (New York, USA), and by donations from Phosplatin (New York, USA), the Luke Heller TECPR2 Foundation (Boston, USA) and Sotio a.s. (Prague, Czech Republic).

Disclosures

L.G. provides remunerated consulting to OmniSEQ (Buffalo, NY, USA), Astra Zeneca (Gaithersburg, MD, USA), Boehringer Ingelheim (Vienna, Austria), VL47 (New York, NY, USA), and the Luke Heller TECPR2 Foundation (Boston, MA, USA), and is member of the Scientific Advisory Committee of OmniSEQ (Buffalo, NY, USA).

References

Alemany, A., Florescu, M., Baron, C. S., Peterson-Maduro, J., & van Oudenaarden, A. (2018). Whole-organism clone tracing using single-cell sequencing. *Nature*, *556*(7699), 108–112. https://doi.org/10.1038/nature25969.

Alley, E. W., Lopez, J., Santoro, A., Morosky, A., Saraf, S., Piperdi, B., et al. (2017). Clinical safety and activity of pembrolizumab in patients with malignant pleural mesothelioma (KEYNOTE-028): Preliminary results from a non-randomised, open-label, phase 1b trial. *The Lancet Oncology*, *18*(5), 623–630. https://doi.org/10.1016/s1470-2045(17)30169-9.

Andre, P., Denis, C., Soulas, C., Bourbon-Caillet, C., Lopez, J., Arnoux, T., et al. (2018). Anti-NKG2A mAb is a checkpoint inhibitor that promotes anti-tumor immunity by unleashing both T and NK cells. *Cell*, *175*(7). 1731–1743.e1713. https://doi.org/10.1016/j.cell.2018.10.014.

Andtbacka, R. H., Kaufman, H. L., Collichio, F., Amatruda, T., Senzer, N., Chesney, J., et al. (2015). Talimogene laherparepvec improves durable response rate in patients with advanced melanoma. *Journal of Clinical Oncology*, *33*(25), 2780–2788. https://doi.org/10.1200/jco.2014.58.3377.

Ansell, S. M., Lesokhin, A. M., Borrello, I., Halwani, A., Scott, E. C., Gutierrez, M., et al. (2015). PD-1 blockade with nivolumab in relapsed or refractory Hodgkin's lymphoma. *The New England Journal of Medicine*, *372*(4), 311–319. https://doi.org/10.1056/NEJMoa1411087.

Antonia, S. J., Villegas, A., Daniel, D., Vicente, D., Murakami, S., Hui, R., et al. (2017). Durvalumab after chemoradiotherapy in stage III non-small-cell lung cancer. *The New England Journal of Medicine*, *377*(20), 1919–1929. https://doi.org/10.1056/NEJMoa1709937.

Apolo, A. B., Infante, J. R., Balmanoukian, A., Patel, M. R., Wang, D., Kelly, K., et al. (2017). Avelumab, an anti-programmed death-ligand 1 antibody, in patients with refractory metastatic urothelial carcinoma: Results from a multicenter, phase Ib study. *Journal of Clinical Oncology*, *35*(19), 2117–2124. https://doi.org/10.1200/jco.2016.71.6795.

Aranda, F., Vacchelli, E., Eggermont, A., Galon, J., Fridman, W. H., Zitvogel, L., et al. (2014). Trial watch: Immunostimulatory monoclonal antibodies in cancer therapy. *Oncoimmunology*, *3*(1), e27297. https://doi.org/10.4161/onci.27297.

Atherton, M. J., Stephenson, K. B., Tzelepis, F., Bakhshinyan, D., Nikota, J. K., Son, H. H., et al. (2018). Transforming the prostatic tumor microenvironment with oncolytic virotherapy. *Oncoimmunology*, 7(7), e1445459. https://doi.org/10.1080/2162402X.2018.1445459.

Balachandran, V. P., Luksza, M., Zhao, J. N., Makarov, V., Moral, J. A., Remark, R., et al. (2017). Identification of unique neoantigen qualities in long-term survivors of pancreatic cancer. *Nature, 551*(7681), 512–516. https://doi.org/10.1038/nature24462.

Balar, A. V., Castellano, D., O'Donnell, P. H., Grivas, P., Vuky, J., Powles, T., et al. (2017). First-line pembrolizumab in cisplatin-ineligible patients with locally advanced and unresectable or metastatic urothelial cancer (KEYNOTE-052): A multicentre, single-arm, phase 2 study. *The Lancet Oncology, 18*(11), 1483–1492. https://doi.org/10.1016/s1470-2045(17)30616-2.

Balar, A. V., Galsky, M. D., Rosenberg, J. E., Powles, T., Petrylak, D. P., Bellmunt, J., et al. (2017). Atezolizumab as first-line treatment in cisplatin-ineligible patients with locally advanced and metastatic urothelial carcinoma: A single-arm, multicentre, phase 2 trial. *Lancet, 389*(10064), 67–76. https://doi.org/10.1016/s0140-6736(16)32455-2.

Bantug, G. R., Galluzzi, L., Kroemer, G., & Hess, C. (2018). The spectrum of T cell metabolism in health and disease. *Nature Reviews. Immunology, 18*(1), 19–34. https://doi.org/10.1038/nri.2017.99.

Bargou, R., Leo, E., Zugmaier, G., Klinger, M., Goebeler, M., Knop, S., et al. (2008). Tumor regression in cancer patients by very low doses of a T cell-engaging antibody. *Science, 321*(5891), 974–977. https://doi.org/10.1126/science.1158545.

Becht, E., McInnes, L., Healy, J., Dutertre, C. A., Kwok, I. W. H., Ng, L. G., et al. (2019). Dimensionality reduction for visualizing single-cell data using UMAP. *Nature Biotechnology, 37*, 38–44. https://doi.org/10.1038/nbt.4314.

Bezu, L., Sauvat, A., Humeau, J., Leduc, M., Kepp, O., & Kroemer, G. (2018). eIF2alpha phosphorylation: A hallmark of immunogenic cell death. *Oncoimmunology*, 7(6). e1431089https://doi.org/10.1080/2162402X.2018.1431089.

Bhutiani, N., Li, Q., Anderson, C. D., Gallagher, H. C., De Jesus, M., Singh, R., et al. (2018). Enhanced gut barrier integrity sensitizes colon cancer to immune therapy. *Oncoimmunology*, 7(11), e1498438. https://doi.org/10.1080/2162402X.2018.1498438.

Borghaei, H., Paz-Ares, L., Horn, L., Spigel, D. R., Steins, M., Ready, N. E., et al. (2015). Nivolumab versus docetaxel in advanced nonsquamous non-small-cell lung cancer. *The New England Journal of Medicine, 373*(17), 1627–1639. https://doi.org/10.1056/NEJMoa1507643.

Borrie, A. E., & Maleki Vareki, S. (2018). T lymphocyte-based cancer immunotherapeutics. *International Review of Cell and Molecular Biology, 341*, 201–276. https://doi.org/10.1016/bs.ircmb.2018.05.010.

Bottomley, A., Coens, C., Suciu, S., Santinami, M., Kruit, W., Testori, A., et al. (2009). Adjuvant therapy with pegylated interferon alfa-2b versus observation in resected stage III melanoma: A phase III randomized controlled trial of health-related quality of life and symptoms by the European Organisation for Research and Treatment of Cancer Melanoma Group. *Journal of Clinical Oncology, 27*(18), 2916–2923. https://doi.org/10.1200/jco.2008.20.2069.

Bouwhuis, M. G., Suciu, S., Testori, A., Kruit, W. H., Sales, F., Patel, P., et al. (2010). Phase III trial comparing adjuvant treatment with pegylated interferon Alfa-2b versus observation: Prognostic significance of autoantibodies—EORTC 18991. *Journal of Clinical Oncology, 28*(14), 2460–2466. https://doi.org/10.1200/jco.2009.24.6264.

Brahmer, J. R., Rodriguez-Abreu, D., Robinson, A. G., Hui, R., Csoszi, T., Fulop, A., et al. (2017). Health-related quality-of-life results for pembrolizumab versus chemotherapy in advanced, PD-L1-positive NSCLC (KEYNOTE-024): A multicentre, international, randomised, open-label phase 3 trial. *The Lancet Oncology, 18*(12), 1600–1609. https://doi.org/10.1016/s1470-2045(17)30690-3.

Brogden, K. A., Parashar, D., Hallier, A. R., Braun, T., Qian, F., Rizvi, N. A., et al. (2018). Genomics of NSCLC patients both affirm PD-L1 expression and predict their clinical responses to anti-PD-1 immunotherapy. *BMC Cancer, 18*(1), 225. https://doi.org/10.1186/s12885-018-4134-y.

Brown, S. D., & Holt, R. A. (2019). Neoantigen characteristics in the context of the complete predicted MHC class I self-immunopeptidome. *Oncoimmunology, 8*(3), 1556080. https://doi.org/10.1080/2162402X.2018.1556080.

Casares, N., Pequignot, M. O., Tesniere, A., Ghiringhelli, F., Roux, S., Chaput, N., et al. (2005). Caspase-dependent immunogenicity of doxorubicin-induced tumor cell death. *The Journal of Experimental Medicine, 202*(12), 1691–1701. https://doi.org/10.1084/jem.20050915.

Cella, D., Grunwald, V., Nathan, P., Doan, J., Dastani, H., Taylor, F., et al. (2016). Quality of life in patients with advanced renal cell carcinoma given nivolumab versus everolimus in CheckMate 025: A randomised, open-label, phase 3 trial. *The Lancet Oncology, 17*(7), 994–1003. https://doi.org/10.1016/s1470-2045(16)30125-5.

Chatterjee, M., Turner, D. C., Felip, E., Lena, H., Cappuzzo, F., Horn, L., et al. (2016). Systematic evaluation of pembrolizumab dosing in patients with advanced non-small-cell lung cancer. *Annals of Oncology, 27*(7), 1291–1298. https://doi.org/10.1093/annonc/mdw174.

Chen, G., Huang, A. C., Zhang, W., Zhang, G., Wu, M., Xu, W., et al. (2018). Exosomal PD-L1 contributes to immunosuppression and is associated with anti-PD-1 response. *Nature, 560*(7718), 382–386. https://doi.org/10.1038/s41586-018-0392-8.

Chen, R., Zinzani, P. L., Fanale, M. A., Armand, P., Johnson, N. A., Brice, P., et al. (2017). Phase II study of the efficacy and safety of pembrolizumab for relapsed/refractory classic Hodgkin lymphoma. *Journal of Clinical Oncology, 35*(19), 2125–2132. https://doi.org/10.1200/jco.2016.72.1316.

Chevrier, S., Levine, J. H., Zanotelli, V. R. T., Silina, K., Schulz, D., Bacac, M., et al. (2017). An immune atlas of clear cell renal cell carcinoma. *Cell, 169*(4), 736–749.e718. https://doi.org/10.1016/j.cell.2017.04.016.

Chow, L. Q. M., Haddad, R., Gupta, S., Mahipal, A., Mehra, R., Tahara, M., et al. (2016). Antitumor activity of pembrolizumab in biomarker-unselected patients with recurrent and/or metastatic head and neck squamous cell carcinoma: Results from the Phase Ib KEYNOTE-012 expansion cohort. *Journal of Clinical Oncology, 34*(32), 3838–3845. https://doi.org/10.1200/jco.2016.68.1478.

Coens, C., Suciu, S., Chiarion-Sileni, V., Grob, J. J., Dummer, R., Wolchok, J. D., et al. (2017). Health-related quality of life with adjuvant ipilimumab versus placebo after complete resection of high-risk stage III melanoma (EORTC 18071): Secondary outcomes of a multinational, randomised, double-blind, phase 3 trial. *The Lancet Oncology, 18*(3), 393–403. https://doi.org/10.1016/s1470-2045(17)30015-3.

Coley, W. B. (1893). The treatment of malignant tumors by repeated inoculations of erysipelas. With a report of ten original cases. *The American Journal of the Medical Sciences, 105*, 487–511.

D'Angelo, S. P., Russell, J., Lebbe, C., Chmielowski, B., Gambichler, T., Grob, J. J., et al. (2018). Efficacy and safety of first-line avelumab treatment in patients with Stage IV metastatic Merkel cell carcinoma: A preplanned interim analysis of a clinical trial. *JAMA Oncology, 4*(9). e180077. https://doi.org/10.1001/jamaoncol.2018.0077.

Dash, P., Fiore-Gartland, A. J., Hertz, T., Wang, G. C., Sharma, S., Souquette, A., et al. (2017). Quantifiable predictive features define epitope-specific T cell receptor repertoires. *Nature, 547*(7661), 89–93. https://doi.org/10.1038/nature22383.

De Henau, O., Rausch, M., Winkler, D., Campesato, L. F., Liu, C., Cymerman, D. H., et al. (2016). Overcoming resistance to checkpoint blockade therapy by targeting PI3Kgamma in myeloid cells. *Nature, 539*(7629), 443–447. https://doi.org/10.1038/nature20554.

Decker, W. K., da Silva, R. F., Sanabria, M. H., Angelo, L. S., Guimaraes, F., Burt, B. M., et al. (2017). Cancer immunotherapy: Historical perspective of a clinical revolution and emerging preclinical animal models. *Frontiers in Immunology*, *8*, 829. https://doi.org/10.3389/fimmu.2017.00829.

Di Trolio, R., Simeone, E., Di Lorenzo, G., Grimaldi, A. M., Romano, A., Ayala, F., et al. (2012). Update on PEG-interferon alpha-2b as adjuvant therapy in melanoma. *Anticancer Research*, *32*(9), 3901–3909.

Diamond, J. M., Vanpouille-Box, C., Spada, S., Rudqvist, N. P., Chapman, J. R., Ueberheide, B. M., et al. (2018). Exosomes shuttle TREX1-sensitive IFN-stimulatory dsDNA from irradiated cancer cells to DCs. *Cancer Immunology Research*, *6*(8), 910–920. https://doi.org/10.1158/2326-6066.cir-17-0581.

Dighe, A. S., Richards, E., Old, L. J., & Schreiber, R. D. (1994). Enhanced in vivo growth and resistance to rejection of tumor cells expressing dominant negative IFN gamma receptors. *Immunity*, *1*(6), 447–456.

Doi, T., Piha-Paul, S. A., Jalal, S. I., Saraf, S., Lunceford, J., Koshiji, M., et al. (2018). Safety and antitumor activity of the anti-programmed death-1 antibody pembrolizumab in patients with advanced esophageal carcinoma. *Journal of Clinical Oncology*, *36*(1), 61–67. https://doi.org/10.1200/jco.2017.74.9846.

Dou, Z., Ghosh, K., Vizioli, M. G., Zhu, J., Sen, P., Wangensteen, K. J., et al. (2017). Cytoplasmic chromatin triggers inflammation in senescence and cancer. *Nature*, *550*(7676), 402–406. https://doi.org/10.1038/nature24050.

Dudley, M. E., Wunderlich, J. R., Robbins, P. F., Yang, J. C., Hwu, P., Schwartzentruber, D. J., et al. (2002). Cancer regression and autoimmunity in patients after clonal repopulation with antitumor lymphocytes. *Science*, *298*(5594), 850–854. https://doi.org/10.1126/science.1076514.

Dunn, G. P., Bruce, A. T., Ikeda, H., Old, L. J., & Schreiber, R. D. (2002). Cancer immunoediting: From immunosurveillance to tumor escape. *Nature Immunology*, *3*(11), 991–998. https://doi.org/10.1038/ni1102-991.

Dunn, G. P., Old, L. J., & Schreiber, R. D. (2004). The three Es of cancer immunoediting. *Annual Review of Immunology*, *22*, 329–360. https://doi.org/10.1146/annurev.immunol.22.012703.104803.

Eggermont, A. M., Bouwhuis, M. G., Kruit, W. H., Testori, A., ten Hagen, T., Yver, A., et al. (2010). Serum concentrations of pegylated interferon alpha-2b in patients with resected stage III melanoma receiving adjuvant pegylated interferon alpha-2b in a randomized phase III trial (EORTC 18991). *Cancer Chemotherapy and Pharmacology*, *65*(4), 671–677. https://doi.org/10.1007/s00280-009-1072-z.

Eggermont, A. M., Chiarion-Sileni, V., Grob, J. J., Dummer, R., Wolchok, J. D., Schmidt, H., et al. (2015). Adjuvant ipilimumab versus placebo after complete resection of high-risk stage III melanoma (EORTC 18071): A randomised, double-blind, phase 3 trial. *The Lancet Oncology*, *16*(5), 522–530. https://doi.org/10.1016/s1470-2045(15)70122-1.

Eggermont, A. M., Chiarion-Sileni, V., Grob, J. J., Dummer, R., Wolchok, J. D., Schmidt, H., et al. (2016). Prolonged survival in stage III melanoma with ipilimumab adjuvant therapy. *The New England Journal of Medicine*, *375*(19), 1845–1855. https://doi.org/10.1056/NEJMoa1611299.

Eggermont, A. M., Suciu, S., Santinami, M., Testori, A., Kruit, W. H., Marsden, J., et al. (2008). Adjuvant therapy with pegylated interferon alfa-2b versus observation alone in resected stage III melanoma: Final results of EORTC 18991, a randomised phase III trial. *Lancet*, *372*(9633), 117–126. https://doi.org/10.1016/s0140-6736(08)61033-8.

Ehrlich, P. (1909). Über den jetzigen stand der karzinomforschung. *Nederlands Tijdschrift voor Geneeskunde*, *5*, 273–290.

El-Khoueiry, A. B., Sangro, B., Yau, T., Crocenzi, T. S., Kudo, M., Hsu, C., et al. (2017). Nivolumab in patients with advanced hepatocellular carcinoma (CheckMate 040): An open-label, non-comparative, phase 1/2 dose escalation and expansion trial. *Lancet*, *389*(10088), 2492–2502. https://doi.org/10.1016/s0140-6736(17)31046-2.

Escudier, B., Motzer, R. J., Sharma, P., Wagstaff, J., Plimack, E. R., Hammers, H. J., et al. (2017). Treatment beyond progression in patients with advanced renal cell carcinoma treated with nivolumab in CheckMate 025. *European Urology*, *72*(3), 368–376. https://doi.org/10.1016/j.eururo.2017.03.037.

Escudier, B., Sharma, P., McDermott, D. F., George, S., Hammers, H. J., Srinivas, S., et al. (2017). CheckMate 025 randomized phase 3 study: Outcomes by key baseline factors and prior therapy for nivolumab versus everolimus in advanced renal cell carcinoma. *European Urology*, *72*(6), 962–971. https://doi.org/10.1016/j.eururo.2017.02.010.

Fehrenbacher, L., Spira, A., Ballinger, M., Kowanetz, M., Vansteenkiste, J., Mazieres, J., et al. (2016). Atezolizumab versus docetaxel for patients with previously treated non-small-cell lung cancer (POPLAR): A multicentre, open-label, phase 2 randomised controlled trial. *Lancet*, *387*(10030), 1837–1846. https://doi.org/10.1016/s0140-6736(16)00587-0.

Ferrari de Andrade, L., Tay, R. E., Pan, D., Luoma, A. M., Ito, Y., Badrinath, S., et al. (2018). Antibody-mediated inhibition of MICA and MICB shedding promotes NK cell-driven tumor immunity. *Science*, *359*(6383), 1537–1542. https://doi.org/10.1126/science.aao0505.

Ferris, R. L., Blumenschein, G., Jr., Fayette, J., Guigay, J., Colevas, A. D., Licitra, L., et al. (2016). Nivolumab for recurrent squamous-cell carcinoma of the head and neck. *The New England Journal of Medicine*, *375*(19), 1856–1867. https://doi.org/10.1056/NEJMoa1602252.

Formenti, S. C., Rudqvist, N. P., Golden, E., Cooper, B., Wennerberg, E., Lhuillier, C., et al. (2018). Radiotherapy induces responses of lung cancer to CTLA-4 blockade. *Nature Medicine*, *24*(12), 1845–1851. https://doi.org/10.1038/s41591-018-0232-2.

Frenel, J. S., Le Tourneau, C., O'Neil, B., Ott, P. A., Piha-Paul, S. A., Gomez-Roca, C., et al. (2017). Safety and efficacy of pembrolizumab in advanced, programmed death ligand 1-positive cervical cancer: Results from the phase Ib KEYNOTE-028 trial. *Journal of Clinical Oncology*, *35*(36), 4035–4041. https://doi.org/10.1200/jco.2017.74.5471.

Fuchs, C. S., Doi, T., Jang, R. W., Muro, K., Satoh, T., Machado, M., et al. (2018). Safety and efficacy of pembrolizumab monotherapy in patients with previously treated advanced gastric and gastroesophageal junction cancer: Phase 2 clinical KEYNOTE-059 trial. *JAMA Oncology*, *4*(5). e180013. https://doi.org/10.1001/jamaoncol.2018.0013.

Fusi, A., Collette, S., Busse, A., Suciu, S., Rietz, A., Santinami, M., et al. (2009). Circulating melanoma cells and distant metastasis-free survival in stage III melanoma patients with or without adjuvant interferon treatment (EORTC 18991 side study). *European Journal of Cancer*, *45*(18), 3189–3197. https://doi.org/10.1016/j.ejca.2009.09.004.

Gabrusiewicz, K., Li, X., Wei, J., Hashimoto, Y., Marisetty, A. L., Ott, M., et al. (2018). Glioblastoma stem cell-derived exosomes induce M2 macrophages and PD-L1 expression on human monocytes. *Oncoimmunology*, 7(4), e1412909. https://doi.org/10.1080/2162402X.2017.1412909.

Galluzzi, L., Bravo-San Pedro, J. M., Demaria, S., Formenti, S. C., & Kroemer, G. (2017). Activating autophagy to potentiate immunogenic chemotherapy and radiation therapy. *Nature Reviews. Clinical Oncology*, *14*(4), 247–258. https://doi.org/10.1038/nrclinonc.2016.183.

Galluzzi, L., Buque, A., Kepp, O., Zitvogel, L., & Kroemer, G. (2015). Immunological effects of conventional chemotherapy and targeted anticancer agents. *Cancer Cell*, *28*(6), 690–714. https://doi.org/10.1016/j.ccell.2015.10.012.

Galluzzi, L., Buque, A., Kepp, O., Zitvogel, L., & Kroemer, G. (2017). Immunogenic cell death in cancer and infectious disease. *Nature Reviews. Immunology*, *17*(2), 97–111. https://doi.org/10.1038/nri.2016.107.

Galluzzi, L., Chan, T. A., Kroemer, G., Wolchok, J. D., & Lopez-Soto, A. (2018). The hallmarks of successful anticancer immunotherapy. *Science Translational Medicine*, *10*(459), eaat7807. https://doi.org/10.1126/scitranslmed.aat7807.

Galluzzi, L., Kepp, O., Chan, F. K., & Kroemer, G. (2017). Necroptosis: Mechanisms and relevance to disease. *Annual Review of Pathology*, *12*, 103–130. https://doi.org/10.1146/annurev-pathol-052016-100247.

Galluzzi, L., Vanpouille-Box, C., Bakhoum, S. F., & Demaria, S. (2018). SnapShot: CGAS-STING signaling. *Cell*, *173*(1), 276–276.e271. https://doi.org/10.1016/j.cell.2018.03.015. .

Galluzzi, L., Vitale, I., Aaronson, S. A., Abrams, J. M., Adam, D., Agostinis, P., et al. (2018). Molecular mechanisms of cell death: Recommendations of the Nomenclature Committee on Cell Death 2018. *Cell Death and Differentiation*, *25*(3), 486–541. https://doi.org/10.1038/s41418-017-0012-4.

Galluzzi, L., Yamazaki, T., & Demaria, S. (2017). Heavy metal to rock the immune infiltrate. *Trends in Immunology*, *38*(8), 539–541. https://doi.org/10.1016/j.it.2017.05.007.

Galluzzi, L., Zitvogel, L., & Kroemer, G. (2016). Immunological mechanisms underneath the efficacy of cancer therapy. *Cancer Immunology Research*, *4*(11), 895–902. https://doi.org/10.1158/2326-6066.CIR-16-0197.

Galon, J., Costes, A., Sanchez-Cabo, F., Kirilovsky, A., Mlecnik, B., Lagorce-Pages, C., et al. (2006). Type, density, and location of immune cells within human colorectal tumors predict clinical outcome. *Science*, *313*(5795), 1960–1964. https://doi.org/10.1126/science.1129139.

Gandhi, L., Rodriguez-Abreu, D., Gadgeel, S., Esteban, E., Felip, E., De Angelis, F., et al. (2018). Pembrolizumab plus chemotherapy in metastatic non-small-cell lung cancer. *The New England Journal of Medicine*, *378*(22), 2078–2092. https://doi.org/10.1056/NEJMoa1801005.

Garcia-Martinez, E., Smith, M., Buque, A., Aranda, F., de la Pena, F. A., Ivars, A., et al. (2018). Trial watch: Immunostimulation with recombinant cytokines for cancer therapy. *Oncoimmunology*, 7(6), e1433982. https://doi.org/10.1080/2162402X.2018.1433982.

Garon, E. B., Rizvi, N. A., Hui, R., Leighl, N., Balmanoukian, A. S., Eder, J. P., et al. (2015). Pembrolizumab for the treatment of non-small-cell lung cancer. *The New England Journal of Medicine*, *372*(21), 2018–2028. https://doi.org/10.1056/NEJMoa1501824.

Geoerger, B., Bergeron, C., Gore, L., Sender, L., Dunkel, I. J., Herzog, C., et al. (2017). Phase II study of ipilimumab in adolescents with unresectable stage III or IV malignant melanoma. *European Journal of Cancer*, *86*, 358–363. https://doi.org/10.1016/j.ejca.2017.09.032.

Glanville, J., Huang, H., Nau, A., Hatton, O., Wagar, L. E., Rubelt, F., et al. (2017). Identifying specificity groups in the T cell receptor repertoire. *Nature*, *547*(7661), 94–98. https://doi.org/10.1038/nature22976.

Gopalakrishnan, V., Spencer, C. N., Nezi, L., Reuben, A., Andrews, M. C., Karpinets, T. V., et al. (2018). Gut microbiome modulates response to anti-PD-1 immunotherapy in melanoma patients. *Science*, *359*(6371), 97–103. https://doi.org/10.1126/science.aan4236.

Graham, J. B., & Graham, R. M. (1959). The effect of vaccine on cancer patients. *Surgery, Gynecology & Obstetrics*, *109*(2), 131–138.

Grobner, S. N., Worst, B. C., Weischenfeldt, J., Buchhalter, I., Kleinheinz, K., Rudneva, V. A., et al. (2018). The landscape of genomic alterations across childhood cancers. *Nature*, *555*(7696), 321–327. https://doi.org/10.1038/nature25480.

Gubin, M. M., Zhang, X., Schuster, H., Caron, E., Ward, J. P., Noguchi, T., et al. (2014). Checkpoint blockade cancer immunotherapy targets tumour-specific mutant antigens. *Nature*, *515*(7528), 577–581. https://doi.org/10.1038/nature13988.

Hagihara, K., Chan, S., Zhang, L., Oh, D. Y., Wei, X. X., Simko, J., et al. (2019). Neoadjuvant sipuleucel-T induces both Th1 activation and immune regulation in localized prostate cancer. *Oncoimmunology, 8*(1), e1486953. https://doi.org/10.1080/2162402X.2018.1486953.

Halperin, S. O., Tou, C. J., Wong, E. B., Modavi, C., Schaffer, D. V., & Dueber, J. E. (2018). CRISPR-guided DNA polymerases enable diversification of all nucleotides in a tunable window. *Nature, 560*(7717), 248–252. https://doi.org/10.1038/s41586-018-0384-8.

Harding, S. M., Benci, J. L., Irianto, J., Discher, D. E., Minn, A. J., & Greenberg, R. A. (2017). Mitotic progression following DNA damage enables pattern recognition within micronuclei. *Nature, 548*(7668), 466–470. https://doi.org/10.1038/nature23470.

Harrington, K. J., Ferris, R. L., Blumenschein, G., Jr., Colevas, A. D., Fayette, J., Licitra, L., et al. (2017). Nivolumab versus standard, single-agent therapy of investigator's choice in recurrent or metastatic squamous cell carcinoma of the head and neck (CheckMate 141): Health-related quality-of-life results from a randomised, phase 3 trial. *The Lancet Oncology, 18*(8), 1104–1115. https://doi.org/10.1016/s1470-2045(17)30421-7.

Heery, C. R., O'Sullivan-Coyne, G., Madan, R. A., Cordes, L., Rajan, A., Rauckhorst, M., et al. (2017). Avelumab for metastatic or locally advanced previously treated solid tumours (JAVELIN Solid Tumor): A phase 1a, multicohort, dose-escalation trial. *The Lancet Oncology, 18*(5), 587–598. https://doi.org/10.1016/s1470-2045(17)30239-5.

Herbst, R. S., Soria, J. C., Kowanetz, M., Fine, G. D., Hamid, O., Gordon, M. S., et al. (2014). Predictive correlates of response to the anti-PD-L1 antibody MPDL3280A in cancer patients. *Nature, 515*(7528), 563–567. https://doi.org/10.1038/nature14011.

Hida, T., Kaji, R., Satouchi, M., Ikeda, N., Horiike, A., Nokihara, H., et al. (2018). Atezolizumab in Japanese patients with previously treated advanced non-small-cell lung cancer: A subgroup analysis of the phase 3 OAK study. *Clinical Lung Cancer, 19*(4), e405–e415. https://doi.org/10.1016/j.cllc.2018.01.004.

Hildner, K., Edelson, B. T., Purtha, W. E., Diamond, M., Matsushita, H., Kohyama, M., et al. (2008). Batf3 deficiency reveals a critical role for CD8alpha+ dendritic cells in cytotoxic T cell immunity. *Science, 322*(5904), 1097–1100. https://doi.org/10.1126/science.1164206.

Hille, F., Richter, H., Wong, S. P., Bratovic, M., Ressel, S., & Charpentier, E. (2018). The biology of CRISPR-Cas: Backward and forward. *Cell, 172*(6), 1239–1259. https://doi.org/10.1016/j.cell.2017.11.032.

Hodi, F. S., Hwu, W. J., Kefford, R., Weber, J. S., Daud, A., Hamid, O., et al. (2016). Evaluation of immune-related response criteria and RECIST v1.1 in patients with advanced melanoma treated with pembrolizumab. *Journal of Clinical Oncology, 34*(13), 1510–1517. https://doi.org/10.1200/jco.2015.64.0391.

Hodi, F. S., O'Day, S. J., McDermott, D. F., Weber, R. W., Sosman, J. A., Haanen, J. B., et al. (2010). Improved survival with ipilimumab in patients with metastatic melanoma. *The New England Journal of Medicine, 363*(8), 711–723. https://doi.org/10.1056/NEJMoa1003466.

Horn, L., Spigel, D. R., Vokes, E. E., Holgado, E., Ready, N., Steins, M., et al. (2017). Nivolumab versus docetaxel in previously treated patients with advanced non-small-cell lung cancer: Two-year outcomes from two randomized, open-label, phase III trials (CheckMate 017 and CheckMate 057). *Journal of Clinical Oncology, 35*(35), 3924–3933. https://doi.org/10.1200/jco.2017.74.3062.

Hoshino, A., Costa-Silva, B., Shen, T. L., Rodrigues, G., Hashimoto, A., Tesic Mark, M., et al. (2015). Tumour exosome integrins determine organotropic metastasis. *Nature, 527*(7578), 329–335. https://doi.org/10.1038/nature15756.

Hsu, C., Lee, S. H., Ejadi, S., Even, C., Cohen, R. B., Le Tourneau, C., et al. (2017). Safety and antitumor activity of pembrolizumab in patients with programmed death-ligand 1-positive nasopharyngeal carcinoma: results of the KEYNOTE-028 study. *Journal of Clinical Oncology*, *35*(36), 4050–4056. https://doi.org/10.1200/jco.2017.73.3675.

Hui, R., Garon, E. B., Goldman, J. W., Leighl, N. B., Hellmann, M. D., Patnaik, A., et al. (2017). Pembrolizumab as first-line therapy for patients with PD-L1-positive advanced non-small cell lung cancer: A phase 1 trial. *Annals of Oncology*, *28*(4), 874–881. https://doi.org/10.1093/annonc/mdx008.

Humbert, M., & Hugues, S. (2019). Warming up the tumor microenvironment in order to enhance immunogenicity. *Oncoimmunology*, *8*(1), e1510710. https://doi.org/10.1080/2162402X.2018.1510710.

Hundal, J., Kiwala, S., Feng, Y. Y., Liu, C. J., Govindan, R., Chapman, W. C., et al. (2019). Accounting for proximal variants improves neoantigen prediction. *Nature Genetics*, *51*(1), 175–179. https://doi.org/10.1038/s41588-018-0283-9.

Iida, N., Dzutsev, A., Stewart, C. A., Smith, L., Bouladoux, N., Weingarten, R. A., et al. (2013). Commensal bacteria control cancer response to therapy by modulating the tumor microenvironment. *Science*, *342*(6161), 967–970. https://doi.org/10.1126/science.1240527.

Jaiswal, S., Jamieson, C. H., Pang, W. W., Park, C. Y., Chao, M. P., Majeti, R., et al. (2009). CD47 is upregulated on circulating hematopoietic stem cells and leukemia cells to avoid phagocytosis. *Cell*, *138*(2), 271–285. https://doi.org/10.1016/j.cell.2009.05.046.

Janikovits, J., Muller, M., Krzykalla, J., Korner, S., Echterdiek, F., Lahrmann, B., et al. (2018). High numbers of PDCD1 (PD-1)-positive T cells and B2M mutations in microsatellite-unstable colorectal cancer. *Oncoimmunology*, 7(2), e1390640. https://doi.org/10.1080/2162402X.2017.1390640.

Joseph, R. W., Elassaiss-Schaap, J., Kefford, R., Hwu, W. J., Wolchok, J. D., Joshua, A. M., et al. (2018). Baseline tumor size is an independent prognostic factor for overall survival in patients with melanoma treated with pembrolizumab. *Clinical Cancer Research*, *24*(20), 4960–4967. https://doi.org/10.1158/1078-0432.ccr-17-2386.

Kalhor, R., Kalhor, K., Mejia, L., Leeper, K., Graveline, A., Mali, P., et al. (2018). Developmental barcoding of whole mouse via homing CRISPR. *Science*, *361*(6405), eaat9804. https://doi.org/10.1126/science.aat9804.

Kantoff, P. W., Higano, C. S., Shore, N. D., Berger, E. R., Small, E. J., Penson, D. F., et al. (2010). Sipuleucel-T immunotherapy for castration-resistant prostate cancer. *The New England Journal of Medicine*, *363*(5), 411–422. https://doi.org/10.1056/NEJMoa1001294.

Kaplan, D. H., Shankaran, V., Dighe, A. S., Stockert, E., Aguet, M., Old, L. J., et al. (1998). Demonstration of an interferon gamma-dependent tumor surveillance system in immunocompetent mice. *Proceedings of the National Academy of Sciences of the United States of America*, *95*(13), 7556–7561.

Karasaki, T., Nagayama, K., Kuwano, H., Nitadori, J. I., Sato, M., Anraku, M., et al. (2017). Prediction and prioritization of neoantigens: Integration of RNA sequencing data with whole-exome sequencing. *Cancer Science*, *108*(2), 170–177. https://doi.org/10.1111/cas.13131.

Kaufman, H. L., Russell, J., Hamid, O., Bhatia, S., Terheyden, P., D'Angelo, S. P., et al. (2016). Avelumab in patients with chemotherapy-refractory metastatic Merkel cell carcinoma: A multicentre, single-group, open-label, phase 2 trial. *The Lancet Oncology*, *17*(10), 1374–1385. https://doi.org/10.1016/s1470-2045(16)30364-3.

Kaufman, H. L., Russell, J. S., Hamid, O., Bhatia, S., Terheyden, P., D'Angelo, S. P., et al. (2018). Updated efficacy of avelumab in patients with previously treated metastatic Merkel cell carcinoma after >/=1 year of follow-up: JAVELIN Merkel 200, a phase 2 clinical trial. *Journal for Immunotherapy of Cancer*, *6*(1), 7. https://doi.org/10.1186/s40425-017-0310-x.

Kazandjian, D., Suzman, D. L., Blumenthal, G., Mushti, S., He, K., Libeg, M., et al. (2016). FDA approval summary: Nivolumab for the treatment of metastatic non-small cell lung cancer with progression on or after platinum-based chemotherapy. *The Oncologist, 21*(5), 634–642. https://doi.org/10.1634/theoncologist.2015-0507.

Kellner, C., Peipp, M., Gramatzki, M., Schrappe, M., & Schewe, D. M. (2018). Perspectives of Fc engineered antibodies in CD19 targeting immunotherapies in pediatric B-cell precursor acute lymphoblastic leukemia. *Oncoimmunology*, 7(8), e1448331. https://doi.org/10.1080/2162402X.2018.1448331.

Kelly, P. N. (2018). The cancer immunotherapy revolution. *Science, 359*(6382), 1344–1345. https://doi.org/10.1126/science.359.6382.1344.

Kiyota, N., Hasegawa, Y., Takahashi, S., Yokota, T., Yen, C. J., Iwae, S., et al. (2017). A randomized, open-label, phase III clinical trial of nivolumab vs. therapy of investigator's choice in recurrent squamous cell carcinoma of the head and neck: A subanalysis of Asian patients versus the global population in checkmate 141. *Oral Oncology, 73*, 138–146. https://doi.org/10.1016/j.oraloncology.2017.07.023.

Koebel, C. M., Vermi, W., Swann, J. B., Zerafa, N., Rodig, S. J., Old, L. J., et al. (2007). Adaptive immunity maintains occult cancer in an equilibrium state. *Nature, 450*(7171), 903–907. https://doi.org/10.1038/nature06309.

Krummel, M. F., & Allison, J. P. (1995). CD28 and CTLA-4 have opposing effects on the response of T cells to stimulation. *The Journal of Experimental Medicine, 182*(2), 459–465.

Lavin, Y., Kobayashi, S., Leader, A., Amir, E. D., Elefant, N., Bigenwald, C., et al. (2017). Innate immune landscape in early lung adenocarcinoma by paired single-cell analyses. *Cell, 169*(4), 750–765.e717. https://doi.org/10.1016/j.cell.2017.04.014.

Le, D. T., Durham, J. N., Smith, K. N., Wang, H., Bartlett, B. R., Aulakh, L. K., et al. (2017). Mismatch repair deficiency predicts response of solid tumors to PD-1 blockade. *Science, 357*(6349), 409–413. https://doi.org/10.1126/science.aan6733.

Le, D. T., Uram, J. N., Wang, H., Bartlett, B. R., Kemberling, H., Eyring, A. D., et al. (2015). PD-1 blockade in tumors with mismatch-repair deficiency. *The New England Journal of Medicine, 372*(26), 2509–2520. https://doi.org/10.1056/NEJMoa1500596.

Leach, D. R., Krummel, M. F., & Allison, J. P. (1996). Enhancement of antitumor immunity by CTLA-4 blockade. *Science, 271*(5256), 1734–1736.

Levy, A., Massard, C., Soria, J. C., & Deutsch, E. (2016). Concurrent irradiation with the anti-programmed cell death ligand-1 immune checkpoint blocker durvalumab: Single centre subset analysis from a phase 1/2 trial. *European Journal of Cancer, 68*, 156–162. https://doi.org/10.1016/j.ejca.2016.09.013.

Li, H., van der Leun, A. M., Yofe, I., Lubling, Y., Gelbard-Solodkin, D., van Akkooi, A. C. J., … Amit, I. (2019). Dysfunctional CD8 T cells form a proliferative, dynamically regulated compartment within human melanoma. Cell, 176(4), 775-789 e718. doi:https://doi.org/10.1016/j.cell.2018.11.043

Lopez-Soto, A., Gonzalez, S., Smyth, M. J., & Galluzzi, L. (2017). Control of metastasis by NK cells. *Cancer Cell, 32*(2), 135–154. https://doi.org/10.1016/j.ccell.2017.06.009.

Mackenzie, K. J., Carroll, P., Martin, C. A., Murina, O., Fluteau, A., Simpson, D. J., et al. (2017). cGAS surveillance of micronuclei links genome instability to innate immunity. *Nature, 548*(7668), 461–465. https://doi.org/10.1038/nature23449.

Majeti, R., Chao, M. P., Alizadeh, A. A., Pang, W. W., Jaiswal, S., Gibbs, K. D., Jr., et al. (2009). CD47 is an adverse prognostic factor and therapeutic antibody target on human acute myeloid leukemia stem cells. *Cell, 138*(2), 286–299. https://doi.org/10.1016/j.cell.2009.05.045.

Mariathasan, S., Turley, S. J., Nickles, D., Castiglioni, A., Yuen, K., Wang, Y., et al. (2018). TGFbeta attenuates tumour response to PD-L1 blockade by contributing to exclusion of T cells. *Nature, 554*(7693), 544–548. https://doi.org/10.1038/nature25501.

Maude, S. L., Laetsch, T. W., Buechner, J., Rives, S., Boyer, M., Bittencourt, H., et al. (2018). Tisagenlecleucel in children and young adults with B-cell lymphoblastic leukemia. *The New England Journal of Medicine*, *378*(5), 439–448. https://doi.org/10.1056/NEJMoa1709866.

McDermott, D., Haanen, J., Chen, T. T., Lorigan, P., & O'Day, S. (2013). Efficacy and safety of ipilimumab in metastatic melanoma patients surviving more than 2 years following treatment in a phase III trial (MDX010-20). *Annals of Oncology*, *24*(10), 2694–2698. https://doi.org/10.1093/annonc/mdt291.

Merchant, M. S., Wright, M., Baird, K., Wexler, L. H., Rodriguez-Galindo, C., Bernstein, D., et al. (2016). Phase I clinical trial of ipilimumab in pediatric patients with advanced solid tumors. *Clinical Cancer Research*, *22*(6), 1364–1370. https://doi.org/10.1158/1078-0432.ccr-15-0491.

Migden, M. R., Rischin, D., Schmults, C. D., Guminski, A., Hauschild, A., Lewis, K. D., et al. (2018). PD-1 blockade with cemiplimab in advanced cutaneous squamous-cell carcinoma. *The New England Journal of Medicine*, *379*(4), 341–351. https://doi.org/10.1056/NEJMoa1805131.

Mirzaei, R., Sarkar, S., Dzikowski, L., Rawji, K. S., Khan, L., Faissner, A., et al. (2018). Brain tumor-initiating cells export tenascin-C associated with exosomes to suppress T cell activity. *Oncoimmunology*, 7(10), e1478647. https://doi.org/10.1080/2162402X.2018.1478647.

Morgan, R. A., Dudley, M. E., Wunderlich, J. R., Hughes, M. S., Yang, J. C., Sherry, R. M., et al. (2006). Cancer regression in patients after transfer of genetically engineered lymphocytes. *Science*, *314*(5796), 126–129. https://doi.org/10.1126/science.1129003.

Motzer, R. J., Escudier, B., McDermott, D. F., George, S., Hammers, H. J., Srinivas, S., et al. (2015). Nivolumab versus everolimus in advanced renal-cell carcinoma. *The New England Journal of Medicine*, *373*(19), 1803–1813. https://doi.org/10.1056/NEJMoa1510665.

Mulqueen, R. M., Pokholok, D., Norberg, S. J., Torkenczy, K. A., Fields, A. J., Sun, D., et al. (2018). Highly scalable generation of DNA methylation profiles in single cells. *Nature Biotechnology*, *36*(5), 428–431. https://doi.org/10.1038/nbt.4112.

Muro, K., Chung, H. C., Shankaran, V., Geva, R., Catenacci, D., Gupta, S., et al. (2016). Pembrolizumab for patients with PD-L1-positive advanced gastric cancer (KEYNOTE-012): A multicentre, open-label, phase 1b trial. *The Lancet Oncology*, *17*(6), 717–726. https://doi.org/10.1016/s1470-2045(16)00175-3.

Nanda, R., Chow, L. Q., Dees, E. C., Berger, R., Gupta, S., Geva, R., et al. (2016). Pembrolizumab in patients with advanced triple-negative breast cancer: Phase Ib KEYNOTE-012 study. *Journal of Clinical Oncology*, *34*(21), 2460–2467. https://doi.org/10.1200/jco.2015.64.8931.

Necchi, A., Joseph, R. W., Loriot, Y., Hoffman-Censits, J., Perez-Gracia, J. L., Petrylak, D. P., et al. (2017). Atezolizumab in platinum-treated locally advanced or metastatic urothelial carcinoma: Post-progression outcomes from the phase II IMvigor210 study. *Annals of Oncology*, *28*(12), 3044–3050. https://doi.org/10.1093/annonc/mdx518.

Neelapu, S. S., Locke, F. L., Bartlett, N. L., Lekakis, L. J., Miklos, D. B., Jacobson, C. A., et al. (2017). Axicabtagene ciloleucel CAR T-cell therapy in refractory large B-cell lymphoma. *The New England Journal of Medicine*, 377(26), 2531–2544. https://doi.org/10.1056/NEJMoa1707447.

Nishimasu, H., Shi, X., Ishiguro, S., Gao, L., Hirano, S., Okazaki, S., et al. (2018). Engineered CRISPR-Cas9 nuclease with expanded targeting space. *Science*, *361*(6408), 1259–1262. https://doi.org/10.1126/science.aas9129.

Obeid, M., Tesniere, A., Ghiringhelli, F., Fimia, G. M., Apetoh, L., Perfettini, J. L., et al. (2007). Calreticulin exposure dictates the immunogenicity of cancer cell death. *Nature Medicine, 13*(1), 54–61. https://doi.org/10.1038/nm1523.

O'Neil, B. H., Wallmark, J. M., Lorente, D., Elez, E., Raimbourg, J., Gomez-Roca, C., et al. (2017). Safety and antitumor activity of the anti-PD-1 antibody pembrolizumab in patients with advanced colorectal carcinoma. *PLoS One, 12*(12), e0189848. https://doi.org/10.1371/journal.pone.0189848.

Ott, P. A., Elez, E., Hiret, S., Kim, D. W., Morosky, A., Saraf, S., et al. (2017). Pembrolizumab in patients with extensive-stage small-cell lung cancer: Results from the phase Ib KEYNOTE-028 study. *Journal of Clinical Oncology, 35*(34), 3823–3829. https://doi.org/10.1200/jco.2017.72.5069.

Ott, P. A., Piha-Paul, S. A., Munster, P., Pishvaian, M. J., van Brummelen, E. M. J., Cohen, R. B., et al. (2017). Safety and antitumor activity of the anti-PD-1 antibody pembrolizumab in patients with recurrent carcinoma of the anal canal. *Annals of Oncology, 28*(5), 1036–1041. https://doi.org/10.1093/annonc/mdx029.

Overman, M. J., McDermott, R., Leach, J. L., Lonardi, S., Lenz, H. J., Morse, M. A., et al. (2017). Nivolumab in patients with metastatic DNA mismatch repair-deficient or microsatellite instability-high colorectal cancer (CheckMate 142): An open-label, multicentre, phase 2 study. *The Lancet Oncology, 18*(9), 1182–1191. https://doi.org/10.1016/s1470-2045(17)30422-9.

Oyer, J. L., Gitto, S. B., Altomare, D. A., & Copik, A. J. (2018). PD-L1 blockade enhances anti-tumor efficacy of NK cells. *Oncoimmunology*, 7(11), e1509819. https://doi.org/10.1080/2162402X.2018.1509819.

Ozcan, M., Janikovits, J., von Knebel Doeberitz, M., & Kloor, M. (2018). Complex pattern of immune evasion in MSI colorectal cancer. *Oncoimmunology*, 7(7), e1445453. https://doi.org/10.1080/2162402X.2018.1445453.

Patnaik, A., Kang, S. P., Rasco, D., Papadopoulos, K. P., Elassaiss-Schaap, J., Beeram, M., et al. (2015). Phase I study of pembrolizumab (MK-3475; Anti-PD-1 monoclonal antibody) in patients with advanced solid tumors. *Clinical Cancer Research, 21*(19), 4286–4293. https://doi.org/10.1158/1078-0432.ccr-14-2607.

Perez-Gracia, J. L., Loriot, Y., Rosenberg, J. E., Powles, T., Necchi, A., Hussain, S. A., et al. (2018). Atezolizumab in platinum-treated locally advanced or metastatic urothelial carcinoma: Outcomes by prior number of regimens. *European Urology, 73*, 462–468. https://doi.org/10.1016/j.eururo.2017.11.023.

Petrella, T. M., Robert, C., Richtig, E., Miller, W. H., Jr., Masucci, G. V., Walpole, E., et al. (2017). Patient-reported outcomes in KEYNOTE-006, a randomised study of pembrolizumab versus ipilimumab in patients with advanced melanoma. *European Journal of Cancer, 86*, 115–124. https://doi.org/10.1016/j.ejca.2017.08.032.

Plimack, E. R., Bellmunt, J., Gupta, S., Berger, R., Chow, L. Q., Juco, J., et al. (2017). Safety and activity of pembrolizumab in patients with locally advanced or metastatic urothelial cancer (KEYNOTE-012): A non-randomised, open-label, phase 1b study. *The Lancet Oncology, 18*(2), 212–220. https://doi.org/10.1016/S1470-2045(17)30007-4.

Pol, J., Kroemer, G., & Galluzzi, L. (2016). First oncolytic virus approved for melanoma immunotherapy. *Oncoimmunology, 5*(1), e1115641. https://doi.org/10.1080/2162402X.2015.1115641.

Pol, J., Levesque, S., Workenhe, S. T., Gujar, S., Le Boeuf, F., D, R. C., et al. (2018). Trial watch: Oncolytic viro-immunotherapy of hematologic and solid tumors. *Oncoimmunology*, 7(12), e1503032. https://doi.org/10.1080/2162402X.2018.1503032.

Pol, J., Vacchelli, E., Aranda, F., Castoldi, F., Eggermont, A., Cremer, I., et al. (2015). Trial watch: Immunogenic cell death inducers for anticancer chemotherapy. *Oncoimmunology, 4*(4). e1008866https://doi.org/10.1080/2162402X.2015.1008866.

Powles, T., O'Donnell, P. H., Massard, C., Arkenau, H. T., Friedlander, T. W., Hoimes, C. J., et al. (2017). Efficacy and safety of durvalumab in locally advanced or metastatic urothelial carcinoma: Updated results from a Phase 1/2 open-label study. *JAMA Oncology*, *3*(9), e172411. https://doi.org/10.1001/jamaoncol.2017.2411.

Raikar, S. S., Fleischer, L. C., Moot, R., Fedanov, A., Paik, N. Y., Knight, K. A., et al. (2018). Development of chimeric antigen receptors targeting T-cell malignancies using two structurally different anti-CD5 antigen binding domains in NK and CRISPR-edited T cell lines. *Oncoimmunology*, 7(3), e1407898. https://doi.org/10.1080/2162402X.2017.1407898.

Rao, S., Gharib, K., & Han, A. (2019). Cancer immunosurveillance by T cells. *International Review of Cell and Molecular Biology*, *342*, 149–173. https://doi.org/10.1016/bs.ircmb.2018.08.001.

Reck, M., Rodriguez-Abreu, D., Robinson, A. G., Hui, R., Csoszi, T., Fulop, A., et al. (2016). Pembrolizumab versus chemotherapy for PD-L1-positive non-small-cell lung cancer. *The New England Journal of Medicine*, *375*(19), 1823–1833. https://doi.org/10.1056/NEJMoa1606774.

Riaz, N., Havel, J. J., Makarov, V., Desrichard, A., Urba, W. J., Sims, J. S., et al. (2017). Tumor and microenvironment evolution during immunotherapy with nivolumab. *Cell*, *171*(4), 934–949.e916. https://doi.org/10.1016/j.cell.2017.09.028.

Ribas, A., Hamid, O., Daud, A., Hodi, F. S., Wolchok, J. D., Kefford, R., et al. (2016). Association of pembrolizumab with tumor response and survival among patients with advanced melanoma. *JAMA*, *315*(15), 1600–1609. https://doi.org/10.1001/jama.2016.4059.

Rittmeyer, A., Barlesi, F., Waterkamp, D., Park, K., Ciardiello, F., von Pawel, J., et al. (2017). Atezolizumab versus docetaxel in patients with previously treated non-small-cell lung cancer (OAK): A phase 3, open-label, multicentre randomised controlled trial. *Lancet*, *389*(10066), 255–265. https://doi.org/10.1016/s0140-6736(16)32517-x.

Rizvi, N. A., Hellmann, M. D., Snyder, A., Kvistborg, P., Makarov, V., Havel, J. J., et al. (2015). Cancer immunology. Mutational landscape determines sensitivity to PD-1 blockade in non-small cell lung cancer. *Science*, *348*(6230), 124–128. https://doi.org/10.1126/science.aaa1348.

Robert, C., Ribas, A., Hamid, O., Daud, A., Wolchok, J. D., Joshua, A. M., et al. (2018). Durable complete response after discontinuation of pembrolizumab in patients with metastatic melanoma. *Journal of Clinical Oncology*, *36*(17), 1668–1674. https://doi.org/10.1200/jco.2017.75.6270.

Robert, C., Ribas, A., Wolchok, J. D., Hodi, F. S., Hamid, O., Kefford, R., et al. (2014). Anti-programmed-death-receptor-1 treatment with pembrolizumab in ipilimumab-refractory advanced melanoma: A randomised dose-comparison cohort of a phase 1 trial. *Lancet*, *384*(9948), 1109–1117. https://doi.org/10.1016/s0140-6736(14)60958-2.

Robert, C., Schachter, J., Long, G. V., Arance, A., Grob, J. J., Mortier, L., et al. (2015). Pembrolizumab versus ipilimumab in advanced melanoma. *The New England Journal of Medicine*, *372*(26), 2521–2532. https://doi.org/10.1056/NEJMoa1503093.

Robert, C., Schadendorf, D., Messina, M., Hodi, F. S., & O'Day, S. (2013). Efficacy and safety of retreatment with ipilimumab in patients with pretreated advanced melanoma who progressed after initially achieving disease control. *Clinical Cancer Research*, *19*(8), 2232–2239. https://doi.org/10.1158/1078-0432.ccr-12-3080.

Rosenberg, J. E., Hoffman-Censits, J., Powles, T., van der Heijden, M. S., Balar, A. V., Necchi, A., et al. (2016). Atezolizumab in patients with locally advanced and metastatic urothelial carcinoma who have progressed following treatment with platinum-based chemotherapy: A single-arm, multicentre, phase 2 trial. *Lancet*, *387*(10031), 1909–1920. https://doi.org/10.1016/s0140-6736(16)00561-4.

Routy, B., Le Chatelier, E., Derosa, L., Duong, C. P. M., Alou, M. T., Daillere, R., et al. (2018). Gut microbiome influences efficacy of PD-1-based immunotherapy against epithelial tumors. *Science*, *359*(6371), 91–97. https://doi.org/10.1126/science.aan3706.

Ruscetti, M., Leibold, J., Bott, M. J., Fennell, M., Kulick, A., Salgado, N. R., et al. (2018). NK cell-mediated cytotoxicity contributes to tumor control by a cytostatic drug combination. *Science*, *362*(6421), 1416–1422. https://doi.org/10.1126/science.aas9090.

Savage, P. A., Vosseller, K., Kang, C., Larimore, K., Riedel, E., Wojnoonski, K., et al. (2008). Recognition of a ubiquitous self antigen by prostate cancer-infiltrating CD8+ T lymphocytes. *Science*, *319*(5860), 215–220. https://doi.org/10.1126/science.1148886.

Schachter, J., Ribas, A., Long, G. V., Arance, A., Grob, J. J., Mortier, L., et al. (2017). Pembrolizumab versus ipilimumab for advanced melanoma: final overall survival results of a multicentre, randomised, open-label phase 3 study (KEYNOTE-006). *Lancet*, *390*(10105), 1853–1862. https://doi.org/10.1016/s0140-6736(17)31601-x.

Schumacher, T. N., & Schreiber, R. D. (2015). Neoantigens in cancer immunotherapy. *Science*, *348*(6230), 69–74. https://doi.org/10.1126/science.aaa4971.

Schuster, S. J., Bishop, M. R., Tam, C. S., Waller, E. K., Borchmann, P., McGuirk, J. P., et al. (2019). Tisagenlecleucel in adult relapsed or refractory diffuse large B-cell lymphoma. *The New England Journal of Medicine*, *380*(1), 45–56. https://doi.org/10.1056/NEJMoa1804980.

Seiwert, T. Y., Burtness, B., Mehra, R., Weiss, J., Berger, R., Eder, J. P., et al. (2016). Safety and clinical activity of pembrolizumab for treatment of recurrent or metastatic squamous cell carcinoma of the head and neck (KEYNOTE-012): An open-label, multicentre, phase 1b trial. *The Lancet Oncology*, *17*(7), 956–965. https://doi.org/10.1016/s1470-2045(16)30066-3.

Shankaran, V., Ikeda, H., Bruce, A. T., White, J. M., Swanson, P. E., Old, L. J., et al. (2001). IFNgamma and lymphocytes prevent primary tumour development and shape tumour immunogenicity. *Nature*, *410*(6832), 1107–1111. https://doi.org/10.1038/35074122.

Sharma, P., Retz, M., Siefker-Radtke, A., Baron, A., Necchi, A., Bedke, J., et al. (2017). Nivolumab in metastatic urothelial carcinoma after platinum therapy (CheckMate 275): A multicentre, single-arm, phase 2 trial. *The Lancet Oncology*, *18*(3), 312–322. https://doi.org/10.1016/s1470-2045(17)30065-7.

Shaverdian, N., Lisberg, A. E., Bornazyan, K., Veruttipong, D., Goldman, J. W., Formenti, S. C., et al. (2017). Previous radiotherapy and the clinical activity and toxicity of pembrolizumab in the treatment of non-small-cell lung cancer: A secondary analysis of the KEYNOTE-001 phase 1 trial. *The Lancet Oncology*, *18*(7), 895–903. https://doi.org/10.1016/s1470-2045(17)30380-7.

Shen, M. W., Arbab, M., Hsu, J. Y., Worstell, D., Culbertson, S. J., Krabbe, O., et al. (2018). Predictable and precise template-free CRISPR editing of pathogenic variants. *Nature*, *563*(7733), 646–651. https://doi.org/10.1038/s41586-018-0686-x.

Spitzer, M. H., & Nolan, G. P. (2016). Mass cytometry: Single cells, many features. *Cell*, *165*(4), 780–791. https://doi.org/10.1016/j.cell.2016.04.019.

Spranger, S., Bao, R., & Gajewski, T. F. (2015). Melanoma-intrinsic beta-catenin signalling prevents anti-tumour immunity. *Nature*, *523*(7559), 231–235. https://doi.org/10.1038/nature14404.

Straussman, R., Morikawa, T., Shee, K., Barzily-Rokni, M., Qian, Z. R., Du, J., et al. (2012). Tumour micro-environment elicits innate resistance to RAF inhibitors through HGF secretion. *Nature*, *487*(7408), 500–504. https://doi.org/10.1038/nature11183.

Tahara, M., Muro, K., Hasegawa, Y., Chung, H. C., Lin, C. C., Keam, B., et al. (2018). Pembrolizumab in Asia-Pacific patients with advanced head and neck squamous cell carcinoma: Analyses from KEYNOTE-012. *Cancer Science*, *109*(3), 771–776. https://doi.org/10.1111/cas.13480.

Tauriello, D. V. F., Palomo-Ponce, S., Stork, D., Berenguer-Llergo, A., Badia-Ramentol, J., Iglesias, M., et al. (2018). TGFbeta drives immune evasion in genetically reconstituted colon cancer metastasis. *Nature*, *554*(7693), 538–543. https://doi.org/10.1038/nature25492.

Vacchelli, E., Eggermont, A., Sautes-Fridman, C., Galon, J., Zitvogel, L., Kroemer, G., et al. (2013). Trial watch: Oncolytic viruses for cancer therapy. *Oncoimmunology*, *2*(6). e24612https://doi.org/10.4161/onci.24612.

van Montfoort, N., Borst, L., Korrer, M. J., Sluijter, M., Marijt, K. A., Santegoets, S. J., et al. (2018). NKG2A blockade potentiates CD8 T cell immunity induced by cancer vaccines. *Cell*, *175*(7), 1744–1755.e1715. https://doi.org/10.1016/j.cell.2018.10.028.

Vanpouille-Box, C., Alard, A., Aryankalayil, M. J., Sarfraz, Y., Diamond, J. M., Schneider, R. J., et al. (2017). DNA exonuclease Trex1 regulates radiotherapy-induced tumour immunogenicity. *Nature Communications*, *8*, 15618. https://doi.org/10.1038/ncomms15618.

Vanpouille-Box, C., Demaria, S., Formenti, S. C., & Galluzzi, L. (2018). Cytosolic DNA sensing in organismal tumor control. *Cancer Cell*, *34*(3), 361–378. https://doi.org/10.1016/j.ccell.2018.05.013.

Vetizou, M., Pitt, J. M., Daillere, R., Lepage, P., Waldschmitt, N., Flament, C., et al. (2015). Anticancer immunotherapy by CTLA-4 blockade relies on the gut microbiota. *Science*, *350*(6264), 1079–1084. https://doi.org/10.1126/science.aad1329.

Viaud, S., Saccheri, F., Mignot, G., Yamazaki, T., Daillere, R., Hannani, D., et al. (2013). The intestinal microbiota modulates the anticancer immune effects of cyclophosphamide. *Science*, *342*(6161), 971–976. https://doi.org/10.1126/science.1240537.

Vitale, I., Sistigu, A., Manic, G., Rudqvist, N. P., Trajanoski, Z., & Galluzzi, L. (2019). Mutational and antigenic landscape in tumor progression and cancer immunotherapy. *Trends in Cell Biology*, *29*, 395–416. https://doi.org/10.1016/j.tcb.2019.01.003.

Vokes, E. E., Ready, N., Felip, E., Horn, L., Burgio, M. A., Antonia, S. J., et al. (2018). Nivolumab versus docetaxel in previously treated advanced non-small-cell lung cancer (CheckMate 017 and CheckMate 057): 3-year update and outcomes in patients with liver metastases. *Annals of Oncology*, *29*(4), 959–965. https://doi.org/10.1093/annonc/mdy041.

Wang, H., Xu, X., Nguyen, C. M., Liu, Y., Gao, Y., Lin, X., et al. (2018). CRISPR-mediated programmable 3D genome positioning and nuclear organization. *Cell*, *175*(5), 1405–1417.e1414. https://doi.org/10.1016/j.cell.2018.09.013.

Weber, J. S., D'Angelo, S. P., Minor, D., Hodi, F. S., Gutzmer, R., Neyns, B., et al. (2015). Nivolumab versus chemotherapy in patients with advanced melanoma who progressed after anti-CTLA-4 treatment (CheckMate 037): A randomised, controlled, open-label, phase 3 trial. *The Lancet Oncology*, *16*(4), 375–384. https://doi.org/10.1016/s1470-2045(15)70076-8.

Weber, J., Mandala, M., Del Vecchio, M., Gogas, H. J., Arance, A. M., Cowey, C. L., et al. (2017). Adjuvant nivolumab versus ipilimumab in resected stage III or IV melanoma. *The New England Journal of Medicine*, *377*(19), 1824–1835. https://doi.org/10.1056/NEJMoa1709030.

Wierz, M., Janji, B., Berchem, G., Moussay, E., & Paggetti, J. (2018). High-dimensional mass cytometry analysis revealed microenvironment complexity in chronic lymphocytic leukemia. *Oncoimmunology*, 7(8), e1465167. https://doi.org/10.1080/2162402X.2018.1465167.

Wu, Y., Williams, E. G., Dubuis, S., Mottis, A., Jovaisaite, V., Houten, S. M., et al. (2014). Multilayered genetic and omics dissection of mitochondrial activity in a mouse reference population. *Cell*, *158*(6), 1415–1430. https://doi.org/10.1016/j.cell.2014.07.039.

Xu, J. X., Maher, V. E., Zhang, L., Tang, S., Sridhara, R., Ibrahim, A., et al. (2017). FDA approval summary: nivolumab in advanced renal cell carcinoma after anti-angiogenic therapy and exploratory predictive biomarker analysis. *The Oncologist*, *22*(3), 311–317. https://doi.org/10.1634/theoncologist.2016-0476.

Yaddanapudi, K., Meng, S., Whitt, A. G., Al Rayyan, N., Richie, J., Tu, A., et al. (2019). Exosomes from GM-CSF expressing embryonic stem cells are an effective prophylactic vaccine for cancer prevention. *Oncoimmunology*, *8*(3), 1561119. https://doi.org/10.1080/2162402X.2018.1561119.

Yan, W. X., Hunnewell, P., Alfonse, L. E., Carte, J. M., Keston-Smith, E., Sothiselvam, S., et al. (2019). Functionally diverse type V CRISPR-Cas systems. *Science*, *363*(6422), 88–91. https://doi.org/10.1126/science.aav7271.

Younes, A., Santoro, A., Shipp, M., Zinzani, P. L., Timmerman, J. M., Ansell, S., et al. (2016). Nivolumab for classical Hodgkin's lymphoma after failure of both autologous stem-cell transplantation and brentuximab vedotin: A multicentre, multicohort, single-arm phase 2 trial. *The Lancet Oncology*, *17*(9), 1283–1294. https://doi.org/10.1016/s1470-2045(16)30167-x.

Zhang, S. Q., Ma, K. Y., Schonnesen, A. A., Zhang, M., He, C., Sun, E., et al. (2018). High-throughput determination of the antigen specificities of T cell receptors in single cells. *Nature Biotechnology*. https://doi.org/10.1038/nbt.4282.

Zhang, L., Zhang, S., Yao, J., Lowery, F. J., Zhang, Q., Huang, W. C., et al. (2015). Microenvironment-induced PTEN loss by exosomal microRNA primes brain metastasis outgrowth. *Nature*, *527*(7576), 100–104. https://doi.org/10.1038/nature15376.

Zheng, C., Zheng, L., Yoo, J. K., Guo, H., Zhang, Y., Guo, X., et al. (2017). Landscape of infiltrating T cells in liver cancer revealed by single-cell sequencing. *Cell*, *169*(7), 1342–1356.e1316. https://doi.org/10.1016/j.cell.2017.05.035.

Zinzani, P. L., Ribrag, V., Moskowitz, C. H., Michot, J. M., Kuruvilla, J., Balakumaran, A., et al. (2017). Safety and tolerability of pembrolizumab in patients with relapsed/refractory primary mediastinal large B-cell lymphoma. *Blood*, *130*(3), 267–270. https://doi.org/10.1182/blood-2016-12-758383.

CHAPTER ONE

Quantitation of calreticulin exposure associated with immunogenic cell death

Peng Liu[a,b,c,d,e,f], **Liwei Zhao**[a,b,c,d,e,f], **Oliver Kepp**[a,b,c,d,e,f,*], **Guido Kroemer**[b,c,d,e,f,g,h,i,*]

[a]Faculty of Medicine, University of Paris Sud, Kremlin-Bicêtre, France
[b]Cell Biology and Metabolomics Platforms, Gustave Roussy Cancer Campus, Villejuif, France
[c]Equipe 11 labellisée Ligue Nationale contre le Cancer, Centre de Recherche des Cordeliers, Paris, France
[d]Institut National de la Santé et de la Recherche Médicale (INSERM), UMR1138, Equipe labellisée Ligue Nationale Contre le Cancer, Paris, France
[e]Université Paris Descartes, Sorbonne Paris Cité, Paris, France
[f]Université Pierre et Marie Curie, Paris, France
[g]Suzhou Institute for Systems Medicine, Chinese Academy of Medical Sciences, Suzhou, China
[h]Pôle de Biologie, Hôpital Européen Georges Pompidou, AP-HP, Paris, France
[i]Department of Women's and Children's Health, Karolinska University Hospital, Stockholm, Sweden
*Corresponding authors: e-mail address: captain.olsen@gmail.com; kroemer@orange.fr

Contents

Abstract

Tumor cells treated by immunogenic cell death (ICD) inducers emit danger associated molecular patterns (DAMP), including but not limited to calreticulin (CALR), which translocates from the ER lumen to the surface of the cellular membrane where it serves as de novo uptake signal for antigen presenting cells of the immune system. CALR is exposed at an early stage of ICD and dictates tumor antigen transfer and therefore the immunogenicity of cancer cell death. Here, we provide a bi-color flow cytometry protocol for the quantification of ICD-associated CALR cell surface exposure in fixed samples. As compared to the detection of surface exposed CALR by confocal microscopy, the present flow cytometry-based analysis is cost-efficient and does not require sophisticated equipment. Moreover, the staining panel can be extended to a multicolor analysis for the parallel assessment of additional parameters.

Methods in Enzymology, Volume 632
ISSN 0076-6879
https://doi.org/10.1016/bs.mie.2019.05.011

1. Introduction

Calreticulin (CALR) is the most abundant calcium-binding chaperone that resides in the endoplasmic reticulum (ER) lumen. It is retained in the ER through a C-terminal KDEL retention signal, which keeps it from exiting via the secretory pathway (Michalak, Groenendyk, Szabo, Gold, & Opas, 2009). In response to the treatment with clinically-approved inducers of immunogenic cell death (ICD) such as anthracyclines, oxaliplatin or crizotinib, CALR can translocate to and is exposed at the surface of the cellular membrane (Garg et al., 2012; Obeid et al., 2007; Panaretakis et al., 2009; van Vliet & Agostinis, 2017). The translocation of CALR is preceded by a split ER stress response elicited by the therapeutic agents. This split ER stress response manifests with the activation of kinases that phosphorylate eIF2α but lacks the activation of other arms of the unfolded stress response (Bezu et al., 2018).

Surface-exposed CALR acts as a de novo uptake signal for dendritic cells (DC) from the immune system. It thus facilitates the recognition of stressed and dying cells by phagocytes expressing the CALR receptor CD91 (Brown & Neher, 2012; Gardai et al., 2005). Thus, CALR-dependent immune recognition of dying tumor cells facilitates the transfer of tumor antigens into dendritic cells (DC), favoring MHC class I-restricted antigen cross-presentation and induction of cancer-specific cytotoxic T lymphocytes (Kroemer, Galluzzi, Kepp, & Zitvogel, 2013). Thus, CALR exposure is an important hallmark of ICD ultimately triggering the induction of T cell-mediated anticancer immune responses (Galluzzi, Buque, Kepp, Zitvogel, & Kroemer, 2017). Here, we detail a simple method for detecting the surface exposure of CALR by immunofluorescence technology.

2. Cell culture

1. Human osteosarcoma U20S cells and murine non-small cell lung cancer (NSCLC) TC1 cells are routinely cultured in Dulbecco's Modified Eagle Medium (DMEM) supplemented with 10% fetal bovine serum (FBS), 100 U/mL penicillin, and 100 μg/mL streptomycin in a humidified atmosphere containing 5% CO_2 at 37 °C (see *Notes 1–3*).
2. When cultured cells reach around 80% confluence, the culture medium is removed by vacuum aspiration, and adherent cells are gently washed

once with pre-warmed PBS (see *Notes 4 and 5*). After removing PBS, cells are detached with 0.25% trypsin and resuspended with complete culture medium. Cells are centrifuged at $500 \times g$ for 5 min, supernatant is discarded and the cellular pellet is resuspended in supplemented medium. Cells are further diluted and reseeded in flasks for maintenance or 6-well plates for experiments (see *Notes 6–10*).

3. For the following treatment schedule, 4×10^5 U2OS cells or 2×10^5 TC-1 cells are seeded in tissue culture 6-well plates, allowed to adapt overnight (see *Notes 11–13*), and processed as detailed in Section 3.

3. Treatments

1. Appropriate confluence is inspected by light microscopy before the cells are subjected to treatment (see *Note 14*).
2. Culture medium is removed and cells are rapidly washed once with pre-warmed medium (2 mL/well).
3. The following controls are prepared (see *Note 15*): 2 mL fresh, complete culture medium (see *Note 16*); 2 mL fresh, complete culture medium supplemented with 150 μM cisplatin (see *Note 17*); 2 mL fresh, complete culture medium supplemented with 300 μM oxaliplatin (see *Note 17*); or 2 mL fresh, complete culture medium supplemented with 10 μM crizotinib (see *Note 18*).
4. Cells are incubated in the above control/treatment culture conditions for 6–8 h (see *Note 19*), then culture medium is removed and cells are rapidly washed once with pre-warmed medium (2 mL/well), and finally supplemented with 1.5 mL fresh complete medium. The plates are incubated for additional 18–24 h (see *Note 20*) in a cell culture incubator at 37 °C.
5. Experimental conditions applied as described above are expected to induce the exposure of CALR in early stages of cell death (Liu et al., 2019; Obeid et al., 2007; Tesniere et al., 2010).

4. Immunostaining and flow cytometric analysis

1. Cells are treated as described above (see *Note 21*), supernatants are collected in 2 mL Eppendorf tubes, and 0.5 mL trypsin is added to each

well (see *Note 22*). Once all cells have detached, a process that is monitored by light microscopy (see *Note 23*), 0.5 mL of the original supernatant is used to flush the wells by pipetting up-and-down several times, and to transfer the cell suspension into pre-labeled 2 mL Eppendorf tubes (see *Note 24*). Cells are centrifuged at 500 × *g* for 5 min at 4 °C to obtain cellular pellets.

2. Supernatant is removed by vacuum aspiration and 200 μL of ice-cold PBS is added to the pellet. Following cells are resuspended by pipetting up-and-down several times. The cell suspension is transferred into 96-well V-shape or U-shape plates (see *Note 25*) and spun at 500 × *g* for 5 min at 4 °C.
3. Supernatant is removed by decantation into an open waste container. Remaining liquid is removed by coherence by bringing it into contact with a disposable paper towel (see *Note 26*). 50 μL of prepared Zombie UV™ Fixable Viability dye solution (Biolegend REF 423108, 1:500 diluted in cold PBS, v:v) is added and the sample is mixed by pipetting up-and-down several times. Following the cells are incubated at 4 °C in the dark for 20 min (see *Note 27*). 200 μL cold PBS is added into each well and the plates are spun at 500 × *g* for 5 min at 4 °C.
4. Supernatant is removed as described before in *step 3*. 50 μL of prepared anti-CALR antibody (solution Abcam REF ab2907, 1:100 diluted in FACS buffer, 1% BSA/PBS, w:v) (see *Note 28*) is added and mixed by pipetting up-and-down several times, and incubated at 4 °C in the dark for 30 min (see *Note 29*). 200 μL cold FACS buffer is added into each well and the plates are spun at 500 × *g* for 5 min at 4 °C.
5. The supernatant is removed as described above in *step 3*. 50 μL of prepared fixation buffer (4% PFA/FACS buffer, v:v) is added and mixed by pipetting up-and-down several times. Cells are incubated at 4 °C in the dark for 15 min (see *Note 30*). And 200 μL of ice-cold FACS buffer into each well before spinning the plates at 500 × *g* for 5 min at 4 °C.
6. The supernatant is removed as described before in *step 3* and 250 μL ice-cold FACS buffer is added into each well before spinning the plates at 500 × *g* for 5 min at 4 °C (see *Note 31*).
7. *Step 6* is repeated once more (see *Note 32*).

8. The supernatant is removed as described before in *step 3*. Add 50 µL prepared Alexa Fluor® 546 goat-anti-rabbit second antibody (Invitrogen REF A-11010, 1:250 diluted in FACS buffer), mix by pipetting up-and-down several times, and incubate at 4 °C in the dark for 30 min (see *Notes 29 and 33*). Add 200 µL cold FACS buffer into each well and spin the plates at 500 × *g* for 5 min at 4 °C.
9. The supernatant is removed as described before in *step 3* and cells are resuspended in 200 µL FACS buffer. Following the plates are kept at 4 °C in the dark until acquisition on a flow cytometry analyzer (see *Note 34*).
10. Samples are diluted in PBS for flow cytometric assessment according to the manufacturer's recommendation (see *Note 35*). Samples are acquired on a FACS LSRFortessa (Becton Dickinson, Franklin Lakes, NJ, USA) or equivalent flow cytometer and data are stored in FCS file format for further analysis (see *Note 36*).
11. FCS files are analyzed by means of the FlowJo Software (FlowJo LLC, Ashland, OR, USA). Cells are gated as explained in Fig. 1 and the percentage of $CALR^+$ $DAPI^-$ cells (see *Note 37*) is measured (Fig. 2; see *Note 38*).

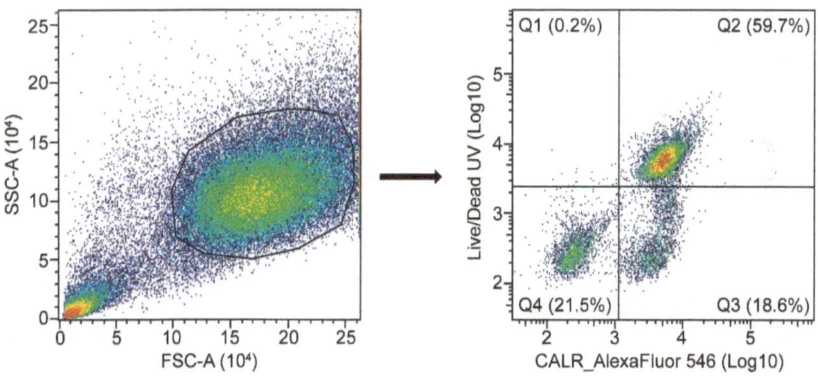

Fig. 1 Gating strategy in FlowJo to analyze CALR exposure. A forward scatter (FSC) versus side scatter (SSC) density plots is gated for a population of cells while debris are excluded (left panel). For the gated cells (circle on the left pane), the CALR-specific fluorescence (PE stands for Alexa Fluor 546) is plotted versus the fluorescence of the exclusion dye (Live/Dead UV). Quadrants placed based on negative controls define the four populations of interest which are $DAPI^-$ $CALR^-$ (Q4), $DAPI^-$ $CALR^+$ (Q3) and $DAPI^+$ (Q1 + Q2).

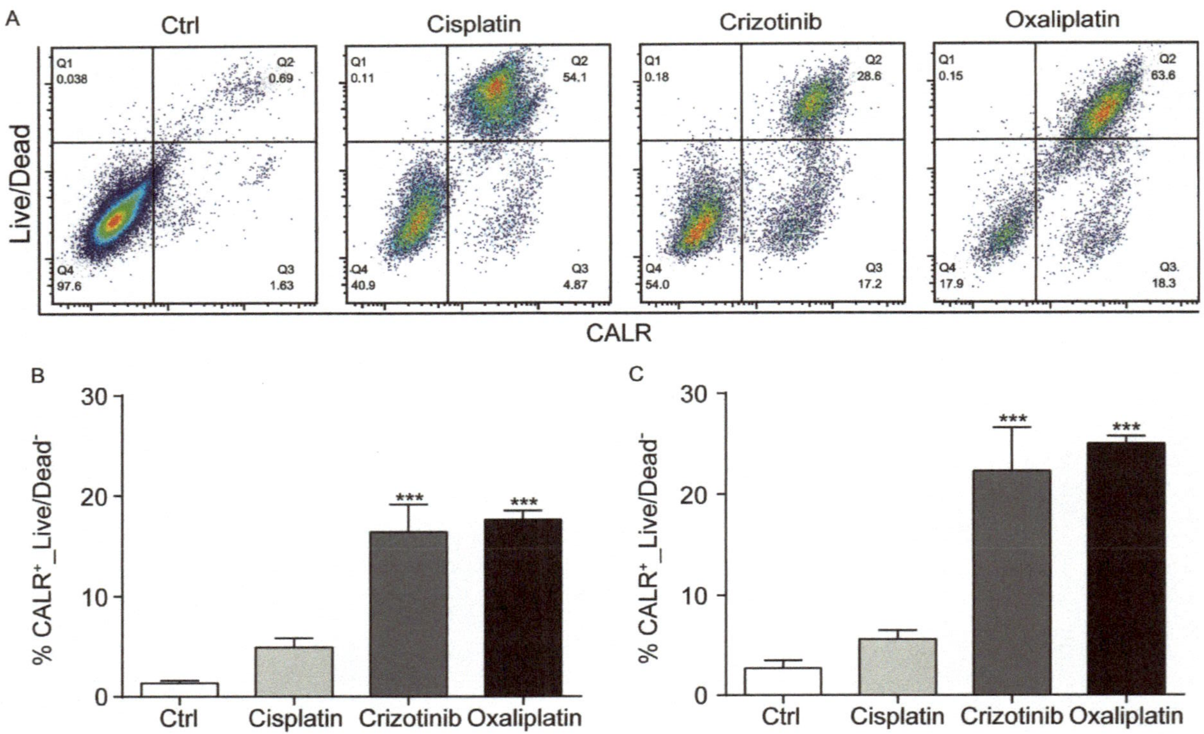

Fig. 2 Oxaliplatin- or crizotinib-induced CALR exposure in U2OS and TC-1 cells. Wild type U2OS or TC-1 cells were seeded in 6-well plates and cultured overnight before treatment with vehicle controls (Ctrl), cisplatin (150 μM), crizotinib (10 μM), or oxaliplatin (300 μM), for an incubation period of 6 h, followed by washing and an additional 18 h incubation in complete medium, as described in Sections 2 and 3. Following cells were collected and subjected to immunostaining and flow cytometric assessment and data analysis as described in Section 4. Representative plots obtained from TC-1 cell line under different treatment conditions are shown in (A), and percentages of $CALR^+$ $Live/Dead^-$ obtained from TC-1 cells (B) and U2OS cells (C) are reported as bar charts (mean ± SD; $n=3$). ***$P<0.001$ as compared to cells maintained in control conditions (Student's *t*-test).

5. Concluding remarks

The protocol presented here describes a reliable methodology to quantify the cell surface exposure of CALR following the treatment with immunogenic cell death inducers (Liu et al., 2019; Tesniere et al., 2010). This method can be used to perform medium-throughput screening campaigns aiming at the identification of bona fide ICD inducers. CALR exposure is one of the decisive hallmarks of ICD, which typically happens at early stages of cell death (Obeid et al., 2007), and can be detected with specific antibodies *before* the (permeabilizing) fixation of cells. In order to exclude permeabilized cells that would manifest with predominantly intracellular CALR staining, an exclusion dye such as DAPI or PI is routinely employed. Neither DAPI nor PI can be fixed. Therefore, this procedure requires immediate flow cytometric assessment of living cells after immunostaining, which limits the time between staining and assessment and thus the number of samples that can be processed per experimental run. In the method presented here, we extended the routine procedure with the usage of a fixable viability dye which allows fixation after the staining procedure and thus increases the number of samples because it allows for long-term storage of the samples following the first antibody staining in *step 4.5* without affecting the specificity of the analysis. This improvement offers a certain flexibility in sample handling and facilitates the adaptation of this methodology to medium- and high-throughput screening approaches.

6. Notes

1. This basic culture medium is different from the official recommendations of the American Type Culture Collection (ATCC, Manassas, VA, USA) and has been optimized according to the lab's expertise.
2. U2OS cells are conveniently maintained in 75 cm^2 tissue culture flasks (with around 1×10^7 cells at full confluence, a subcultivation ratio of 1:5 is recommended 2 times per week), while TC-1 cells are preferably maintained in 175 cm^2 tissue culture flasks (with around 2–3 $\times 10^7$ cells at full confluence, a subcultivation ratio of 1:20 is recommended 2 times per week) as they proliferate more rapidly.
3. Optimal culture media/conditions may differ for other cell lines.
4. If cells can't reach desired confluences, the cells should not be detached for passaging but only the media should be replenished until cellular

confluence reaches 80%. Similarly, cells should not overgrow. Excessively dense cultures should be discarded and new cultures should be started from frozen stocks.

5. Typically, 10 mL for 75 cm^2 flasks and 20 mL for 175 cm^2 flasks. TC1 cells are loosely adherent cells, so a direct flush of PBS on the cell culture layer should be avoided.
6. Commercially available trypsin products such as TrypLE™ Express (from Gibco®-Thermo Fisher Scientific, Carlsbad, CA, USA) are recommended if possible as they are generally more stable and more efficient than self-prepared trypsin solutions.
7. Multiple factors can lead to variable time for detachment. According to our experience, untreated U2OS cells at 70–80% confluence in 75 cm^2 flasks are generally incubated for 2 min in 3 mL TrypLE™ Express, at 37 °C in a cell culture incubator; while the detachment of TC1 cells only takes <30 s in 3 mL TrypLE™ Express at room temperature.
8. It is good practice to keep the cap tightly closed during the incubation of cells in trypsin and visually inspect the procedure under a light microscopy. When most of the cells round up, gently tap the flasks to support complete cellular detachment before adding complete growth medium to stop the enzymatic activity of trypsin. Typically, a twofold volume of complete medium is recommended to stop trypsinization.
9. Cell suspension can be centrifuged at 300–500 × *g* for 5 min at room temperature. Centrifugation is not obligatory if cells are only passaged for maintenance cultures, yet it is strongly recommended if they are used for seeding plates for experimental procedures.
10. A U2OS cell culture at 80% confluence passaged 1:5 reaches 80% confluence in approximately 2–3 days. TC-1 cultures that have reached 80% confluence must be divided as soon as possible as prolonged culture will lead to an aggregation of cells and causes cell death and cellular detachment (see also *Note 2*).
11. For subsequent experimental procedures the cellular pellet after centrifugation shall be resuspended and diluted in media supplemented only with FBS but not with antibiotics.
12. Cells are generally stained with trypan blue to exclude damaged/dead cells in the cell counting step. Cells in desired numbers are resuspended and seeded in 6-well plates (typically 2 mL/well). The plates can be optionally shaken several times in a "cross" order (i.e., 5 times moving back and forth, plus 5 times moving left and right) to support a

homogenous distribution of cells. Avoid agitating the plates in a spiraling movement, which will lead to an aggregation of cells in the center of the well.

13. The time required for cellular adherence also depends on cell type. U2OS cells show a relatively short adaptation time of 12 h. Conversely, the TC1 cells need 16–24 h to completely adhere and spread over the cell culture substrate. Of note, incompletely adapted cells may display increased sensitivity to experimental interventions.

14. It is recommended to check cellular confluence before starting the treatment as it is important to the response of cells to the treatments, especially for cells with a high proliferation rate, like TC-1. A high number of cells from the start should be avoided as it decreases the ratio of drug molecules/cell and would lead to over confluence and thus cellular stress in untreated controls before harvest, which could imply on CALR exposure.

15. To compensate for inter-well variability and facilitate statistical analysis, it is recommended to establish all control/experimental conditions in at least three replicates per experimental condition.

16. Control conditions should match the vehicles of desired chemicals, in this protocol, the control condition we used is 2 mL fresh complete medium plus 40 μL DPBS (2%) and 2 μL DMSO (0.1%) (see *Notes 17 and 18*).

17. Stock solution of oxaliplatin (Sigma REF Y0000271) is prepared at 15 mM in DPBS, stock solution of cisplatin (Sigma REF C2210000) is prepared at 1 mM in DPBS, and both of them are aliquoted in 1.5 mL Eppendorf tubes which can be stored for 12 months at −20 °C. Aliquots should be used only once and freeze-thaw cycles need to be avoided by any means. Do not dissolve oxaliplatin or cisplatin in DMSO (Hall et al., 2014). The working concentrations for oxaliplatin/cisplatin (Tesniere et al., 2010) and crizotinib (Liu et al., 2019) are chosen based on the lab's experience. Both chemically heterogenous agents that functionally converge on CALR exposure have been used as positive controls to underline the complexity of this pathway. Please explore the best working concentrations for other treatments by testing a series of different concentrations based on their LD50 values.

18. Crizotinib (Sigma REF PZ0191) stock solution is prepared at 10 mM in DMSO and aliquoted to 0.5 mL Eppendorf tubes that can be stored for 12 months at −20 °C. Less than 20 freeze-thaw cycles do not limit efficacy.

19. The treatment time applied here is optimized for oxaliplatin and crizotinib (Liu et al., 2019; Tesniere et al., 2010), which may differ when using other chemicals/treatments. Please explore the best treatment conditions for different chemicals (see also *Notes 17 and 20*).
20. This incubation period in fresh medium is flexible and can be adapted between overnight (~12 h) to 1 day (~24 h). For treatments that exhibit low toxicity the washing step might be avoided and the cells can be incubated in the treatment conditions for 24 h, which need to be explored according to the experimental conditions.
21. It is recommended to visually check the cell killing induced by the treatment. Excessive cytotoxicity that manifests with cellular detachment and aggregation should be avoided. In such case treatment conditions that would lead to a significant proportion of dying (yet not dead) cells should be (re-)established by reducing the concentration of the drug utilized for cell killing (see also *Notes 17 and 20*).
22. Please collect the supernatant as it contains dead cells. It is not mandatorily necessary to do additional washing steps before adding trypsin, yet it should be ensured that no cells culture supernatant is left in the wells before trypsinization.
23. Cells following certain treatments (e.g., oxaliplatin here) may become more resistant to detachment, putting the plates in 37 °C is recommended to support the enzymatic activity of trypsin.
24. The original cell culture supernatant can be used to stop the trypsin digestion. The maximum capacity of a 2 mL Eppendorf tube is about 2.1 mL, and should thus suffice to hold 1.5 mL supernatant plus 0.5 mL trypsin.
25. This step is optional but strongly recommend if larger numbers of samples must be stained in one experiment. Staining performed in 96-well plates are more homogeneous as the handling interval between samples are massively shortened. In addition, it saves time and labor. If testing less samples, the staining procedures can be directly done in 2 mL Eppendorf tubes and the resuspension volume can be adjusted to 1 mL.
26. This handling procedure can be applied to all the centrifugation-removing supernatant steps in the following process. It is important to always remove as much supernatant as possible.
27. The stock solution of Zombie UV™ Fixable Viability dye is prepared according to the manufacture's protocol. The concentration and incubation periods are adapted from the manufacturer's instructions to get best staining effects for U2OS and TC-1 cells collected from 6-well

plates. The Zombie UV™ Fixable Viability kit can be replaced with any other fixable viability dyes, such as the LIVE/DEAD Cell Viability probes from Thermo Fisher Scientific, according to availability. However, the staining condition might need to be adapted accordingly.

28. This antibody has been validated for immunostaining of both human and murine cell lines, other antibodies may be employed yet would need to be tested by the experimentator. FACS buffer can be 0.5–2% BSA dissolved in PBS, filtered through a 0.22 μm filter and kept at 4 °C for up to 3 months.

29. This concentration and incubation period is adapted from the manufacturer's instructions to get best staining effect in this protocol, for U2OS and TC-1 cells collected from 6-wells plates. We tested the incubation period with or without plate shaking with no major difference in the results.

30. After adding the 50 μL fixation buffer, resuspend the cell pellets immediately by pipetting. We recommend to fix the cells row by row to avoid any delay in mixing cellular suspensions.

31. Washing is mandatory as remaining PFA would negatively affect the subsequent immunostaining. It is not necessary to resuspend cells by pipetting during the washing procedure.

32. After the second wash, cells can be optionally resuspended in 200 μL FACS buffer and kept at 4 °C for up to 2 weeks until staining with the fluorescence-labeled secondary antibody right before flow cytometric acquisition.

33. Alexa Fluor® 546 can be excited with a 488 nm (or a 532 nm laser) and fits to the PE channel that is commonly installed on many flow cytometric analyzers.

34. Samples that have been stained with the fluorescence-labeled secondary antibody should be analyzed within a week as the fluorescence intensity will decrease even in the absence of extensive exposure to light. If samples need to be kept longer before acquisition, they should be stored directly after fixation (see also *Note 32*).

35. Generally, the FACS buffer mentioned above is suitable for most flow cytometers. The dilution factor will dependent on the specifications of the cytometer, but generally $1–10 \times 10^6$ cells/mL in the final suspension will be fine.

36. If using the staining panel Zombie UV™ Fixable Viability dye + Alexa Fluor® 546, it is not necessary to consider fluorescence compensation (see also *Notes 27 and 33*).

37. Here the "CALR" stands for CALR_ Alexa Fluor® 546 channel, while "DAPI" stands for the channel used to acquire the Zombie UV™ Fixable Viability dye. Restricting the analysis to DAPI negative cells is crucial here, as DAPI positive cells are permeabilized which allow for the staining of intracellular CALR epitopes, leading to false positive events (Gardai et al., 2005).
38. The DAPI channel can be also used as a parameter to estimate cytotoxicity of the treatment. If >50% of cell become DAPI positive, it may indicate a relatively high killing effect of the treatment and reducing the applied concentrations of drugs may help to improve the assessment of CALR exposure.

Acknowledgments

G.K. is supported by the Ligue contre le Cancer (équipe labellisée); Agence National de la Recherche (ANR)—Projets blancs; ANR under the frame of *E*-Rare-2, the ERA-Net for Research on Rare Diseases; Association pour la recherche sur le cancer (ARC); Cancéropôle Ile-de-France; Chancelerie des universités de Paris (Legs Poix), Fondation pour la Recherche Médicale (FRM); a donation by Elior; European Research Area Network on Cardiovascular Diseases (ERA-CVD, MINOTAUR); Gustave Roussy Odyssea, the European Union Horizon 2020 Project Oncobiome; Fondation Carrefour; High-end Foreign Expert Program in China (GDW20171100085), Institut National du Cancer (INCa); Inserm (HTE); Institut Universitaire de France; LeDucq Foundation; the LabEx Immuno-Oncology; the RHU Torino Lumière; the Seerave Foundation; the SIRIC Stratified Oncology Cell DNA Repair and Tumor Immune Elimination (SOCRATE); and the SIRIC Cancer Research and Personalized Medicine (CARPEM).

References

Bezu, L., Sauvat, A., Humeau, J., Gomes-da-Silva, L. C., Iribarren, K., Forveille, S., et al. (2018). eIF2alpha phosphorylation is pathognomonic for immunogenic cell death. *Cell Death and Differentiation*, *25*, 1375–1393.

Brown, G. C., & Neher, J. J. (2012). Eaten alive! Cell death by primary phagocytosis: 'Phagoptosis'. *Trends in Biochemical Sciences*, *37*(8), 325–332.

Galluzzi, L., Buque, A., Kepp, O., Zitvogel, L., & Kroemer, G. (2017). Immunogenic cell death in cancer and infectious disease. *Nature Reviews. Immunology*, *17*(2), 97–111.

Gardai, S. J., McPhillips, K. A., Frasch, S. C., Janssen, W. J., Starefeldt, A., Murphy-Ullrich, J. E., et al. (2005). Cell-surface calreticulin initiates clearance of viable or apoptotic cells through trans-activation of LRP on the phagocyte. *Cell*, *123*(2), 321–334.

Garg, A. D., Krysko, D. V., Verfaillie, T., Kaczmarek, A., Ferreira, G. B., Marysael, T., et al. (2012). A novel pathway combining calreticulin exposure and ATP secretion in immunogenic cancer cell death. *The EMBO Journal*, *31*(5), 1062–1079.

Hall, M. D., Telma, K. A., Chang, K. E., Lee, T. D., Madigan, J. P., Lloyd, J. R., et al. (2014). Say no to DMSO: Dimethylsulfoxide inactivates cisplatin, carboplatin, and other platinum complexes. *Cancer Research*, *74*(14), 3913–3922.

Kroemer, G., Galluzzi, L., Kepp, O., & Zitvogel, L. (2013). Immunogenic cell death in cancer therapy. *Annual Review of Immunology*, *31*, 51–72.

Liu, P., Zhao, L., Pol, J., Levesque, S., Petrazzuolo, A., Pfirschke, C., et al. (2019). Crizotinib-induced immunogenic cell death in non-small cell lung cancer. *Nature Communications*, *10*, 1486.

Michalak, M., Groenendyk, J., Szabo, E., Gold, L. I., & Opas, M. (2009). Calreticulin, a multi-process calcium-buffering chaperone of the endoplasmic reticulum. *The Biochemical Journal*, *417*(3), 651–666.

Obeid, M., Tesniere, A., Ghiringhelli, F., Fimia, G. M., Apetoh, L., Perfettini, J. L., et al. (2007). Calreticulin exposure dictates the immunogenicity of cancer cell death. *Nature Medicine*, *13*(1), 54–61.

Panaretakis, T., Kepp, O., Brockmeier, U., Tesniere, A., Bjorklund, A. C., Chapman, D. C., et al. (2009). Mechanisms of pre-apoptotic calreticulin exposure in immunogenic cell death. *The EMBO Journal*, *28*(5), 578–590.

Tesniere, A., Schlemmer, F., Boige, V., Kepp, O., Martins, I., Ghiringhelli, F., et al. (2010). Immunogenic death of colon cancer cells treated with oxaliplatin. *Oncogene*, *29*(4), 482–491.

van Vliet, A. R., & Agostinis, P. (2017). PERK and filamin A in actin cytoskeleton remodeling at ER-plasma membrane contact sites. *Molecular & Cellular Oncology*, *4*(5), e1340105.

CHAPTER TWO

Side-by-side comparison of flow cytometry and immunohistochemistry for detection of calreticulin exposure in the course of immunogenic cell death

Lenka Kasikova[a,b], Iva Truxova[a,b], Isabelle Cremer[c,d], Catherine Sautes-Fridman[c,d], Oliver Kepp[e,f,g], Guido Kroemer[e,f,g,h,i], Radek Spisek[a,b], Jitka Fucikova[a,b,*]

[a]Department of Immunology, Charles University, 2nd Faculty of Medicine and University Hospital Motol, Prague, Czech Republic
[b]Sotio, Prague, Czech Republic
[c]Team Inflammation, Complement and Cancer, INSERM, Centre de Recherche des Cordeliers, Paris, France
[d]Sorbonne Université, Paris, France
[e]Centre de Recherche des Cordeliers, Equipe labellisée par la Ligue contre le cancer, Université de Paris Descartes, Université Sorbonne Paris Cité, Université Paris Diderot, Sorbonne Université, INSERM, Paris, France
[f]Metabolomics and Cell Biology Platforms, Institut Gustave Roussy, Villejuif, France
[g]Pôle de Biologie, Hôpital Européen Georges Pompidou, AP-HP, Paris, France
[h]Suzhou Institute for Systems Biology, Chinese Academy of Sciences, Suzhou, China
[i]Department of Women's and Children's Health, Karolinska Institute, Karolinska University Hospital, Stockholm, Sweden
*Corresponding author: e-mail address: fucikova@sotio.com

Contents

Methods in Enzymology, Volume 632
ISSN 0076-6879
https://doi.org/10.1016/bs.mie.2019.05.025

Abstract

Immunogenic cell death (ICD), a functionally peculiar type of apoptosis, represents a unique way to deliver danger-associated molecular patterns (DAMPs) to the tumor microenvironment. Once emitted by dying cancer cells, DAMPs orchestrate antigen-specific immune responses by acting on both innate and adaptive components of the immune system. Accumulating preclinical and clinical evidence indicates that one of these DAMPs, calreticulin (CALR) represents a novel powerful prognostic biomarker, reflecting the activation of a clinically relevant anticancer immune response in different cancer malignancies. Therefore, the assessment of CALR emission can provide a therapeutic tool for the stratification of cancer patients and the identification of individuals that are intrinsically capable to respond to a particular treatment. Here we describe methods for the quantification of CALR exposure in the tumor microenvironment of cancer patients by flow cytometry and immunohistochemistry.

1. Introduction

Malignant cells responding to distinct types of treatment, including some chemotherapeutics and physical modalities, undergo an immunogenic form of cell death that is associated with exposure (on their surface) and release (in the tumor microenvironment) of danger signals in the form of danger-associated molecular patterns (DAMPs) (Galluzzi, Buque, Kepp, Zitvogel, & Kroemer, 2017; Kroemer, Galluzzi, Kepp, & Zitvogel, 2013). Immunogenic cell death (ICD)-relevant DAMPs encompass endoplasmic reticulum (ER) chaperones including calreticulin (CALR), heat-shock proteins (HSP70 and HSP90), the non-histone chromatin-binding nuclear protein high mobility group box 1 (HMGB1), nucleic acids, small metabolites such as ATP, as well as type I interferons (Garg et al., 2010; Krysko et al., 2012; Scaffidi, Misteli, & Bianchi, 2002; Sistigu et al., 2014). In physiological conditions, CALR functions as a Ca^{2+} binding chaperone with multiple functions inside and outside the ER, which goes largely beyond the scope of this paper and can be found in Bedard, Szabo, Michalak, and Opas, (2005) and Kroemer et al. (2013). Once translocated to the cell surface of dying tumor cells that undergo ICD, CALR can be recognized by dendritic cells of the immune system via pattern recognition receptors and contribute to the activation of a therapeutically relevant immune response (Obeid et al., 2007). Corroborating the clinical relevance of ICD, CALR exposure has been attributed prognostic and predictive values in distinct solid cancers and hematologic malignancies (Fucikova, Becht, et al., 2016; Fucikova, Truxova, et al., 2016; Fucikova et al., 2015; Wemeau et al., 2010). Here, we provide a detailed description of methods for the quantification of CALR exposure in the tumor microenvironment of cancer patients by flow cytometry and immunohistochemistry (IHC).

2. Methods to monitor CALR exposure: Principles

CALR has been shown to play a critical role in activation of both innate and adaptive immune response (Gold et al., 2010; Kroemer et al., 2013). Therefore, monitoring CALR exposure can provide a therapeutic tool for the clinical management of cancer patients. Different methods have been used to monitor CALR exposure and they were comprehensively summarized by Kepp et al. (2014). (1) IHC is widely used to analyze the presence of CALR in the tumor microenvironment. The main advantage of this technique is its applicability for the retrospective monitoring of CALR expression profiles in large cohorts of patients. However, using IHC analyses, it is not possible to efficiently distinguish between intracellular and surface-exposed CALR (Fucikova, Becht, et al., 2016; Hsu et al., 2005; Muth et al., 2016; Peng et al., 2010). (2) Flow cytometric detection of CALR exposure by the staining of non-permeabilized cells with an anti-CALR specific antibody represents the gold standard (Fucikova, Truxova et al., 2016; Obeid et al., 2007; Wemeau et al., 2010). The use of exclusion dyes such as DAPI or PI helps to subtract the population of permeabilized cells from the analysis, to avoid the staining of intracellularly located CALR. Flow cytometry can be widely used to analyze the exposure of CALR on primary tumor cells isolated from freshly resected tumor tissue.

3. Methods to monitor CALR exposure: Protocols

3.1 IHC detection and analysis

3.1.1 Deparaffinization and Rehydration

(1) Formalin-fixed paraffin embedded tissue blocks are sectioned into thin slices (3–5 μm) by microtome.

(2) To remove paraffin wax, immerse sections three times with xylen for 5 min each time.

(3) To start rehydration, immerse sections once in 100% ethanol for 5 min.

(4) Immerse sections once in 96% ethanol for 5 min.

(5) Immerse sections in 70% ethanol for 5 min.
(6) Immerse sections in 50% ethanol for 5 min.
(7) Immerse sections in dH20 for 5 min (see Note 1).

3.1.2 Target retrieval

(8) To restore epitopes which were crosslinked during the process of fixation, immerse section in citrate buffer pH 6 (Epitope Retrieval Solution, Novocastra; Leica Biosystems, RE7113) at 97 °C for 30 min.
(9) Cool section on bench top for 30 min.
(10) Rinse section with dH_20 once for 5 min.

3.1.3 Blocking endogenous peroxidase

(11) Use a hydrophobic pen (DAKO, S2002) to circumscribe the tissue (see Note 2).
(12) To block endogenous peroxidase activity, incubate the section with 200 μL of 3% hydrogen peroxide for 15 min at room temperature in a slide staining tray.
(13) Wash section with 1 × TBS buffer once for 5 min.
(14) To prevent non-specific binding of the antibody to the tissues, block each section with 1–3 drops of serum blocking reagent (DAKO, X0909) for 15 min in a slide staining trays (see Note 3).

3.1.4 Primary antibody incubation

(15) Dilute mouse anti-CALR primary antibody (Table 1) with Antibody Dilution Buffer (DAK0, S2022).
(16) Add 200 μL of diluted primary antibody to each section and incubate at room temperature for 2 h in a slide staining tray.
(17) Rinse the excess antibody from the section with one gentle movement and wash section with 1 × TBS buffer with 0.04% Tween 20 once for 5 min.

Table 1 Primary and secondary antibodies used for the immunohistochemistry detection of CALR expression.

Antibody	Cat. no.	Source	Producer	Clone	Dilution
Anti-CALR	ab22683	mouse	Abcam	FMC 75	1:200
Anti-mouse	715-066-150	donkey	Jackson ImmunoResearch	–	1:500

3.1.5 *Secondary antibody and streptavidin-HRP*

(18) Dilute biotin-labeled secondary anti-mouse antibody (Table 1) with 1× TBS.

(19) Add 200μL of diluted biotin-labeled secondary anti-mouse antibody to each section and incubate at room temperature for 30min in a slide staining tray.

(20) Rinse the excess antibody from the section with one gentle movement and wash section with 1× TBS buffer 0.5% Tween 20 once for 5min.

(21) Dilute streptavidin-HRP solution (DAKO, P0397) with 1× TBS (1:300).

(22) Add 200μL of diluted streptavidin-HRP to each section and incubate at room temperature for 30min in a slide staining tray.

(23) Rinse the excess streptavidin-HRP from the section with one gentle movement and wash section with 1× TBS buffer with 0.5% Tween 20 once for 5min.

3.1.6 *DAB staining*

(24) Add 200μL of DAB Reagent (DAKO, K3468) to each section and react for 2min to stain.

(25) Rinse sections with 1× TBS buffer to stop reaction.

(26) Rinse in dH_20 and drain the section.

3.1.7 *Hematoxylin staining and mounting on coverslips*

(27) Counterstain sections in hematoxylin reagent (DAKO, S3309).

(28) Mount coverslips with Glycergel mounting medium (DAKO, C0563).

3.1.8 *Image analysis*

The expression level of CALR for each section was determined as the score of positive tumor cells as published previously (Fucikova, Becht, et al., 2016). The score index was calculated for 10 independent fields at 20× magnification under a light microscope (DM2000LED; Leica) and classified into five categories (score 1: <10% positive cells, score 2: 10–25% positive cells, score 3: 26–50% positive cells, score 4: 51–75% positive cells, score 5: >75% positive cells) (Fig. 1).

3.2 Flow cytometric detection and analysis

(1) Resect the fresh tumor tissue and cut it into small pieces.

(2) Transfer the tissue into the gentleMACS C Tube containing 10mL of PBS buffer.

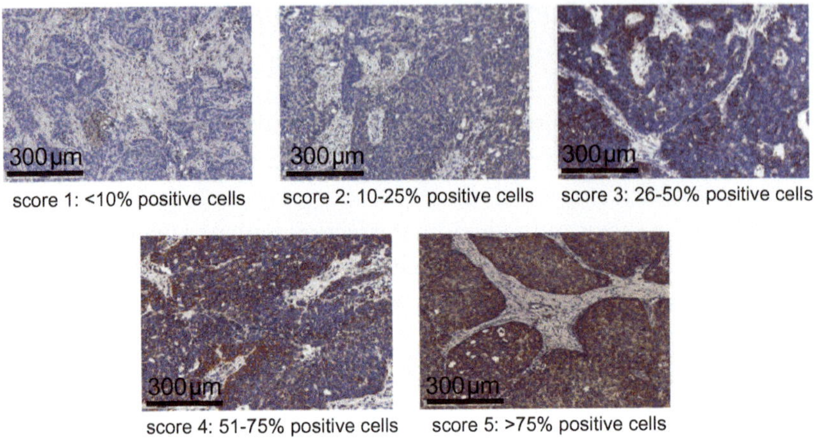

Fig. 1 Representative images of CALR immunostaining. Scale bar = 300 µm.

(3) Insert the tube into gentleMACS™ Dissociator and run the gentleMACS Program Tumor_01.

(4) Add 1 mg/mL of Collagenase II and 100 µg/mL of DNase I (see Note 4).

(5) Incubate the tube on a rotating platform (18 rpm) in an incubator (37 °C, 5% CO_2) for 30 min.

(6) Transfer gentleMACS C Tube into gentleMASC™ dissociator and run program gentleMACS Program Tumor_01).

(7) Strain the cell suspension with a 70 µm cell strainer into a new 50 mL conical tube and wash cells with PBS.

(8) Centrifuge collected cell suspension at 270 ×*g* at room temperature for 10 min, discard supernatant.

(9) Wash 50 mL of PBS through the 70 µm strainer into the new tube.

(10) Transfer cell suspension into the 96-well V-bottom plate.

(11) Dilute mouse anti-CALR primary antibody with PBS (Table 2)

(12) Add diluted mouse anti-CALR primary antibody and incubate at 4 °C for 20 min.

(13) Wash the cells twice with PBS. Centrifuge at 423 ×*g* for 5 min at room temperature. Discard supernatant.

(14) Dilute APC-labeled secondary antibody with PBS (Table 2)

(15) Add secondary antibody to the cell suspension and incubate at 4 °C for 20 min

(16) Wash the cells twice with PBS. Centrifuge at 423 ×*g* for 5 min at room temperature. Discard supernatant.

Table 2 Antibodies used for the flow cytometry detection of CALR expression.

Antibody	Cat. no.	Source	Producer	Clone	Fluorochrome	Dilution
Calreticulin	ADI-SPA-601-F	Mouse	Enzo Life Sciences, Inc.	FMC 75	–	2.4:100
Allophycocyanin-conjugated anti-mouse	115–136-068	Goat	Jackson ImmunoResearch	–	APC	3:100
Anti-Pan cytokeratin	53–9003-82	Murine	eBioscience	AE1/AE3	A488	2:100
Epitelial antigen	F0860	Mouse	DAKO	Ber-EP4	FITC	2:100
Epcam	324204	Mouse	BioLegend	9C4	FITC	2:100
CD227	559774	Mouse	BD Biosciences	HMPV	FITC	2:100
CD45	ED7029	Mouse	EXBIO	MEM-28	PerCP	4:100

(17) Add the mixture of primary antibodies: anti-EpCAM, anti-CD227, anti-Epithelial antigen, anti-Pan cytokeratin, anti-CD45 and incubate at 4 °C for 20 min (Table 2) (see Note 5).

(18) Wash the cells twice with PBS. Centrifuge at 423 ×*g* for 5 min at room temperature. Discard supernatant

(19) Use Annexin V/DAPI staining kit according to the manufacturer instructions (eBioscience, Thermo Fisher) to determine the population of apoptotic/necrotic cells (see Note 6).

(20) Analyze the cell suspension by flow cytometry (see Note 7) (Fig. 2)

3.2.1 Analysis

Expression of ecto-CALR was analyzed on both living and early apoptotic (DAPI$^-$) tumor cells identified as a population of CD45$^-$, EpCAM$^+$,

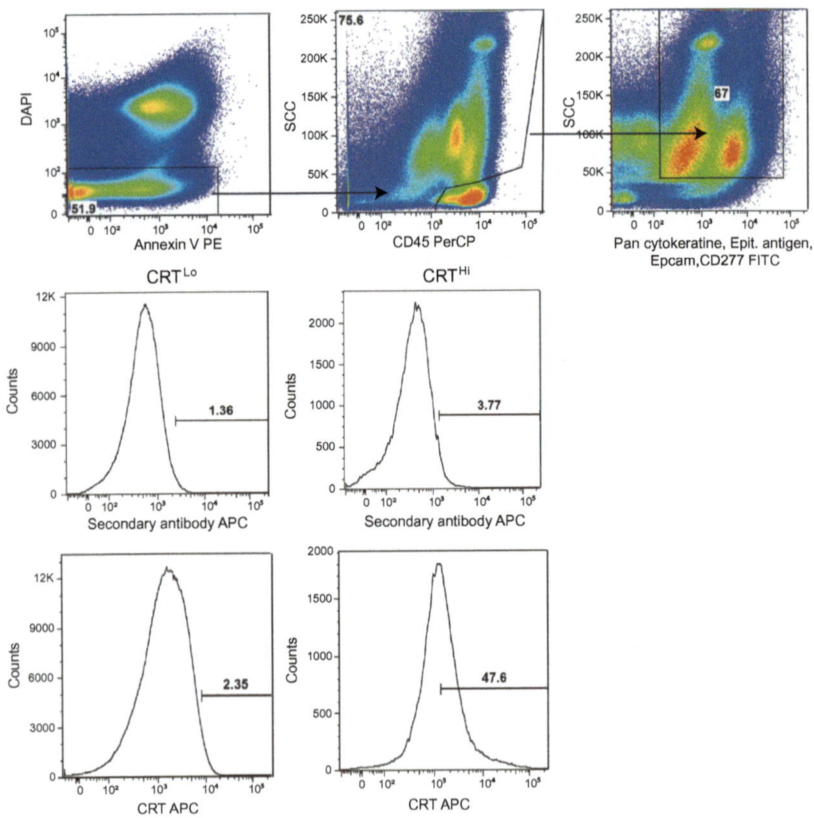

Fig. 2 Flow cytometry-assisted quantification of surface exposed CALR.

$CD227^+$, epithelial antigen$^+$ and pan-cytokeratin$^+$ as indicated in the Table 2. A representative gating strategy of ecto-CALR for $CALR^{Lo}$ and $CALR^{Hi}$ patients is indicated in the Fig. 2.

4. Concluding remarks

CALR represents a multi-functional chaperon protein, which is involved in a wide spectrum of cellular processes, including calcium homeostasis, folding of ER proteins, cell adhesion and migration (Bedard et al., 2005; Gold et al., 2010). CALR is best characterized for its function as an eat-signal, leading to the maturation of dendritic cells and activation of adaptive immune response against tumors (Obeid, Tesniere, et al., 2007). The accumulating preclinical and clinical evidences indicate that CALR represents a novel powerful prognostic biomarker for cancer patients (Fucikova et al., 2015). Therefore, monitoring CALR exposure may provide powerful clinical tool for stratifying cancer patients into different risk groups and identifying the fraction of patients suitable for the particular therapeutic regimen. Here we provide a detailed protocol of two major assays (flow cytometry and immunohistochemistry) for the detection of CALR exposure in the tumor microenvironment.

5. Notes

(1) Do not allow sections to dry at any time until ready to perform antigen retrieval. Drying will cause non-specific antibody binding and therefore high background staining.

(2) Avoid touching the tissue during the staining procedure, which creates a hydrophobic boundary on the sample.

(3) Drain the slides and wipe away any excess blocking reagent before proceeding to the next step.

(4) Collagenase and DNase enzyme are used for chemical or enzymatic digestion of tissue.

(5) Composition of antibodies for the identification of tumor population is dependent on cancer type.

(6) Annexin V is commonly used to stain phosphatidylserine residues exposed during apoptosis. DAPI is a cell-impermeant dye that stains double-strand nucleic acids and allows for the identification of necrotic cell population.

(7) Cells should be analyzed within 2 h after the initial incubation period. For prolonged periods, store at 4 °C and protect from light until analysis.

Acknowledgments

This study was exclusively sponsored by Sotio, Prague, Czech Republic.

References

Bedard, K., Szabo, E., Michalak, M., & Opas, M. (2005). Cellular functions of endoplasmic reticulum chaperones calreticulin, calnexin, and ERp57. *International Review of Cytology*, *245*, 91–121.

Fucikova, J., Becht, E., Iribarren, K., Goc, J., Remark, R., Damotte, D., et al. (2016). Calreticulin expression in human non-small cell lung cancers correlates with increased accumulation of antitumor immune cells and favorable prognosis. *Cancer Research*, *76*, 1746–1756.

Fucikova, J., Moserova, I., Urbanova, L., Bezu, L., Kepp, O., Cremer, I., et al. (2015). Prognostic and predictive value of DAMPs and DAMP-associated processes in cancer. *Frontiers in Immunology*, *6*, 402.

Fucikova, J., Truxova, I., Hensler, M., Becht, E., Kasikova, L., Moserova, I., et al. (2016). Calreticulin exposure by malignant blasts correlates with robust anticancer immunity and improved clinical outcome in AML patients. *Blood*, *128*, 3113–3124.

Galluzzi, L., Buque, A., Kepp, O., Zitvogel, L., & Kroemer, G. (2017). Immunogenic cell death in cancer and infectious disease. *Nature Reviews Immunology*, *17*, 97–111.

Garg, A. D., Nowis, D., Golab, J., Vandenabeele, P., Krysko, D. V., & Agostinis, P. (2010). Immunogenic cell death, DAMPs and anticancer therapeutics: An emerging amalgamation. *Biochimica et Biophysica Acta*, *1805*, 53–71.

Gold, L. I., Eggleton, P., Sweetwyne, M. T., Van Duyn, L. B., Greives, M. R., Naylor, S. M., et al. (2010). Calreticulin: Non-endoplasmic reticulum functions in physiology and disease. *The FASEB Journal*, *24*, 665–683.

Hsu, W. M., Hsieh, F. J., Jeng, Y. M., Kuo, M. L., Chen, C. N., Lai, D. M., et al. (2005). Calreticulin expression in neuroblastoma—A novel independent prognostic factor. *Annals of Oncology: Official Journal of the European Society for Medical Oncology*, *16*, 314–321.

Kepp, O., Senovilla, L., Vitale, I., Vacchelli, E., Adjemian, S., Agostinis, P., et al. (2014). Consensus guidelines for the detection of immunogenic cell death. *Oncoimmunology*, *3*, e955691.

Kroemer, G., Galluzzi, L., Kepp, O., & Zitvogel, L. (2013). Immunogenic cell death in cancer therapy. *Annual Review of Immunology*, *31*, 51–72.

Krysko, D. V., Garg, A. D., Kaczmarek, A., Krysko, O., Agostinis, P., & Vandenabeele, P. (2012). Immunogenic cell death and DAMPs in cancer therapy. *Nature Reviews. Cancer*, *12*, 860–875.

Muth, C., Rubner, Y., Semrau, S., Ruhle, P. F., Frey, B., Strnad, A., et al. (2016). Primary glioblastoma multiforme tumors and recurrence: Comparative analysis of the danger signals HMGB1, HSP70, and calreticulin. *Strahlentherapie und Onkologie: Organ der Deutschen Rontgengesellschaft [et al]*, *192*, 146–155.

Obeid, M., Panaretakis, T., Joza, N., Tufi, R., Tesniere, A., van Endert, P., et al. (2007). Calreticulin exposure is required for the immunogenicity of gamma-irradiation and UVC light-induced apoptosis. *Cell Death and Differentiation*, *14*, 1848–1850.

Obeid, M., Tesniere, A., Ghiringhelli, F., Fimia, G. M., Apetoh, L., Perfettini, J. L., et al. (2007). Calreticulin exposure dictates the immunogenicity of cancer cell death. *Nature Medicine*, *13*, 54–61.

Peng, R. Q., Chen, Y. B., Ding, Y., Zhang, R., Zhang, X., Yu, X. J., et al. (2010). Expression of calreticulin is associated with infiltration of T-cells in stage IIIB colon cancer. *World Journal of Gastroenterology*, *16*, 2428–2434.

Scaffidi, P., Misteli, T., & Bianchi, M. E. (2002). Release of chromatin protein HMGB1 by necrotic cells triggers inflammation. *Nature*, *418*, 191–195.

Sistigu, A., Yamazaki, T., Vacchelli, E., Chaba, K., Enot, D. P., Adam, J., et al. (2014). Cancer cell-autonomous contribution of type I interferon signaling to the efficacy of chemotherapy. *Nature Medicine*, *20*, 1301–1309.

Wemeau, M., Kepp, O., Tesniere, A., Panaretakis, T., Flament, C., De Botton, S., et al. (2010). Calreticulin exposure on malignant blasts predicts a cellular anticancer immune response in patients with acute myeloid leukemia. *Cell Death & Disease*, *1*, e104.

CHAPTER THREE

Quantitative determination of phagocytosis by bone marrow-derived dendritic cells via imaging flow cytometry

Giulia Cerrato[a,b,c,d], Peng Liu[a,b,c,d], Isabelle Martins[a,b,c,d,e], Oliver Kepp[a,b,c,d,e,*], Guido Kroemer[a,b,c,d,e,f,g,h,i,*]

[a]Metabolomics and Cell Biology Platforms, Gustave Roussy Cancer Campus, Villejuif, France
[b]INSERM, U1138, Paris, France
[c]Equipe 11 labellisée par la Ligue Nationale contre le Cancer, Centre de Recherche des Cordeliers, Paris, France
[d]Université Paris Descartes/Paris V, Sorbonne Paris Cité, Paris, France
[e]Université Pierre et Marie Curie/Paris VI, Paris, France
[f]Faculté de Medicine, Université Paris Saclay/Paris XI, Le Kremlin-Bicêtre, France
[g]Suzhou Institute for Systems Medicine, Chinese Academy of Medical Sciences, Suzhou, China
[h]Pôle de Biologie, Hôpital Européen Georges Pompidou, AP-HP, Paris, France
[i]Department of Women's and Children's Health, Karolinska University Hospital, Stockholm, Sweden
*Corresponding authors: e-mail address: captain.olsen@gmail.com; kroemer@orange.fr

Contents

Abstract

Immunogenic cell death (ICD), induced by certain anticancer chemotherapeutics, leads to the emission of danger associated molecular patterns (DAMP) by cancer cells, which

Methods in Enzymology, Volume 632
ISSN 0076-6879
https://doi.org/10.1016/bs.mie.2019.07.021

facilitates the attraction, activation and maturation of dendritic cells (DC) as well as the subsequent priming of effector T cells. In this context calreticulin (CALR) exposed at an early stage of ICD at the surface of the cancer cells serves as phagocytic signal and triggers the formation of immunological synapses between malignant cells and DC. Subsequent phagocytosis facilitates the transfer of tumor associated antigen and thus depicts a fundamental step in the generation of anticancer immunity. Here we provide an imaging flowcytometric protocol for the quantification of ICD-associated DC phagocytosis of cancer cells. As compared to the traditional flowcytometry-based analysis, the presented method offers additional means of differentiation between the transient formation of immunological synapses and the final DC-mediated phagocytosis of cancer cells.

1. Introduction

Long-lasting therapeutic effects of anticancer regimen beyond treatment discontinuation depend on the stimulation of tumor associated antigen-specific immunity (Galluzzi, Chan, Kroemer, Wolchok, & Lopez-Soto, 2018; Kroemer, 2017; Senovilla et al., 2012; Vacchelli et al., 2015). Quality and quantity of tumor neoantigens have been associated with the therapeutic success of anticancer immunotherapies (Galluzzi et al., 2018). One way to trigger adaptive anticancer immunity and to generate immunological memory is the induction of immunogenic cell death (ICD). Thus, exposing cancer cells to specific cytotoxicants, such as anthracyclins, oxaliplatin or crizotinib stimulate the emission of danger associated molecular patterns (DAMPs) that operate as adjuvants and boost anticancer immune responses (Galluzzi, Buque, Kepp, Zitvogel, & Kroemer, 2017; Kepp, Loos, Liu, & Kroemer, 2017; Liu, Zhao, Kepp, & Kroemer, 2019; Pitt, Kroemer, & Zitvogel, 2017).The mechanisms that are at the basis of ICD include lysosomal ATP secretion, translocation of calreticulin (CALR) to the cell surface (Obeid et al., 2007), production of type-1 interferon (IFN) responses (Sistigu et al., 2014), as well as the exodus of annexin A1 (ANXA1) and nuclear high mobility group box 1 (HMGB1) (Apetoh et al., 2007; Vacchelli et al., 2015).

Dendritic cells are attracted into the tumor microenvironment by a gradient of ATP that ligates P2Y2 purinergic receptors on DC. ANXA1 in this context facilitates the final DC homing *via* binding to formyl peptide receptor 1 (FPR1), and CALR acts as eat me signal on CD91 for antigen transfer into DC, subsequently HMGB1 ligates TLR4 and triggers the maturation of DC, and type I interferon facilitates the stimulation of T cells. The uptake of (parts of the) dying tumor cells by phagocytic DC provides the basis

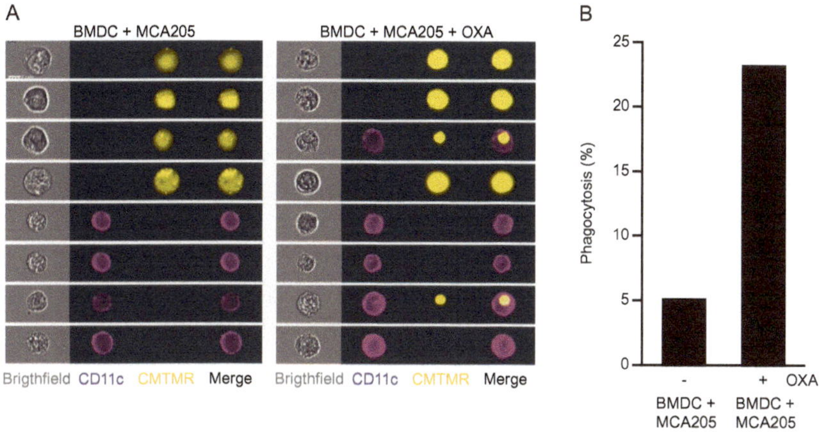

Fig. 1 Dendritic cell phagocytosis stimulated by oxaliplatin. (A and B) Bone marrow-derived dendritic cells (BMDC) were harvested from femur and tibia of C57BL/6 mice and were differentiated for 7 days in the presence of GM-CSF. Syngeneic murine fibrosarcoma MCA205 cells were stained with 0.5 μM CellTracker Orange (CMTMR) and were then treated for 24 h with 500 μM oxaliplatin (OXA). Then MCA205 were co-cultured with differentiated BMDC for 4 h. Cells were collected and BMDC were immunostained with Pacific Blue-labeled anti-mouse CD11c antibody before fixation and analysis by imaging flow cytometry (ImageStream, Amnis). (A) Representative images of MCA205 untreated and treated with OXA (brightfield and PE channel) and BMDC (brightfield and Pacifc Blue channel) are depicted. The scale bar equals to 7 μm. (B) The ratio of $CMTMR^{+}CD11c^{+}$ cells among $CD11c^{+}$ BMDC is depicted.

for tumor-associated antigen transfer from malignant cells to the immune system. This process favors MHC class I-restricted antigen cross-presentation and eventually leads to the induction of cancer-specific cytotoxic T lymphocytes (Kroemer, Galluzzi, Kepp, & Zitvogel, 2013). DC-mediate phagocytosis of cancer cells thus constitutes a quintessential step in the generation of anticancer immunity (Galluzzi et al., 2017) (Fig. 1).

2. Material

2.1 Disposables

1. Polystyrene flasks (75 and 25 cm^2) and 6-well plates for cell culture
2. Cell scrapers and cell lifters
3. Conical polypropylene centrifuge tubes (15 and 50 mL)
4. Pipet-aid and serological pipettes (5 and 10 mL)
5. Micropipettes (0.2–2 μL, 1–20 μL, 20–200 μL, 200–1000 μL) and tips
6. Cell counting slides
7. 2 mL microcentrifuge tubes

3. Equipment

1. Humidified cell culture incubator allowing for standard cell culture conditions (37 °C, 5% CO_2)
2. Standard bench top centrifuge
3. Cell culture laminar flow hood safety level II
4. Imaging flow cytometer with the appropriate software, *e.g.*, Amnis (ImageStream, Seattle, WA, USA) with IDEAS software (Luminex Corporation, Austin, TX, USA)

4. Reagents

1. Adequate culture medium supplemented with serum (*e.g.*, 10% v/v fetal bovine serum) as well as 1% v/v non-essential amino acids, 1 mM sodium pyruvate, 10 mM 2-[4-(2-hydroxyethyl)piperazin-1-yl]ethanesulfonic acid (HEPES) buffer and antibiotics (*e.g.*, 100 U/mL penicillin G and 100 μg/mL streptomycin sulfate) if required. Store at 4 °C
2. 0.05–0.25% Trypsin solution with EDTA. Store at 4 °C
3. 0.4% Trypan blue solution
4. 1 × Phosphate-buffered saline (PBS) at pH 7.4
5. Red blood cell lysis buffer (*e.g.*, RBC Lysis Buffer, BioLegend, San Diego, CA, USA)
6. CellTracker Dye (*e.g.*, CellTracker Orange CMTMR Dye, Invitrogen, Carlsbad, CA, USA). Store at −20 °C under protection from light
7. Anti-mouse CD11c antibody (*e.g.*, Pacific Blue anti-mouse CD11c antibody, BioLegend, San Diego, CA, USA). Store at 4 °C under protection from light
8. *Fixation solution*: 4% PFA v/v in PBS. Store at 4 °C
9. *FACS buffer:* 1% BSA w/v in PBS. Store at 4 °C

5. Animals

1. Six-week-old female C57BL/6 mice were purchased from Envigo France (Huntingdon, UK) and were housed in the animal facility at the Gustave Roussy Campus Cancer in pathogen–free and environmental-controlled cages with 12 h day, 12 h night cycles and received food and water *ad libitum*.

6. Fibrosarcoma cell culture

1. Murine fibrosarcoma MCA205 cells are routinely maintained in standard culture conditions (at 37 °C, under 5% CO_2) and cultured in Roswell Park Memorial Institute (RPMI)-1640 medium supplemented with 10% fetal bovine serum (FBS), 1 mM sodium pyruvate and 10 mM HEPES buffer (see **Note 1**). After thawing, the cells are maintained in culture in complete growth medium at least 1 week.
2. Once maintenance cultures reach 75–80% confluence, exhausted culture medium is discarded (see **Notes 2 and 3**). Thereafter, the cell monolayer is detached with 0.25% trypsin/EDTA, diluted and reseeded (see **Notes 4 and 5**). This provides cells for maintenance culture as well as for experimental procedures.
3. 4×10^5 fibrosarcoma MCA205 cells are seeded in standard 25 cm^2 polystyrene flasks for cell culture, allowed to adapt and recover standard proliferation for 24 h, and processed as detailed in Section 5.

7. Bone marrow-derived dendritic cells (BMDC) culture

1. BMDC are isolated postmortem from C57BL/6 using a sterile scalpel. Femur and the tibia are surgically removed above the hip joint. Tissue is carefully removed and femur and tibia are kept in a sterile tube filled with PBS on ice (see **Notes 6–8**).
2. The exterior of both femur and tibia are sterilized (see **Note 9**) by passing them through 70% ethanol for 15 s. Following femur and tibia are washed with sterile PBS.
3. The extremities of both femur and tibia are cut with sterile scissors as close as possible to the joints. Bones are purged with PBS supplemented with 1% BSA with a syringe inserted into the bone and the bone marrow is collected in sterile centrifuge tubes. Bones are repeatedly purged until all bone marrow has been collected (see **Note 10**). Remaining cellular aggregates are dissolved by pipetting.
4. Cells are centrifuged at 200*g* for 5 min and supernatant is discarded
5. Red blood cells are lysed by resuspending cells in 1 mL RBC lysis buffer per bone, incubation for 3 min, followed by the addition of an excess volume of (10 ×) of RPMI-1640 complete medium. The cell suspension is passed through a 70 μm cell strainer and centrifuged at 200*g* for 5 min (see **Note 11**) and the supernatant is discarded.

6. The cell pellet is resuspended in RPMI-1640 complete medium supplemented with 50 nM β-mercaptoethanol and 20 ng/mL GM-CSF. Cellular viability is evaluated by trypan blue staining in a hemocytometer using a light microscope. Following cells are seeded in 6-well plates at a density of 1.5×10^6 viable cells in 2 mL medium per well (see **Note 12**) and incubated at 37 °C, 5% CO_2.
7. At day 3 additional 1 mL of RPMI-1640 complete medium supplemented with 20 ng/mL GM-CSF and 50 nM β-mercaptoethanol are added. At day 6 half of the supernatant is removed, centrifuged at 200*g* for 5 min, resuspended in fresh RPMI-1640 complete medium supplemented with 20 ng/mL GM-CSF and 50 nM β-mercaptoethanol and added back to the original culture (see **Notes 13 and 14**). Cells can be harvested on day 7 (see **Note 15**).

8. Labelling and chemotherapeutic treatment of fibrosarcoma

1. Appropriate cell confluence is frequently inspected *via* light microscopy (see **Note 16**)
2. Stock solution of CellTracker™ Orange CMTMR (5-(and-6)-(((4-chloromethyl)benzoyl)amino) tetramethylrhodamine (Thermo Fisher Scientific, Waltham, Massachusetts, USA) is diluted to a final working concentration of 0.5 μM in serum-free medium and pre-warmed to 37 °C (see **Note 17**).
3. Exhausted cell culture medium is removed and the pre-warmed CellTracker™ working solution is added (1 mL per flask). Cells are incubated for 30 min at 37 °C.
4. In the meantime oxaliplatin is diluted to 500 μM final concentration in fresh complete culture medium and vortexed (see **Note 18**).
5. CellTracker™ working solution is removed and 5 mL of oxaliplatin solution is added. Cells are incubated at 37 °C, 5% CO_2 for additional 24 h.

9. Fibrosarcoma and dendritic cell co-culture

1. MCA205 cells are harvested by collecting the cell culture supernatant containing dead cells as well as adherent cells that have been detached with 0.25% trypsin/EDTA into a sterile tube. Cells are centrifuged at 200*g* for 5 min and resuspended in PBS, centrifuged again at 200*g* for 5 min and eventually resuspended in RPMI-1640 complete medium.

2. BMDC are harvested by collecting cell culture supernatant containing non-adherent cells in a sterile tube together with less-adherent cells obtained by washing the cell layer with medium and by gently tapping the flask (see **Note 19**). Cells are centrifuged at 200*g* for 5 min.
3. MCA205 cells treated or not and BMDC are co-cultured in 6-well plates in a 1:4 ratio (BMDC:MCA205) (*e.g.*, 2×10^5 BMDC and 8×10^5 MCA205 cells) (see **Note 20**). Cells are separated by pipetting up and down and plates are incubated at 37 °C, 5% CO_2 for 4 h.

10. Sample preparation and immunostaining

1. Following the co-culture, cell culture supernatants and cells carefully detached with a cell scraper are collected into sterile centrifuge tubes. Tubes are centrifuged at 200*g* for 5 min, medium is discarded and cells are re-suspended in FACS buffer (see **Note 21**).
2. FACS buffer is discarded, 100 µL of CD11c anti-mouse antibody (N418) Pacific Blue diluted 1:200 in FACS buffer is added and samples are incubated at 4 °C for 30 min (see **Note 22**)
3. Three subsequent washing steps are performed by adding FACS buffer and centrifuging at 200 g for 5 min.
4. An additional washing step is performed by adding FACS buffer and centrifuging at 200 g for 5 min. Two-hundred µL of 4% PFA in FACS buffer are added in each tube and the cells are incubated for 20 min at 4 ° C (see **Notes 23 and 24**). Two subsequent washings step are performed as described in step 3.

11. Imageflowcytometric acquisition and data analysis

1. Samples are prepared in microcentrifuge tubes. 50 µL of cell suspension containing around 1×10^6 cells are diluted in FACS buffer (see **Note 25**).
2. Fluorescent channels of the imaging cytometer (ImageStream, Amnis, Seattle, USA) are chosen according to fluorochromes. Ten-thousand images are acquired in using the bright field channel and the fluorescence emission channels 7 (bandpass filter 430–505 nm) and 3 (bandpass filter 560–595 nm) for each sample. Laser intensities are set to 70 mW for 375 nm laser, 200 mW for 561 nm.
3. Unlabeled cells as well as cells labeled with only one fluorochrome are used to adjust spectral compensation.

4. Samples are passed through the AMNIS imaging flow cytometer using a 60 × objective with a velocity of 44.0 (mm/ms) (see **Note 26**).
5. Analysis: The acquired images are analyzed with IDEAS software. First, cells in focus are selected by gating the gradient RMS (root means square for image sharpness) histogram. Second, a single cell gating is performed by means of a scatter plot of aspect ratio/area. Image analysis is performed on single cells in focus to determine the percentage of CMTMR and Pacific Blue double positive cells.

12. Concluding remarks

The protocol presented here describes a novel methodology to monitor dendritic cell (DC)-mediated phagocytosis of cancer cells by imaging flow cytometry. The procedure offers the possibility to perform low-throughput secondary screening campaigns aiming at the identification of novel ICD inducers from collections of drugs pre-selected from primary phenotypic screens. Imaging flow cytometry is a sophisticated technology and, apart from the initial investment, requires experienced handling for the generation of solid data. Nevertheless, compared to traditional flow cytometry-based methods, imaging flow-cytometry offers the advantage of a clear distinction between transient cell-to-cell contact and cell-in-cell phagocytosis. It therefore helps to distinguish distinct biological phenomena and to establish a chronology of biologically distinct events. Moreover, the procedure presented here is rather flexible and can be extended by the additional assessment of stress markers in cancer cells and activation and maturation markers in DC.

13. Notes

1. MCA205 cells can be conveniently maintained in 25-, 75- or 175 cm^2 flasks for cell culture.
2. Other cell lines might have different culture conditions.
3. For maintaining optimal culture conditions, MCA205 cells can be split at 1:15–1:20 every 2–3 days to reach approximately 80% confluence. Cell culture density is a major parameter to prevent metabolic, nutritional and/or oxidative stress in case of growth >80% or genetic drifts of the original cell population due to excessively low confluence (<30%).

4. Multiple factors play a role in the detachment time, *e.g.*, cell type, confluence, presence of experimental agents in the culture medium. <1 min is required to efficiently detach MCA205 cells at 70–80% confluence. A visual inspection by light microscopy is recommended.
5. Cells should be washed once with PBS and then incubated with trypsin/EDTA for a sufficient time to detach the cellular monolayer. An extended time of incubation could have negative impact on viability.
6. To minimize impact on cellular viability, all reagents and samples need to be kept on ice during the procedure.
7. Fractured bones need to be excluded as bone marrow sterility might be affected.
8. The femur generally contains more bone marrow cells than the tibia.
9. The time of immersion in ethanol is crucial as a prolonged exposure could affect the viability of the bone marrow cells.
10. Purged bones will appear completely white. Repeat the procedure for all the bones.
11. The cell pellet should have a white appearance and a red ring of blood-cell debris should be visible around the outer surface of the pellet.
12. Cells can also be cultured in flasks for cell culture. As there is a risk of contamination, it is preferable to subdivide cells into different wells or flasks.
13. Gently add and remove medium to avoid affecting cellular homeostasis.
14. At day 3 the dendritic cells are only loosely adherent.
15. Procedure of day 6 can be repeated to harvest cells as late as day 9.
16. Underconfluence should be avoided as excessively subconfluent cells generally display increased sensitivity to various forms of stress. Subconfluent murine fibrosarcoma MCA205 cells exposed to treatment will die faster as compared to cells at optimal confluence. Along similar lines overconfluence needs to be strictly avoided as baseline viability will be affected.
17. Cellular staining should be chosen according to the spectral properties of the drug used for treatment. In addition, any kind of spectral overlap has to be avoided. Other fluorescent probes are commercially available to perform this step, *e.g.*, CellTracker™ Blue, Violet, Deep Red, Green CMFDA.
18. Any other immunogenic cell death inducer can be used at this step to treat the cells.
19. Dendritic cells (DC) are mostly non-adherent or loosely adherent, while the majority of adherent cells are macrophages. In tissue-treated plates and flasks DC can also be adherent.

20. The ratio of MCA205 to BMDC as well as the incubation time and the cellular numbers can be adjusted according to the experimental conditions. Four hours of co-culture allows for up to 25% of phagocytosis. If cell-to-cell interaction needs to be studied a shorter incubation time would be advisable.
21. This additional washing step limits drug carryover that could prolong the treatment duration.
22. Other anti-mouse CD11c antibody are commercially available and the fluorescent conjugate has to be chosen to avoid overlap with MCA205 staining (*e.g.* Pacific Blue-CMTMR, CFSE-APC, *etc.*).
23. As it is a volatile carcinogen, PFA has to be manipulated under a laminar flow and disposed in appropriate reservoirs.
24. Samples can be stored up to 2 weeks at 4 °C.
25. It is important not to have bubbles inside the tubes as this could potentially interfere with the acquisition and it is a good practice to minimize aggregation by straining the sample through a mesh strainer.
26. A compensation matrix has to be built, even if fluorochrome emission wavelengths do not overlap. The IDEAS software automatically detects samples that have a single stain and calculates compensation
27. In the method presented here murine fibrosarcoma MCA205 cells have been utilized. Alternatively, other murine syngeneic systems could be used for DC phagocytosis studies. Uptake efficacy might vary and needs to established before pharmacological or genetic interventions are tested

Acknowledgments

GK is supported by the Ligue contre le Cancer (équipe labellisée); Agence National de la Recherche (ANR) – Projets blancs; ANR under the frame of E-Rare-2, the ERA-Net for Research on Rare Diseases; Association pour la recherche sur le cancer (ARC); Cancéropôle Ile-de-France; Institut National du Cancer (INCa); Fondation Bettencourt-Schueller; Fondation de France; Fondation pour la Recherche Médicale (FRM); the European Commission (ArtForce); the European Research Council (ERC); the LabEx Immuno-Oncology; the SIRIC Stratified Oncology Cell DNA Repair and Tumor Immune Elimination (SOCRATE); the SIRIC Cancer Research and Personalized Medicine (CARPEM); the Swiss Bridge Foundation, ISREC, and the Paris Alliance of Cancer Research Institutes (PACRI).

References

Apetoh, L., Ghiringhelli, F., Tesniere, A., Obeid, M., Ortiz, C., Criollo, A., et al. (2007). Toll-like receptor 4-dependent contribution of the immune system to anticancer chemotherapy and radiotherapy. *Nature Medicine*, *13*(9), 1050–1059.

Galluzzi, L., Buque, A., Kepp, O., Zitvogel, L., & Kroemer, G. (2017). Immunogenic cell death in cancer and infectious disease. *Nature Reviews. Immunology*, *17*(2), 97–111.

Galluzzi, L., Chan, T. A., Kroemer, G., Wolchok, J. D., & Lopez-Soto, A. (2018). The hallmarks of successful anticancer immunotherapy. *Science Translational Medicine*, *10*(459), eaat7807.

Kepp, O., Loos, F., Liu, P., & Kroemer, G. (2017). Extracellular nucleosides and nucleotides as immunomodulators. *Immunological Reviews*, *280*(1), 83–92.

Kroemer, G. (2017). Death, danger & immunity: Fundamental mechanisms linking pathogenic or iatrogenic cell death events to immune responses. *Immunological Reviews*, *280*(1), 5–7.

Kroemer, G., Galluzzi, L., Kepp, O., & Zitvogel, L. (2013). Immunogenic cell death in cancer therapy. *Annual Review of Immunology*, *31*, 51–72.

Liu, P., Zhao, L., Kepp, O., & Kroemer, G. (2019). Crizotinib—A tyrosine kinase inhibitor that stimulates immunogenic cell death. *Oncoimmunology*, *8*(7), 1596652.

Obeid, M., Tesniere, A., Ghiringhelli, F., Fimia, G. M., Apetoh, L., Perfettini, J. L., et al. (2007). Calreticulin exposure dictates the immunogenicity of cancer cell death. *Nature Medicine*, *13*(1), 54–61.

Pitt, J. M., Kroemer, G., & Zitvogel, L. (2017). Immunogenic and non-immunogenic cell death in the tumor microenvironment. *Advances in Experimental Medicine and Biology*, *1036*, 65–79.

Senovilla, L., Vitale, I., Martins, I., Tailler, M., Pailleret, C., Michaud, M., et al. (2012). An immunosurveillance mechanism controls cancer cell ploidy. *Science*, *337*(6102), 1678–1684.

Sistigu, A., Yamazaki, T., Vacchelli, E., Chaba, K., Enot, D. P., Adam, J., et al. (2014). Cancer cell-autonomous contribution of type I interferon signaling to the efficacy of chemotherapy. *Nature Medicine*, *20*(11), 1301–1309.

Vacchelli, E., Ma, Y., Baracco, E. E., Sistigu, A., Enot, D. P., Pietrocola, F., et al. (2015). Chemotherapy-induced antitumor immunity requires formyl peptide receptor 1. *Science*, *350*(6263), 972–978.

CHAPTER FOUR

Cytofluorometric assessment of dendritic cell-mediated uptake of cancer cell apoptotic bodies

Martina Musella[a,b,†], Gwenola Manic[a,†], Claudia Galassi[c], Ilio Vitale[d,e,*,‡], Antonella Sistigu[a,c,*,‡]

[a]Unit of Tumor Immunology and Immunotherapy, Department of Research, Advanced Diagnostics, and Technological Innovation, IRCCS Regina Elena National Cancer Institute, Rome, Italy
[b]Department of Molecular Medicine, University "La Sapienza", Rome, Italy
[c]Istituto di Patologia Generale, Università Cattolica del Sacro Cuore, Rome, Italy
[d]Italian Institute for Genomic Medicine (IIGM), Turin, Italy
[e]Candiolo Cancer Institute—FPO, IRCCS, Candiolo, Italy
*Corresponding authors: e-mail address: iliovit@gmail.com; antonella.sistigu@gmail.com

Contents

Abstract

Dendritic cells (DCs) are specialized antigen presenting cells (APCs) able to intake and crosspresent antigens (Ags) on major histocompatibility complex (MHC) class I and II molecules to T cells thus initiating primary and memory immune responses. DC-mediated Ag uptake and crosspresentation represent crucial steps toward cancer recognition and eventually elimination. Cytofluorometry is a standardized procedure

† Equally contributed to this article.
‡ Share senior co-authorship.

Methods in Enzymology, Volume 632
ISSN 0076-6879
https://doi.org/10.1016/bs.mie.2019.08.005

to study phagocytosis. By fast and reproducible single cell measurements, flow cytometry allows for simultaneous biochemical and functional analyses of Ag intake. In this chapter, we discuss a two-color flow cytometric analysis of DC-mediated uptake of apoptotic bodies. We also show data on the adjuvanticity of Type-I-interferons (Type-I-IFNs) during Ag retention as we offer a guideline and a range of advice on sample preparation and acquisition.

Abbreviations

Ags	antigens
APCs	antigen presenting cells
ATP	adenosine triphosphate
cDCs	conventional DCs
CFSE	5(6)-carboxyfluorescein diacetate N-succinimidyl ester
DAMPs	damage-associated molecular patterns
DAPI	4′,6-diamidino-2-phenylindole
DCs	dendritic cells
EDTA	ethylenediaminetetraacetic acid
ERAAP	endoplasmic reticulum aminopeptidase
FBS	fetal bovine serum
ICD	immunogenic cell death
Ig	immunoglobulin
OXP	oxaliplatin
PBS	phosphate-buffered saline
PE	phycoerythrin
PI	propidium iodide
PS	phosphatidylserine
RBC	red blood cell
RPMI	Roswell Park Memorial Institute
RT	room temperature
TME	tumor microenvironment
Type-I-IFNs	Type-I-interferons

1. Introduction

Effective cancer cell killing by immune cells strictly requires the occurrence—and iterative amplification—of a sequence of events in a process known as the Cancer-Immunity Cycle (Chen & Mellman, 2013). Dendritic cells (DCs) are professional innate antigen presenting cells (APCs) lying at the apex of these stepwise cascade. DCs have indeed the unique ability to induce primary immune responses by capturing and processing antigens (Ags) for their crosspresentation to T cells and by providing additional signals that either shape responses toward immunity [e.g., Type-I-interferons (Type-I-IFNs) and interleukins (IL)], or promote immune tolerance [e.g., tumor growth factor β 1 (TGFB1, also known as TGF-β)] (Banchereau et al., 2000; Mildner & Jung, 2014). Since their discovery by Ralph Steinman in 1973, DCs have been extensively studied over the last years and a plethora of different cell subsets have been identified (Steinman, 2012). The so-called "conventional" or "classical" DCs (cDCs) are the prototype of highly phagocytic immune sentinels (Mildner & Jung, 2014; Sichien, Lambrecht, Guilliams, & Scott, 2017). Immature cDCs strategically reside at body barriers where constantly sense and sample Ags. Moreover, DCs are well equipped with Ag capturing and processing machineries which includes (i) cytosolic and nuclear proteasomes and the endoplasmic reticulum aminopeptidase (ERAP), which are involved, respectively, in peptide generation and trimming, (ii) the transporter 1, ATP binding cassette subfamily B member (TAP1) and TAP2, which translocate proteasomal derived peptides from the cytoplasm to the endoplasmic reticulum, and (iii) the chaperone proteins calreticulin (CALR), protein disulfide isomerase family A member 3 (PDIA3, best known as ERp57) and tapasin, which form the peptide-loading complex. Once having captured and processed Ags and then loaded derived antigenic epitopes on major histocompatibility complex class I (MHC-I), mature cDCs migrate from peripheral sites to lymphoid organs. In these sites, cDCs initiate immune responses by confining and concentrating rare Ags and crosspriming naïve T cells (Steinman, 2012). Related to such a unique ability to induce primary immunity, cDCs also favor the establishment of immunological memory (Banchereau & Steinman, 1998; Bell, Young, & Banchereau, 1999; Troy & Hart, 1997).

There is compelling evidence that for anticancer immune responses to efficiently make inroad, neoplastic cells must express and present Ags (antigenicity) and, at the same time, must emit a set of known immunogenic danger signals (adjuvanticity), both of which influence the breath and depth of the response. In this context, the pressure the immune system exerts on cancer cells together with the functional interactions between malignant and tumor-associated stromal cells continuously shape the architecture and composition of the tumor system, which in turn drives tumor Ag evolution and promotes tumor progression (Vitale, Manic, Galassi, & Galluzzi, 2019; Vitale, Sistigu, et al., 2019). It is well demonstrated that apoptotic cancer cell death triggered in response to a set of anticancer therapeutic approaches, including some types of chemotherapies as well as radiation, photodynamic, and oncolytic viral therapies, is associated with the emission of highly immunogenic signals, the so-called damage-associated molecular patterns (DAMPs) (Apetoh et al., 2007; Michaud et al., 2011; Obeid et al., 2007; Sistigu et al., 2014; Vacchelli et al., 2015). Together with Ag recognition by T cells, such emitted DAMPs trigger a stable and long-lasting antitumor response (Galluzzi, Buque, Kepp, Zitvogel, & Kroemer, 2017; Galluzzi et al., 2018). This concept, pioneered by the labs of Guido Kroemer and Laurence Zitvogel, has been termed immunogenic cell death (ICD) (Bloy et al., 2017; Galluzzi et al., 2017; Green, Ferguson, Zitvogel, & Kroemer, 2009; Kepp et al., 2014; Kroemer, Galluzzi, Kepp, & Zitvogel, 2013). At the basis of ICD there is evidence that apoptotic bodies generated by ICD-inducing agents are surrounded by a plethora of "find-me" and "eat-me" signals emitted by dying cells in a spatiotemporally coordinated fashion. Such DAMPs help elevate apoptotic bodies recognition and uptake by cDCs through distinct mechanisms. Specifically, *pre-mortem* release of Type-I-IFNs and adenosine triphosphate (ATP) favors the chemotaxis of inflammatory monocytes and DCs within the tumor microenvironment (TME) (Kepp et al., 2014; Sistigu et al., 2014; Michaud et al., 2011). Moreover, surface exposure of the lipocortin annexin A1 (ANXA1) and of the endoplasmic reticulum resident CALR stimulates, respectively, the approaching and engulfing of dying cancer cells (Kepp et al., 2014; Vacchelli et al., 2015; Obeid et al., 2007). Beyond ANXA1 and CALR, the inner-membrane lipid phosphatidylserine (PS), which is exposed at the cell surface during apoptosis, can also operate as "eat-me" signal (Fadok et al., 2000), although with immunosuppressive effect. Finally, the *post-mortem* release of the non-histone chromatin-binding nuclear protein high-mobility group box 1 (HMGB1) inhibits the fusion between lysosomes and Ag-containing phagosomes, therefore ensuring efficient Ag presentation to T cells (Apetoh et al., 2007).

Reportedly, phagosomal and endosomal pH have been shown to play a modulatory role in DC-mediated Ag crosspresentation (Mantegazza et al., 2008; Savina et al., 2006, 2009). In particular, as discovered by Sebastian Amigorena's group, alkalinization of the phagocytic and endocytic pathways in DCs delays lysosomal enzyme proteolytic activity, thus enhancing Ag retention and presentation to cytotoxic T cells (Mantegazza et al., 2008; Savina et al., 2006, 2009). Notably, beyond licensing DCs during ICD, Type-I-IFNs have been shown to increase DC phagosomal pH and thus enhance their Ag crosspresentation potential (Lorenzi et al., 2011). Along with this, Type-I-IFNs (i) sustain the survival of Ag-loaded DCs by selectively regulating the apoptosis-related genes B-cell lymphoma 2 (BCL2) and BCL2 like 1 (BCL2L1, best known as BCL-XL) (Lorenzi et al., 2011), (ii) promote DC activation and maturation (Le Bon & Tough, 2008; Montoya et al., 2002), (iii) upregulate MHC-I (Kobayashi & van den Elsen, 2012), and (iv) exert a multipronged regulatory role in innate and adaptive immunity leading to efficient CD8 T cell crosspriming (Musella, Manic, De Maria, Vitale, & Sistigu, 2017).

Evaluating the phagocytic uptake of apoptotic bodies by DCs *in vitro* is a preliminary standard approach to detect ICD (Schiavoni et al., 2011). Over the last years, flow cytometry has replaced to some extent fluorescence and confocal microscopy (Lehmann, Sornes, & Halstensen, 2000; Steinkamp, Wilson, Saunders, & Stewart, 1982) as this procedure allows for an easy and quick monitoring of the uptake of fluorescent targets (e.g., beads, bacteria, or apoptotic bodies). Moreover, flow cytometry gives the opportunity to perform multiparametric analyses of a large number of samples, favoring its application to both experimental and clinical testing.

This chapter describes a two-color flow cytometric method to quantify the *in vitro* uptake of fluorescent apoptotic bodies by murine DCs in the presence or absence of Type-I-IFN stimulation.

2. Materials

2.1 Disposable

- 75-cm^2 cell culture treated flasks (Corning®, New York, NY)
- 6-well plates (Corning®)
- 40 μm cell strainer (Falcon, Corning®)
- Cell counting slides
- 15 and 50 mL tubes (Falcon, Corning®)
- 5 mL polystyrene tubes (Falcon, Corning®)
- 1.5 mL Eppendorf microtubes (Eppendorf®, Hamburg, Germany)
- Pipet aid and serological pipettes (2 mL, 5 mL, 10 mL)
- Pipettes and sterile tips (1–10 μL, 20–200 μL, 1000 μL)
- 10 mL syringe [BD Plastipak™; BD Biosciences, Becton, Dickinson and Company (BD), Franklin Lakes, NJ]
- MACS separation columns (LS, MS columns; Miltenyi Biotec, Bergisch Gladbach, Germany)
- MACS Separators (MiniMACS™, MidiMACS™ separators; Miltenyi Biotec)

- MACS Multistand (Miltenyi Biotec)
- Aluminum foil
- Ice
- Timer
- Sterile scissors

2.2 Equipment

- Humidified cell culture incubator at 37 °C with 5% CO_2
- Safety cabinet (Class II)
- Optical microscope
- Refrigerable centrifuge
- Laboratory refrigerator (4 °C)
- Ice machine
- Thermostatic bath
- Cytofluorometer (such as BD FACSCanto II, with up to three lasers: blue (488 nm), red (633 nm), and violet (405 nm); BD Biosciences)
- Software for flow cytometry analysis [such as FlowJo software (FlowJo, LLC, Becton, Dickinson and Company)]
- Software for statistical analysis (such as GraphPad Prism; GraphPad Software, San Diego, CA)

2.3 Reagents

- MCA.205 murine fibrosarcoma cells
- Roswell Park Memorial Institute (RPMI) 1640 medium supplemented with 10% (v/v) fetal bovine serum (FBS), 2 mM L-glutamine, 100 IU/mL penicillin G sodium salt, and 100 µg/mL streptomycin sulfate (growth medium, all from Gibco-Thermo Scientific; Thermo Fisher Scientific Inc., Waltham, MA)
- Phosphate-buffered saline (PBS, Gibco-Thermo Scientific)
- Trypsin/ethylenediaminetetraacetic acid (EDTA) solution (Gibco-Thermo Scientific)
- Spleen from immunocompetent histocompatible ($H\text{-}2^b$) mice (C57Bl/6J)
- Red blood cell (RBC) Lysis Buffer (eBioscience™, Thermo Scientific)
- EDTA, 0.1 M, pH 7.2 (Sigma-Aldrich; Millipore-Sigma, Saint-Louis, MO)
- Histodenz (1.077 g/mL, Sigma-Aldrich)
- FcR Blocking Reagent (CD16/32 antibody; Miltenyi Biotec)
- Separation buffer (Miltenyi Biotec)

- CD11c MicroBeads UltraPure (Miltenyi Biotec)
- 5(6)-Carboxyfluorescein diacetate N-succinimidyl ester (CFSE, Sigma-Aldrich)
- 1% FBS in cold PBS
- Oxaliplatin (OXP, Sigma-Aldrich)
- Purified Type-I-IFNs
- Anti-mouse phycoerythrin (PE)-CD11c (clone N418, eBioscience™, Thermo Scientific) and IgG isotype control (eBioscience™, ThermoFisher)
- 4′,6-diamidino-2-phenylindole (DAPI, ThermoFisher)

3. Methods

3.1 Cancer cell culture

1. MCA.205 murine fibrosarcoma cells are routinely maintained at 37 °C, under 5% CO_2 in Roswell Park Memorial Institute (RPMI) 1640 medium supplemented with 10% (v/v) fetal bovine serum (FBS), 2 mM L-glutamine, 100 IU/mL penicillin G sodium salt, and 100 μg/mL streptomycin sulfate (growth medium) (*see* **Notes 1, 2**).
2. For optimal culture conditions, MCA.205 cells are placed in 75-cm^2 flasks at an approximate density of 5×10^6 cells/flask (*see* **Note 3**) in 15 mL of growth medium.
3. When maintenance cultures reach 75–80% confluence (*see* **Notes 3, 4**), growth medium is discarded from the cell monolayer, and cells are first washed with prewarmed phosphate-buffered saline (PBS) to completely remove FBS, which contains protease inhibitors that inactivate EDTA enzymatic activity, and then incubated with pre-warmed trypsin/EDTA (see **Note 5**) for 1–2 min at 37 °C to detach cells (*see* **Notes 6, 7**).
4. Detached cells are washed in growth medium for 5 min at 1100 *g* at room temperature (RT), and reseeded upon dilution in growth medium (*see* **Note 8**). This provides cells for maintenance culture and experimental procedures.
5. For coculture experiments, 1×10^6 cells are seeded in 6-well plates in 3 mL of growth medium and processed as detailed in Section 4.

3.2 Spleen processing and dendritic cell isolation

1. Spleens previously explanted from immunocompetent mice [in our case histocompatible (H-2^b) either xenografted or naïve C57Bl/6]

are placed into 40 μm cell strainer on the top of 50 mL tube and crushed with the plunger of a 10 mL syringe (*see* **Note 9**).

2. Cell strainer is then rinsed twice with 10 mL of growth medium, and the obtained splenocyte cell suspension collected in 50 mL tube is washed/centrifuged twice by adding each time 10 mL of growth medium prior to centrifugation at 1100 *g* for 5 min (at RT).
3. Cell pellet (containing approximately 1×10^8 cells) is then resuspended in 2 mL of RBC Lysis Buffer and incubated for 1–3 min at RT (*see* **Note 10**).
4. Thereafter, splenocytes are washed/centrifuged twice with 10 mL growth as in step 2, and the resulting RBC-free white pellet is resuspended with 10 mL of growth medium followed by centrifugation at 750 *g* for 10 min (at RT) to discard platelets.
5. Cell pellet is then incubated for 5 min at RT with EDTA to allow disruption of DC-T cell complexes. The obtained cell suspension (containing approximately 5×10^7 cells upon cell doublet exclusion) is washed/centrifuged twice with 10 mL of growth medium as in step 2.
6. Cell pellet is resuspended in 6 mL of Histodenz (1.077 g/mL) (*see* **Note 11**), overlaid on an additional layer of 6 mL Histodenz, and centrifuged at 1700 *g* for 20 min at 4 °C. To avoid disturbing the gradients, acceleration and deceleration should be as slowly as possible and without brakes.
7. The low-density fraction enriched for DCs is then collected and washed once with cold PBS prior to centrifugation at 1100 *g* for 5 min (at 4 °C) and cell count.
8. DC-enriched cell fraction is incubated with FcR Blocking Reagent (CD16/32 antibody) for 20 min at 4 °C (*see* **Note 12**) followed by two consecutive washes in 10 mL of separation buffer.
9. For DC purification, cell pellet is resuspended in CD11c MicroBeads UltraPure in a volume of 100 μL per 1×10^8 total cells and incubated for 10 min at 4 °C in the dark.
10. After a wash in 10 mL of separation buffer followed by a centrifugation at 1100 *g* for 10 min (at RT), cell pellet is resuspended in 500 μL of separation buffer (up to 1×10^8 cells *see* **Note 13**).
11. Columns are rinsed with separation buffer before applying cell suspension.
12. Once cells have been transferred, columns are washed 3 times with separation buffer to allow unlabeled cells to flow-through, and placed on suitable collection tubes (*e.g.*, 15 mL tubes).

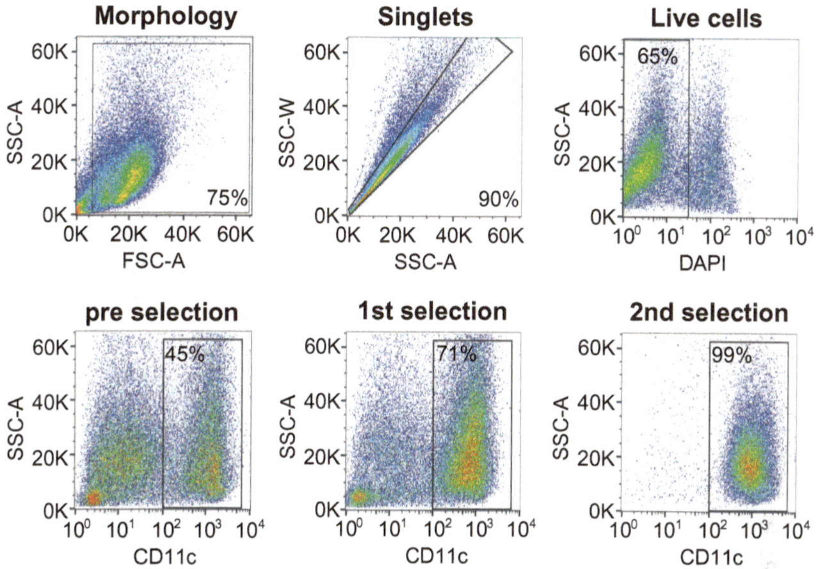

Fig. 1 Cytofluorometric assessment of CD11c-expressing cells. DCs are purified from immunocompetent histocompatible C57Bl/6J mice spleens by means of Histodenz-based density gradient and two consecutive beads-based positive selections. Samples are first gated for morphology, by plotting forward scatter area (FCS-A, x-axis) and side scatter area (SSC-A, y-axis). Thereafter, singlets are gated by plotting SSC-A (x-axis) and side scatter widths (SSC-W, y-axis). Dead cells are excluded by means of DAPI staining, by plotting blue laser area (x-axis) and SSC-A (y-axis). Finally, the percentage of CD11c^{pos} cells are analyzed after one or two consecutive positive selections, by plotting PE-CD11c area (x-axis) and SSC-A (y-axis).

13. Labeled DCs are eluted by adding 500 μL of separation buffer and firmly pushing plungers into the columns.
14. A second separation procedure with a new column is performed to increase the purity of CD11c^{pos} cells.
15. The enriched CD11c^{pos} fraction is then analyzed for purity by a BD FACSCanto II flow cytometer as described in Section 3.3 (steps 8–11) and as illustrated in Fig. 1.

3.3 Cancer cell—DC coculture and uptake analysis

1. MCA.205 murine fibrosarcoma cells are resuspended in PBS supplemented with 1% (v/v) FBS at a concentration of 1×10^7 cells/mL and then incubated at 37 °C for 10 min in the dark with CFSE at a final concentration of 1 μM.

2. Upon incubation with CFSE, cells are washed twice by resuspending them in 10 mL of 1% FBS in cold PBS prior to centrifugation at 1100 *g* for 5 min at 4 °C. From this step onward, cells should be protected from light to preserve fluorochrome signal and avoid photobleaching.
3. CFSE-stained cells are counted and then seeded in 6-well plates at a density of 1×10^6 cancer cells per well in 3 mL of growth medium.
4. Once adherent to the support (i.e., 16–24 h after seeding, according to cell type), cancer cells are treated for 6–8 h with OXP in growth medium at a final concentration of 300 μM (*see* **Note 14**).
5. Cancer cells are then washed twice in growth medium and cultivated for 24–48 h in standard culture conditions to let them die (*see* **Note 15**).
6. Upon delicate removal of 1 mL of growth medium from the culture, DCs isolated as described in Section 3.2 are added to the system at a 1:4 ratio with dying tumor cells (2.5×10^5 DCs for 1×10^6 tumor cells) in a volume of 1 mL of fresh growth medium (*see* **Notes 16, 17**).
7. Purified Type-I-IFNs are then added to the cancer cell/DC coculture to a final concentration of 5×10^3 IU/mL (*see* **Notes 18, 19**), and cells are incubated at 37 °C. As experimental controls cells can be incubated at 4 °C (a condition in which phagocytosis is hampered) or administered with the phagocytic inhibitor cytochalasin D to a final concentration of 20 μM.
8. After 3–18 h of incubation, DCs are purified by means of Histodenz as described in Section 3.2 (step 6–12).
9. For cell surface staining, DCs are resuspended in 100 μL of cold PBS solution containing 1% FBS and 0.25 μg of either IgG isotype control or anti-mouse CD11c.
10. Upon incubation for 20 min at 4 °C in the dark to avoid the loss of signal due to photobleaching of fluorochromes, cell are washed twice in 1% FBS in cold PBS and incubated with the vitality dye DAPI at a final concentration of 1 μM (*see* **Note 20**) prior to flow cytometric analysis of CD11c and CFSE positivity by means of a BD FACSCanto II flow cytometer.
11. Data analysis are performed using the FlowJo software.
12. During flow cytometer acquisition and data analysis, samples are sequentially gated (1) for morphology, by plotting forward scatter area (FSC-A, x-axis) and side scatter area (SSC-A, y-axis) (gate 1); (2) for singlets, by plotting SSC-A (x-axis) and side scatter width (SSC-W, y-axis) (gate 2); and (3) for viability, by plotting blue laser area (emission 461 nm, x-axis) and SSC-A (y-axis) (*see* **Note 21**) (gate 3). Based on

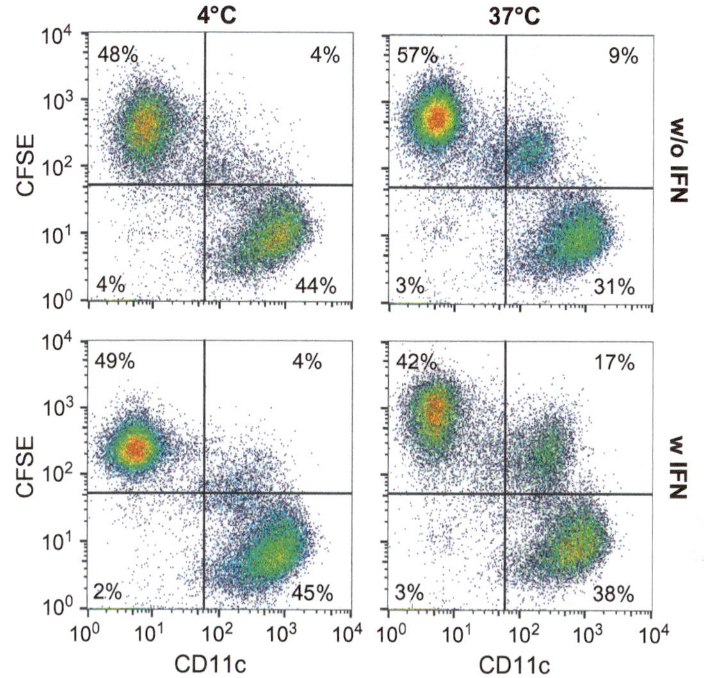

Fig. 2 Cytofluorometric analysis of CFSE positivity in CD11c-expressing cells in the presence or not of Type-I-IFNs at 37 and 4 °C. DCs, enriched from coculture with dying cancer cells, were stained with a PE-conjugated anti-CD11c antibody and analyzed for CFSE intake. Coculture at 4 °C was used as a negative control of uptake. Plots depict the percentages of cells previously gated on morphology, singlets and viability. CFSE-CD11c double positive cells at 37 °C (plots on the right) represent the uptake of apoptotic bodies, while at 4 °C CFSE positivity on CD11c cells is due to juxtaposition of cells.

gate 3, cells are finally analyzed for CFSE, by plotting green laser area (emission 521 nm, x-axis), and for PE-CD11c, by plotting red laser area (emission 573 nm, y-axis) (*see* **Note 22**) (Fig. 2).

4. Discussion and concluding remarks

In this chapter, we first isolated DCs from the spleen explanted by histocompatible immunocompetent mice and then assessed DC purity by using a PE-conjugated antibody recognizing the DC marker CD11c. As shown in the plots in Fig. 1, this approach ensures efficient enrichment of DCs from mouse spleen. Moreover, we analyzed the capability of DCs to engulf apoptotic bodies in the presence and absence of Type-I-IFNs, which reportedly favor antitumor immune responses via the release of chemokine

C-X-C motif chemokine ligand 10 (CXCL10) into the TME (Michaud et al., 2011; Sistigu et al., 2014). To this aim, isolated DCs were cocultured with cancer cells previously treated with OXP, a known ICD inducer via regulated cell death, *de facto* resulting in abundant generation of apoptotic bodies. Cocultured cells were left untreated or incubated with exogenous Type-I-IFNs during cancer cell ICD induction. To assess DC uptake of apoptotic bodies, we made use of CFSE as an apoptotic body tracker in combination with the PE-conjugated anti-CD11c antibody. CFSE was chosen as it guarantees an easy and efficient labeling with an emission wavelength well separated from the PE-conjugated anti-CD11c antibody (CFSE: 521 nm; PE: 573 nm), simplifying the compensation settings for the two fluorophores. This allows for an easy, rapid and reliable quantification of DC-mediated engulfment of apoptotic bodies. Through this approach, we could confirm that Type-I-IFNs enhance the intake of cancer cell-derived apoptotic bodies by DCs (Fig. 2). As expected, such effect was more evident when cancer cells and DCs were cocultured at 37 °C than 4 °C, in line with the fact that the former is the optimal, physiological experimental condition. We surmise that the positive impact of Type-I-IFNs on the uptake of apoptotic cells by DCs can arise from the previously reported capability of Type-I-IFNs to prolong the intracellular persistence of antigenic particles engulfed by DCs (Lorenzi et al., 2011). This seems to occur through a mechanism dependent on pH alkalinization, resulting in enhanced Ag crosspresentation to CD8 T cells. Further experiments are required to prove this hypothesis.

In conclusion, we demonstrated the feasibility and suitability of this *in vitro* approach, which should be adopted by researchers dealing with ICD detection and, in particular, with the assessment of the phagocytic uptake of apoptotic bodies by DCs. In this context, we strongly recommend the use of additional procedures in order to confirm and extend experimental results, such as *in vitro* and *in vivo* validation of the immunogenicity of cell death as well as assessment of the ability of "Ag-loaded" DCs to efficiently prime cytotoxic T cells.

5. Notes

1. Optimal culture conditions may vary depending on the cell line of choice.
2. MCA.205 cells can conveniently grow in 25-, 75-, or 175-cm^2 flasks. As a guideline, MCA.205 cells have a duplication rate of 18–24 h and

reach 75–80% confluence at a density of approximately $5–10 \times 10^6$, $20–30 \times 10^6$, $40–60 \times 10^6$ cells within 25-, 75-, or 175-cm^2 flasks, respectively.

3. The number of cells seeded could be different across cell lines usually depending on cell size (and thus ploidy) and replication rate.
4. Passaging of cells at suboptimal confluence may influence cell state quite dramatically. Confluence levels >85% can cause cell stress by starvation, conversely confluence levels <30% can favor clone selection, thus changing the genetic and cytogenetic background.
5. Optimum trypsin activity is achieved at 37 °C, so pre-warmed trypsin speeds up cell detachment.
6. Trypsinization time varies depending on the cell type. However, as a general guideline, long term incubation should be avoided as it damages cells by striping cell surface proteins and inducing cell stress and cell death.
7. At this step, it is a good practice to inspect cell detachment by light microscopy.
8. For MCA.205 cells, 1:20 dilution is recommended to reach 75–80% confluence rates within 3 days. Optimal cell dilutions depend on the cell line of choice.
9. As an alternative, mechanical and enzymatic dissociation can be performed. Briefly, spleens are placed in 1.5 mL Eppendorf tubes with 1 mL of RPMI 1640 medium supplemented with 400 U/mL Collagenase IV and 200 U/mL DNase I (digestion buffer). Spleens are cut with sterile scissors and incubated for 10 min at 37 °C. After the incubation with digestion buffer, spleens are cut again, transferred into 50 mL tubes through 40 μm cell strainers and washed twice for 5 min at 1200 rpm at RT in growth medium before cell count.
10. Lysis of RBCs improves leukocyte enrichment. We recommend the use of this buffer as it ensures optimal removal of RBCs with minimal effects on white blood cells.
11. As an alternative, Nycodenz (1.077 g/mL) can be used.
12. In order to obtain high purities of DCs (≥90%), FcR-blocking should be performed with either an anti-CD16/32 antibody or a mouse Ig, before proceeding with magnetic labeling.
13. Buffer volume should be scaled up for higher cell numbers, according to manufacturer's instructions.
14. A few hour treatment pulse ensures the induction of cancer cell death at later time points. In our hands, a 6–8 h pulse of 300 μM OXP induces

40% or 80% of MCA.205 tumor cell death upon 24 h or 48 h incubation, respectively. At this step a flow cytometry analysis of annexin V-propidium iodide should be performed.

15. It can be especially useful to wash cells before apoptotic bodies develop so to preserve an optimal cancer cells:DCs ratio for coculture.
16. If possible, we recommend to not dilute the coculture so as to favor tumor cell–DC interactions and ensure optimal phagocytic conditions. Moreover, we recommend not removing >1 mL of culture medium before adding DCs, as higher volume can lead to loss of floating apoptotic bodies.
17. In case human DCs are obtained by differentiating peripheral blood monocytes (monocyte-derived DCs), cell numbers can be scaled down to a minimum of 2.5×10^4 DCs for 1×10^5 tumor cells. Moreover, monocyte-derived DCs are phenotypically heterogeneous and therefore a spectrum of phagocytic ability may result.
18. Purified Type-I-IFNs can be added to a different final concentration, as determined by previous dose-response studies.
19. We recommend thawing Type-I-IFNs on ice and preparing small aliquots to avoid freeze-thaw cycles.
20. Alternatively, other cell viability dyes with emission wavelengths far from 521 to 573 nm and thus not overlapping with CFSE or PE-CD11c fluorescences [e.g., 7-amino-actinomycin D or vitality fixable dead-cell either blue or violet or aqua] can be used.
21. The emission filter of the flow cytometer can be adjusted accordingly to the spectral properties of the vital dye chosen in the study.
22. It is crucial to employ fluorescences not overlapping and, so, easily discriminated during cytofluorometric analysis.

Acknowledgments

A.S. is supported by the Associazione Italiana per la Ricerca sul Cancro (AIRC, Start-Up 2016 #18418) and Ministero Italiano della Salute (RF_GR-2013-02357273). I.V. is supported by the AIRC (IG 2017 grant number 20417) and a startup grant from the Italian Institute for Genomic Medicine (Turin, Italy).

References

Apetoh, L., Ghiringhelli, F., Tesniere, A., Obeid, M., Ortiz, C., Criollo, A., et al. (2007). Toll-like receptor 4-dependent contribution of the immune system to anticancer chemotherapy and radiotherapy. *Nature Medicine*, *13*(9), 1050–1059. https://doi.org/10.1038/nm1622.

Banchereau, J., Briere, F., Caux, C., Davoust, J., Lebecque, S., Liu, Y. J., et al. (2000). Immunobiology of dendritic cells. *Annual Review of Immunology*, *18*, 767–811. https://doi.org/10.1146/annurev.immunol.18.1.767.

Banchereau, J., & Steinman, R. M. (1998). Dendritic cells and the control of immunity. *Nature*, *392*(6673), 245–252. https://doi.org/10.1038/32588.

Bell, D., Young, J. W., & Banchereau, J. (1999). Dendritic cells. *Advances in Immunology*, *72*, 255–324.

Bloy, N., Garcia, P., Laumont, C. M., Pitt, J. M., Sistigu, A., Stoll, G., et al. (2017). Immunogenic stress and death of cancer cells: Contribution of antigenicity vs adjuvanticity to immunosurveillance. *Immunological Reviews*, *280*(1), 165–174. https://doi.org/10.1111/imr.12582.

Chen, D. S., & Mellman, I. (2013). Oncology meets immunology: The cancer-immunity cycle. *Immunity*, *39*(1), 1–10. https://doi.org/10.1016/j.immuni.2013.07.012.

Fadok, V. A., Bratton, D. L., Rose, D. M., Pearson, A., Ezekewitz, R. A., & Henson, P. M. (2000). A receptor for phosphatidylserine-specific clearance of apoptotic cells. *Nature*, *405*(6782), 85–90. https://doi.org/10.1038/35011084.

Galluzzi, L., Buque, A., Kepp, O., Zitvogel, L., & Kroemer, G. (2017). Immunogenic cell death in cancer and infectious disease. *Nature Reviews. Immunology*, *17*(2), 97–111. https://doi.org/10.1038/nri.2016.107.

Galluzzi, L., Vitale, I., Aaronson, S. A., Abrams, J. M., Adam, D., Agostinis, P., et al. (2018). Molecular mechanisms of cell death: Recommendations of the nomenclature committee on cell death 2018. *Cell Death and Differentiation*, *25*(3), 486–541. https://doi.org/10.1038/s41418-017-0012-4.

Green, D. R., Ferguson, T., Zitvogel, L., & Kroemer, G. (2009). Immunogenic and tolerogenic cell death. *Nature Reviews. Immunology*, *9*(5), 353–363. https://doi.org/10.1038/nri2545.

Kepp, O., Senovilla, L., Vitale, I., Vacchelli, E., Adjemian, S., Agostinis, P., et al. (2014). Consensus guidelines for the detection of immunogenic cell death. *Oncoimmunology*, *3*(9), e955691. https://doi.org/10.4161/21624011.2014.955691.

Kobayashi, K. S., & van den Elsen, P. J. (2012). NLRC5: A key regulator of MHC class I-dependent immune responses. *Nature Reviews. Immunology*, *12*(12), 813–820. https://doi.org/10.1038/nri3339.

Kroemer, G., Galluzzi, L., Kepp, O., & Zitvogel, L. (2013). Immunogenic cell death in cancer therapy. *Annual Review of Immunology*, *31*, 51–72. https://doi.org/10.1146/annurev-immunol-032712-100008.

Le Bon, A., & Tough, D. F. (2008). Type I interferon as a stimulus for cross-priming. *Cytokine & Growth Factor Reviews*, *19*(1), 33–40. https://doi.org/10.1016/j.cytogfr.2007.10.007.

Lehmann, A. K., Sornes, S., & Halstensen, A. (2000). Phagocytosis: Measurement by flow cytometry. *Journal of Immunological Methods*, *243*(1–2), 229–242.

Lorenzi, S., Mattei, F., Sistigu, A., Bracci, L., Spadaro, F., Sanchez, M., et al. (2011). Type I IFNs control antigen retention and survival of CD8alpha(+) dendritic cells after uptake of tumor apoptotic cells leading to cross-priming. *Journal of Immunology*, *186*(9), 5142–5150. https://doi.org/10.4049/jimmunol.1004163.

Mantegazza, A. R., Savina, A., Vermeulen, M., Perez, L., Geffner, J., Hermine, O., et al. (2008). NADPH oxidase controls phagosomal pH and antigen cross-presentation in human dendritic cells. *Blood*, *112*(12), 4712–4722. https://doi.org/10.1182/blood-2008-01-134791.

Michaud, M., Martins, I., Sukkurwala, A. Q., Adjemian, S., Ma, Y., Pellegatti, P., et al. (2011). Autophagy-dependent anticancer immune responses induced by chemotherapeutic agents in mice. *Science*, *334*(6062), 1573–1577. https://doi.org/10.1126/science.1208347.

Mildner, A., & Jung, S. (2014). Development and function of dendritic cell subsets. *Immunity*, *40*(5), 642–656. https://doi.org/10.1016/j.immuni.2014.04.016.

Montoya, M., Schiavoni, G., Mattei, F., Gresser, I., Belardelli, F., Borrow, P., et al. (2002). Type I interferons produced by dendritic cells promote their phenotypic and functional activation. *Blood*, *99*(9), 3263–3271.

Musella, M., Manic, G., De Maria, R., Vitale, I., & Sistigu, A. (2017). Type-I-interferons in infection and cancer: Unanticipated dynamics with therapeutic implications. *Oncoimmunology*, *6*(5), e1314424. https://doi.org/10.1080/2162402X.2017.1314424.

Obeid, M., Tesniere, A., Ghiringhelli, F., Fimia, G. M., Apetoh, L., Perfettini, J. L., et al. (2007). Calreticulin exposure dictates the immunogenicity of cancer cell death. *Nature Medicine*, *13*(1), 54–61. https://doi.org/10.1038/nm1523.

Savina, A., Jancic, C., Hugues, S., Guermonprez, P., Vargas, P., Moura, I. C., et al. (2006). NOX2 controls phagosomal pH to regulate antigen processing during crosspresentation by dendritic cells. *Cell*, *126*(1), 205–218. https://doi.org/10.1016/j.cell.2006.05.035.

Savina, A., Peres, A., Cebrian, I., Carmo, N., Moita, C., Hacohen, N., et al. (2009). The small GTPase Rac2 controls phagosomal alkalinization and antigen crosspresentation selectively in CD8(+) dendritic cells. *Immunity*, *30*(4), 544–555. https://doi.org/10.1016/j.immuni.2009.01.013.

Schiavoni, G., Sistigu, A., Valentini, M., Mattei, F., Sestili, P., Spadaro, F., et al. (2011). Cyclophosphamide synergizes with type I interferons through systemic dendritic cell reactivation and induction of immunogenic tumor apoptosis. *Cancer Research*, *71*(3), 768–778. https://doi.org/10.1158/0008-5472.CAN-10-2788.

Sichien, D., Lambrecht, B. N., Guilliams, M., & Scott, C. L. (2017). Development of conventional dendritic cells: From common bone marrow progenitors to multiple subsets in peripheral tissues. *Mucosal Immunology*, *10*(4), 831–844. https://doi.org/10.1038/mi.2017.8.

Sistigu, A., Yamazaki, T., Vacchelli, E., Chaba, K., Enot, D. P., Adam, J., et al. (2014). Cancer cell-autonomous contribution of type I interferon signaling to the efficacy of chemotherapy. *Nature Medicine*, *20*(11), 1301–1309. https://doi.org/10.1038/nm.3708.

Steinkamp, J. A., Wilson, J. S., Saunders, G. C., & Stewart, C. C. (1982). Phagocytosis: Flow cytometric quantitation with fluorescent microspheres. *Science*, *215*(4528), 64–66.

Steinman, R. M. (2012). Decisions about dendritic cells: Past, present, and future. *Annual Review of Immunology*, *30*, 1–22. https://doi.org/10.1146/annurev-immunol-100311-102839.

Troy, A. J., & Hart, D. N. (1997). Dendritic cells and cancer: Progress toward a new cellular therapy. *Journal of Hematotherapy*, *6*(6), 523–533. https://doi.org/10.1089/scd.1.1997.6.523.

Vacchelli, E., Ma, Y., Baracco, E. E., Sistigu, A., Enot, D. P., Pietrocola, F., et al. (2015). Chemotherapy-induced antitumor immunity requires formyl peptide receptor 1. *Science*, *350*(6263), 972–978. https://doi.org/10.1126/science.aad0779.

Vitale, I., Manic, G., Galassi, C., & Galluzzi, L. (2019). Stress responses in stromal cells and tumor homeostasis. *Pharmacology & Therapeutics*, *200*, 55–68. https://doi.org/10.1016/j.pharmthera.2019.04.004.

Vitale, I., Sistigu, A., Manic, G., Rudqvist, N. P., Trajanoski, Z., & Galluzzi, L. (2019). Mutational and antigenic landscape in tumor progression and cancer immunotherapy. *Trends in Cell Biology*, *29*(5), 396–416. https://doi.org/10.1016/j.tcb.2019.01.003.

CHAPTER FIVE

Methods to assess DC-dependent priming of T cell responses by dying cells

Anne-Marie Chalupova[b,†], Sarka Vosahlikova[b,†], Daniela Rozkova[b], Klara Sochorova[b], Lenka Palova-Jelinkova[a,b], Jirina Bartunkova[a,b], Radek Spisek[a,b], Jitka Fucikova[a,b,*]

[a]Department of Immunology, Charles University, 2nd Faculty of Medicine and University Hospital Motol, Prague, Czech Republic
[b]Sotio, Prague, Czech Republic
*Corresponding author: e-mail address: fucikova@sotio.com

Contents

Abstract

Dendritic cells have been widely investigated in cancer immunotherapy clinical trials for the last two decades mainly due to their robust ability to elicit an adaptive anticancer immune response of the cellular and humoral types. Immature DCs can be easily loaded with desired antigens. However, to become efficient antigen-presenting cells, DCs must first undergo a process of maturation. Protocols for the generation of DCs for use in cancer immunotherapy, including the generation of a large number of immature DCs for antigen pulsing and the selection of a well-defined immunostimulatory agent to achieve complete and reproducible maturation, which is a crucial step for further

† Contributed equally.

Methods in Enzymology, Volume 632
ISSN 0076-6879
https://doi.org/10.1016/bs.mie.2019.05.045

stimulation of T cell activation, must carefully consider the characteristics of DC physiology. In this report, we provided a detailed protocol for DC generation, pulsation and activation with the subsequent induction of T cell-specific immune responses.

1. Introduction

The main aim of immunotherapy for human cancer is to induce a tumor-specific and long-lasting immune response that slows the growth of tumor cells. The induction of anti-tumor immunity includes the presentation of tumor antigens by dendritic cells (DCs), priming and activation of a specific T cell response with the subsequent homing of effector T cells to the site of malignant cells, leading to their final elimination (Truxova et al., 2017). DCs are highly specialized antigen-presenting cells (APCs) with a unique ability to establish and control primary immune responses (Banchereau et al., 2000). Therefore, DCs have been extensively used in immunotherapy for their properties of initiation and modulation of the immune response for two decades (Palucka, Ueno, Roberts, Fay, & Banchereau, 2011; Vacchelli et al., 2013). Since the discovery of DCs in 1970, autologous DCs have been employed in immunotherapy for several tumor types, including melanoma, lymphoma, prostate cancer, renal cell carcinoma, and glioblastoma, with limited clinical success (Anguille et al., 2015; Anguille, Smits, Lion, van Tendeloo, & Berneman, 2014; Cao et al., 2014; Fucikova et al., 2018; Hsu et al., 1996; Nakai, Hartmann, Kishimoto, & Katoh, 2010; Podrazil et al., 2015; Vacchelli et al., 2013). The food and drug administration (FDA) approval of Sipuleucel-T, which is a cellular product based on enriched blood APCs cultured with the fusion protein prostatic acid phosphatase (PAP) and granulocyte-macrophage colony-stimulating factor (GM-CSF), for the treatment of metastatic prostate cancer has provided promise for cancer immunotherapy based on DCs (Kantoff et al., 2010). The main strategies for the development of DC-based anticancer vaccines are as follows: (1) ex vivo generation of DC vaccines, (2) targeting DC receptors by tumor-associated antigens in vivo, and (3) DC-derived exosomes (Boczkowski, Nair, Nam, Lyerly, & Gilboa, 2000; Galluzzi et al., 2014; Viaud et al., 2010). Ex vivo DCs are mainly manufactured through the in vitro differentiation of peripheral blood mononuclear cells (PBMCs) or previously separated monocytes in the presence of GM-CSF and interleukin 4 (IL-4) (Curti, Ferri, Pandolfi, Isidori, & Lemoli, 2004; Fucikova et al., 2011). After expansion, DCs are loaded with tumor antigens, activated and reintroduced into patients to efficiently generate anti-tumor immunity. Protocols for the generation of a large number of DCs for use in cancer immunotherapy trials must carefully consider these characteristics of DC physiology. Therefore, crucial steps for the large-scale production of DCs for clinical trials involve generating large numbers of immature DCs, the selection of an appropriate source of tumor antigens and a well-defined maturation stimulus to facilitate reproducible maturation. In this report, we provide a description of several assays that are crucial for DC preparation, pulsation and activation and DC-dependent priming of T cell-specific responses.

2. Generation of dendritic cells: Principles

Dendritic cells (DCs) are a heterogeneous cell population of bone marrow-derived cells that are present in all tissues with the ability to accumulate at sites facing intense antigenic challenges (lymphoid organs, mucosal surfaces, skin, etc.) (Banchereau et al., 2000). DCs are categorized on the basis of their origin, surface receptor expression and functions as myeloid DCs (mDCs) and plasmacytoid DCs (pDCs). As APCs, DCs regulate both the innate and adaptive arms of the immune system and exhibit plasticity in response to pathogens (Huang et al., 2001). DCs specialize in foreign entity uptake and in processing and presenting antigens to T lymphocytes, consequently activating naïve T-lymphocytes and establishing immunological memory. These functions provide DCs with the unique ability to induce a primary immune response (Banchereau et al., 2000). Circulating blood DCs are rare (accounting for <1% of human PBMCs) and are difficult to maintain in culture (Stockwin, McGonagle, Martin, & Blair, 2000). Although the in vivo expansion of blood DCs using hematopoietic growth factors and GM-CSF is a widely used strategy to increase the yield of isolated cells, most experimental and clinical studies currently rely on the in vitro development of DCs from $CD34^{+}$ progenitor cells or blood monocytes (Nair, Archer, & Tedder, 2012). According to a standard protocol, monocytes are cultured for 5–7 days with GM-CSF and IL-4 to generate immature DCs (Curti et al., 2004). To obtain fully mature DCs with a T cell stimulatory ability, immature DCs must be cultured for 1–2 days with microbial, pro-inflammatory or T-cell derived stimuli. The ability to prepare DCs from blood mononuclear cells with added stimulating cytokines has enabled cancer immunotherapy and facilitated the use of immunotherapy vaccines containing DCs pulsed with tumor antigens (Anguille et al., 2014; Vacchelli et al., 2013).

3. Generation of dendritic cells: Protocol

(1) For peripheral blood mononuclear cell (PBMC) isolation, dilute peripheral blood or the buffy coat with PBS + EDTA (*Lonza BioScience, Belgium*) (see Notes 1 and 2). Isolate peripheral blood mononuclear cells (PBMCs) using Ficoll Paque™ Plus (GE HealthCare, Sweden).

(2) Separate $CD14^+$ cells from PBMCs using the EasySep™ Human CD14 Positive Selection Kit II according to the manufacturer's instructions (*StemCell Technologies, Canada*). Desired cells are targeted with tetrameric antibody complexes recognizing CD14 and dextran-coated magnetic particles. Labeled cells are separated using an EasySep™ magnet without the use of columns.

(3) Wash the isolated $CD14^+$ cells with PBS, and after a second wash step, resuspend the isolated cells in culture medium enriched with IL-4 (*Gentaur, Belgium*) and granulocyte-macrophage colony-stimulating factor (GM-CSF; *Gentaur, Belgium*) and incubate the cells at 37 °C with 5% CO_2 for 6 days (see Note 3).

(4) Freeze the $CD14^-$ fraction from the separation step (T lymphocytes). Centrifuge the fraction at 1200 rpm for 10 min at 4 °C. Resuspend the pellet in 1 mL of Cryostor (BioLife Solutions, USA). Freeze the cell suspension in a Mr. Frosty container at −80 °C and transfer the cell vial into liquid nitrogen after 24 h.

4. Phagocytosis of apoptotic tumor cells and subsequent DC maturation: Principles

The main role of immature DCs is to efficiently and continuously sample the antigenic content in their environment by phagocytosis of particles, high-volume fluid phase micropinocytosis, receptor-mediated endocytosis, and direct contact with apoptotic or infected cells (Nagl et al., 2002). Unlike professional phagocytes, their main task does not involve clearing pathogens but rather enabling the presentation of antigens to the immune system on their cell surface via major histocompatibility complexes (MHC classes I and II) (Cabral et al., 2013). After receiving appropriate signals either directly from pathogens or via inflammatory stimuli, DCs alter the expression of chemokine receptors, allowing them to leave the peripheral tissue and migrate to draining lymphoid organs to undergo a process of maturation, which includes various functional and phenotypic changes from their immature to mature stage (Banchereau et al., 2000). Mature DCs are characterized by decreased endocytic activity, upregulation of major histocompatibility complex (MHC) class I and II molecules and co-stimulatory molecules (CD86 and CD80), and responsiveness to inflammatory chemokines. Importantly, mature DCs synthesize high levels of IL-12 that enhance innate (natural killer cells) and adaptive (B and T cells) immunity. Simultaneously, DCs are very effective in presenting captured antigens to T cells in lymphoid tissue, leading to the induction of an anti-tumor immune response (Banchereau et al., 2000).

5. Phagocytosis of apoptotic tumor cells and subsequent DC maturation: Protocol

1. LNCaP human prostate cells are routinely maintained at 37 °C (with 5% CO_2) in RPMI-1640 medium supplemented with 10% FBS, 2 mmol/L L-glutamine and 100 U/mL penicillin (see Notes 4–6).
2. Assess the viability and immunogenicity of treated tumor cells using Annexin V/DAPI, as previously described (see Note 7) (Fucikova et al., 2014).
3. For subsequent analyses of the phagocytosis rate, add 5 μL of Vybrant DiO cell labelling solution to 1×10^6 iDCs and Vybrant DiD cell labelling solution to 1×10^6 LNCaP cells for 30 min at 37 °C.
4. Cultivate iDCs with previously UV-B-treated LNCaP cells at a 5:1 ratio for 24 h at 37 °C.
5. Analyze the percentage of phagocytosed tumor cells by iDCs using flow cytometry analysis of single-positive and double-positive LNCaP cells and iDCs (Fig. 1).
6. To fully activate DCs, use the appropriate maturation stimulus (either the TLR3 ligand poly I:C, the TLR4 ligand LPS or a standard maturation cocktail (TNF, IL-1β, IL-6 and PGE2)).
7. To analyze the DC maturation status, collect DCs and add a mixture of antibodies against co-stimulatory and surface molecules, including those described in detail in Table 1.
8. Analyze the expression profile of costimulatory and surface molecules, including CD80, CD83, CD86 and HLA-DR, on the cell surfaces of iDCs and activated pulsed $CD11c^+$ DCs (Note 8).

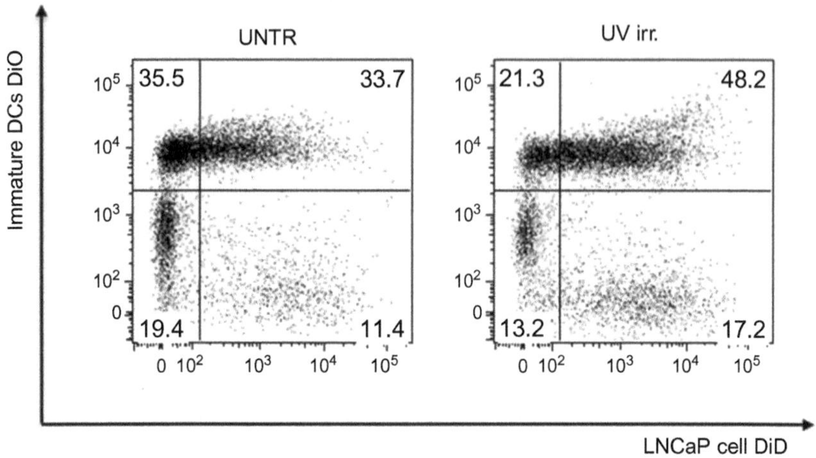

Fig. 1 The rate of phagocytosis of LNCaP cells by DCs. The gating strategy for detecting the rate of phagocytosis of DiD-stained LNCaP cells by DiO-stained iDCs.

Table 1 The antibodies used for flow cytometry analysis of the DC phenotype.

Antibody	**Fluorophore**	**Clone**	**Manufacturer**
CD80	FITC	MAB104	Beckman Coulter
CD86	PE	B7-2	Beckman Coulter
CD14	PE-Dy594	MEM-15	Beckman Coulter
CD83	PC5	HB15a	BD Biosciences
HLA-DR	PC7	L243	Exbio
CD11C	APC	BU15	Exbio

6. The role of DCs in T cell activation: Principles

The maturation status of DCs is crucial for adequate T cell recruitment, activation, expansion and differentiation (Banchereau et al., 2000). DCs produce large amounts of inflammatory chemokines that contribute to the recruitment of DC precursors in inflamed tissues, thus sustaining antigen sampling in peripheral tissue and presentation to T cells in lymph nodes (Savina & Amigorena, 2007). The initiation of an immune response requires at least two signals for the activation of naïve T cells by APCs, including T cell receptor (TCR) signaling and CD28 co-stimulation. T-cell receptors interact with foreign antigens in the context of MHC recognition, providing the first signal. However, a second signal, which is crucial for final T cell activation, occurs when the CD28 receptor is constitutively expressed on T cells and binds to B7-1 (D80) and B7-2 (CD86) molecules that are expressed on APCs. Then, T cells interact with APCs to receive both signals for appropriate T-cell activation, leading to cytokine production, proliferation and induction of antigen-specific T cell responses. DCs play a crucial role in every aspect of the T cell response, including the deletion of self-reactive thymocytes, generation of effector and memory cells and induction of peripheral tolerance (Banchereau et al., 2000).

7. The role of DCs in T cell activation: Protocols

(1) Thaw the $CD14^-$ fraction and collect the cells into RPMI culture medium enriched with 5% AB serum, 1% GlutaMAX, 1% nonessential amino acids and 1% β-mercaptoethanol (Life Technologies; USA) and incubate the cells at 37 °C for 12h.
(2) Cultivate T cells with activated DCs at a 5:1 ratio in a 96-well plate at 37 °C for 10 days (see Note 9).
(3) Add IL-2 (*PeproTech; USA*) to the culture every third day at a final concentration of 20U/mL.
(4) To determine the population of activated T cells, add a mixture of monoclonal extracellular antibodies, as described in detail in Table 2.
(5) Incubate at 4 °C for 25min.
(6) Wash cells twice in PBS.
(7) Add fixation buffer (see Note 10), which is prepared as follows: fixation/permeabilization concentrate and fixation/permeabilization diluent at a 1:3 ratio (*eBioscience; USA*).
(8) Add permeabilization buffer diluted in distilled water at a 1:9 ratio.

Table 2 The antibodies used for flow cytometry detection of tumor-specific T cells.

Antibody	Fluorophore	Clone	Manufacturer
CD3	A700	17AT2	BioLegend
CD4	ECD	SFCI12T4D11	Beckman Coulter
CD8	HV500	RPA-T8	BD Biosciences
IFN-γ (intracellular)	PC7	XMG1.2	BioLegend
CD154 (intracellular)	PE	24–31	BioLegend

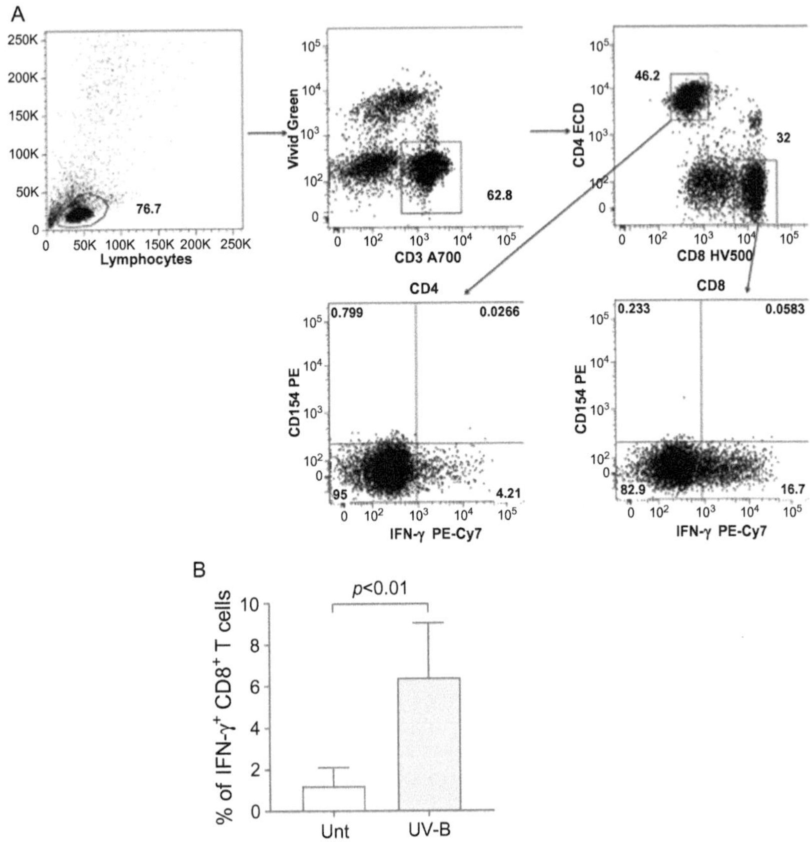

Fig. 2 The induction of tumor-specific T cells by DCs pulsed by UV-B-treated cells. (A) The gating strategy for detecting IFN-γ^+ CD8$^+$ T cells. (B) The percentage of IFN-γ^+ CD8$^+$ T cells after cultivation with DCs pulsed with UV-B-treated LNCaP cells.

(9) Add a mixture of intracellular antibodies and incubate for 30 min at 4 °C, as indicated in Table 2.

(10) Perform the final analysis of IFN-γ- and CD154-producing CD8$^+$ T cells, as indicated by the gating strategy (Fig. 2).

8. Concluding remarks

Over the past decade, increasing knowledge of DC biology and the role of DCs in the coordination of the immune response have made DCs a promising target for cancer immunotherapy (Anguille et al., 2014; Vacchelli et al., 2013). A broad panel of protocols have been developed over the past two decades to harness the potent immunostimulatory activity of DCs against cancer. However, only one cellular product (Sipuleucel-T) involving DCs is currently approved for the treatment of asymptomatic and minimally symptomatic metastatic castration-refractory prostate cancer (Kantoff et al., 2010). Despite this achievement, the development of clinically useful DC-based immunotherapeutic regimens is limited by several obstacles, particularly including the isolation and expansion of autologous DCs that are loaded with TAAs and optionally activated in GMP-compliant facilities. In this report, we provided a detailed protocol for DC generation, pulsation and activation and the subsequent induction of T cell-specific immune responses.

9. Notes

(1) The final purity of mononuclear cells can be improved by diluting blood samples.

(2) The peripheral blood and buffy coat should not be older than 12 h and supplemented with anticoagulants (e.g., heparin, EDTA, citrate or citrate phosphate dextrose).

(3) Performing this step improves the purity of target cells within the separation procedure.

(4) This step is conducted according to an official recommendation by the American Type Culture Collection (ATCC, Manassas, VA, USA) (http://www.lgcstandards-atcc.org/products/all/CRL-1740.aspx?geo_country=cz).

(5) Optimal culture conditions may differ for other cell lines.

(6) Cells should not be detached for passaging at an excessively low (<30%) confluence because this may favor the emergence of subclones, resulting in genetic drift of the original cell population. Similarly, cells should not grow beyond 80% confluence because excessively dense cultures may suffer from metabolic, nutritional, and/or oxidative stress.

(7) Annexin V is a cellular protein that is commonly used to characterize cellular apoptosis. DAPI is a cell-permeant dye that stains double-strand nucleic acids and allows for the identification of necrotic cell populations.

(8) To determine the phenotypic and functional activation profiles of DCs, we recommend analyzing the cytokine profile, including IL-1β, IL-6, IL-8, IL-12 and TNF-α, using an ELISA kit.

(9) Repeat this step to increase the frequency of antigen-specific T cells in the culture.

(10) The Intracellular Fixation & Permeabilization Buffer Set is designed for use in intracellular staining and flow cytometric analysis, and it has been specially formulated to reduce non-specific staining of fluorochrome-labeled antibodies and increase fluorescence signal to noise ratios. In the first step, live cells can be fixed with fixation buffer, which crosslinks proteins. The second step (the use of permeabilization buffer) creates holes in the membrane, thereby allowing the intracellular staining antibodies to effectively enter the cells.

Acknowledgments

This study was exclusively sponsored by Sotio, Prague, Czech Republic.

Reference

Anguille, S., Smits, E. L., Bryant, C., Van Acker, H. H., Goossens, H., Lion, E., et al. (2015). Dendritic cells as pharmacological tools for cancer immunotherapy. *Pharmacological Reviews*, *67*, 731–753.

Anguille, S., Smits, E. L., Lion, E., van Tendeloo, V. F., & Berneman, Z. N. (2014). Clinical use of dendritic cells for cancer therapy. *The Lancet Oncology*, *15*, e257–e267.

Banchereau, J., Briere, F., Caux, C., Davoust, J., Lebecque, S., Liu, Y.-J., et al. (2000). Immunobiology of dendritic cells. *Annual Review of Immunology*, *18*, 767–811.

Boczkowski, D., Nair, S. K., Nam, J. H., Lyerly, H. K., & Gilboa, E. (2000). Induction of tumor immunity and cytotoxic T lymphocyte responses using dendritic cells transfected with messenger RNA amplified from tumor cells. *Cancer Research*, *60*, 1028–1034.

Cabral, M. G., Silva, Z., Ligeiro, D., Seixas, E., Crespo, H., Carrascal, M. A., et al. (2013). The phagocytic capacity and immunological potency of human dendritic cells is improved by α2,6-sialic acid deficiency. *Immunology*, *138*, 235–245.

Cao, J. X., Zhang, X. Y., Liu, J. L., Li, D., Li, J. L., Liu, Y. S., et al. (2014). Clinical efficacy of tumor antigen-pulsed DC treatment for high-grade glioma patients: Evidence from a meta-analysis. *PLoS One*, *9*, e107173.

Curti, A., Ferri, E., Pandolfi, S., Isidori, A., & Lemoli, R. M. (2004). Dendritic cell differentiation. *Journal of Immunology*, *172*, 3 (author reply-4).

Fucikova, J., Moserova, I., Truxova, I., Hermanova, I., Vancurova, I., Partlova, S., et al. (2014). High hydrostatic pressure induces immunogenic cell death in human tumor cells. *International Journal of Cancer*, *135*, 1165–1177.

Fucikova, J., Podrazil, M., Jarolim, L., Bilkova, P., Hensler, M., Becht, E., et al. (2018). Phase I/II trial of dendritic cell-based active cellular immunotherapy with DCVAC/PCa in patients with rising PSA after primary prostatectomy or salvage radiotherapy for the treatment of prostate cancer. *Cancer Immunology, Immunotherapy: CII*, *67*, 89–100.

Fucikova, J., Rozkova, D., Ulcova, H., Budinsky, V., Sochorova, K., Pokorna, K., et al. (2011). Poly I: C-activated dendritic cells that were generated in CellGro for use in cancer immunotherapy trials. *Journal of Translational Medicine*, *9*, 223.

Galluzzi, L., Vacchelli, E., Bravo-San Pedro, J. M., Buque, A., Senovilla, L., Baracco, E. E., et al. (2014). Classification of current anticancer immunotherapies. *Oncotarget*, *5*, 12472–12508.

Hsu, F. J., Benike, C., Fagnoni, F., Liles, T. M., Czerwinski, D., Taidi, B., et al. (1996). Vaccination of patients with B-cell lymphoma using autologous antigen-pulsed dendritic cells. *Nature Medicine*, *2*, 52–58.

Huang, Q., Liu, D., Majewski, P., Schulte, L. C., Korn, J. M., Young, R. A., et al. (2001). The plasticity of dendritic cell responses to pathogens and their components. *Science*, *294*, 870.

Kantoff, P. W., Higano, C. S., Shore, N. D., Berger, E. R., Small, E. J., Penson, D. F., et al. (2010). Sipuleucel-T immunotherapy for castration-resistant prostate cancer. *The New England Journal of Medicine*, *363*, 411–422.

Nagl, M., Kacani, L., Müllauer, B., Lemberger, E.-M., Stoiber, H., Sprinzl, G. M., et al. (2002). Phagocytosis and killing of bacteria by professional phagocytes and dendritic cells. *Clinical and Diagnostic Laboratory Immunology*, *9*, 1165–1168.

Nair, S., Archer, G. E., & Tedder, T. F. (2012). Isolation and generation of human dendritic cells. *Current Protocols in Immunology / Ed John E Coligan [et al]*, Chapter 7: Unit 7.32.

Nakai, N., Hartmann, G., Kishimoto, S., & Katoh, N. (2010). Dendritic cell vaccination in human melanoma: Relationships between clinical effects and vaccine parameters. *Pigment Cell & Melanoma Research*, *23*, 607–619.

Palucka, K., Ueno, H., Roberts, L., Fay, J., & Banchereau, J. (2011). Dendritic cell subsets as vectors and targets for improved cancer therapy. *Current Topics in Microbiology and Immunology*, *344*, 173–192.

Podrazil, M., Horvath, R., Becht, E., Rozkova, D., Bilkova, P., Sochorova, K., et al. (2015). Phase I/II clinical trial of dendritic-cell based immunotherapy (DCVAC/PCa) combined with chemotherapy in patients with metastatic, castration-resistant prostate cancer. *Oncotarget*, *6*, 18192–18205.

Savina, A., & Amigorena, S. (2007). Phagocytosis and antigen presentation in dendritic cells. *Immunological Reviews*, *219*, 143–156.

Stockwin, L. H., McGonagle, D., Martin, I. G., & Blair, G. E. (2000). Dendritic cells: Immunological sentinels with a central role in health and disease. *Immunology and Cell Biology*, *78*, 91–102.

Truxova, I., Hensler, M., Skapa, P., Halaska, M. J., Laco, J., Ryska, A., et al. (2017). Rationale for the combination of dendritic cell-based vaccination approaches with chemotherapy agents. *International Review of Cell and Molecular Biology*, *330*, 115–156.

Vacchelli, E., Vitale, I., Eggermont, A., Fridman, W. H., Fucikova, J., Cremer, I., et al. (2013). Trial watch: Dendritic cell-based interventions for cancer therapy. *Oncoimmunology*, *2*, e25771.

Viaud, S., Thery, C., Ploix, S., Tursz, T., Lapierre, V., Lantz, O., et al. (2010). Dendritic cell-derived exosomes for cancer immunotherapy: What's next? *Cancer Research*, *70*, 1281–1285.

CHAPTER SIX

Isolation of mononuclear phagocytes from the mouse gut

Pailin Chiaranunt, Kyle Burrows, Louis Ngai, Arthur Mortha*
Department of Immunology, University of Toronto, Toronto, ON, Canada
*Corresponding author: e-mail address: arthur.mortha@utoronto.ca

Contents

Abstract

The intestinal tract is home to trillions of microbes that make up the gut microbiota and is a major source of environmental antigens that can be derived from food, commensal microorganisms, and potential pathogens. Amidst this complex environment, myeloid cells, including macrophages (MPs) and dendritic cells (DCs), are key immunological sentinels that locally maintain both tissue and immune homeostasis. Recent research has revealed substantial functional and developmental heterogeneity within the intestinal DC and MP compartments, with evidence pointing to their regulation by the microbiota. DCs are classically divided into three subsets based on their CD103 and CD11b expression: $CD103^+CD11b^-$($XCR1^+$) cDC1s, $CD103^+CD11b^+$ cDC2s, and $CD103^-CD11b^+$ cDC2s. Meanwhile, mature gut MPs have recently been classified by their expression of Tim-4 and CD4 into a long-lived, self-maintaining Tim-$4^+$$CD4^+$ population and short-lived, monocyte-derived Tim-$4^-$$CD4^+$ and Tim-$4^-$$CD4^-$ populations. In this chapter, we provide experimental procedures to classify and isolate these myeloid subsets from the murine intestinal lamina propria for functional characterization.

Methods in Enzymology, Volume 632
ISSN 0076-6879
https://doi.org/10.1016/bs.mie.2019.10.004

1. Introduction

Phagocytosis is a fundamental biological process used for nutrient uptake and sensing of the extracellular environment (Gordon, 2016). Organisms as primitive as amoeba phagocytose microbial prey, while higher order cell systems have evolved variations of this process to orchestrate more complex interactions within their cellular communities (Yutin, Wolf, Wolf, & Koonin, 2009). The largest density of foreign antigens and microbes is believed to be found within the intestinal tract (Grainger, Askenase, Guimont-Desrochers, da Fonseca, & Belkaid, 2014). In a highly diverse cellular environment, environmental sampling is an essential part of immune surveillance and is mandatory for the recognition and elimination of invading pathogens or clearance of dying and infected cells. Highly endocytic cells, found within the non-hematopoietic, epithelial compartment of the gut (i.e., Microfold (M) cells and goblet cells), form the first line of defense against potential harmful microbes residing in our intestinal tract (Gunther & Seyfert, 2018; Knoop, Miller, & Newberry, 2013). Furthermore, nutrients, commensal microbes, and pathogens can all become potential antigens (Grainger et al., 2014). Thus, phagocytic cells that can sample the environment and communicate with non-phagocytic immune cells are essential to preventing invasion and inflammation at the intestinal barrier site. The communication networks that exist between the epithelial barrier, phagocytes, and non-phagocytic immune cells is therefore of great importance to our immune system (Mazzini, Massimiliano, Penna, & Rescigno, 2014; McDole et al., 2012; Ohno, 2016).

Mononuclear phagocytes (MNPs) are myeloid cells specialized in the uptake, processing, and presentation of antigens to prime immune cells across all mammalian tissues (Merad, Sathe, Helft, Miller, & Mortha, 2013). These cells are the central element integrating antigenic information into adaptive immunity and are vital to sustain intestinal and peripheral immune homeostasis and host defense (Merad et al., 2013). MNPs are a heterogeneous group of cells composed of monocytes (MOs), macrophages (MPs) and dendritic cells (DCs). To accurately differentiate between MPs and DCs, it is necessary to examine their distinct origins (Fig. 1). To start, MPs are the first immune cells to develop in the embryo, while DC development is only fully established during late fetal development (De Kleer, Willems, Lambrecht, & Goriely, 2014). Early embryonic MPs are generated from Erythroid-Myeloid Precursor cells in the blood islands

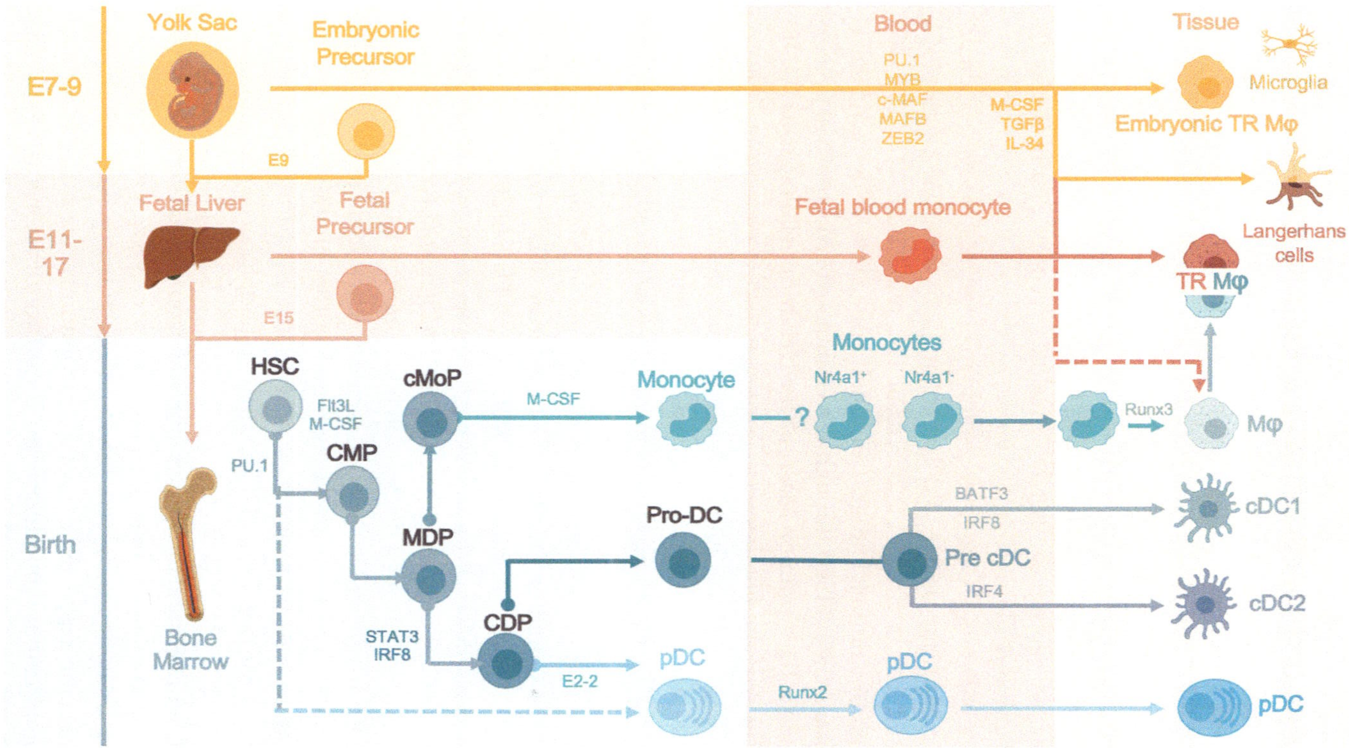

Fig. 1 Development of mononuclear phagocytes. The diagram shows different waves of myelopoiesis across mouse development. The yolk sac, fetal liver, and bone marrow are reservoirs for tissue-resident macrophage precursors, which seed tissues at different stages of embryonic development. Critical growth and differentiation factors support macrophage and dendritic cell development. Key transcription factors regulating lineage specifications are shown. Dotted lines indicate possible differentiation pathways.

of the yolk sac (YS) and seed the early developing brain and skin during the first primitive wave of hematopoiesis. Brain MPs, also termed microglia, and skin MPs, termed Langerhans cells, are long-lived, tissue-resident cells that locally self-renew in the steady state and depend on the growth factors Macrophage-Colony Stimulating Factor (M-CSF) and Interleukin (IL)-34 (Ginhoux & Guilliams, 2016). During a second wave of primitive hematopoiesis, fetal liver (FL)-derived myeloid precursors seed developing tissues to differentiate into tissue-resident MPs. These MPs, like their embryonic counterparts, are long-lived and locally proliferating cells but, in contrast to microglia, are replaceable by adult myeloid precursor cells in the steady state (Perdiguero & Geissmann, 2016).

In adults, MPs and DCs are rarely found in circulation. However, circulating blood MOs are capable of differentiating into tissue MPs, while DCs are generated from a circulating precursor DC pool. Both MPs and DCs arise through two distinct lineages from adult bone marrow-derived Hematopoietic Stem Cells (HSC) through the Granulocyte Macrophage Precursor, diverging at the Common Macrophage Dendritic cell Precursor (MDP) stage into the MP and DC lineages (Fogg et al., 2006; Geissmann, Jung, & Littman, 2003; Liu et al., 2009). Myeloid commitment at the HSC stage requires the transcriptional activity of PU.1 and instructive signaling through the Csf1 receptor (Csf1R) (Mossadegh-Keller et al., 2013). Orchestrated activation of STAT3 and IRF8 further promotes MDP differentiation. MDPs either differentiate into the Common Monocyte Precursor (CMoP) or the Common Dendritic cell Precursor (CDP) within the bone marrow (Becker et al., 2012; Laouar, Welte, Fu, & Flavell, 2003; Naik et al., 2007; Onai et al., 2007). Prior to leaving the bone marrow, CMoPs differentiate into M-CSF-dependent MOs, while FMS-related tyrosine kinase 3 ligand (Flt3L)-dependent CDPs develop into common DC precursors (pre-cDCs) and plasmacytoid DCs (pDCs), respectively (Hettinger et al., 2013; Liu et al., 2009). The latter are rarely found in the intestinal tract but are highly prominent in circulation.

Circulating MOs and pre-cDCs migrate to peripheral organs and differentiate into MPs or DCs (Liu et al., 2009). Pre-cDCs express transcription factors that predetermine their adaptation to become cDC1s or cDC2s, allowing for differentiation into pre-cDC1 or pre-cDC2 (Schlitzer et al., 2015). The transcription factors Interferon Regulatory Factor 8 (IRF8) and Basic Leucine Zipper ATF-Like Transcription Factor 3 (Batf3) regulate cDC1 differentiation, while Interferon Regulatory Factor 4 (IRF4) regulates cDC2 differentiation (Schlitzer et al., 2015). Common surface markers that allow for the discrimination of cDC1 and cDC2 include XCR1 (receptor for lymphotactin-1), CD11b, and CD172a (SIRPa). Further subdivision into cDC1 and cDC2 subsets can be made using combinations of the markers, CD24, CD103, and CD11b (Guilliams et al., 2016). A detailed overview of DC differentiation has recently been provided by Scott and colleagues (Sichien, Lambrecht, Guilliams, & Scott, 2017).

Tissue MPs can arise from yolk sac-derived erythromyeloid precursors, fetal liver, or bone marrow-derived circulating MOs (Ginhoux & Guilliams, 2016). Blood MOs comprise of at least two different subsets that can be distinguished by their high and low expression of the surface marker, Ly6C, or expression of the chemokine receptors, CCR2 and CX3CR1 (Geissmann et al., 2003). Ly6C^{high} MOs are CCR2^{high} CX3CR1^{-}, while Ly6C^{lo} MOs are CCR2^{lo} CX3CR1^{high}. Ly6C^{high} MOs differentiate into Ly6C^{lo} MOs in a manner dependent on the transcription factor Nuclear Receptor Subfamily 4 Group A Member 1 (NR4A1) (Hanna et al., 2011). Ly6C^{high} MO development from CMoPs requires the transcription factors CCAAT/enhancer-binding protein (C/EBP) alpha (a) and beta (b), IRF8, and PU.1 (Zhu, Thomas, & Hedrick, 2016). Upon entering a tissue, MOs respond to the local environment and adapt a transcriptional profile as they differentiate into tissue MPs (Lavin et al., 2014). The transition of MOs into MPs within tissues is easily visualized using a flow cytometry-based gating strategy that segregates CD11b-expressing cells into a "waterfall" continuum, in which high expression of Ly6C indicates MO identity while MHCII expression correlates with maturation into MPs (Tamoutounour et al., 2012). Tissue-resident MPs depend on continuous organ-specific signals to regulate a large set of their transcriptome, while other parts of their transcriptome remain expressed even if these MPs are moving into a different tissue. These data collectively suggest a role for epigenetic imprinting of tissue-specific MP identity depending on environmental signals (Lavin, Mortha, Rahman, & Merad, 2015).

Within the adult intestinal tract, MPs can be subdivided into at least three subsets based on their expression of CD4 and Tim-4. Tim-4^{+}CD4^{+} MPs arise from fetal monocytes and are the earliest MPs found in the intestinal tract. This population of cells has a limited ability to self-maintain through proliferation and experiences slow but steady replacement by circulating MOs (Liu et al., 2019; Shaw et al., 2018). Tim-4^{-}CD4^{+} MPs arise later during postnatal life, followed by Tim-4^{-}CD4^{-} MPs, both of which get permanently replaced by circulating adult monocytes. A new, elegant transgenic mouse system, tracking the history of *Ms4a3* expression (a gene exclusively expressed in adult MOs), revealed that the Tim-4^{-}CD4^{-} and Tim-4^{-}CD4^{+} MP populations require 6–8 weeks to be entirely replaced by infiltrating MOs in the steady state. In turn, fetal-derived Tim-4^{+}CD4^{+} MPs self-maintained for over 8 months prior to their complete replacement by adult MOs, collectively suggesting differential contribution of adult MOs to the gut MP pool (Liu et al., 2019). A recent characterization of these MP populations also uncovered distinct gene expression profiles associated with specialized functions that support homeostasis of local microanatomic structures and processes within the intestinal tract (De Schepper et al., 2018). These findings raise the question as to whether there exist environmentally controlled plasticity and dynamics between these MP populations that may impact intestinal tissue homeostasis, similar to those described for liver MPs (Bonnardel et al., 2019). Meanwhile, cDC1 and cDC2 within the intestinal tract arise from pre-committed circulatory pre-cDC1 and pre-cDC2 precursors, which undergo terminal differentiation within the gut. In contrast to tissue MPs, DCs lack tissue-specific transcription profiles, much in line with their conserved functions across the entire body (Heidkamp et al., 2016).

Both DCs and MPs are surveilling cells, a function that requires them to migrate to and within tissues and to localize within microanatomic areas for antigen sampling. The chemokine receptors CCR9 and CCR10, in conjunction with the alpha 4 beta 7 (α4β7) integrin complex, determine homing of lymphocytes and leukocytes to the small and large intestinal mucosa (Habtezion, Nguyen, Hadeiba, & Butcher, 2016). Chemokines, such as CX3CL1, produced by intestinal epithelial cells position MPs in close proximity to the epithelium to directly sample luminal content (Kim et al., 2011; Niess et al., 2005). These luminal antigens can also be sampled and transferred to MNPs via M cells and goblet cells either through transcytosis or goblet cell associated antigen passages (GAP) (Knoop et al., 2013; Ohno, 2016). Observations of antigen transfer between MPs and DCs hint at their division of labor in the context of foreign antigen presentation (Fig. 2) (Mazzini et al., 2014).

Antigen sampling by DCs and MPs occurs inside the tissue and involves a broad spectrum of direct and indirect mechanisms. Phagocytosis by MPs of intestinal microbes or antigens induces the MP-specific release of IL-10, which in turn sustains the local pool of suppressive T effector cells and supports maintenance of oral tolerance (Hadis et al., 2011; Jinushi et al., 2007; Kim et al., 2018). However, priming of T cells rarely requires MPs and their migration to lymph nodes. Instead, DCs, upon antigen engulfment or

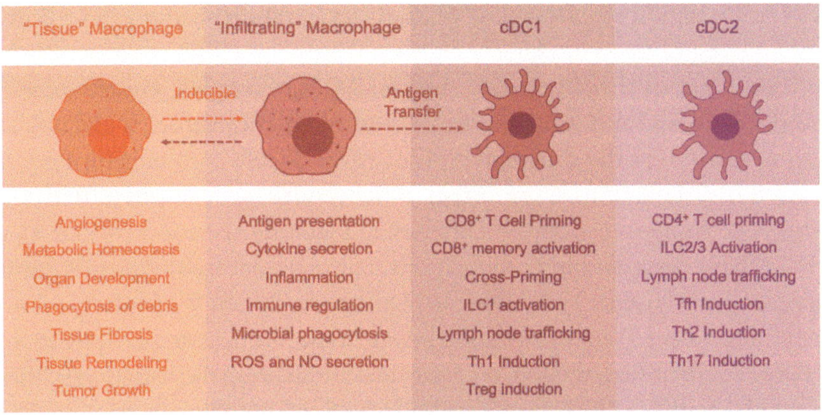

Fig. 2 Physiological roles of mononuclear phagocytes. Macrophages and dendritic cells are functionally distinct. Embryonic and fetal derived "tissue" macrophages regulate local tissue homeostasis with minimal contribution to immunity. "Infiltrating" adult macrophages contribute to host defense and local immune responses. Dendritic cells are specialized in their interaction with adaptive and innate lymphoid cells and initiate tailored immune responses.

antigen-transfer by MPs, upregulate the chemokine receptor CCR7 and migrate to tissue draining mesenteric lymph nodes carrying their engulfed antigenic cargo (Forster et al., 1999; Jang et al., 2006). Within the lymph node, DCs present antigen to naïve T cells, facilitating their priming, proliferation, and adaptation of effector functions. Post DC-encounter, effector T cells migrate back into the mucosal tissue as differentiated helper or cytotoxic T cells.

While antigen presentation to T cells is important for the initiation of an inflammatory response toward pathogens, immunologic tolerance to most luminal antigens is preferable during homeostasis. Regulatory T cells (Tregs) are generated by MHCII-mediated antigen presentation by DCs within gut-draining lymph nodes, while tissue-resident MPs are involved in the local maintenance of Tregs (Klebanoff et al., 2013). Consequently, the loss of MHCII on all CD11c-expressing DCs and MPs in an otherwise healthy mouse results in the loss of immune tolerance and the development of microbiota-dependent intestinal inflammation (Loschko et al., 2016). Similarly, deficiency in CCR7 results in the loss of T cell and B cell homeostasis due to the migratory defects in DCs (Forster et al., 1999). Interestingly, the loss of MHCI does not affect intestinal immune tolerance but impairs the capacity of the immune system to clear virally infected cells and tumor cells (Kurts, Cannarile, Klebba, & Brocker, 2001; Zijlstra et al., 1990). Collectively, antigen presentation by

MNPs is a critical regulator of T cell activation, differentiation, and adaptive immune responses to intestinal antigens.

Importantly, in addition to antigen-MHC complex presentation by MNPs, cytokine secretion and expression of co-stimulatory molecules are also necessary for the differentiation of naïve T cells into effector lineages (i.e., Th1, Th2, Th17, or Treg) (Guermonprez, Valladeau, Zitvogel, Thery, & Amigorena, 2002). Differential cytokine secretion by cDC1s or cDC2s have been demonstrated to prime tailored responses to distinct groups of pathogens (Hildner et al., 2008; Williams et al., 2013). While the cytokines required to drive T cell polarization toward a specific T helper cell phenotype have been identified, the signaling cascades stimulating the secretion of these cytokines within cDC subsets remain to be fully uncovered.

The exact role of DCs during embryogenesis remains a vastly understudied field from the perspective of the embryo as opposed to the maternal counterpart. Considering the absence of pathogen-specific reactivity within the early days of embryonic development, MPs arising during this stage may serve as regulators of tissue homeostasis in addition to their postnatal role in tissue immunity. This becomes particularly obvious when comparing tissue cell development in the absence of MPs. Mice lacking growth factors for MPs display severe development defects in their bones, teeth, mammary glands, muscles, and brains, while deficiencies in cDC subsets do not have a major impact on organogenesis (Dai et al., 2002; Gouon-Evans, Rothenberg, & Pollard, 2000; Horsley, Jansen, Mills, & Pavlath, 2003; McKenna et al., 2000; Michaelson et al., 1996). MP-derived growth and differentiation factors for tissue cells, alongside the machinery responsible for the clearance of dying tissue cells, are essential components in controlling proper development. In particular, clearance of dead cell debris, a crucial function executed by MPs, prevents the onset of autoinflammatory pathologies (Hanayama et al., 2004). Interestingly, phagocytosis of apoptotic cells primed an immunosuppressive effector program in MPs rather than unleashing the production of inflammatory mediators (Jinushi et al., 2007). With MPs being present in all tissues since birth, this important caretaker function by MPs primes the tissue landscape for newly infiltrating immune cells during pre- and post-natal life. MPs therefore serve as multi-modular regulators of tissue and immune homeostasis, while bridging tissue damage and infection to adaptive immune regulation through DC independent and dependent pathways. Notably, drastic functional changes in MPs are seen during chronic and acute inflammation. In response to inflammatory signals from the damaged or infected tissue, MPs mediate the recruitment of MOs and neutrophils. In particular, experimental models of intestinal inflammation reveal changes in the composition and function of gut MPs alongside increased abundance of MOs and neutrophils (Griseri, McKenzie, Schiering, & Powrie, 2012; Zigmond et al., 2012). Thus, analysis of the gut myeloid landscape should take these alterations into consideration.

MNPs are critical to tumor immunity, with cDC1s, adult monocyte-derived MPs, and fetal tissue-resident MPs playing opposing roles (Bottcher & Reis, 2018; Franklin et al., 2014; Lavin et al., 2017). Cytotoxic $CD8^+$ T cells and Natural Killer (NK) cells are essential components in the rejection of malignant cells as they initiate crosstalk with cDC1s to eliminate and clear tumors (Bottcher et al., 2018). However, while a tissue immune profile rich in $CD8^+$ T cells is a good predictor of disease outcome in colorectal cancer (CRC), experimental evidence for cDC1s as regulators of colorectal tumor rejection remains sparse (Bindea et al., 2013). Interestingly, data from non-intestinal solid tumors strongly suggest that targeting cDC1 function for the improvement of anti-tumor immunity may be a promising option for the treatment of tumors (Barry et al., 2018; Salmon et al., 2016). On the other hand, MPs were found to play opposing roles in anti-tumor immunity. While fetal tissue-resident MPs were reported to foster tumor growth, infiltrating adult monocyte-derived MPs are generally associated with strong anti-inflammatory responses and immunosuppression (Engblom, Pfirschke, & Pittet, 2016). In line with these observations, experimental mouse models of CRC revealed that CCR2-dependent MOs differentiate into MPs within the intestinal tract and contributed to exacerbated tumor growth due to elevated release of proinflammatory cytokines (Jung et al., 2017; Popivanova et al., 2009). Despite being contradictory to results observed in extra-intestinal solid tumors, functional reprogramming or selective depletion of MPs within tumors is a potential target for the improvement of anti-tumor immunity. However, the opposing secretion of these inflammatory cytokines by gut MPs is a critical factor, requiring strategic consideration when developing therapies targeting MNPs for the treatment of intestinal or extra-intestinal tumors. Epithelial cell homeostasis, for example, depends on MP-derived inflammatory cytokine stimulation to achieve balanced proliferation, differentiation, and cell death. The secretion of many of these "homeostatic" cytokines acting to support healthy epithelial turnover is produced and induced by MNPs in the intestinal tract and may support the exacerbated growth and survival of malignant cells as well, collectively suggesting that targeting steady-state functions of MPs within the gut may require careful evaluation (Kirchberger et al., 2013).

In essence, MNPs are a central element controlling intestinal immunity. Here, we provide a fast and cost-efficient protocol for the reproducible isolation of small and large intestinal MNPs in mice. In addition, we provide a set of surface marker flow cytometry stains that easily allow for the classification and isolation of highly purified gut DC subsets and MP subsets on five-laser cytometer setups. This protocol has been used in the laboratory to characterize gene expression in FACS-sorted MNPs and to analyze the production of chemokines and cytokines following intracellular cytokine staining. We hope this protocol will be a useful recipe for researchers interested in studying intestinal MNPs during either steady state or inflammation.

2. Identification of gut lamina propria myeloid cell populations by flow cytometry

2.1 Equipment

- Flow cytometer (BD LSR Fortessa X20, five lasers (355/405/488/561/640))
- 5 mL polystyrene tubes (FACS tubes)
- Pipettes, 0.5–10 μL (10 μL), 2–20 μL (20 μL), 20–200 (200 μL), and 100–1000 (1000 μL)
- Serological pipettes, 5, 10, 25, and 50 mL
- Pipette controller (PIPETBOY acu, INTEGRA Biosciences)
- Conical tubes, 15 and 50 mL (Falcon)
- Microcentrifuge tubes, 1.5 mL (Eppendorf)
- Centrifuge (Eppendorf 5810 R)
- Inverted phase microscope (Fisher Scientific)
- Cell strainers, 70 μm (Corning, cat. no. CLS431751)
- Fine curved scissors
- Fine forceps
- Petri dishes, 100 mm × 15 mm (Fisherbrand, cat. no. FB0875712)
- 6-well plates (Corning, cat. no. 3506)
- Incubator shaker (New Brunswick Scientific, cat. no. I24/I24R)

2.2 Mice

C57BL/6 mice, male or female, 8–10 weeks old (The Jackson Laboratory, cat. no. 000644).

2.3 Chemicals/reagents

- Deoxyribonuclease I (DNase I) from bovine pancreas (SIGMA Life Science, cat. no. DN25)
- Collagenase from Clostridium histolyticum (SIGMA Life Science, cat. no. 5138)
- Hank's balanced salt solution without calcium/magnesium (HBSS −/−) (Gibco, cat. no. 14175)
- HBSS with calcium and magnesium (HBSS+/+) (Gibco, cat. no. 14025)
- Percoll (GE Healthcare, cat. no. 17-0891-01) (*see Notes*)
- HEPES (1 M) (Gibco, cat. no. 15630)
- Ethylenediaminetetraacetic acid (EDTA) (Bioshop, cat. no. EDT111)
- Fetal bovine serum (Gibco, cat. no. 12483), heat inactivated at 56 °C for at least 30 min, aliquoted and stored sterilely at −20 °C
- Dulbecco's Phosphate Buffered Saline (Modified, without calcium chloride and magnesium chloride, liquid, sterile-filtered, suitable for cell culture) (SIGMA Life Science, cat. no. D8537)
- Dulbecco's Modified Eagle's Medium (DMEM)—high glucose, with 4500 mg/L glucose, L-glutamine, and sodium bicarbonate, without sodium pyruvate, liquid, sterile-filtered, suitable for cell culture (SIGMA Life Science, cat. no. D5976)
- Penicillin-Streptomycin (100 ×) (SIGMA Life Science, cat. no. P4333)
- 10 × Dulbecco's phosphate buffered saline (DPBS; Invitrogen, cat. no. D1408)
- Sodium pyruvate (100 mM) (Gibco, cat. no. 11360)
- *Dissociation buffer*: HBSS −/− supplemented with 5 mM EDTA and 10 mM HEPES (*see Notes*)
- *Digestion buffer*: HBSS+/+ containing 4% FBS, 10 mM HEPES, 0.5 mg/mL collagenase, and 0.5 mg/mL DNase I
- *D-10 cell culture medium*: High-glucose DMEM, supplemented with 10% FBS, 200 mg/L glutamine, 50,000 U/L penicillin, 1 mM sodium pyruvate (Invitrogen). Stored at 4 °C
- *FACS buffer:* 1 × DPBS supplemented with 2% FBS and 5 mM EDTA. Stored at 4 °C
- eBioscience™ FOXP3/Transcription Factor Staining Buffer Set (ThermoFisher, cat. no. 00-5523-00)

2.4 Antibodies

Antibody	Clone	Fluorophore	Source	Catalog number	Stock concentration (mg/mL)
Anti-CD103	2E7	PE	BioLegend	121406	0.2
Anti-CD11b	M1/70	Alexa Fluor 488	eBioscience	53-0112-82	0.5
Anti-CD11c	N418	PE-Cy7	BioLegend	117318	0.2
Anti-CD16/CD32 (Fc block)	2.4G2	Purified	BD Pharmingen	553141	0.5
Anti-CD4	GK1.5	FITC	BioLegend	100406	0.5
Anti-CD45.2	104	APC-Cy7	BioLegend	109824	0.2
Anti-CD64	X54-5/7.1	Alexa Fluor 647	BioLegend	139322	0.5
Anti-CD64	X54-5/7.1	Brilliant Violet 605	BioLegend	139323	0.2
Anti-IRF4	3E4	FITC	eBioscience	11-9858-82	0.5
Anti-IRF8	V3GYWCH	APC	eBioscience	17-9852-82	0.2
Anti-Ly-6C	HK1.4	PerCP-Cy5.5	eBioscience	45-5932-82	0.2
Anti-MHCII (I-A/I-E)	M5/114.15.2	Pacific Blue	BioLegend	107620	0.5
Anti-CD172a (SIRPa)	P84	APC	BioLegend	144014	0.2
Anti-Tim-4	F31-5G3	PE	BioLegend	129906	0.2
Anti-XCR1	ZET	Alexa Fluor 647	BioLegend	148214	0.5

3. Protocol

3.1 Lamina propria lymphocyte isolation

Time is of the essence! Lamina propria cells die rapidly once removed from the mouse; small intestinal cells are more sensitive and should be processed before the colon.

Prepare buffers beforehand.

1. Euthanize mouse according to institutional guidelines. Open the peritoneal cavity, pull the small intestine below the stomach using forceps and cut at the duodenum/stomach junction. Carefully pull the intestine out to not damage the tissue, sliding it through the scissors to trim out the mesentery, then cut at the cecum to fully remove the small intestine. Next, hold the large intestine using forceps, cut out the cecum, then pull the intestine and cut as close to the rectum as possible to remove the colon.
2. Place small intestine and colon into separate petri dishes containing ice-cold D-10 media and keep moist at all times.
3. Using forceps and scissors, remove the mesentery, residual fat, and connective tissue. A well-cleaned tissue will lead to a cleaner lymphocyte preparation, which will give better staining.
4. Remove the Peyer's patches of the small intestine by cutting out small sections with curved scissors. Peyer's patches can be recognized by their opaque color and nodal structure; there are 7–10 Peyer's patches per small intestine. The Peyer's patches of the distal end are small and more difficult to visualize, but they become larger toward the proximal end of the small intestine. Peyer's patches are not found on the colon.
5. Cut open the intestine longitudinally using scissors while gently holding the tissue with forceps or a gloved hand. Once opened, gently shake the intestine in the petri dish containing ice-cold D-10 in order to remove the fecal material. Next, gently scrape the luminal side of the intestine with curved forceps to remove any remaining fecal matter and excess mucus. For the small intestine, cut the sample into four equal parts.
6. Place intestine into a 50 mL conical tube containing dissociation buffer (10 mL for colon, 15 mL for small intestine), which will remove the epithelial layer containing both epithelial cells and intra-epithelial lymphocytes.
7. Incubate at 37 °C for 15 min with slow rotation at 100 rpm in a shaker.

8. Vortex vigorously for 30 s, and discard the supernatant containing epithelial cells and intra-epithelial lymphocytes.
9. Collect the tissue, gently scrape both sides with the forceps to remove any remaining epithelium and mucus, and rinse in a petri dish containing HBSS +/+. Then place the tissue in a 1.5 mL microcentrifuge tube and mince into fine pieces using scissors. The minced tissue should be able to pass through a 1000 μL pipette tip with ease.
10. Transfer tissue into a 50 mL conical tube with digestion buffer (10 mL for colon, 15 mL for small intestine) and incubate at 37 °C for 20 min with slow rotation at 100 rpm in a shaker (*see Notes*).
11. Vortex well for 30 s, then collect the supernatant by filtering through a 70 μm cell strainer into a 50 mL conical tube.
12. Spin down the cells at 500 × *g* for 10 min.
13. Prepare Percoll solutions (*see Notes*).
14. Resuspend the cells from step 12 in 5 mL of 40% Percoll.
15. Fill a fresh 15 mL centrifuge tube with 5 mL 80% Percoll and overlay the 40% Percoll containing the cells on top of the 80%. A slow and constant pace while pipetting is key to preventing mixing while overlaying the Percoll.
16. Centrifuge the tubes at 1300 × *g* for 20 min at 20 °C with brakes off.
17. Remove and aspirate out the top layer of epithelial cells.
18. Leukocytes will form an opaque layer at the interface between 40% and 80% Percoll. Collect these cells using a fresh serum pipette (*see Notes*).
19. Transfer the leukocytes to a fresh 15 mL conical tube and fill to 10 mL with FACS buffer. Spin down the cells at 500 × *g* for 7 min, then re-suspend in FACS buffer and keep cells on ice. Save a small aliquot for cell counting (*see Notes*). Cells are now ready for staining.

3.2 Staining procedure for monocytes and macrophages

1. Wash cells with PBS. Incubate in 100 μL of Live/Dead Aqua (1:1000 of stock in PBS) for 20 min on ice (*see Notes*).
2. Wash cells twice with FACS buffer.
3. Incubate cells with a 100 μL cocktail of surface staining, fluorescently labeled antibodies, including Fc block (1:200), for 20 min on ice. For monocytes/macrophages, we have found that by minimally co-staining with anti-CD45.2 APC-Cy7 (1:200) as our hematopoietic marker, anti-MHCII Pacific Blue (1:2000) as our marker for antigen presenting cells, and anti-CD64 APC (1:100) and anti-Ly6C PerCP-Cy5.5

(1:1000) as our macrophage and monocyte markers, respectively, we can utilize many combinations of other FITC-, PE-, and PE-Cy7-labeled antibodies to assess further surface markers on lamina propria cells (Fig. 3).

4. Wash stained cells twice with FACS buffer. Analyze on a flow cytometer.

3.3 Staining procedure for cDC1s and cDC2s

3.3.1 Surface markers staining

1. Wash cells with PBS. Incubate in 100 μL of Live/Dead Aqua (1:2000 of stock in PBS) for 20 min on ice.
2. Wash cells twice with FACS buffer.
3. Incubate cells with a 100 μL cocktail of surface staining, fluorescently labeled antibodies, including Fc block (1:200), for 20 min on ice. For DCs, we have found that by minimally co-staining with anti-CD45.2 APC-Cy7 (1:200) as our hematopoietic marker, anti-MHCII Pacific Blue (1:2000) as our marker for antigen presenting cells, anti-CD64 BV605 (1:100) as our marker to exclude macrophages, and anti-CD11c PE-Cy7 (1:1000) as the integrin marker for DCs (and some macrophages), we can utilize many combinations of other FITC-, PE-, and APC-labeled antibodies to assess further surface markers on lamina propria cells. Importantly, identification of DC subsets will require additional staining with anti-CD11b FITC (1:1000) or anti-CD172a (SIRPa) APC (1:100) and anti-CD103 PE (1:500) (Fig. 4) (*see Notes*).
4. Wash stained cells twice with FACS buffer. Analyze on a flow cytometer.

3.3.2 Transcription factors staining

1. After the last wash post-surface staining, fix and permeabilize cells for at least 20 min using the fixation/permeabilization buffer from the eBioscience FOXP3 Staining Kit according to the manufacturer's protocol.
2. Wash 2 × with the kit's wash/permeabilization buffer.
3. Intracellular staining for DCs is performed with anti-IRF4 FITC (1:100) and anti-IRF8 APC (1:100) diluted in wash/permeabilization buffer solution for 1 h at 4 °C in the dark (Fig. 4).
4. Wash 2 × with the kit's wash/permeabilization buffer.
5. Resuspend stained cells in FACS buffer. Analyze on a flow cytometer.

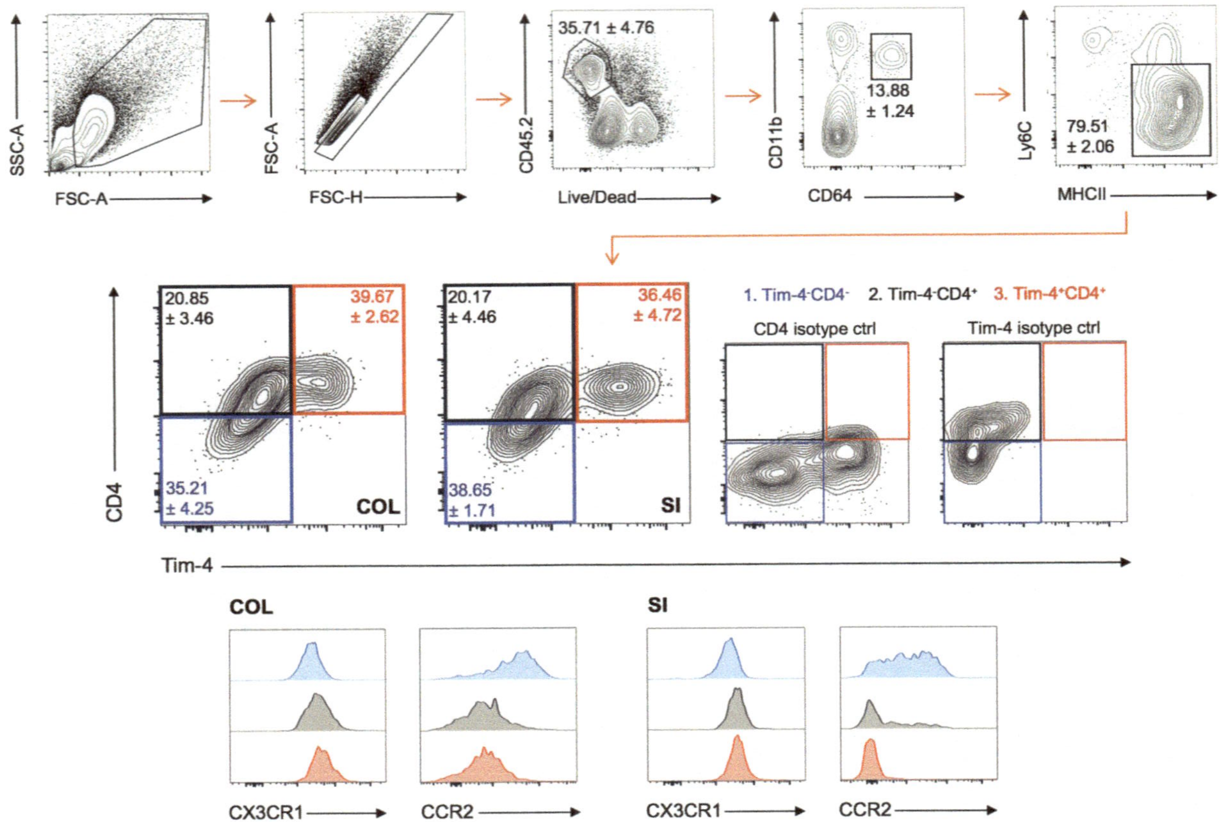

Fig. 3 Identification of intestinal macrophage subsets. Gating strategy to identify intestinal macrophages via flow cytometry. Lamina propria leukocytes were stained with a dead cell exclusion dye and antibodies against CD45, CD11b, CD64, CD4, Tim-4, MHCII, and Ly6C. Following exclusion of dead cells and gating on $CD45^+$ cells, $CD11b^+CD64^+$ macrophages are identified and further subdivided based on Ly6C and MHCII expression. $Ly6C^+MHCII^{-/lo}$ cells contain infiltrating monocytes and immature macrophages. $Ly6C^{-/lo}MHCII^+$ mature macrophages were then subdivided into three additional subsets based on the expression of Tim-4 and CD4. $Tim\text{-}4^-CD4^-$ macrophages (blue gate), $Tim\text{-}4^-CD4^+$ macrophages (black gate) and $Tim\text{-}4^+CD4^+$ macrophages (red gate) can be identified. Isotype controls for CD4 and Tim-4 are shown. Macrophage subsets can additionally be discriminated based on their expression of CX3CR1 and CCR2. Numbers adjacent to gates represent percentages of each population ± SEM.

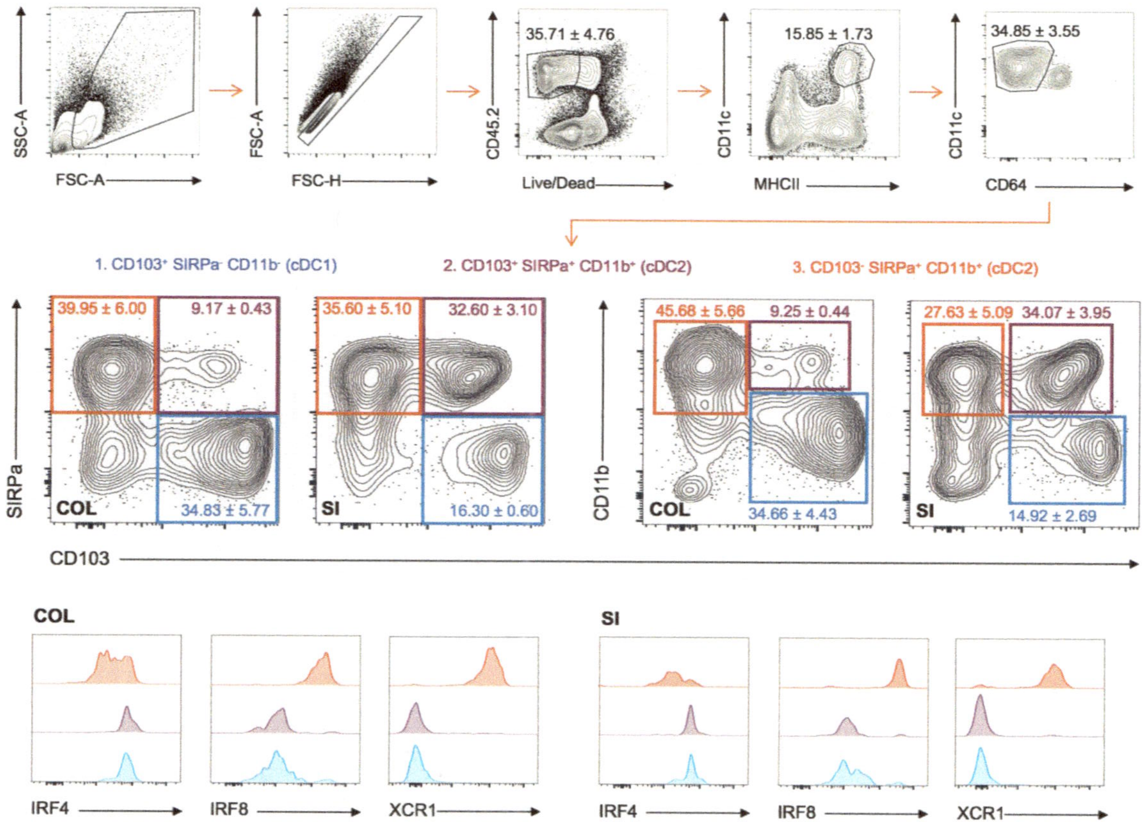

Fig. 4 Identification of intestinal dendritic cell subsets. Gating strategy to identify intestinal dendritic cells via flow cytometry. Lamina propria leukocytes were stained with a dead cell exclusion dye and antibodies against CD45, CD11c, MHCII, CD64, CD172a, and CD103. Following exclusion of dead cells and gating on $CD45^+$ cells, $CD11c^{hi}MHCII^+$ dendritic cells were identified and further exclusion of $CD64^+$ macrophages performed. Dendritic cell subsets are then identified based on the expression of CD103 and CD172a or CD11b. $CD103^+CD172a^-CD11b^-$ dendritic cells (red gate), $CD103^+CD172a^+CD11b^+$ dendritic cells (violet gate), and $CD103^-CD172a^+CD11b^+$ dendritic cells (blue gate) can be identified. Dendritic cell subsets express their lineage defining transcription factors IRF4 (cDC2) and IRF8 (cDC1) as well as the CD172a substitute, CD11b, and the cDC1 marker, XCR1. Numbers adjacent to gates represent percentages of each population ± SEM.

Notes

It is very important to use Ca/Mg-free HBSS as Ca/Mg-containing HBSS will not allow for separation of epithelial cells and intra-epithelial lymphocytes. Excessive epithelial cell contamination during digestion results in decreased viability of lamina propria lymphocytes.

Our protocol involves one round of digestion in DNAse and collagenase, while other laboratories may perform up to three rounds of digestion and may substitute collagenase for liberase or dispase. We prefer the simple use of collagenase and single round of digestion as it is less time-consuming and more cost-effective while yielding sufficient cell numbers. Furthermore, in our experience with dispase-containing digestion buffers, decreased effectiveness of antibody stains, particularly CD4, is to be anticipated due to dispase.

Due to large numbers of non-hematopoietic cells following digestion of the lamina propria, our method highly recommends the Percoll gradient enrichment step to clean up the sample. Percoll is not an isotonic solution. Thus, 1 vol. of 10 × DPBS is added to 9 vol. of Percoll in order to obtain an isotonic 90% Percoll solution. Dilute 90% Percoll with the appropriate amount of cell culture medium in order to obtain 40% and 80% Percoll solutions used for the separation gradient.

Using a Percoll gradient will remove neutrophils, stromal cells, platelets, epithelial cells, and erythrocytes, thus decreasing the overall cell yield. In our hands, the DC and MP pools, as well as overall viability, remain unaltered after Percoll enrichment. Analysis of neutrophil numbers, often required in experimental models of intestinal inflammation, should be performed on preparations prior to the Percoll step. However, skipping the Percoll gradient often results in decreased cell viability due to an increase in dead non-hematopoietic cells. These cells often aggregate and clump and may negatively impact leukocyte viability and yield, increase acquisition time at the flow cytometer, and cause clotting of nozzles on high speed cell sorters. For sorting strategies and downstream RNASeq analysis, researchers should follow a stringent doublet exclusion and inclusion of a "dump/lineage" channel, to avoid confounding contaminations by B or T lymphocytes.

Typical cell yields for colons and small intestines of wild-type C57BL/6 male or female, 8- to 10-week-old mice housed in our facility are $1\text{–}2 \times 10^6$ and $3\text{–}5 \times 10^6$ cells (50–60% viable), respectively. We have previously performed this protocol in animal facilities across three different countries on animals housed under different grades of pathogen exclusion. The cell yields reported here may vary with those observed at different institutions and may even reach higher numbers when performed on mice housed under reduced pathogen exclusion criteria. The exclusion of segmented filamentous bacteria (SFB), mouse norovirus (MNV), mouse hepatitis virus (MHV), *Helicobacter* spp., and intestinal protozoa in our institute seems to be particularly important factors. These pathogens are not commonly excluded across animal facilities and will result in higher yields of gut $CD45^+$ cells.

Live/Dead Aqua is preferred over other viability dyes, such as DAPI or 7AAD/PI, as its excitation/emission spectra does not overlap with commonly used fluorophores (Pacific Blue/BV421/eFluor450 or PerCP-Cy5.5, respectively). An additional advantage of this protocol is the routine incorporation of a fixable dye for Live/Dead cell exclusion, which has proven valuable for our analysis of intracellular markers, including cytokines, chemokines, and transcription factors in MPs and DCs.

Notably, for the analysis of MPs and DCs, CD11b as marker for myeloid cells can be exchanged with CD172a. Both markers are suitable for the identification and discrimination of the cDC1 and cDC2 subsets and MPs (Fig. 4). CD172a has been shown to be the superior marker due to its cross-species applicability (Guilliams et al., 2016).

Acknowledgments

A.M. is supported by CIHR-Project Grant (388337) and NSERC-Discovery Grant (RGPIN-2019-04521). A.M. is the Tier 2 Canadian Research Chair in Mucosal Immunology and supported by the Tier 2 CRC-CIHR program. Figs. 1 and 2 were "Created with BioRender."

References

Barry, K. C., Hsu, J., Broz, M. L., Cueto, F. J., Binnewies, M., Combes, A. J., et al. (2018). A natural killer-dendritic cell axis defines checkpoint therapy-responsive tumor microenvironments. *Nature Medicine*, *24*(8), 1178–1191. https://doi.org/10.1038/s41591-018-0085-8.

Becker, A. M., Michael, D. G., Satpathy, A. T., Sciammas, R., Singh, H., & Bhattacharya, D. (2012). IRF-8 extinguishes neutrophil production and promotes dendritic cell lineage commitment in both myeloid and lymphoid mouse progenitors. *Blood*, *119*(9), 2003–2012. https://doi.org/10.1182/blood-2011-06-364976.

Bindea, G., Mlecnik, B., Tosolini, M., Kirilovsky, A., Waldner, M., Obenauf, A. C., et al. (2013). Spatiotemporal dynamics of intratumoral immune cells reveal the immune landscape in human cancer. *Immunity*, *39*(4), 782–795. https://doi.org/10.1016/j.immuni.2013.10.003.

Bonnardel, J., T'Jonck, W., Gaublomme, D., Browaeys, R., Scott, C. L., Martens, L., et al. (2019). Stellate cells, hepatocytes, and endothelial cells imprint the Kupffer cell identity on monocytes colonizing the liver macrophage niche. *Immunity*, *51*, 638–654. https://doi.org/10.1016/j.immuni.2019.08.017.

Bottcher, J. P., Bonavita, E., Chakravarty, P., Blees, H., Cabeza-Cabrerizo, M., Sammicheli, S., et al. (2018). NK cells stimulate recruitment of cDC1 into the tumor microenvironment promoting cancer immune control. *Cell*, *172*(5). https://doi.org/10.1016/j.cell.2018.01.0041022–1037.e1014.

Bottcher, J. P., & Reis, E. S. C. (2018). The role of type 1 conventional dendritic cells in Cancer immunity. *Trends in Cancer*, *4*(11), 784–792. https://doi.org/10.1016/j.trecan.2018.09.001.

Dai, X. M., Ryan, G. R., Hapel, A. J., Dominguez, M. G., Russell, R. G., Kapp, S., et al. (2002). Targeted disruption of the mouse colony-stimulating factor 1 receptor gene results in osteopetrosis, mononuclear phagocyte deficiency, increased primitive progenitor cell frequencies, and reproductive defects. *Blood*, *99*(1), 111–120. https://doi.org/10.1182/blood.v99.1.111.

De Kleer, I., Willems, F., Lambrecht, B., & Goriely, S. (2014). Ontogeny of myeloid cells. *Frontiers in Immunology*, *5*, 423. https://doi.org/10.3389/fimmu.2014.00423.

De Schepper, S., Verheijden, S., Aguilera-Lizarraga, J., Viola, M. F., Boesmans, W., Stakenborg, N., et al. (2018). Self-maintaining gut macrophages are essential for intestinal homeostasis. *Cell*, *175*(2). https://doi.org/10.1016/j.cell.2018.07.048400-415.e413.

Engblom, C., Pfirschke, C., & Pittet, M. J. (2016). The role of myeloid cells in cancer therapies. *Nature Reviews. Cancer*, *16*(7), 447–462. https://doi.org/10.1038/nrc.2016.54.

Fogg, D. K., Sibon, C., Miled, C., Jung, S., Aucouturier, P., Littman, D. R., et al. (2006). A clonogenic bone marrow progenitor specific for macrophages and dendritic cells. *Science*, *311*(5757), 83–87. https://doi.org/10.1126/science.1117729.

Forster, R., Schubel, A., Breitfeld, D., Kremmer, E., Renner-Muller, I., Wolf, E., et al. (1999). CCR7 coordinates the primary immune response by establishing functional microenvironments in secondary lymphoid organs. *Cell*, *99*(1), 23–33. https://doi.org/10.1016/s0092-8674(00)80059-8.

Franklin, R. A., Liao, W., Sarkar, A., Kim, M. V., Bivona, M. R., Liu, K., et al. (2014). The cellular and molecular origin of tumor-associated macrophages. *Science*, *344*(6186), 921–925. https://doi.org/10.1126/science.1252510.

Geissmann, F., Jung, S., & Littman, D. R. (2003). Blood monocytes consist of two principal subsets with distinct migratory properties. *Immunity*, *19*(1), 71–82.

Ginhoux, F., & Guilliams, M. (2016). Tissue-resident macrophage ontogeny and homeostasis. *Immunity*, *44*(3), 439–449. https://doi.org/10.1016/j.immuni.2016.02.024.

Gordon, S. (2016). Phagocytosis: An immunobiologic process. *Immunity*, *44*(3), 463–475. https://doi.org/10.1016/j.immuni.2016.02.026.

Gouon-Evans, V., Rothenberg, M. E., & Pollard, J. W. (2000). Postnatal mammary gland development requires macrophages and eosinophils. *Development*, *127*(11), 2269–2282.

Grainger, J. R., Askenase, M. H., Guimont-Desrochers, F., da Fonseca, D. M., & Belkaid, Y. (2014). Contextual functions of antigen-presenting cells in the gastrointestinal tract. *Immunological Reviews*, *259*(1), 75–87. https://doi.org/10.1111/imr.12167.

Griseri, T., McKenzie, B. S., Schiering, C., & Powrie, F. (2012). Dysregulated hematopoietic stem and progenitor cell activity promotes interleukin-23-driven chronic intestinal inflammation. *Immunity*, *37*(6), 1116–1129. https://doi.org/10.1016/j.immuni.2012.08.025.

Guermonprez, P., Valladeau, J., Zitvogel, L., Thery, C., & Amigorena, S. (2002). Antigen presentation and T cell stimulation by dendritic cells. *Annual Review of Immunology*, *20*, 621–667. https://doi.org/10.1146/annurev.immunol.20.100301.064828.

Guilliams, M., Dutertre, C. A., Scott, C. L., McGovern, N., Sichien, D., Chakarov, S., et al. (2016). Unsupervised high-dimensional analysis aligns dendritic cells across tissues and species. *Immunity*, *45*(3), 669–684. https://doi.org/10.1016/j.immuni.2016.08.015.

Gunther, J., & Seyfert, H. M. (2018). The first line of defence: Insights into mechanisms and relevance of phagocytosis in epithelial cells. *Seminars in Immunopathology*, *40*(6), 555–565. https://doi.org/10.1007/s00281-018-0701-1.

Habtezion, A., Nguyen, L. P., Hadeiba, H., & Butcher, E. C. (2016). Leukocyte trafficking to the small intestine and Colon. *Gastroenterology*, *150*(2), 340–354. https://doi.org/10.1053/j.gastro.2015.10.046.

Hadis, U., Wahl, B., Schulz, O., Hardtke-Wolenski, M., Schippers, A., Wagner, N., et al. (2011). Intestinal tolerance requires gut homing and expansion of FoxP3 + regulatory T cells in the lamina propria. *Immunity*, *34*(2), 237–246. https://doi.org/10.1016/j.immuni.2011.01.016.

Hanayama, R., Tanaka, M., Miyasaka, K., Aozasa, K., Koike, M., Uchiyama, Y., et al. (2004). Autoimmune disease and impaired uptake of apoptotic cells in MFG-E8-deficient mice. *Science*, *304*(5674), 1147–1150. https://doi.org/10.1126/science.1094359.

Hanna, R. N., Carlin, L. M., Hubbeling, H. G., Nackiewicz, D., Green, A. M., Punt, J. A., et al. (2011). The transcription factor NR4A1 (Nur77) controls bone marrow differentiation and the survival of $Ly6C^-$ monocytes. *Nature Immunology*, *12*(8), 778–785. https://doi.org/10.1038/ni.2063.

Heidkamp, G. F., Sander, J., Lehmann, C. H. K., Heger, L., Eissing, N., Baranska, A., et al. (2016). Human lymphoid organ dendritic cell identity is predominantly dictated by ontogeny, not tissue microenvironment. *Science Immunology*, *1*(6). https://doi.org/10.1126/sciimmunol.aai7677.

Hettinger, J., Richards, D. M., Hansson, J., Barra, M. M., Joschko, A. C., Krijgsveld, J., et al. (2013). Origin of monocytes and macrophages in a committed progenitor. *Nature Immunology*, *14*(8), 821–830. https://doi.org/10.1038/ni.2638.

Hildner, K., Edelson, B. T., Purtha, W. E., Diamond, M., Matsushita, H., Kohyama, M., et al. (2008). Batf3 deficiency reveals a critical role for CD8alpha + dendritic cells in cytotoxic T cell immunity. *Science*, *322*(5904), 1097–1100. https://doi.org/10.1126/science.1164206.

Horsley, V., Jansen, K. M., Mills, S. T., & Pavlath, G. K. (2003). IL-4 acts as a myoblast recruitment factor during mammalian muscle growth. *Cell*, *113*(4), 483–494. https://doi.org/10.1016/s0092-8674(03)00319-2.

Jang, M. H., Sougawa, N., Tanaka, T., Hirata, T., Hiroi, T., Tohya, K., et al. (2006). CCR7 is critically important for migration of dendritic cells in intestinal lamina propria to mesenteric lymph nodes. *Journal of Immunology*, *176*(2), 803–810. https://doi.org/10.4049/jimmunol.176.2.803.

Jinushi, M., Nakazaki, Y., Dougan, M., Carrasco, D. R., Mihm, M., & Dranoff, G. (2007). MFG-E8-mediated uptake of apoptotic cells by APCs links the pro- and antiinflammatory activities of GM-CSF. *The Journal of Clinical Investigation*, *117*(7), 1902–1913. https://doi.org/10.1172/JCI30966.

Jung, K., Heishi, T., Incio, J., Huang, Y., Beech, E. Y., Pinter, M., et al. (2017). Targeting CXCR4-dependent immunosuppressive Ly6C(low) monocytes improves anti-angiogenic therapy in colorectal cancer. *Proceedings of the National Academy of Sciences of the United States of America*, *114*(39), 10455–10460. https://doi.org/10.1073/pnas.1710754114.

Kim, M., Galan, C., Hill, A. A., Wu, W. J., Fehlner-Peach, H., Song, H. W., et al. (2018). Critical role for the microbiota in CX3CR1(+) intestinal mononuclear phagocyte regulation of intestinal T cell responses. *Immunity*, *49*(1). https://doi.org/10.1016/j.immuni.2018.05.009151-163.e155.

Kim, K. W., Vallon-Eberhard, A., Zigmond, E., Farache, J., Shezen, E., Shakhar, G., et al. (2011). In vivo structure/function and expression analysis of the CX3C chemokine fractalkine. *Blood*, *118*(22), e156–e167. https://doi.org/10.1182/blood-2011-04-348946.

Kirchberger, S., Royston, D. J., Boulard, O., Thornton, E., Franchini, F., Szabady, R. L., et al. (2013). Innate lymphoid cells sustain colon cancer through production of interleukin-22 in a mouse model. *The Journal of Experimental Medicine*, *210*(5), 917–931. https://doi.org/10.1084/jem.20122308.

Klebanoff, C. A., Spencer, S. P., Torabi-Parizi, P., Grainger, J. R., Roychoudhuri, R., Ji, Y., et al. (2013). Retinoic acid controls the homeostasis of pre-cDC-derived splenic and intestinal dendritic cells. *The Journal of Experimental Medicine*, *210*(10), 1961–1976. https://doi.org/10.1084/jem.20122508.

Knoop, K. A., Miller, M. J., & Newberry, R. D. (2013). Transepithelial antigen delivery in the small intestine: Different paths, different outcomes. *Current Opinion in Gastroenterology*, *29*(2), 112–118. https://doi.org/10.1097/MOG.0b013e32835cf1cd.

Kurts, C., Cannarile, M., Klebba, I., & Brocker, T. (2001). Dendritic cells are sufficient to cross-present self-antigens to CD8 T cells in vivo. *Journal of Immunology*, *166*(3), 1439–1442. https://doi.org/10.4049/jimmunol.166.3.1439.

Laouar, Y., Welte, T., Fu, X. Y., & Flavell, R. A. (2003). STAT3 is required for Flt3L-dependent dendritic cell differentiation. *Immunity*, *19*(6), 903–912.

Lavin, Y., Kobayashi, S., Leader, A., Amir, E. D., Elefant, N., Bigenwald, C., et al. (2017). Innate immune landscape in early lung adenocarcinoma by paired single-cell analyses. *Cell*, *169*(4). https://doi.org/10.1016/j.cell.2017.04.014, 750-765.e717.

Lavin, Y., Mortha, A., Rahman, A., & Merad, M. (2015). Regulation of macrophage development and function in peripheral tissues. *Nature Reviews Immunology*, *15*(12), 731–744. https://doi.org/10.1038/nri3920.

Lavin, Y., Winter, D., Blecher-Gonen, R., David, E., Keren-Shaul, H., Merad, M., et al. (2014). Tissue-resident macrophage enhancer landscapes are shaped by the local microenvironment. *Cell*, *159*(6), 1312–1326. https://doi.org/10.1016/j.cell.2014.11.018.

Liu, Z., Gu, Y., Chakarov, S., Bleriot, C., Kwok, I., Chen, X., et al. (2019). Fate mapping via Ms4a3-expression history traces monocyte-derived cells. *Cell*, *178*(6). https://doi.org/10.1016/j.cell.2019.08.009, 1509-1525.e1519.

Liu, K., Victora, G. D., Schwickert, T. A., Guermonprez, P., Meredith, M. M., Yao, K., et al. (2009). In vivo analysis of dendritic cell development and homeostasis. *Science*, *324*(5925), 392–397. https://doi.org/10.1126/science.1170540.

Loschko, J., Schreiber, H. A., Rieke, G. J., Esterhazy, D., Meredith, M. M., Pedicord, V. A., et al. (2016). Absence of MHC class II on cDCs results in microbial-dependent intestinal inflammation. *The Journal of Experimental Medicine*, *213*(4), 517–534. https://doi.org/10.1084/jem.20160062.

Mazzini, E., Massimiliano, L., Penna, G., & Rescigno, M. (2014). Oral tolerance can be established via gap junction transfer of fed antigens from CX3CR1(+) macrophages to CD103(+) dendritic cells. *Immunity*, *40*(2), 248–261. https://doi.org/10.1016/j.immuni.2013.12.012.

McDole, J. R., Wheeler, L. W., McDonald, K. G., Wang, B., Konjufca, V., Knoop, K. A., et al. (2012). Goblet cells deliver luminal antigen to CD103+ dendritic cells in the small intestine. *Nature*, *483*(7389), 345–349. https://doi.org/10.1038/nature10863.

McKenna, H. J., Stocking, K. L., Miller, R. E., Brasel, K., De Smedt, T., Maraskovsky, E., et al. (2000). Mice lacking flt3 ligand have deficient hematopoiesis affecting hematopoietic progenitor cells, dendritic cells, and natural killer cells. *Blood*, *95*(11), 3489–3497.

Merad, M., Sathe, P., Helft, J., Miller, J., & Mortha, A. (2013). The dendritic cell lineage: Ontogeny and function of dendritic cells and their subsets in the steady state and the inflamed setting. *Annual Review of Immunology*, *31*, 563–604. https://doi.org/10.1146/annurev-immunol-020711-074950.

Michaelson, M. D., Bieri, P. L., Mehler, M. F., Xu, H., Arezzo, J. C., Pollard, J. W., et al. (1996). CSF-1 deficiency in mice results in abnormal brain development. *Development*, *122*(9), 2661–2672.

Mossadegh-Keller, N., Sarrazin, S., Kandalla, P. K., Espinosa, L., Stanley, E. R., Nutt, S. L., et al. (2013). M-CSF instructs myeloid lineage fate in single haematopoietic stem cells. *Nature*, *497*(7448), 239–243. https://doi.org/10.1038/nature12026.

Naik, S. H., Sathe, P., Park, H. Y., Metcalf, D., Proietto, A. I., Dakic, A., et al. (2007). Development of plasmacytoid and conventional dendritic cell subtypes from single precursor cells derived in vitro and in vivo. *Nature Immunology*, *8*(11), 1217–1226. https://doi.org/10.1038/ni1522.

Niess, J. H., Brand, S., Gu, X., Landsman, L., Jung, S., McCormick, B. A., et al. (2005). CX3CR1-mediated dendritic cell access to the intestinal lumen and bacterial clearance. *Science*, *307*(5707), 254–258. https://doi.org/10.1126/science.1102901.

Ohno, H. (2016). Intestinal M cells. *Journal of Biochemistry*, *159*(2), 151–160. https://doi.org/10.1093/jb/mvv121.

Onai, N., Obata-Onai, A., Schmid, M. A., Ohteki, T., Jarrossay, D., & Manz, M. G. (2007). Identification of clonogenic common Flt3 + M-CSFR + plasmacytoid and conventional dendritic cell progenitors in mouse bone marrow. *Nature Immunology*, *8*(11), 1207–1216. https://doi.org/10.1038/ni1518.

Perdiguero, E. G., & Geissmann, F. (2016). The development and maintenance of resident macrophages. *Nature Immunology*, *17*(1), 2–8. https://doi.org/10.1038/ni.3341.

Popivanova, B. K., Kostadinova, F. I., Furuichi, K., Shamekh, M. M., Kondo, T., Wada, T., et al. (2009). Blockade of a chemokine, CCL2, reduces chronic colitis-associated carcinogenesis in mice. *Cancer Research*, *69*(19), 7884–7892. https://doi.org/10.1158/0008-5472.CAN-09-1451.

Salmon, H., Idoyaga, J., Rahman, A., Leboeuf, M., Remark, R., Jordan, S., et al. (2016). Expansion and activation of CD103(+) dendritic cell progenitors at the tumor site enhances tumor responses to therapeutic PD-L1 and BRAF inhibition. *Immunity*, *44*(4), 924–938. https://doi.org/10.1016/j.immuni.2016.03.012.

Schlitzer, A., Sivakamasundari, V., Chen, J., Sumatoh, H. R., Schreuder, J., Lum, J., et al. (2015). Identification of cDC1- and cDC2-committed DC progenitors reveals early lineage priming at the common DC progenitor stage in the bone marrow. *Nature Immunology*, *16*(7), 718–728. https://doi.org/10.1038/ni.3200.

Shaw, T. N., Houston, S. A., Wemyss, K., Bridgeman, H. M., Barbera, T. A., Zangerle-Murray, T., et al. (2018). Tissue-resident macrophages in the intestine are long lived and defined by Tim-4 and CD4 expression. *The Journal of Experimental Medicine*, *215*(6), 1507–1518. https://doi.org/10.1084/jem.20180019.

Sichien, D., Lambrecht, B. N., Guilliams, M., & Scott, C. L. (2017). Development of conventional dendritic cells: From common bone marrow progenitors to multiple subsets in peripheral tissues. *Mucosal Immunology*, *10*(4), 831–844. https://doi.org/10.1038/mi.2017.8.

Tamoutounour, S., Henri, S., Lelouard, H., de Bovis, B., de Haar, C., van der Woude, C. J., et al. (2012). CD64 distinguishes macrophages from dendritic cells in the gut and reveals the Th1-inducing role of mesenteric lymph node macrophages during colitis. *European Journal of Immunology*, *42*(12), 3150–3166. https://doi.org/10.1002/eji.2012 42847.

Williams, J. W., Tjota, M. Y., Clay, B. S., Vander Lugt, B., Bandukwala, H. S., Hrusch, C. L., et al. (2013). Transcription factor IRF4 drives dendritic cells to promote Th2 differentiation. *Nature Communications*, *4*, 2990. https://doi.org/10.1038/ncomms3990.

Yutin, N., Wolf, M. Y., Wolf, Y. I., & Koonin, E. V. (2009). The origins of phagocytosis and eukaryogenesis. *Biology Direct*, *4*, 9. https://doi.org/10.1186/1745-6150-4-9.

Zhu, Y. P., Thomas, G. D., & Hedrick, C. C. (2016). 2014 Jeffrey M. Hoeg award lecture: Transcriptional control of monocyte development. *Arteriosclerosis, Thrombosis, and Vascular Biology*, *36*(9), 1722–1733. https://doi.org/10.1161/ATVBAHA.116.304054.

Zigmond, E., Varol, C., Farache, J., Elmaliah, E., Satpathy, A. T., Friedlander, G., et al. (2012). Ly6C hi monocytes in the inflamed colon give rise to proinflammatory effector cells and migratory antigen-presenting cells. *Immunity*, *37*(6), 1076–1090. https://doi.org/10.1016/j.immuni.2012.08.026.

Zijlstra, M., Bix, M., Simister, N. E., Loring, J. M., Raulet, D. H., & Jaenisch, R. (1990). Beta 2-microglobulin deficient mice lack CD4-8+ cytolytic T cells. *Nature*, *344*(6268), 742–746. https://doi.org/10.1038/344742a0.

CHAPTER SEVEN

Generation of mouse bone marrow-derived macrophages using tumor coculture assays to mimic the tumor microenvironment

Esra Güç[a,†], Demi Brownlie[a,†], Carolina Rodriguez-Tirado[b], Takanori Kitamura[a], Jeffrey W. Pollard[a,b,*]

[a]MRC Centre for Reproductive Health, Queen's Medical Research Institute, The University of Edinburgh, Edinburgh, United Kingdom

[b]Department of Developmental and Molecular Biology, Albert Einstein College of Medicine, New York, NY, United States

*Corresponding author: e-mail address: jeff.pollard@ed.ac.uk

Contents

† Equal contribution.

Methods in Enzymology, Volume 632
ISSN 0076-6879
https://doi.org/10.1016/bs.mie.2019.11.014

Abstract

Macrophages are one of the key immune cells within the tumor microenvironment that encourage the growth of tumors at the primary site as well as contributing to all parts of the metastatic cascade. Although it is possible to isolate macrophages directly from the tumor, this can be a laborious process and due to their plasticity, it is not possible to maintain their in vivo phenotype in vitro. For this reason, differentiating macrophages from bone marrow is an attractive alternative. Here we present robust methods to study in vitro derived macrophages including (i) the isolation and generation of macrophages from bone marrow, (ii) differentiation/characterization of classically activated, alternatively activated and tumor-conditioned macrophages, as well as (iii) in vitro co-culturing assays for tumor cell-macrophage interaction/transmigration.

1. Introduction

Macrophages in vivo can be derived from bone marrow (BM), fetal liver and from yolk-sac-progenitors. They are usually defined as mononuclear phagocytic cells that reside in all tissues and are involved in a wide range of processes such as in immune responses and in the regulation of tissue growth and homeostasis (Geissmann et al., 2010; Pollard, 2009). In addition, they are involved in almost all pathologies either in their resolution or in their acerbation (Wynn, Chawla, & Pollard, 2013). In the context of this series, macrophages have been shown in most cases to exert pro-tumoral effects within the tumor microenvironment (Cassetta & Pollard, 2018; De Palma & Lewis, 2013; Mantovani, Marchesi, Malesci, Laghi, & Allavena, 2017). Macrophages in the tumor microenvironment show a different phenotype to inflammatory and homeostatic cells (Qian & Pollard, 2010) and exert functions that enhance cancer malignancy at every step of the metastatic cascade. These functions include the induction of angiogenesis, enhancement of tumor cell dissemination to metastatic sites where they also enhance extravasation and metastatic cell survival and growth (Kitamura, Qian, & Pollard, 2015). They also produce an immunosuppressive microenvironment which contributes to immunotherapy/chemotherapy resistance (Cassetta & Pollard, 2018; DeNardo & Coussens, 2007). The link between poor prognosis and high macrophage infiltration has also been established in many but not all human cancers (Jochems & Schlom, 2011; Yang, McKay, Pollard, & Lewis, 2018). Besides pro-tumoral macrophages, the release of inflammatory factors in the tumor microenvironment such as IFN-γ, can activate tumoricidal macrophages that enhance anti-tumor immune responses, however, the persistence of this phenotype within an immunosuppressive tumor microenvironment is unclear (Mantovani & Sica, 2010). Recent studies have shown that macrophages can be therapeutically reprogrammed using agonistic or antagonistic drugs and thereby be diverted to fight against cancer. (Cassetta & Pollard, 2018).

In mice, bone marrow-derived $CD11b^{+}CD11c^{-}Ly6G^{-}Ly6C^{high/low}$ circulatory monocytes are recruited to the tissue and differentiate into $Ly6C^{-}F4/80^{+}CD11b^{+}$ macrophages. These cells populate inflammatory tissue and are present in large numbers in the tumor microenvironment (Epelman, Lavine, & Randolph, 2014) as well as many tissues during homeostasis such as the intestine, replacing yolk sac derived macrophages. Macrophages are highly versatile cells that display a wide spectrum of phenotypes. Despite this fact they are often over-simplistically classified as classically activated macrophages that play an immune effector/anti-tumoral role, and alternatively activated macrophages that play an immunoregulatory/pro-tumoral role (Gordon & Taylor, 2005; Mosser & Edwards, 2008) This nomenclature has also been superimposed upon the classification of "M1 or M2" based observations in different mouse strains with different susceptibilities to cancer (Mantovani & Sica, 2010) to reflect Th1 and Th2 responses in vivo. However, this classification is more representative of in vitro derived macrophages using cytokine-mediated differentiation rather than their phenotype in vivo. To clarify these responses, Murray and colleagues suggest a nomenclature that links to the cytokines used for the activation such as M(IL-4), M(IL-10) (Murray et al., 2014) at least in vitro. But the situation in vivo is likely to represent a much more mixed phenotype given the plethora of signals such cells respond to at any given time (Cassetta and Pollard, 2018).

Although it is possible to sort macrophages from digested tissues, samples can be scarce, and this can be a time consuming and expensive process with high variability between subjects/animals. Consequently, the development of means to isolate macrophages from bone marrow originally perfected by Stanley (Stanley, 1985, 1997; Tushinski et al., 1982) has provided a boon to studies of macrophage biology. This is because large numbers of cells at greater than 95% purity can be obtained from a small number of mice (approximately 10–15 million bone marrow cells can be isolated from a mouse femur) and the isolation protocol and differentiation is straightforward and cost-effective. These cells respond by proliferation to the growth factors Colony stimulating factor (CSF)1, CSF2 and IL-3 and they can be polarized in vitro with a wide range of factors. At least 95–98% of these cells are dependent on CSF1 for viability (Stanley, 1985; Tushinski et al., 1982). Furthermore, the ability to isolate them from any mutant mouse that contains a null or conditional allele of any gene of interest allows studies on growth factor signaling and macrophage function. In addition, there are studies where these bone marrow-derived macrophages (BMDMs) or their conditioned medium (CM) can be transferred to recipient mice and functions studied in vivo (Saha et al., 2016; Thomas et al., 2011).

Although BMDM isolation is a widely used protocol for in vitro macrophage studies there is a diversity of methods for their maturation and differentiation. Therefore, in this chapter we aim to demonstrate a standard BMDM isolation, methods for differentiation with cytokines/growth factors and for studying the roles of macrophages in tumor biology.

2. Isolation and differentiation of mouse BMDM

2.1 Introduction

BMDM isolation shows methodological differences among various studies but as yet it is unclear whether those differences affect functional outcomes. Some important steps in this protocol should be noted to reduce the variation among different isolation methods. For example, to increase the purity of hematopoietic stem cell isolation, we used tissue culture plates on the first day of isolation to allow mesenchymal stem cells, fibroblasts, mature macrophages and other adherent cells to attach but allow hematopoietic cells to remain within the medium. We then transfer the cell suspension into a low adhesive petri plate to continue their differentiation. In addition, different BMDM seeding densities are used among various methods and it has been shown that higher seeding densities might affect the macrophage phenotype (Lee & Hu, 2013). In this protocol, seeding 3×10^6 cell per $10\,cm^2$ dish gives a pure $Ly6C^-Ly6G^-$ $CD11b^+F4/80^+$ macrophage population as shown in Fig. 1. Both human and mouse recombinant CSF1 (historically often called macrophage-CSF, M-CSF) can be used for differentiation. Human recombinant CSF1 has been shown to be fully effective for mouse BMDM differentiation (Stanley, 1985, 1997). We analyzed BMDM phenotype on day 8, although BMDMs can be used for up to 3 weeks after differentiation as previously reported (Davies & Gordon, 2005). They can also be frozen down for use later.

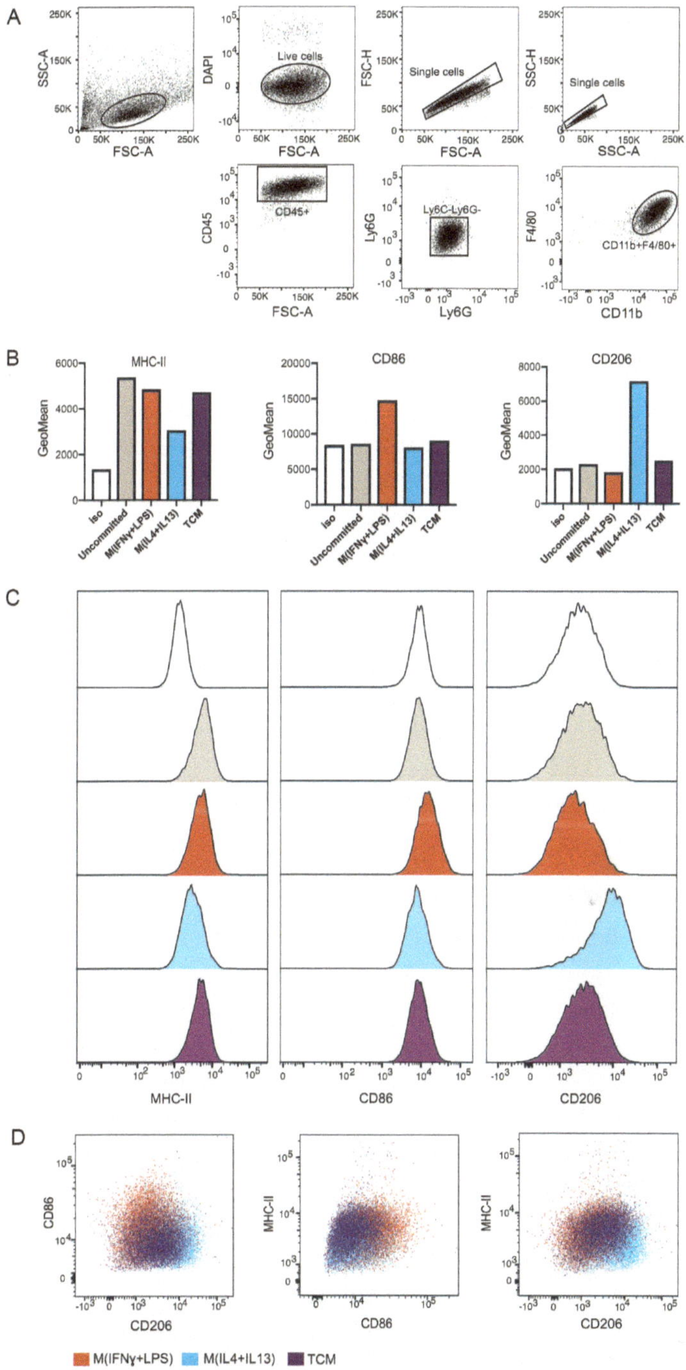

Fig. 1 (A) Flow cytometry gating strategy to identify BMDM. (B) Geometric mean Bar charts. (C) Histograms showing the expression of BMDM markers exposed to M(IFNγ+LPS), M(IL4+IL13), TCM polarizing conditions. The color legends are same in (B) and (C). (D) Flow cytometry dot plots of M(IFNγ+LPS), M(IL4+IL13), TCM showing the expression of different BMDM markers.

2.2 Collection of mouse femur and tibia

1. Euthanize the mouse according to the approved animal license procedure.
2. Place the mouse in supine position and disinfect the legs and the hips by spraying with 70% (v/v) Ethanol.
3. Using scissors cut the skin along the inguinal area and pull back the entire leg skin.
4. Remove the quadriceps and the tibias anterior muscle with forceps or a blade, until the femur and tibia are visible.
5. Using forceps twist the femur and dislocate the femoral head from the hip with scissors. Make sure the femur is not damaged.

   Note: Bones can be stored or shipped in 45 mL of fetal bovine serum free DMEM medium supplemented with 1% (v/v) Penicillin Streptomycin at 4 °C for 1–3 days. The effect of storage on BMDM isolation is characterized by Papazian, Kfoury, and Scadden (2017).
6. After the femur is completely dislocated from the cadaver, using forceps twist the knee joint between femur and tibia. Cut the knee joint and connective tissue. Make sure the tibia is not damaged.
7. Repeat the same procedure to dislocate distal tibia from the talus.
8. Remove the excess muscle and connective tissue using sterile 70% ethanol-soaked tissue wipes and a blade.
9. Place the femurs and tibias in a petri dish and place them on ice.

2.3 Bone marrow isolation

1. Under the sterile laminar flow hood, hold the femur or tibia using forceps and gently cut the tip of the proximal and distal epiphysis of the bone. Make sure metaphysis at both ends are exposed.
2. Hold the bone in a vertical position and gently insert a $27^{1/2}$G needle from the proximal metaphysis. Using a 5 mL syringe, flush 5 mL of cold phosphate buffered saline (PBS), (without calcium chloride or magnesium chloride, pH 7.4) into a 50 mL empty falcon tube.
3. Rotate the bone upside down and repeat the same procedure from the distal metaphysis side.
4. Using a 1 mL pipette, mix the bone marrow cell suspension sufficiently to homogenize into a single cell suspension.
5. Pass the suspension through a 70 μm cell strainer to remove any bone fragments and blood clots.
6. By using the cap of a 1 mL plunger, homogenize any bone marrow fragments that remain on 70 μm cell strainer. Wash the filter with 5 mL more PBS.

7. Centrifuge at $400 \times g$ for 5 min at 4 °C.
8. Discard the supernatant and resuspend the pellet in 30 mL of PBS.
9. Take 20 μL of homogenized cell suspension and count the cell number using a manual or automated cell counter.
10. Centrifuge at $400 \times g$ for 5 min at 4 °C.
11. Discard the supernatant.

2.4 BMDM maturation and differentiation

1. Resuspend the cells with alpha-MEM medium (Gibco, Cat #: 12571063) supplemented with 10% v/v fetal bovine serum (FBS), (Gibco, Cat #: 11573397) and 1% v/v penicillin streptomycin (PS) (Gibco, Cat #: 15140122) at a concentration of 3×10^5 cell/mL.
2. Split 10 mL of cell suspension to each tissue culture treated 10 cm^2 plate.
3. Add 4.4 nM (10^4 U/mL) recombinant CSF1 to each plate.
4. Incubate cells in a humidified incubator with 5% CO_2 at 37 °C.
5. 24 h later, transfer non-adherent cells into a non-tissue culture treated petri dish. *Note*: This procedure is required to remove adherent mesenchymal stem cells.
6. Replace the medium and 4.4 nM (10^4 U/mL) of recombinant CSF1 every 2–3 days.

   Note: CSF1 consumption rate/cell density is described in detail by Tushinski et al. (1982).
7. For the first 3 days cells will not be adherent, therefore add fresh CSF1 and top up to 10 mL final volume.

   Note: Bone marrow cells that are 3–4 days in culture can be stored in 10% v/v DMSO FBS solution at −196 °C for the long term.
8. For subsequent days aspirate the media and add fresh CSF1 and medium to adherent cells on the plate.
9. On day 7 collect the cells for analysis or for differentiation.
10. To collect cells, place the cell culture plate on ice, aspirate the medium and wash the plates with ice cold PBS. Add 5 mL alpha-MEM medium and using a cell scraper gently scrape the bottom of the plate and collect the macrophages.
11. Collect the cell suspension into a 50 mL falcon tube, centrifuge at $400 \times g$ for 5 min at 4 °C.
12. Process cells for analysis or seed 2×10^5 cells per 6-well plate with 3 mL alpha-MEM full medium with 4.4 nM CSF1 for differentiation.
13. To differentiate them into different macrophage subtypes, M(IFNγ + LPS), M(IL4 + IL13) and tumor conditioned macrophages (TCM) use the reagent mix shown in Tables 1 and 2.

Table 1 BEDIM differentiation reagents.

Macrophage subtype	Reagents
M(IFNγ + LPS) or "Activated"	IFN-γ + LPS
M(IL4 + IL13) or "Alternatively activated"	IL-4 + IL-13
Tumor conditioned macrophages (TCM)	20% v/v Tumor-conditioned medium

Table 2 Differentiation reagents.

Differentiation reagents	Final concentration	Catalog number	Supplier
IFN-γ	20 ng/mL	315-05	Peprotech
LPS	100 ng/mL	297-473-0	Sigma-Aldrich
IL-13	20 ng/mL	210-13	Peprotech
IL-4	20 ng/mL	214-14	Peprotech

Note: Tumor-conditioned medium (CM) can be collected from any murine or human tumor cell line, not forgetting species difference in response to factors. In this assay we used the murine mammary carcinoma E0771 cell line. Briefly, seed 1×10^6 tumor cells into 75 cm^2 flask using DMEM medium supplemented with 10% v/v FBS. 24 h later when cells are 70–80% confluent, collect the medium and centrifuge at $400 \times g$ for 5 min. Filter the medium using 0.44 μm low protein binding syringe filter. Transfer the supernatant into a new Falcon tube. Use the CM fresh or freeze it in aliquots at −80 °C. Use 20% v/v of this medium for BMDMs differentiation.

14. 24 h after differentiation aspirate the medium and wash the plates with PBS.
15. Aspirate PBS and add 1 mL of Accutase® (Sigma-Aldrich, Cat # A6964) to each well.

    Note: Because differentiated cells are more adherent, accutase will help cells to detach from the plate.
16. Incubate cells for 15–20 min at 37 °C.
17. Place the cell culture plate on ice and using a cell scraper gently scrape the bottom of the plate and collect macrophages.

3. Characterization of mouse BMDM phenotype

3.1 Introduction

While BMDMs are an incredibly useful model to study macrophage function in vitro, for some of the reasons mentioned above, it's important to

Table 3 Suggested cell surface markers for identification of BMDM subtype via flow cytometry.

BMDM subtype	Markers
Pan-macrophage (lineage)	$CD45^+$, $Ly6C^-$, $Ly6G^-$, $F4/80^+$, $CD115^+ CD11b^+$
M(IFNγ + LPS) or "Activated"	$CD80^+$, $CD86^+$, $MHC\text{-}II^{high}$, $TLR\text{-}4^+$
M(IL4 + IL13) or "Alternatively activated" and TCM	$CD206^+$, $Arg1^+$, $CD163^+$, $CD204^+$, $CD209^+ CD169^+$ $MHCII^{high/low}$

establish standard characterization methods to reduce the variability between different in vitro experiments and to validate their phenotypic similarities to in vivo macrophage subtypes.

There are a variety of ways to do this, one of which is flow cytometry. Flow cytometry is a convenient and effective method allowing a researcher to accurately and simultaneously analyze several cell surface and intracellular markers by immunolabeling cells, allowing the identification of sub-populations that may exist within the BMDM preparation (e.g., mesenchymal stem cells, dendritic cells etc.), while gating out any other cell population of interest that could be cocultured with BMDM.

It is important to select markers that successfully distinguish different macrophage subtypes clearly. There are a number of cell surface markers that can be used for phenotypic characterization using flow cytometry. With no exception they will all show $CSF1R^+$ (CD115) and F4/80 positivity (Table 3). To identify different subtypes, here we have selected CD86, CD206 and MHC-II markers. CD86 is a co-stimulatory molecule that is the most upregulated in classically activated macrophages (Gensel, Kopper, Zhang, Orr, & Bailey, 2017) or M(IFNγ + LPS) and successfully distinguishes them from both alternatively activated macrophages or M(IL4 + IL13) and those treated with TCM (Fig. 1C). CD206 is a mannose receptor that is upregulated on M(IL4 + IL13) (Stein, Keshav, Harris, & Gordon, 1992) and TCM (Fig. 1C). As well as these markers, MHC-II can be used in combination with the above to distinguish TCM from the other macrophage subtypes. For example, M(IFNγ + LPS) are $CD86^{hi}$, $CD206^{lo}$, $MHC\text{-}II^{hi}$ while M(IL4 + IL13) are $CD86^{lo}$, $CD206^{hi}$ and $MHC\text{-}II^{lo}$ and TCMs are $CD86^{lo}$, $CD206^{hi}$ and $MHC\text{-}II^{hi}$.

3.2 Staining and controls

1. Collect cells from petri plate as described in Section 2.4. Centrifuge them at $400 \times g$ for 5 min at 4 °C.

2. Discard supernatant and homogenize pellet with 5 mL of flow cytometry buffer (5% w/v bovine serum albumin in 1 × PBS, pH 7.4).
3. Centrifuge at 400 × *g* for 5 min at 4 °C.
4. Discard supernatant and resuspend with flow cytometry buffer on ice.
5. Pass suspension through a 40 μm cell strainer to ensure a single cell suspension.
6. Count cells and aliquot 0.5–1 × 10^6 cells/100 μL to flow cytometry tubes on ice-label appropriately (samples and controls listed in Table 4).
7. Add 1 μL of Fc block™ (CD16/CD32) (BD Pharmingen, Cat #: 564219) per 1 × 10^6 cells on ice for 10 min.
8. Add antibody cocktail (Table 4–5) and incubate on ice for 30–60 min.
9. Add 1–2 mL flow cytometry buffer to each tube and centrifuge at 400 × *g* for 5 min at 4 °C. Discard the supernatant and repeat the wash step one more time.
10. Resuspend in 300 μL flow cytometry buffer and add 0.3 μL of 100 μg/mL DAPI before running the samples on the flow cytometer. Make sure samples are thoroughly mixed.

3.3 Data acquisition and gating strategy (DIVA software, BD 6-laser-LSR)

1. Acquire unstained cells to set up the correct voltages for forward scatter (FSC) and side scatter (SSC). Make sure all cells are on scale and the main cell population is distinct from cell debris.
2. Prior to fluorescence compensation setup, run an unstained and a fully stained sample to make sure that none of the colors are off-scale.

Table 4 Samples staining and controls for flow cytometry.

Tube name	Samples to be stained	Ab's to be used
Unstained	Cells	No Ab's – just buffer and Fc Block
Full stain	Cells	All Ab's + DAPI
Isotype controls	Cells	All Ab's, swap one Ab for its isotype control instead + DAPI
Fluorescence-minus one controls (FMO)	Cells	All Ab's apart from Ab of interest + DAPI
Compensation controls	Beads	Single Ab

Table 5 Antibody list for immunolabeling.

Antibody	Dilution	Cat. No.	Company
CD45-PerCPCy5.5	1/100	109827	BioLegend®
Ly6C-BV711	1/100	128037	BioLegend®
Ly6G-BV510	1/200	127633	BioLegend®
CD11b-BV605	1/200	101237	BioLegend®
F4/80-APC	1/50	MCA497–A647	Bio-Rad
CD86-APCCy7	1/100	105029	BioLegend®
MHC-II-FITC	1/00	107605	BioLegend®
CD206-PECy7	1/100	141719	BioLegend®
Isotype ctrl for CD86	1/100	400523	BioLegend®
Isotype ctrl for MHC-II	1/100	400633	BioLegend®
Isotype ctrl for CD206	1/100	40052	BioLegend®

The unstained histogram should be around 10^2 in logarithmic scale and fully stained should be under 10^5.

3. Run compensation beads (eBioscience) stained with a single antibody. This will result in a proportion of stained and unstained beads. Follow the software's usual method for compensation.
4. Start acquiring your samples. When acquiring data, set the "stopping gate" to the $CD11b^+$, $F4/80^+$ gate (see Fig. 1A) and collect at least 10,000 events within this gate. These are the macrophages. These gates can be set up by gating on positive cells compared to their respective FMO.
5. Save all of your data into appropriate software/hardware for further analysis on your preferred analysis software.

4. BMDM cancer cell coculture assays

4.1 Introduction

Macrophages induce tumor cell progression, survival and migration through reciprocal cell–cell interactions in vivo. Several of these activities can be mimicked in cell culture thus allowing exploration of mechanisms that in turn can be reexplored within in vivo models. We have performed a number of these assays over the years describing, for example, a paracrine loop between TAMs and tumor cells involving tumor synthesized CSF1 and macrophage produced EGF ligands (Condeelis & Pollard, 2006). Most recently we have studied the mechanism whereby macrophages promote metastasis via their stimulation of extravasation and tumor cell survival at metastatic sites (Kitamura & Pollard, 2015). To this end we developed relatively simple assays involving two- or three-cell types that have high fidelity when translated back in vivo. These assays allow screening using mutant BMDM and inhibitors to establish function. One such assay we use is the "In vitro transendothelial migration" (eTEM) assay which is designed to mimic the extravasation of tumor cells and thus combines tumor and endothelial cells with macrophages. A similar assay was also developed for intravasation (Wyckoff et al., 2004). For the eTEM assay, macrophages are seeded on the underside of a transwell, while the upper-side is coated with matrix and endothelial cells. Cancer cell extravasation from the apical side of endothelial cells to the macrophage rich underside is quantified. This assay has previously been used to demonstrate that inflammatory monocytes as well as VEGF and CCL3 expression in metastasis-associated macrophages promotes cancer cell transmigration (Kitamura et al., 2015; Qian et al., 2011). Time-lapse imaging in vitro is a second method used in the same study to demonstrate tumor cell-macrophage cell–cell interactions and thus showed that CCR1 expression is required by macrophages to interact with cancer cells (Kitamura, Qian, Soong, et al., 2015). Below is a detailed description of these two assays. It should be acknowledged that there are many other types of assays published that look at tumor cell–macrophage interactions often using sophisticated microfluidic devices see for example (Zervantonakis et al., 2012) or spheroid type assays. Readers are referred to this large literature to select the assay most suitable to their needs.

4.2 In vitro transendothelial migration assay

1. Two days before the experiment: For migration assays use a growth factor reduced matrigel coated 8 μm pore size transwell insert (Corning, Cat # 354483). Thaw the inserts at room temperature for about 10 min.
2. Preincubate the transwells with DMEM (Corning, Cat # 10-013CVR) medium and 10% v/v FBS by adding 300 μL on upper chamber and 500 μL on bottom chamber. Incubate at 37 °C for 2 h.
3. Preparation of 3B-11 cells (SV40 transformed mouse endothelial cell line):
   a. Culture 3B-11 cells in DMEM medium supplemented with 10% v/v FBS in a T75 cm^2 flask.

   b. Aspirate the medium and wash cells with 1 × PBS. Aspirate the buffer.
   c. Add 3 mL Trypsin-EDTA (0.25%) (Gibco) per T75 cm^2 flask.
   d. Incubate for 5 min at 37 °C. Stop the enzymatic process by adding 3 mL of DMEM complete medium. Collect detached cells into a 15 mL falcon tube.
   e. Centrifuge cells at 400 × *g* for 5 min. Discard the supernatant.
   f. Resuspend cells in DMEM complete medium. Adjust the cell concentration to 10^5 cells per/mL.

4. Aspirate the medium on both sides of the transwells. Make sure the matrigel and membrane are not damaged.

5. Add 200 μL of the 3B-11 cell suspension on top of the matrigel and add 750 μL DMEM medium on the bottom chamber. Incubate cells for 48 h at 37 °C in 5% CO_2 until cells form a monolayer over matrigel.
   Note: Incubation times can vary according to standardization methods described at Section 4.3.

6. On the experiment day: Prepare and label E0771 cells (mouse breast adenocarcinoma cell line, cells are cultured in the same condition as 3B-11 cells):
   a. Aspirate the cell culture medium from T75 cm^2 flask and wash cells with prewarmed PBS.
   b. Add 10 mL of prewarmed serum free DMEM and 1 μM Cell Tracker Green CMFDA Dye (Thermo Fisher). Incubate cells for 15 min at 37 °C in 5% CO_2.
   c. First check the staining under the microscope.
   d. Aspirate the medium and repeat step 3 (a–e) of Section 4.2 to collect cells.
   e. Centrifuge cells at 400 × *g* for 5 min. Discard the supernatant and wash cells with 1 × PBS.
   f. Repeat the centrifuge step. Discard the supernatant. Resuspend the cells in DMEM medium with 0.5% v/v FBS and 4.4 nM CSF1. Adjust the cell concentration to 2×10^5 cells. Keep the cells on ice until migration assay is ready to start.

7. Preparation of macrophages
   a. Use BMDMs 7 days after maturation and collect macrophages from tissue culture plates as indicated in Section 2.4.
   b. Centrifuge macrophages at 400 × *g* for 5 min, discard the supernatant.
   c. Resuspend cells in DMEM complete medium with 4.4 nM CSF1 and adjust the concentration to 5×10^5 cells/mL.

d. Aspirate the medium from transwells and make sure endothelial cells, matrigel and transwells are not damaged. Using a sterile tweezer flip the membrane upside down on 24-well plate-lid.

e. Seed 20 μL of macrophage suspension on the underside of the transwell and let the macrophages attach for 15 min under the hood at room temperature.

Note: Make sure there are no bubbles and the solution is dispersed evenly on the membrane.

f. Place the inserts back into the empty 24 wells and let the cells attach for another 15 min in the hood.

Note 1: Since endothelial cells will be attached on the matrigel, they won't get dry during the 30 min incubation.

Note 2: Step 6 and step 7 can be initiated at the same time and preparation can be coordinated between the waiting time of each step. However as soon as macrophages are attached to the transwell the transmigration assay needs to be started. Tumor cells should be kept on ice until they are ready to be placed on transwells for the transmigration.

8. Move transwells to a 12-well plate with 750 μL of 4.4 nM CSF1 and DMEM complete medium.

9. Add 200 μL of tumor cell suspension on top of the transwell.

10. 36 h later, aspirate the medium from top and bottom chambers and fix transwells with 4% w/v paraformaldehyde (PFA) solution for 15 min at room temperature.

11. Remove PFA and wash transwells twice with 1 × PBS. To remove non-migrated cells, scrub the upper side of the transwells with a cotton swab.

12. Transfer the wells into a 35 mm Mat-Tex glass bottom dish. Using an inverted light microscope equipped with 20 × (NA + 0.40 air) objective and cooled CCD camera, take 20 × images from 6 fields. Quantify the number of migrated cells using Fiji/Image J software. The images from multiple fields can be analyzed in batches. Convert images into a grayscale, and by using the threshold tool, create binary images (positive pixels in black, background white). By using analyze particle feature, select the average size of a cells and count the number of the cells on the membrane. (If the number of migrated cells per 20 × is low, cells can also be counted manually).

13. To obtain more accurate results, the whole transmembrane can be imaged using a whole slide scanner microscope. To block the autofluorescence caused by the transmembrane, use a Matrigel coated membrane that blocks transmission of light between 400 and 700 nm (Fluoroblok™, Corning, 354165)

a. Repeat the same procedure until step 12.
b. Cut the membranes from the transwell by using a scalpel. Using fine tweezers transfer the membranes onto a glass slide.
c. Place each membrane on a glass slide with the underside of the transwell membrane facing up. Wash membranes with 1 × PBS.
d. Aspirate PBS buffer and counterstain by adding 100 μL of 0.5 μg/mL DAPI in PBS. Incubate at room temperature for 10 min.
e. Aspirate the DAPI solution, wash membranes twice with PBS.
f. Aspirate PBS and mount membranes using mounting medium (Prolong Diamond, Thermo Fisher) and seal with coverslip.
g. After 24 h when mounting medium is polymerized and coverslip is stabilized, image the entire membrane using a fluorescent whole slide scanner with 20 × magnification. Acquire whole slide fluorescence images within 358 nm excitation and 461 nm emission for DAPI and 492 nm excitation and 517 emission for cell tracker green. (AxioScan Z1 Slide scanner, 20 × magnification).

14. Process auto-stitched tile images using an image processor (i.e., Definiens AG, Tissue Studio). Detect cell tracker positive tumor cells by using a nuclear detection and cell simulation algorithm. Detect DAPI and Cell Tracker positive cells on the entire membrane and run the batch analysis for all samples.

    Note: This algorithm can be already built-in to the image processor software such as in Definiens AG, Tissue Studio, or new algorithms could be created using average nucleus/cell size and staining thresholding parameters.

4.3 Standardization methods to obtain endothelial cell monolayer for eTEM assay

The methods described below are used to determine an optimum incubation time for endothelial cells to form a confluent monolayer.

1. Repeat step 4.2 (1–4). Incubate endothelial cells for 24, 48, 72 or 96 h at 37 °C in 5% CO_2. A transwell insert with no cells must also be included.
2. Trans-endothelial resistance:
   a. Replace DMEM complete medium with DMEM 0.5% v/v FBS, 1 mM sodium pyruvate, 1% v/v PS.

   b. Two hours later measure the transendothelial resistance using EndOhm-6 chamber and $EVOM^2$ resistance meter (World Precision Instruments) or equivalent.
   c. Express resistance values (Ω) recorded in function of the effective surface area of the filter membrane ($\Omega \times cm^2$).

3. Endothelial permeability:
   a. Repeat step 4.2 (1–4).
   b. Replace the upper and bottom chamber medium with phenol red free-DMEM, 0.5% v/v FBS, 1 mM sodium pyruvate, 1% v/v PS and 100 μL/mL 70 kDa Rhodamine B Isothiocyanate-Dextran (TRITC), (Sigma-Aldrich) and incubate the cells for 36 h.
   c. Take 100 μL of samples from the top and bottom chamber. Dilute samples collected from the top chamber 1:5 in phenol-red free DMEM.
   d. Read the fluorescence at emission at 586 nm (excitation T 557 nm) using a microplate reader.
   e. Calculate the TRITC concentration based on a standard curve for TRITC.
   f. Determine the optimum incubation time for endothelial cells when no TRITC is detected on the bottom chamber and TER has the highest value compared to the other incubation time.

4.4 2D coculture time-lapse imaging

1. Preparation of Matrigel coated chamber slides.
   a. Add 300 μL of 5 mg/mL Matrigel (Corning) in alpha-MEM into each well of Lab Tek II 4-well Chamber Slide (Nunc).
   b. Incubate the slides for 1 h at 37 °C.

2. During the incubation time of Step 1b, start preparation and labeling of macrophages:
   a. Prepare bone marrow-derived macrophages as shown in procedure 2.4 (step 3–6).
   b. Replace the medium to prewarmed serum free alpha-MEM medium with 1.25 μm Cell Tracker Orange CMTMR Dye (Thermo Fisher).
   c. Incubate cells for 15 min at 37 °C in 5% v/v CO_2.

   Note: During this incubation time, tumor cells can be prepared (step 3).

   d. Isolate the labeled macrophages as described in procedure Section 2.4 (step 10).

   e. Centrifuge cells at 400 × *g* for 5 min. Discard the supernatant and wash cells with 1 × PBS.
   f. Repeat the centrifuge step. Discard the supernatant. Resuspend the cells in alpha-MEM complete medium with 4.4 nM CSF1.
3. Preparation of E0771 cells.
   a. After repeating procedure 4.2 step 6 (a–e), discard the supernatant and resuspend the cells in alpha-MEM complete medium.
4. Aspirate the medium and add 10^4 E0071 cell and 10^4 BMDMs in total 1 mL of alpha-MEM complete medium with 4.4 nm CSF1, and gently add to Matrigel coated chamber slides. Let them attach to the plate for 1 h at 37 °C in 5% CO_2.
5. Use an inverted microscope with 10 × magnification, motorized stage, CCD camera sensitive for time-lapse imaging (Axiovert 200, Zeiss) and equipped with a microscope cage incubator. Set the incubator for 5% v/v CO_2 and 37 °C. Acquire brightfield and fluorescent images every 10 min on 10 randomly selected fields.

   Note: In the current staining protocol the staining intensity and exposure time is kept minimal to reduce phototoxicity. Any modification of the current protocol requires optimization to keep the cell status healthy.
6. Process movies in Fiji/ImageJ software and track macrophage-tumor interaction using a manual-tracking tool in the software. Transfer the data to a spreadsheet and calculate the mean of interaction (in minutes) and the percentage of interacting cells.

   Note: The examples of the movies and quantification results are in Kitamura, Qian, Soong, et al. (2015).

5. Concluding remarks

In this study we have shown a reproducible macrophage differentiation and characterization method using murine bone marrow-derived macrophages and introduced two coculture methods to study macrophage–tumor interactions. Cytokine doses used for macrophage differentiation have been tested and characterized in many other studies (Cho et al., 2014; Zajac et al., 2013; Zhao et al., 2017), however, the dose of conditioned medium used for tumor-conditioned macrophages needs to be optimized for each experiment. The confluency of tumor cells, the origin of cell lines, the type of the medium, the batch of FBS and storage conditions might affect the quality of the conditioned medium and macrophage differentiation. To improve the reproducibility, these parameters need to be kept the same between each experiment.

We demonstrated three of the most commonly studied macrophage subtypes in vitro. Although macrophages are differentiated through LPS stimulation do not represent a tumor-associated macrophage phenotype but instead a bactericidal phenotype, reprogramming alternatively activated macrophages into classically activated macrophages has been shown to promote an anti-tumoral effect and macrophage reprogramming has been studied in vitro (Genard, Lucas, & Michiels, 2017). Using LPS or IFNγ stimulation decreases CSF1 receptor level of BMDMs in vitro therefore macrophage polarization should be induced after BMDMs are grown in CSF1 rich medium (Stanley, 1985). In addition, BMDMs derived from C3H/HeJ and C57BL/10Scr strains are not responsive to LPS stimulation and this allows for testing of reagents that may be contaminated by endotoxin (Guilbert & Stanley, 1984; Poltorak et al., 1998). It is also possible to use TNF-α rather than IFN-γ for activated macrophage differentiation (Martinez, Sica, Mantovani, & Locati, 2008). Thus, it is absolutely necessary to define treatments rather than give generic descriptions to macrophage types.

Cytokine mediated differentiation causes changes in expression of a wide variety of cell-surface and intracellular markers, and flow cytometry is a robust method to characterize and monitor those changes. In our study we selected a set of markers that shows the evident differences between M(IFNγ + LPS), M(IL4 + IL13) and TCM, however marker choice can be enriched by using other markers (Table 3).

We describe two tumor-macrophage coculture assays that have been validated in our earlier studies. Transmigration models of extravasation could be adapted for tumor intravasation (iTEM) by changing the order of cells seeded in the transwells such that the macrophages and tumor cells are cocultured together on top of the filter with CSF1 in the lower chamber (Pignatelli et al., 2014). 2D macrophage-tumor coculture assays are a robust method to study the direct interaction of those cells, yet to mimic the complexity of tumor microenvironment more advanced 3D models can be also used (Rebelo et al., 2018).

It is important to note that in vitro macrophage differentiation is in fact an oversimplification of the actual in vivo situation. Macrophages are highly susceptible to changes within the extracellular environment and they display a wide spectrum of subtypes (Mosser & Edwards, 2008) particularly in the tumor microenvironment (Condeelis & Pollard, 2006). Indeed, the exact mechanisms of in vivo differentiation and macrophage ontogeny in some tissues are still not well known and this makes it difficult to define a standard method for macrophage characterization in vivo. Nevertheless in vitro macrophage differentiation and in vitro coculture studies continues play an essential role to reveal fundamental mechanisms.

Acknowledgments

Dr. Jeffrey Pollard acknowledges Dr. Richard Stanley's (Albert Einstein College of Medicine, NY) generosity in teaching him the BMDM isolation methods. We thank all the members of the Pollard lab who over the years have contributed to these methods especially Dr. Bin-Zhi Qian. This manuscript preparation was supported by grants from the Wellcome Trust (101067/Z/13/Z), MRC Centre grant MR/N022556/1 and CRUK C17950/A26783 to J.W.P.

References

Cassetta, L., & Pollard, J. W. (2018). Targeting macrophages: Therapeutic approaches in cancer. *Nature Reviews. Drug Discovery*, *17*, 887–904.

Cho, D. I., Kim, M. R., Jeong, H. Y., Jeong, H. C., Jeong, M. H., Yoon, S. H., et al. (2014). Mesenchymal stem cells reciprocally regulate the M1/M2 balance in mouse bone marrow-derived macrophages. *Experimental & Molecular Medicine*, *46*, e70.

Condeelis, J., & Pollard, J. W. (2006). Macrophages: Obligate partners for tumor cell migration, invasion, and metastasis. *Cell*, *124*(2), 263–266.

Davies, J. Q., & Gordon, S. (2005). Isolation and culture of murine macrophages. *Methods in Molecular Biology*, *290*, 91–103.

De Palma, M., & Lewis, C. E. (2013). Macrophage regulation of tumor responses to anticancer therapies. *Cancer Cell*, *23*(3), 277–286.

DeNardo, D. G., & Coussens, L. M. (2007). Inflammation and breast cancer. Balancing immune response: Crosstalk between adaptive and innate immune cells during breast cancer progression. *Breast Cancer Research*, *9*(4) 212.

Epelman, S., Lavine, K. J., & Randolph, G. J. (2014). Origin and functions of tissue macrophages. *Immunity*, *41*(1), 21–35.

Geissmann, F., Manz, M. G., Jung, S., Sieweke, M. H., Merad, M., & Ley, K. (2010). Development of monocytes, macrophages, and dendritic cells. *Science*, *327*(5966), 656–661.

Genard, G., Lucas, S., & Michiels, C. (2017). Reprogramming of tumor-associated macrophages with anticancer therapies: Radiotherapy versus chemo- and immunotherapies. *Frontiers in Immunology*, *8*, 828.

Gensel, J. C., Kopper, T. J., Zhang, B., Orr, M. B., & Bailey, W. M. (2017). Predictive screening of M1 and M2 macrophages reveals the immunomodulatory effectiveness of post spinal cord injury azithromycin treatment. *Scientific Reports*, 7 40144.

Gordon, S., & Taylor, P. R. (2005). Monocyte and macrophage heterogeneity. *Nature Reviews. Immunology*, *5*(12), 953–964.

Guilbert, L. J., & Stanley, E. R. (1984). Modulation of receptors for the colony-stimulating factor, CSF-1, by bacterial lipopolysaccharide and CSF-1. *Journal of Immunological Methods*, *73*(1), 17–28.

Jochems, C., & Schlom, J. (2011). Tumor-infiltrating immune cells and prognosis: The potential link between conventional cancer therapy and immunity. *Experimental Biology and Medicine (Maywood, N.J.)*, *236*(5), 567–579.

Kitamura, T., & Pollard, J. W. (2015). Therapeutic potential of chemokine signal inhibition for metastatic breast cancer. *Pharmacological Research*, *100*, 266–270.

Kitamura, T., Qian, B. Z., & Pollard, J. W. (2015). Immune cell promotion of metastasis. *Nature Reviews. Immunology*, *15*(2), 73–86.

Kitamura, T., Qian, B. Z., Soong, D., Cassetta, L., Noy, R., Sugano, G., et al. (2015). CCL2-induced chemokine cascade promotes breast cancer metastasis by enhancing retention of metastasis-associated macrophages. *The Journal of Experimental Medicine*, *212*(7), 1043–1059.

Lee, C. M., & Hu, J. (2013). Cell density during differentiation can alter the phenotype of bone marrow-derived macrophages. *Cell & Bioscience*, *3*, 30.

Mantovani, A., Marchesi, F., Malesci, A., Laghi, L., & Allavena, P. (2017). Tumour-associated macrophages as treatment targets in oncology. *Nature Reviews. Clinical Oncology*, *14*(7), 399–416.

Mantovani, A., & Sica, A. (2010). Macrophages, innate immunity and cancer: Balance, tolerance and diversity. *Current Opinion in Immunology*, *22*(2), 231–237.

Martinez, F. O., Sica, A., Mantovani, A., & Locati, M. (2008). Macrophage activation and polarization. *Frontiers in Bioscience*, *13*, 453–461.

Mosser, D. M., & Edwards, J. P. (2008). Exploring the full spectrum of macrophage activation. *Nature Reviews. Immunology*, *8*(12), 958–969.

Murray, P. J., Allen, J. E., Biswas, S. K., Fisher, E. A., Gilroy, D. W., Goerdt, S., et al. (2014). Macrophage activation and polarization: Nomenclature and experimental guidelines. *Immunity*, *41*(1), 14–20.

Papazian, A. E., Kfoury, Y. S., & Scadden, D. T. (2017). Shipping mouse bone marrow. Keep it in the bone. *Experimental Hematology*, *49*, 68–72.

Pignatelli, J., Goswami, S., Jones, J. G., Rohan, T. E., Pieri, E., Chen, X., et al. (2014). Invasive breast carcinoma cells from patients exhibit MenaINV- and macrophage-dependent transendothelial migration. *Science Signaling*, 7(353), ra112.

Pollard, J. W. (2009). Trophic macrophages in development and disease. *Nature Reviews Immunology*, *9*(4), 259–270.

Poltorak, A., He, X., Smirnova, I., Liu, M. Y., Van Huffel, C., Du, X., et al. (1998). Defective LPS signaling in C3H/HeJ and C57BL/10ScCr mice: Mutations in Tlr4 gene. *Science*, *282*(5396), 2085–2088.

Qian, B. Z., Li, J., Zhang, H., Kitamura, T., Zhang, J., Campion, L. R., et al. (2011). CCL2 recruits inflammatory monocytes to facilitate breast-tumour metastasis. *Nature*, *475*(7355), 222–225.

Qian, B. Z., & Pollard, J. W. (2010). Macrophage diversity enhances tumor progression and metastasis. *Cell*, *141*(1), 39–51.

Rebelo, S. P., Pinto, C., Martins, T. R., Harrer, N., Estrada, M. F., Loza-Alvarez, P., et al. (2018). 3D-3-culture: A tool to unveil macrophage plasticity in the tumour microenvironment. *Biomaterials*, *163*, 185–197.

Saha, S., Aranda, E., Hayakawa, Y., Bhanja, P., Atay, S., Brodin, N. P., et al. (2016). Macrophage-derived extracellular vesicle-packaged WNTs rescue intestinal stem cells and enhance survival after radiation injury. *Nature Communications*, 7 13096.

Stanley, E. R. (1985). The macrophage colony-stimulating factor, CSF-1. *Methods in Enzymology*, *116*, 564–587.

Stanley, E. R. (1997). Murine bone marrow-derived macrophages. *Methods in Molecular Biology*, *75*, 301–304.

Stein, M., Keshav, S., Harris, N., & Gordon, S. (1992). Interleukin 4 potently enhances murine macrophage mannose receptor activity: A marker of alternative immunologic macrophage activation. *Journal of Experimental Medicine*, *176*(1), 287–292.

Thomas, J. A., Pope, C., Wojtacha, D., Robson, A. J., Gordon-Walker, T. T., Hartland, S., et al. (2011). Macrophage therapy for murine liver fibrosis recruits host effector cells improving fibrosis, regeneration, and function. *Hepatology*, *53*(6), 2003–2015.

Tushinski, R. J., Oliver, I. T., Guilbert, L. J., Tynan, P. W., Warner, J. R., & Stanley, E. R. (1982). Survival of mononuclear phagocytes depends on a lineage-specific growth factor that the differentiated cells selectively destroy. *Cell*, *28*(1), 71–81.

Wyckoff, J., Wang, W., Lin, E. Y., Wang, Y., Pixley, F., Stanley, E. R., et al. (2004). A paracrine loop between tumor cells and macrophages is required for tumor cell migration in mammary tumors. *Cancer Research*, *64*(19), 7022–7029.

Wynn, T. A., Chawla, A., & Pollard, J. W. (2013). Macrophage biology in development, homeostasis and disease. *Nature*, *496*(7446), 445–455.

Yang, M., McKay, D., Pollard, J. W., & Lewis, C. E. (2018). Diverse functions of macrophages in different tumor microenvironments. *Cancer Research*, *78*(19), 5492–5503.

Zajac, E., Schweighofer, B., Kupriyanova, T. A., Juncker-Jensen, A., Minder, P., Quigley, J. P., et al. (2013). Angiogenic capacity of M1- and M2-polarized macrophages is determined by the levels of TIMP-1 complexed with their secreted proMMP-9. *Blood*, *122*(25), 4054–4067.

Zervantonakis, I. K., Hughes-Alford, S. K., Charest, J. L., Condeelis, J. S., Gertler, F. B., & Kamm, R. D. (2012). Three-dimensional microfluidic model for tumor cell intravasation and endothelial barrier function. *Proceedings of the National Academy of Sciences of the United States of America*, *109*(34), 13515–13520.

Zhao, Y. L., Tian, P. X., Han, F., Zheng, J., Xia, X. X., Xue, W. J., et al. (2017). Comparison of the characteristics of macrophages derived from murine spleen, peritoneal cavity, and bone marrow. *Journal of Zhejiang University. Science. B*, *18*(12), 1055–1063.

CHAPTER EIGHT

Methods for macrophage differentiation and *in vitro* generation of human tumor associated-like macrophages

Esther Heideveld[a,†], Marta Horcas-Lopez[a,†], Martha Lopez-Yrigoyen[b], Lesley M. Forrester[b], Luca Cassetta[a,*,†], Jeffrey W. Pollard[a,*,†]

[a]MRC Centre for Reproductive Health, Queen's Medical Research Institute, The University of Edinburgh, Edinburgh, United Kingdom
[b]MRC Centre for Regenerative Medicine, University of Edinburgh, Edinburgh, United Kingdom
*Corresponding authors: e-mail address: luca.cassetta@ed.ac.uk; jeff.pollard@ed.ac.uk

Contents

† Equal contribution.

Methods in Enzymology, Volume 632
ISSN 0076-6879
https://doi.org/10.1016/bs.mie.2019.10.005

Abstract

Tumor-associated macrophages (TAMs) are becoming a promising target for cancer immunotherapy. Significant efforts have been made to study the detrimental role of TAMs both *in vivo* and *in vitro*. However, it remains challenging to isolate these macrophages to study their function in human cancers and there is the need to seek alternatives to address these limitations. In this review, we will focus on the three most relevant approaches to obtain *in vitro* fully differentiated macrophages *i.e.* peripheral blood, immortalized cell lines such as THP-1 or human induced pluripotent stem cells. We will also provide protocols for the polarization of human macrophages to a TAM-like cells *in vitro*.

1. Introduction

Macrophages due to their dynamic nature and their phenotypic and functional diversity play fundamental roles in development and in homeostatic processes such as tissue surveillance and repair, as well as in immunity. Consequently, macrophages are involved in essentially every disease process. These include defense against pathogens and repair of damaged tissue; and at the same time, exacerbation of persistent inflammatory conditions such as arthritis, dementia and cancer (Pollard, 2009). Recent studies highlighted the important role played by tumor-associated macrophages (TAMs) in human cancers and their ability to interfere with T cell immunotherapy (Cassetta & Pollard, 2018). Human TAMs are becoming a promising target for cancer therapy, but we still do not have enough information to develop and screen successful TAM-targeting drugs. For this reason there is the need to either extract human macrophages from solid tumors or cultivate human macrophages and polarize them into a TAM-like phenotype to perform *in vitro* functional studies.

Several methods have described the isolation of TAMs from human biopsies (Cassetta et al., 2016, 2019). However, fresh samples and sufficient numbers of macrophages can be difficult to obtain from *ex vivo* isolated human samples. Therefore, macrophages have been obtained and cultured from other sources. These include monocytes (monocyte-derived macrophages; MDMs), monocytic cell lines such as THP-1, or human induced pluripotent stem cell lines (hiPSC) that can be used to reach the required macrophage numbers for *in vitro* studies. Macrophages from these sources can be polarized into various macrophage populations, including TAM-like cells, by treatment with cytokine cocktails, chemokines and/or conditioned medium from other cell lines. However, as there are different approaches and research purposes, every method for generating macrophages need to be tailored for the scientific question to be addressed.

Here we describe and compare several methods for generating human macrophages and TAM-like cells *in vitro*. In addition, we will discuss their characteristic phenotype, function and whether they recapitulate their counterparts *in vivo*. Furthermore, we will describe methods to polarize monocytic cells to TAM-like cells and finally summarize the advantages and limitations of cultured macrophages used for *in vitro* functional assays.

2. Monocyte-derived macrophages

Human peripheral blood mononuclear cell (PBMC) isolation is the most common source for MDMs. There are several "no-touch" procedures (*e.g.* elutriation, RosetteSep, plastic-adherence) to isolate monocytes without altering and/or activating the cells, yet, the purity of these populations is often reduced due to contamination by other cells. Density gradient centrifugation by counterflow elutriation can be used to obtain large numbers of monocytes, even within a closed system for clinical-scale generation of cells (Berger et al., 2005). The purity of these monocytes is relatively high (~90–95%; (Heideveld et al., 2015), although there might be some contamination with monocyte-derived dendritic cells and plasmacytoid dendritic cells that have a similar density to monocytes. As macrophages are adherent cells, they can also be isolated based on their adherence to plastic or glass. Although this method is cost-efficient, monocytes are pre-activated and their purity (85–95%) can be reduced compared to other methods due to contamination with dendritic cells and lymphocytes. Therefore, some protocols add a two-step gradient centrifugation with a Percoll step after Histopaque separation (Repnik, Knezevic, & Jeras, 2003). Alternatively, fluorescent activated cell sorting (FACS) can also be used for purification of monocytes and even specific subset isolation if needed. However, this procedure has a low cell recovery rate, high costs/effort and cells can get activated by the antibodies used for the isolation. Finally, highly pure monocytes can be isolated using magnetic beads based on magnetic activated cell sorting (MACS) either by positive or negative isolation. Although positive isolation will activate the monocytes and result in cells full of ingested beads after their isolation, this is still the most commonly used method. As an alternative, negative MACS isolation can be used resulting in non-activated monocytes, although cell numbers and purity are reduced compared to positive selection. In the next section we will focus on monocytes isolation using adherence and magnetic beads (positive and negative selection).

2.1 PBMC isolation

1. Collect peripheral blood in EDTA, heparin or coagulation tubes and dilute whole blood 1:2 with phosphate-buffered saline (PBS).
2. Add 15 mL of Ficoll-Paque 1.077 g/mL (Sigma) to a 50 mL Falcon tube and slowly overlay a maximum of 30 mL blood/PBS mixture. Spin down at room temperature (RT) for 30 min, 400 g, acceleration 1, deceleration 0.
3. Gently remove the upper plasma layer and collect the ring fraction containing PBMCs. Discard the remaining red blood cells and neutrophils.
4. Transfer the ring fraction to a new 50 mL falcon tube (maximum 20 mL) and fill up until 50 mL with PBS.
5. Centrifuge at RT for 8 min at 300 g and discard the supernatant containing platelets.
6. Resuspend the pellet in PBS and count the cells using a Casy Counter or counting chamber.
7. Red cell lysis buffer can be used to remove remaining red blood cells following the manufacturers' protocol (Biolegend).

2.2 Monocytic enrichment

2.2.1 Adherence protocol

1. Plate PBMC at 8×10^6 cells in 1 mL of serum free DMEM in a 12 well-plate.
2. Incubate in a humidified incubator at 37 °C and 5% v/v CO_2 for 1 h.
3. Discard non-adherent cells and wash twice with PBS.
4. Add 2 mL of DMEM supplemented with 10% v/v FBS and 25 ng/mL recombinant human Colony Stimulating Factor-1 (CSF1, PeproTech).
5. Proceed to step 2.3.

2.2.2 Pan monocyte purification

1. MACS buffer preparation: PBS supplemented with 0.5% w/v bovine serum albumin (BSA) and 2 mM EDTA, pH 7.2.
2. Resuspend PBMC in 80 μL pre-cooled MACS buffer per 10^7 total cells maximum.

3a. Positive isolation: add 20 µL CD14 MicroBeads (#130-050-201; Miltenyi Biotech) per 10^7 total cells.

3b. Negative selection: add 20 µL MicroBeads per 10^7 total cells.

4. Mix and incubate for 15–30 min at 4 °C, preferably on a rocking plate or tube roller.

5. Fill up with pre-cooled MACS buffer and spin down for 5 min, 4 °C, 300 g.

6. According to cell numbers, use either a MS Column (#130-042-201; up to 10^7 magnetically labeled cells from up to 2×10^8 total cells) or LS Column (#130–042-401; up to 10^8 magnetically labeled cells from up to 2×10^9 total cells).

7. Assemble a pre-separation filter (#130–041-407) onto the column and prime with 500 µL (MS) or 3 mL (LS) of MACS buffer.

8. Resuspend up to 10^8 cells in 500 µL pre-cooled MACS buffer. Scale up for higher cell numbers.

9. Load cells onto the pre-separation filters over the column and collect unlabelled cells that pass through.

10. Wash three times with 500 µL (MS) or 3 mL (LS) pre-cooled MACS buffer.

11a. For positive selection: discard the flow-through containing PBMC depleted for CD14+ cells, remove the column and place it on a collection tube. Purge with 1 mL (MS) or 5 mL (LS) pre-cooled MACS buffer. These cells are CD14+ cells.

11b. For negative selection: discard the column and continue with the flow-through containing CD14+ cells.

12. Add 2 mL of DMEM supplemented with 10% v/v FBS and 25 ng/mL recombinant human CSF1 (PeproTech).

2.3 Monocyte differentiation to macrophages

1. Replace half of the media every 3 days. MDMs will be differentiated after 6–7 days (Fig. 1).

2a. For classically-activated macrophages: stimulate MDMs with 50–100 ng/mL LPS (Peprotech) and 20 ng/mL IFNγ (R&D systems) overnight (O/N).

2b. For alternatively-activated macrophages: stimulate MDMs with 20 ng/mL IL4 (R&D systems) and 20 ng/mL IL13 (R&D systems) O/N.

3. Wash cells with cytokine-free culture medium for further analysis.

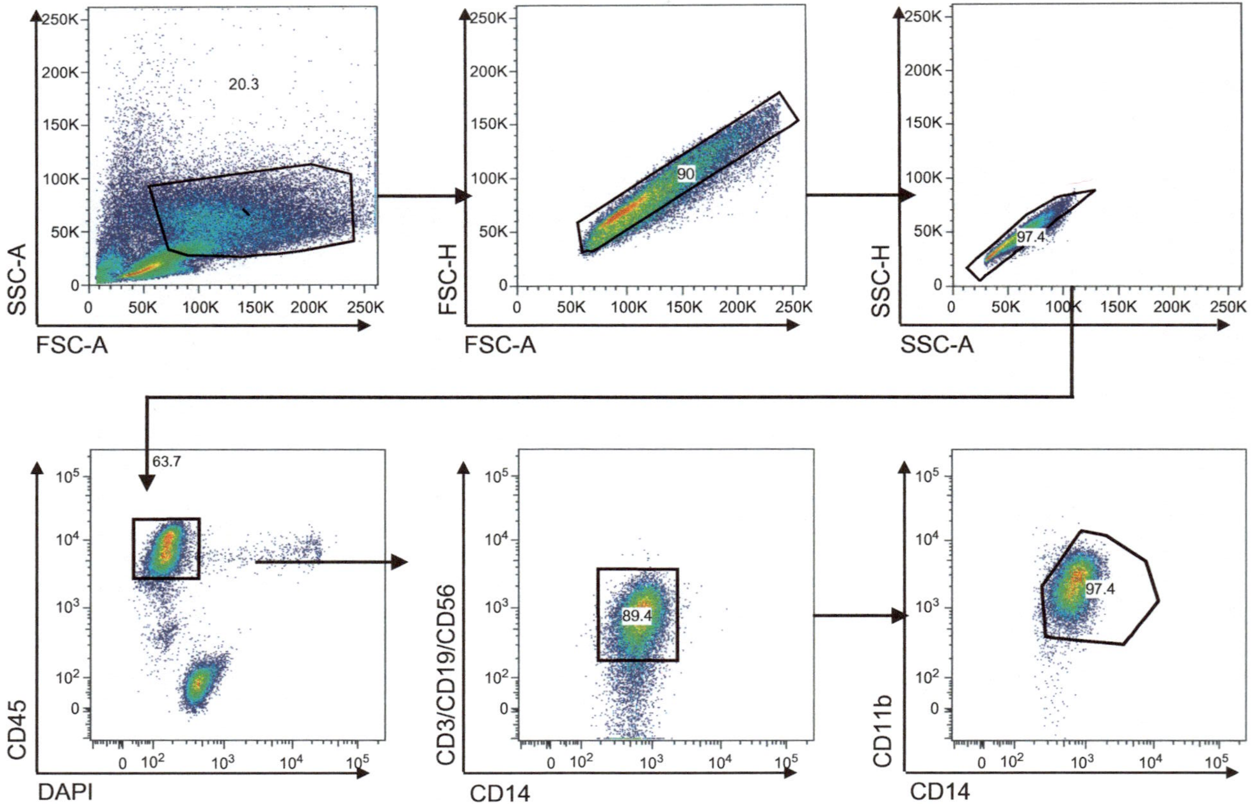

Fig. 1 Gating strategy for human MDMs. Cells are first gated by morphology to exclude any debris and non-desired cells, followed by gating out doublets using FSC-H vs. FSC-A and SSC-H vs. SSC-A. Live myeloid cells are defined as $CD45^+$ $DAPI^-$ cells and $CD14^+$ $CD3/CD19/CD56^-$ (Lineage negative; markers for T cells, B cells and NK cells). Finally, MDMs are isolated as $CD11b^+$ $CD14^+$.

3. THP-1-derived macrophages

Human macrophages can also be obtained by differentiating immortalized monocytic cell lines. Although multiple cell lines are available, the THP-1 cell line is the most widely used and therefore recognized as a reasonable model. This human monocytic leukemia cell line was established in 1980 and resembles primary monocytes (Tsuchiya et al., 1980). These cells have a homogeneous genetic background, which confers a minimal degree of variability. Moreover, THP-1 are non-adherent cells that can be differentiated to both macrophages as well as dendritic cells. Most protocols use phorbol 12-myristate 13-acetate (PMA) to differentiate THP-1 cells toward macrophages. Because of discrepancies in the literature regarding the PMA concentration required to differentiate THP-1 cells towards mature macrophages, we recommend using low concentrations of PMA which have been reported to have a less pronounced effect regarding non-specific gene expression (Park et al., 2007).

3.1 Cell line maintenance

1. Thaw a THP-1 vial in a water bath at 37 °C.
2. Add cells to 10 mL of complete growth medium containing RPMI 1640 supplemented with 10% v/v heat inactivated FBS.
3. Spin down at 125 g for 5 min.
4. Resuspend cell pellet to a concentration of 2–4 × 10^5 cells/mL.
5. Culture cells in T75 flasks at 37 °C, 5% v/v CO_2 (Fig. 2A).

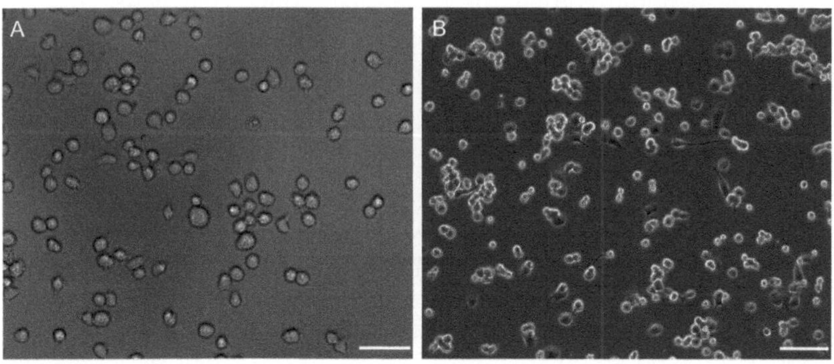

Fig. 2 THP-1 differentiation to macrophage-like cells. (A) Representative image of THP-1 cells in culture. (B) Morphology of THP-1 cells treated with 5 ng/mL PMA for 24 h. Scale Bar = 100 μm.

6. Refresh media every 2–3 days.
7. Split cells when reaching a density of 8×10^5 cells/mL.
8. Alternatively, cells can be frozen in complete growth medium containing 5% v/v DMSO.

3.2 PMA-induced macrophages

1. Treat 2×10^5 THP-1 cells with 5 ng/mL PMA (Sigma) for 48 h.
2. Check adherence after 24 h (Fig. 2B) and macrophage morphology after 48 h.

3a. For classically-activated macrophages: stimulate THP-1 cells with 50–100 ng/mL LPS (Peprotech) and 20 ng/mL IFNγ (R&D systems) O/N.

3b. For alternatively-activated macrophages: stimulate THP-1 cells with 20 ng/mL IL-4 (R&D systems) and 20 ng/mL IL-13 (R&D systems) O/N.

4. Wash cells with cytokine-free culture medium for further analysis

4. Human induced pluripotent stem cell-derived macrophages

Human pluripotent stem cells, including embryonic stem cells and hiPSC, have revolutionized medical research. They represent an inexhaustible cell source that is infection free and amenable to a vast arsenal of genetic manipulation technologies (*e.g.* Zinc Finger nuclease, TALENS and CRISPR-Cas9) (Ackermann et al., 2017; Inoue, Nagata, Kurokawa, & Yamanaka, 2014; Lopez-Yrigoyen et al., 2018, 2019). These technologies enable genome editing for functional studies and gene correction for potential therapies. Genetic alteration of hiPSC and their subsequent differentiation into a desired cell type such as macrophages, is a particularly attractive strategy when the primary cell of interest has proven difficult to genetically manipulate and/or to expand *in vitro* such as is the case for macrophages. hiPSC circumvent the ethical limitations of embryonic stem cells, and as hiPSC can be derived from any somatic cell, they provide a viable strategy for the development of personalized medicine approaches for example, the derivation of patient-specific disease models and personalized drug testing platforms; and the development of autologous cell therapies. Autologous cell therapies circumvent risks associated with immune rejection and donor transmitted infections (Lachmann et al., 2015; Narsinh, Plews, & Wu, 2011).

There are a number of published protocols describing the production of macrophages from hiPSC. Most of these procedures consist of a three-step process: (1) Haemogenic endothelial specification; (2) Myeloid progenitor expansion; and (3) Terminal macrophage maturation. In the first step, BMP4, SCF and VEGF are added to drive mesoendodermal commitment and support to emerging haemogenic endothelial cells (Douvaras et al., 2017; Lopez-Yrigoyen et al., 2018; Pandya et al., 2017). Haemogenic endothelial cells also require signaling from the WNT and Activin-Nodal pathways. In some protocols, this signaling can be achieved by the addition of exogenous WNT or activin agonists such as Activin A (Douvaras et al., 2017; Pandya et al., 2017), or by the activation of endogenous signaling *via* the formation of 3D aggregates of pluripotent stem cells, called embryoid bodies (EBs) (Haideri et al., 2017; Lopez-Yrigoyen et al., 2018). The nature of the second stage depends on the conditions of the first step. If exogenous WNT or Activin agonists are added, a cell monolayer is allowed to develop but if the first step involves EBs formation, these EBs are then plated on an adherent surface. Both monolayer cultures or EBs are then exposed to a cocktail of cytokines to promote the emergence of hematopoietic and myeloid progenitors that are released into suspension. Most protocols use IL3 to support hematopoietic progenitors and CSF1 to drive myeloid specification. In the third stage of the protocol, these myeloid progenitors are harvested and plated onto low attachment plates in the presence of CSF1, which supports macrophage development and maturation.

Here, we describe a protocol for the maintenance of hiPSC as a self-renewing population and for their differentiation into functional macrophages (Haideri et al., 2017; Lopez-Yrigoyen et al., 2018). We focus on serum- and feeder-free protocols that, although more costly, have proven to be more reproducible.

4.1 Cell line maintenance

1. Coat one well of a 6-well plate with CELLstart (Gibco) for 1 h.
2. Incubate frozen vial of cells at 37 °C until thawed, then transfer cells into 5 mL StemPro hESC SFM media, prepared by supplementing DMEM/F12 with Glutamax (Gibco), 1.8% w/v BSA (Gibco) and 0.1 mM β-mercaptoethanol (Gibco).
3. Centrifuge at 100 g for 3 min.
4. Resuspend cells in 1.5 mL of StemPro hESC supplemented with 20 ng/mL bFGF and 10 μM Rock Inhibitor (Merck) and transfer to the CELLstart coated well and culture for 24 h.

5. Change media to StemPro hESC supplemented with 20 ng/mL bFGF (no Rock inhibitor). Change media every day until they reach 80% confluency.
6. To passage the cells, replace the media in the confluent well and cut the cell monolayer using the StemPro EZPassage™ tool (Thermofisher Scientific).
7. Transfer cells at a 1:4 ratio onto pre-coated wells (see step 1) to a final media volume (StemPro hESC supplemented with 20 ng/mL bFGF) of 1.5 mL per well.
8. To freeze cells, treat a 70–80% confluent well of a 6-well plate with 10 μM Rock Inhibitor (Merck) for 1 h.
9. Cut cell colonies with the StemPro EZPassage™ tool (Thermofisher Scientific), and transfer to a centrifuge tube.
10. Centrifuge at 100 g for 3 min.
11. Resuspend in 1 mL of CryoStor Cell cryopreservation media (Sigma) and divide equally into two cryovials.
12. Place cryovials into a pre-chilled "Mr. Frosty" and store at −80 °C for 24–48 h.
13. Transfer vials either to −135 °C freezer or to a liquid nitrogen tank.

4.2 Differentiation to macrophages

Stage 1: Haemogenic endothelial specification

1. Replace maintenance media on one confluent well of a 6-well plate with 1.5 mL of Stage1 media (StemPro hESC SFM (Gibco) supplemented with 50 ng/mL BMP4 (R&D), 50 ng/mL VEGF (R&D) and 20 ng/mL SCF (Life Technologies) (Fig. 3A).
2. Cut colonies using the EZPassage™ tool and transfer into two wells of an Ultralow Attachment 6-well plate (Greiner) in 2.25 mL of Stage1 media.
3. Two days later (day 2), top-up cytokines to a final concentration of 50 ng/mL BMP4 (R&D), 50 ng/mL VEGF (R&D) and 20 ng/mL SCF (Life Technologies) using 0.5 mL of StemPro hESC SFM media.

Stage 2: Myeloid progenitor expansion

4. On day 4, coat a tissue culture 6-well plate with 0.1% w/v gelatin for at least 10 min.
5. Collect EBs and transfer them into a 50 mL Falcon tube.
6. Allow EBs to settle to the bottom by gravity and remove media carefully.
7. Transfer 10–15 EBs to a gelatin coated well containing 3 mL of Stage 2 media: X-VIVO15 (Lonza) supplemented with 100 ng/mL CSF1

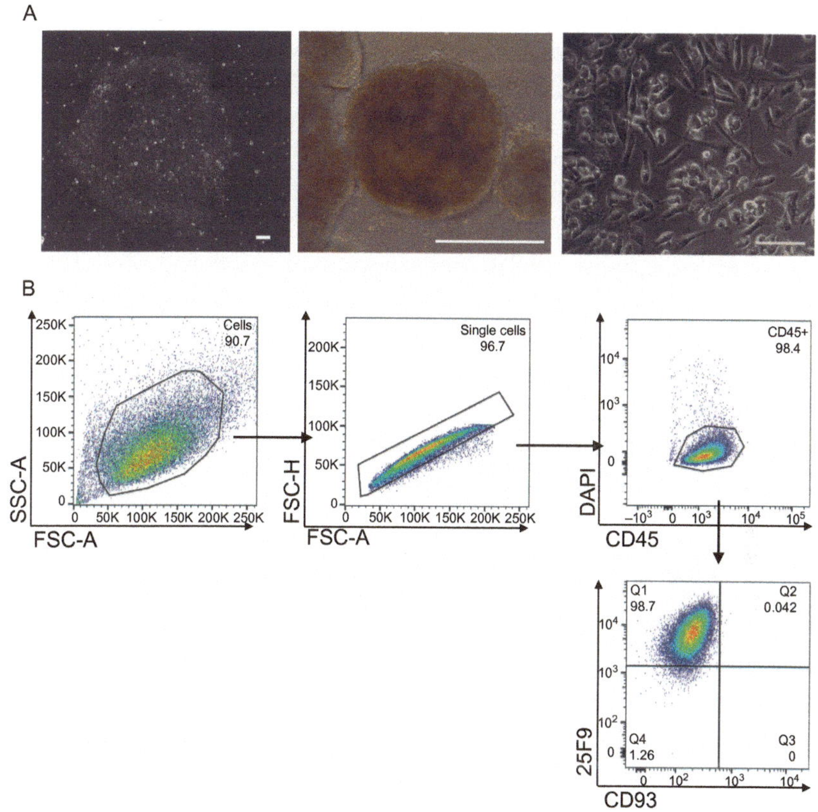

Fig. 3 hiPSCs-derived macrophages. (A) Cellular morphology of hiPSCs differentiation into macrophages. hiPSCs colony culture (left panel), EB formation (mid panel) and fully differentiated hiPSC-derived macrophages (right panel). (B) Gating strategy for hiPSC-derived macrophage maturation. Cells are first gated based on morphology to exclude any debris, followed by gating out doublets using FSC-H vs. FSC-A. Myeloid-like cells are $CD45^+$ $DAPI^-$. Mature hiPSC-derived macrophages are characterized for being $25F9^+$ $CD93^-$. Scale Bar = 100 μm.

(Biolegend), 25 ng/mL IL3 (Preprotech), 2 mM Glutamax (Gibco), 1% Penicillin-Streptomycin (Life Technologies) and 0.055 mM β-mercaptoethanol (Gibco) (Fig. 3A).

8. Change media every 3–4 days for 2.5–3 weeks.
9. After 2.5–3 weeks, EBs non-adherent monocyte-like precursors are released into suspension.

Stage 3: Terminal macrophage maturation

10. Harvest suspension monocyte-like precursors and replace media (Stage 2 media) on EB plate.

11. Centrifuge monocyte precursors at 120g for 3 min.
12. Resuspend in Stage 3 media: X-VIVO15 supplemented with 100 ng/mL CSF1, 2 mM Glutamax and 1% Penicillin-Streptomycin.
13. Plate cells onto untreated plastic bacteriological-grade plates or uncoated 6-well plates at a density of 0.2×10^6 cells/mL.
14. Keep cells in Stage 3 media for 7–11 days, changing media every 3–4 days (Fig. 3A).

15a. For classically-activated macrophages: stimulate cells with 50–100 ng/mL LPS (Peprotech) and 20 ng/ml IFNγ (R&D systems) for 24–48 h.

15b. For alternatively-activated macrophages: stimulate cells with 20 ng/mL IL4 (R&D systems) and 20 ng/mL IL13 (R&D systems) for 24–48 h.

Note: Stages 10–14 can be repeated every 3–4 days and monocyte precursors harvested from the original EB plate for up to 3 month.

5. Phenotyping cultured human macrophages

5.1 Protein level

Flow cytometry remains the most common way to phenotype human macrophages. Macrophages should be firstly characterized as having a positive expression for the lineage markers CD45, CD14, CD11b, CD163 and 25F9, a novel human macrophage marker recently discovered (Pilling, Fan, Huang, Kaul, & Gomer, 2009).

This panel can be followed by the polarization markers that describe either classical or alternative activated macrophages. Alternative activated macrophages are characterized by having high levels of CD169, CD204 and CD206, while classical activated macrophages are positive for the markers CD25, CD68, CD86, CD127 and CD215 (Table 1).

5.2 Gene expression

Gene expression analysis is also a useful tool to test the polarization status of macrophages. Upon LPS and IFNγ stimulation, macrophages will

Table 1 List of lineage and polarization macrophage markers.

Lineage	CD45, CD11b, CD14, CD163, 25F9
Alternative activated macrophages	CD169, CD204, CD206
Classical activated macrophages	CD25, CD68, CD86, CD127, CD215

upregulate mRNA expression of the genes *CD40*, *CD80*, *VCAM1* and *TNFA*. Upon IL4 stimulation, cells upregulate mRNA expression of the genes *CD84*, *CD68*, *TGM2* and *MRC1* (Lopez-Yrigoyen et al., 2018).

6. Culturing and phenotyping tumor-associated macrophages *in vitro*

Finding reliable protocols for human TAM cultivation and phenotyping can be challenging as the literature is populated with contradictory information. TAM origin and phenotype is much more complex than hypothesized in the past, with studies published every day adding more complexity to the picture (Cassetta & Pollard, 2018).

Before starting a project which involves human TAM culture *in vitro*, we suggest to take in consideration the following statements and adjust your scientific plans accordingly:

1. Human TAM do not show a unique and uniform phenotype but they are influenced by the presence of cancer-specific stimuli which drives their polarization (Cassetta et al., 2019; Zilionis et al., 2019);
2. The M1/M2 dichotomy is challenged by several studies suggesting that human TAM transcriptional profiles are more complex than what was initially proposed by *in vitro* polarization studies (Cassetta et al., 2019; Nahrendorf & Swirski, 2016)
3. Human TAM extraction from a tumor specimen (tissue digestion + flow sorting + cultivation in a plate) can be challenging due to tumor size, TAM yield and phenotype. Moreover, appropriate controls are needed to perform a fair comparison (such as resident macrophages extracted from healthy tissue) and this is not always possible (Cassetta et al., 2016).
4. Due to the extreme level of plasticity of macrophages, *ex vivo* extracted TAM cultivated *in vitro* can easily change their phenotype, especially when being switched from 3D to a 2D environment.

6.1 Polarization of human macrophages to TAM-like cells

Alternatively to *ex vivo* TAM extraction, human macrophages can be extracted using methods previously described in this chapter and polarized *in vitro* to a TAM-like phenotype using cancer cell conditioned medium (CM) and/or specific cytokine cocktails (Cassetta et al., 2019; Chen, Ying, Wang, Wu, & Zhu, 2017; Cho et al., 2016; Hsieh, Tai, & Yang, 2018; Solinas et al., 2010; Sousa et al., 2015; Zhang et al., 2016). However, it is important to use these strategies only to recapitulate a phenotype observed *ex vivo* on primary TAM as both cancer cell CM and cytokines will only partially recapitulate human *in vivo* TAM phenotypes.

An alternative to the methods just described is to establish 3D cultures where macrophages can infiltrate tumor organoids somewhat mimicking a real tumor microenvironment; a recent study by Rebelo et al. proposed a protocol for the generation of non-small cell lung carcinoma tumor spheroids where monocytes and fibroblasts can be co-cultivated in a 3D environment with cancer cells. Results showed successful myeloid tumor infiltration and macrophages exhibited a TAM-like phenotype; moreover, TAMs were susceptible to chemo- and immunotherapeutic agents such as CSF1R inhibitors (Rebelo et al., 2018). Patient-derived tumor organoids are widely used around the world to run drug screenings and evaluate the response to therapies (Aberle et al., 2018); therefore the protocol described by Rebelo et al. could be potentially useful for TAM *in vitro* screening of new drugs.

6.2 Example of breast TAM-like cells polarization using cytokines and cancer cell conditioned medium

In a recent publication we profiled by RNA sequencing the transcriptome of human breast cancer TAM (Cassetta et al., 2019); using different validation techniques and patients cohorts we identified several markers, which define a breast cancer associated phenotype of TAMs. In order to perform *in vitro* studies and understand the regulation of SIGLEC1 in TAMs, we incubated THP-1, MDM and hiPSC derived macrophages with cytokines and cancer cell CM derived from triple negative breast cancer cell lines (MDA-MB-231, MDA-MB-468).

To obtain MDA-MB-231 and MDA-MB -468 CM:

1. Tumor cells are initially resuspended in culture medium, seeded at a density of 1×10^5 cells/mL in 2.5 mL culture medium in 6-well plates and cultured O/N at 37 °C in a 5% v/v CO_2 atmosphere before changing medium to match the macrophage cultures (see 2, 3 and 4)
2. For CM exposure on PMA-THP-1 monocytes, cancer-cell culture medium was replaced with 10% v/v FBS RPM1640 medium.
3. For CM exposure on human MDMs, cancer cell culture medium was replaced with 10% v/v FBS DMEM supplemented with 5% v/v human serum.
4. For CM exposure on human hiPSC derived macrophages, cancer cell culture medium was replaced with 10% v/v FBS DMEM.

5. After medium change, cells were cultured for an additional 24h with fresh medium and thereafter, cell free supernatants were harvested and directly used for the experiment.
6. Conditioned medium was collected, filtered using a 0.22 μm filter to remove any cellular debris and used immediately.
7. CM is then exposed on macrophages at different percentages (20–50%) and cells are incubated for 24–48h before performing macrophage phenotyping or functional assays.

7. Troubleshooting and useful tips

As with any scientific protocol there are caveats and important considerations to take into account in order to avoid artifacts and the generation of unreliable results. Here below a list of potential pitfalls and useful tips.

7.1 MDM

a) It is essential to bring all reagents to room temperature before starting the protocol; temperature changes Ficoll density so if Ficoll is used at 4 °C the isopycnic centrifugation will not generate a correct PBMC ring in the vial and cellular yield will be negatively affected;
b) Blood stratification needs to be performed very slowly, decreasing to minimum the pipetman speed; if blood and Ficoll mixes during the stratification, the resulting ring quality will be affected;
c) Ring collection is a critical step in this protocol; make sure to collect only the ring and not the Ficoll below it, being conservative as much as possible; a Pasteur pipette can be used to collect the cell ring.

7.2 THP-1

a) Make sure to check cells for Mycoplasma infection;
b) If cells are not imported from certified suppliers, it is suggested to confirm cell line identity using appropriate services provided by multiple companies.

7.3 hiPSC-derived macrophages

a) hiPSC, if extensively cultivated, can partially differentiate and lose their stem cell properties with a consequent reduction in EBs and progenitors yield; cell morphology of the colonies is a very important indicator of hiPSC differentiation status, as shown in Fig. 4.

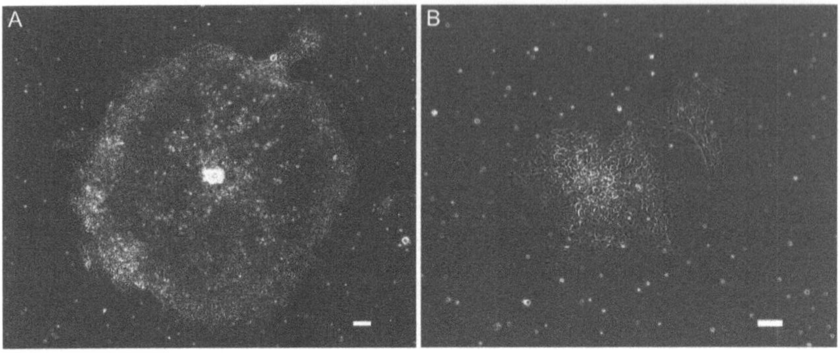

Fig. 4 hiPSC culture colonies troubleshooting. (A) Representative image of a undifferentiated hiPSC colony. (B) Example of partially differentiated hiPSC colony.

b) EB seeding (see Section 4.2, Stage 2) is a very critical part of the protocol; EBs need to be seeded carefully in order not to have all of them at the centre of the well, as they will attach to the bottom of the plate and form huge cellular aggregates; it is suggested to seed them and leave the plate for 10–15 min under the hood to promote EBs fall by gravity onto the bottom of the plate.

c) Macrophage progenitor production should be constantly monitored in order to evaluate the EB status and plan the next hiPSC accordingly; it is suggested to keep a detailed record of number of progenitors collected every 3 days in order to predict when the EBs will be exhausted and will stop producing viable progenitors.

8. Discussion and concluding remarks

This review has detailed the most common methods to obtain human macrophages and polarize them into TAM-like cells *in vitro*. There are several parameters that need to be considered before selecting the correct macrophage source for a project. Table 2 summarizes the essential characteristics of each source to aid with this process. For instance, MDMs are the most clinically relevant source of macrophages to validate targets, however, they might not be ideal for experiments that require large cell numbers with consistent macrophage populations and phenotypes nor for genetic manipulation. THP-1 cells allow the generation of large numbers of macrophages that can also be genetically modified. The use of this source may be beneficial for screening purposes; however, these cells are not dependent on CSF1 and are derived from a cancer, questioning their clinical relevance. Finally, hiPSC-derived macrophage can also be used for screening

Table 2 Summary of parameters for optimal macrophage source selection.

	MDMs	THP-1	hiPSC
Clinical Relevance	Y	N	Y
CSF1 dependency	Y	N	Y
Proliferation	–	+++	++
Time	++	+	+++
Reproducibility	+	+++	+++
Costs	++	+	+++
Transducibility	+	+++	++
Donor independent	N	Y	Y /N
Known genetic background	N	Y	Y
Suitable for genetic modifications	+	+++	+++
Origin	Bone Marrow	Tumor	Dermal Fibroblasts or peripheral blood derived iPSCs[a]

[a]These cells resemble Yolk Sac derived resident macrophages due to their MYB independence.

experiments, they depend on CSF1 and are also amenable to genetic manipulations when at the iPSC stage; however, this macrophage source is labour intensive and expensive with regards to iPSC maintenance and differentiation.

We therefore recommend the investigator to carefully select the proper macrophage source based on the needs of each project (Table 2).

Acknowledgments

This manuscript was supported by Wellcome Trust (101067/Z/13/Z), MRC Centre grant MR/N022556/1 and CRUK C17950/A26783 to JWP. MHL was supported by the Wellcome Trust grant 215151/Z/18/Z. M.L.Y. was supported by CONACYT.

Authors contribution

E.H., M.H.L., M.L.Y., L.M.F., L.C. and J.W.P. wrote the manuscript.

L.C. and J.W.P. conceptualized the manuscript and proofread the final version.

References

Aberle, M. R., Burkhart, R. A., Tiriac, H., Olde Damink, S. W. M., Dejong, C. H. C., Tuveson, D. A., et al. (2018). Patient-derived organoid models help define personalized management of gastrointestinal cancer. *The British Journal of Surgery*, *105*, e48–e60.

Ackermann, M., Kuhn, A., Kunkiel, J., Merkert, S., Martin, U., Moritz, T., et al. (2017). Ex vivo generation of genetically modified macrophages from human induced pluripotent stem cells. *Transfusion Medicine and Hemotherapy*, *44*, 135–142.

Berger, T. G., Strasser, E., Smith, R., Carste, C., Schuler-Thurner, B., Kaempgen, E., et al. (2005). Efficient elutriation of monocytes within a closed system (Elutra) for clinical-scale generation of dendritic cells. *Journal of Immunological Methods*, *298*, 61–72.
Cassetta, L., Fragkogianni, S., Sims, A. H., Swierczak, A., Forrester, L. M., Zhang, H., et al. (2019). Human tumor-associated macrophage and monocyte transcriptional landscapes reveal cancer-specific reprogramming, biomarkers, and therapeutic targets. *Cancer Cell*, *35*(4), 588–602.
Cassetta, L., Noy, R., Swierczak, A., Sugano, G., Smith, H., Wiechmann, L., et al. (2016). Isolation of mouse and human tumor-associated macrophages. *Advances in Experimental Medicine and Biology*, *899*, 211–229.
Cassetta, L., & Pollard, J. W. (2018). Targeting macrophages: Therapeutic approaches in cancer. *Nature Reviews. Drug Discovery*, *17*(12), 887–904.
Chen, X., Ying, X., Wang, X., Wu, X., & Zhu, Q. (2017). Exosomes derived from hypoxic epithelial ovarian cancer deliver microRNA-940 to induce macrophage M2 polarization. *Oncology Reports*, *38*, 522–528.
Cho, S. W., Kim, Y. A., Sun, H. J., Oh, B. C., Yi, K. H., Park, D. J., et al. (2016). CXCL16 signaling mediated macrophage effects on tumor invasion of papillary thyroid carcinoma. *Endocrine-Related Cancer*, *23*, 113–124.
Douvaras, P., Sun, B., Wang, M., Kruglikov, I., Lallos, G., Zimmer, M., et al. (2017). Directed differentiation of human pluripotent stem cells to microglia. *Stem Cell Reports*, *8*, 1516–1524.
Haideri, S. S., McKinnon, A. C., Taylor, A. H., Kirkwood, P., Starkey Lewis, P. J., O'Duibhir, E., et al. (2017). Injection of embryonic stem cell derived macrophages ameliorates fibrosis in a murine model of liver injury. *NPJ Regenerative Medicine*, *2*, 14.
Heideveld, E., Masiello, F., Marra, M., Esteghamat, F., Yağcı, N., von Lindern, M., et al. (2015). CD14+ cells from peripheral blood positively regulate hematopoietic stem and progenitor cell survival resulting in increased erythroid yield. *Haematologica*, *100*, 1396–1406.
Hsieh, C. H., Tai, S. K., & Yang, M. H. (2018). Snail-overexpressing Cancer cells promote M2-like polarization of tumor-associated macrophages by delivering MiR-21-abundant exosomes. *Neoplasia*, *20*, 775–788.
Inoue, H., Nagata, N., Kurokawa, H., & Yamanaka, S. (2014). iPS cells: A game changer for future medicine. *The EMBO Journal*, *33*, 409–417.
Lachmann, N., Ackermann, M., Frenzel, E., Liebhaber, S., Brennig, S., Happle, C., et al. (2015). Large-scale hematopoietic differentiation of human induced pluripotent stem cells provides granulocytes or macrophages for cell replacement therapies. *Stem Cell Reports*, *4*, 282–296.
Lopez-Yrigoyen, M., Fidanza, A., Cassetta, L., Axton, R. A., Taylor, A. H., Meseguer-Ripolles, J., et al. (2018). A human iPSC line capable of differentiating into functional macrophages expressing ZsGreen: A tool for the study and. *Philosophical Transactions of the Royal Society of London. Series B, Biological Sciences*, *373*(1750). https://doi.org/10.1098/rstb.2017.0219.
Lopez-Yrigoyen, M., Yang, C. T., Fidanza, A., Cassetta, L., Taylor, A. H., McCahill, A., et al. (2019). Genetic programming of macrophages generates an in vitro model for the human erythroid island niche. *Nature Communications*, *10*, 881.
Nahrendorf, M., & Swirski, F. K. (2016). Abandoning M1/M2 for a network model of macrophage function. *Circulation Research*, *119*, 414–417.
Narsinh, K. H., Plews, J., & Wu, J. C. (2011). Comparison of human induced pluripotent and embryonic stem cells: Fraternal or identical twins? *Molecular Therapy*, *19*, 635–638.
Pandya, H., Shen, M. J., Ichikawa, D. M., Sedlock, A. B., Choi, Y., Johnson, K. R., et al. (2017). Differentiation of human and murine induced pluripotent stem cells to microglia-like cells. *Nature Neuroscience*, *20*, 753–759.

Park, E. K., Jung, H. S., Yang, H. I., Yoo, M. C., Kim, C., & Kim, K. S. (2007). Optimized THP-1 differentiation is required for the detection of responses to weak stimuli. *Inflammation Research*, *56*, 45–50.

Pilling, D., Fan, T., Huang, D., Kaul, B., & Gomer, R. H. (2009). Identification of markers that distinguish monocyte-derived fibrocytes from monocytes, macrophages, and fibroblasts. *PLoS One*, *4*, e7475.

Pollard, J. W. (2009). Trophic macrophages in development and disease. *Nature Reviews. Immunology*, *9*, 259–270.

Rebelo, S. P., Pinto, C., Martins, T. R., Harrer, N., Estrada, M. F., Loza-Alvarez, P., et al. (2018). 3D-3-culture: A tool to unveil macrophage plasticity in the tumour microenvironment. *Biomaterials*, *163*, 185–197.

Repnik, U., Knezevic, M., & Jeras, M. (2003). Simple and cost-effective isolation of monocytes from buffy coats. *Journal of Immunological Methods*, *278*, 283–292.

Solinas, G., Schiarea, S., Liguori, M., Fabbri, M., Pesce, S., Zammataro, L., et al. (2010). Tumor-conditioned macrophages secrete migration-stimulating factor: A new marker for M2-polarization, influencing tumor cell motility. *Journal of Immunology*, *185*, 642–652.

Sousa, S., Brion, R., Lintunen, M., Kronqvist, P., Sandholm, J., Mönkkönen, J., et al. (2015). Human breast cancer cells educate macrophages toward the M2 activation status. *Breast Cancer Research*, *17*, 101.

Tsuchiya, S., Yamabe, M., Yamaguchi, Y., Kobayashi, Y., Konno, T., & Tada, K. (1980). Establishment and characterization of a human acute monocytic leukemia cell line (THP-1). *International Journal of Cancer*, *26*, 171–176.

Zhang, W., Chen, L., Ma, K., Zhao, Y., Liu, X., Wang, Y., et al. (2016). Polarization of macrophages in the tumor microenvironment is influenced by EGFR signaling within colon cancer cells. *Oncotarget*, *7*, 75366–75378.

Zilionis, R., Engblom, C., Pfirschke, C., Savova, V., Zemmour, D., Saatcioglu, H. D., et al. (2019). Single-cell transcriptomics of human and mouse lung cancers reveals conserved myeloid populations across individuals and species. *Immunity*, *50*(5), 1317–1334.

CHAPTER NINE

In vitro assessment of cancer cell-induced polarization of macrophages

Danique E.M. Duits, Max D. Wellenstein, Karin E. de Visser*

Division of Tumor Biology & Immunology, Oncode Institute, Netherlands Cancer Institute, Amsterdam, The Netherlands

*Corresponding author: e-mail address: k.d.visser@nki.nl

Contents

Abstract

The progression of cancer is strongly influenced by the crosstalk between cancer cells and immune cells. Immune cells can have both pro- and anti-tumor functions depending on the signals present in the environment. A significant proportion of the immune compartment of most solid tumors consists of tumor-associated macrophages. Although their abundance has been associated with poor prognosis in many solid tumor types, the molecular mechanisms by which cancer cells influence macrophage phenotype and function are largely unknown. In this chapter, we provide a detailed description of in vitro assays to study the impact of cancer cells on macrophages. We provide protocols to obtain macrophages from murine bone marrow and human peripheral blood, and to expose these macrophages to cancer cell-derived secreted molecules using conditioned medium from cancer cells. We describe several assays to assess cancer cell-induced polarization of macrophages. This experimental set-up can be utilized to gain molecular insights into how cancer cells influence macrophages.

Methods in Enzymology, Volume 632
ISSN 0076-6879
https://doi.org/10.1016/bs.mie.2019.06.011

1. Introduction

The development and progression of cancer strongly depends on interactions between cancer cells and their microenvironment (Quail & Joyce, 2013). Besides cancer cells, the tumor microenvironment (TME) consists of host stromal cells, extracellular matrix, vascular endothelial cells and immune cells. The immune landscape of tumors can differ tremendously between and within cancer types (Gentles et al., 2015; Thorsson et al., 2018). While intratumoral presence of some immune cell types can be favorable for disease prognosis, infiltration by others can promote tumor progression. For instance, increased influx of $CD8^+$ T cells is indicative of an anti-tumor immune response and as such is frequently associated with improved overall survival, whereas influx of regulatory T cells (Tregs) or myeloid cells provides an immunosuppressive environment which may promote tumorigenesis (Gentles et al., 2015; Thorsson et al., 2018).

The cancer-immune cell crosstalk is a dynamic process that is influenced by multiple factors and determines whether a pro- or anti-inflammatory environment is established at the tumor site. The state of immune cells can dictate cancer cell behavior, but the phenotype of cancer cells also influences the composition and function of tumor-associated immune cells. A major part of the immune compartment of solid tumors comprises tumor-associated macrophages (TAMs), which play a key role in regulating inflammation, angiogenesis, remodeling of the extracellular matrix and cancer cell migration (Mantovani, Marchesi, Malesci, Laghi, & Allavena, 2017). Due to their plasticity, TAMs can exert pro- or anti-tumorigenic functions depending on the cues they receive from their environment (Lewis & Pollard, 2006; Mantovani et al., 2017). TAMs are able to dampen anti-tumor immune responses by inducing an immunosuppressive environment via multiple mechanisms. Direct interaction of TAMs with T cells via immune checkpoints, such as PD1-PDL1, results in immunosuppression by blocking anti-tumor immunity (DeNardo & Ruffell, 2019). Indirectly, TAM-derived cytokines can induce Tregs or inhibit immune cell populations that are involved in anti-tumor immunity, such as dendritic cells (DCs) and $CD8^+$ T cells, thus leading to an immunosuppressive microenvironment (DeNardo & Ruffell, 2019). On the other hand, phagocytosis and antigen presentation can be key anti-tumorigenic features of macrophages, although these functions are infrequently observed in TAMs

(DeNardo & Ruffell, 2019; Mantovani et al., 2017). Altogether, these findings indicate a crucial role for cancer-immune cell crosstalk in cancer progression. Better understanding of TAM regulation and biology could provide new targets to block pro-tumorigenic processes or to re-educate TAMs toward an anti-tumorigenic phenotype.

The clinical relevance of macrophages in cancer progression has been demonstrated in studies examining the correlation between macrophage abundance and disease outcome. In the majority of solid cancer types, such as breast, gastric, urogenital and head and neck cancer, as well as hematological cancers, high macrophage levels correlate with poor prognosis (Mahmoud et al., 2012; Medrek, Ponten, Jirstrom, & Leandersson, 2012; Steidl et al., 2010; Tan et al., 2012; Zhang et al., 2012). This suggests that in most cancer types, these cells play an adverse role in disease progression. Interestingly, in colorectal cancer patients, high macrophage infiltrate is associated with increased overall survival (Zhang et al., 2012), underscoring the functional plasticity of these cells in different contexts and anatomical locations. In addition, recent evidence suggests that disease stage is another important factor in TAM phenotype (Singhal et al., 2019). Accumulation of macrophages has also been associated with poor response to certain anti-cancer therapeutics, such as chemotherapy (DeNardo et al., 2011), indicating that these cells can influence both cancer progression and therapy response. Therefore, macrophage-targeting agents may prove beneficial for the efficacy of conventional anti-cancer drugs (Salvagno et al., 2019). These findings demonstrate that studying the crosstalk between macrophages and cancer cells will aid the development and improvement of agents targeting macrophages and their associated activating molecules, of which some are currently being tested in clinical trials (Brown, Recht, & Strober, 2017; Mantovani et al., 2017).

Despite the apparent crucial role for macrophages in cancer progression, the molecular mechanisms of how cancer cells influence macrophage phenotype and function are largely unresolved. The *in vitro* generation of macrophages from human peripheral blood and mouse bone marrow to study cancer-immune cell interactions provides a relatively quick and easy approach to gain molecular insights into the crosstalk between cancer and immune cells. We here describe an *in vitro* assay focusing on how cancer cell-derived molecules can impact the phenotype and functional state of macrophages. By utilizing conditioned medium from cancer cell lines, we mimic paracrine signaling between cancer cells and TAMs occurring in the TME.

In this chapter, we provide a detailed overview of several *in vitro*-based approaches to study the impact of cancer cells on macrophages. Macrophages can be differentiated in a 7-day protocol from murine bone marrow or human peripheral blood mononuclear cells (PBMCs) to be used in a variety of experiments. In one to several days following differentiation, the impact of the cancer cell secretome on macrophage phenotype can be examined by exposing macrophages to the conditioned medium from cancer cells, followed by several assays to determine macrophage activation. By blocking or genetically disrupting the production and secretion of mediators of interest (e.g., cytokines or growth factors) by cancer cells, the impact of specific cancer cell-derived proteins on macrophage polarization can be studied. Although we describe a powerful method, it is limited to the assessment of paracrine signaling from cancer cells to macrophages. The complexity of an *in vivo* situation where cancer cell-macrophage interactions and paracrine signaling between both cell types impact cancer progression cannot be studied by solely using cancer cell conditioned medium. We will provide examples of experiments showing how cancer cells can influence macrophage phenotype in terms of activation state, intracellular pathway activity and expression of chemokines. In summary, our *in vitro* system can be utilized to gain molecular insights into cancer cell-induced polarization of macrophages.

2. Protocol

2.1 Isolation of murine bone marrow and differentiation into bone marrow-derived macrophages (BMDMs)

Researchers working with mouse models to study cancer can make use of the murine set-up of the culture method described in this chapter. In the murine setting, wild-type non-tumor-bearing mice are used to isolate the bone marrow to be differentiated into bone marrow-derived macrophages (BMDMs) by recombinant murine M-CSF stimulation for 7 days (Fig. 1A and B). We refer to Section 2.2 for the isolation of macrophages from human peripheral blood. An overview of the average cell yield and the corresponding number of experiments that can be performed per mouse can be found in Table 1.

2.1.1 Materials

- Mouse bones (tibiae and femora)
- Disposable scalpels (Swann-Morton)
- Clean paper towels

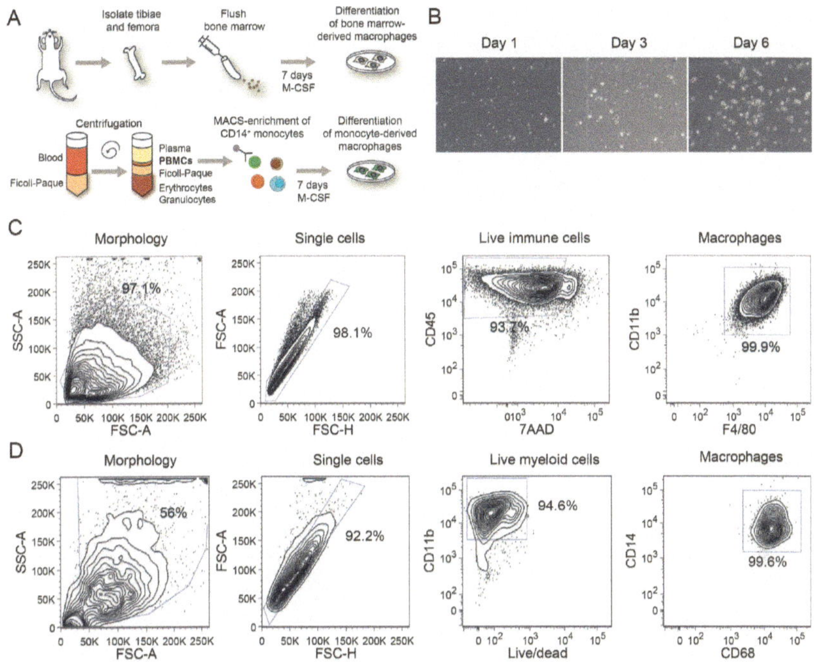

Fig. 1 Isolation and purity of murine bone marrow-derived and human monocyte-derived macrophages. (A) Schematic representation of isolation and differentiation protocol for murine bone marrow-derived macrophages (BMDMs) or human peripheral blood monocyte-derived macrophages (MDMs). For BMDMs, bone marrow is flushed from tibiae and femora and cultured for 7 days in presence of recombinant murine M-CSF. For MDMs, peripheral blood mononuclear cells (PBMCs) are isolated using centrifugation over a Ficoll-Paque layer followed by enrichment of $CD14^+$ monocytes using magnetic beads, which are cultured for 7 days in recombinant human M-CSF. (B) Microscopic images of differentiating BMDMs at day 1, day 3 and day 6. Morphological changes can be seen around day 6. (C) Representative flow cytometry plots of differentiated BMDMs. After gating on single cells, live immune cells are identified as 7-AAD^-CD45^+, and macrophages are identified as $CD11b^+F4/80^+$ cells. (D) Representative flow cytometry plots of differentiated MDMs. After gating on single cells, live myeloid cells are identified as Live/dead-eFluor-780^--$CD11b^+$, and macrophages are identified as $CD14^+CD68^+$ cells.

- 70% Ethanol (diluted from 100% ethanol in H_2O)
- 10 mL syringe (BD Emerald)
- 27G × ½″ 0.4 mm × 13 mm needles (BD Microlance 3)
- 70 μm cell strainer (Falcon)
- 50 mL tubes (Falcon)
- Full RPMI medium: RPMI medium 1640 (Gibco) supplemented with 10% Fetal Bovine Serum (Sigma), 100 IU mL^{-1} penicillin and 100 μg mL^{-1} streptomycin (Roche)

Table 1 Indicative cell numbers upon differentiation.

	Average cell yield	Number of conditions: Plated 300,000 cells per well in Section 2.4.2	Number of conditions: Plated 500,000 cells per well in Section 2.4.2
Murine bone marrow			
One leg	7–10 × 10^6 BMDMs	22–32 conditions	14–20 conditions
Two legs	15–20 × 10^6 BMDMs	50–65 conditions	30–40 conditions
Human blood			
50 mL of blood (buffy coat)	0.5–0.8 × 10^9 PBMCs = 50–80 × 10^6 $CD14^+$ cells = 25–40 × 10^6 MDMs	83–133 conditions	50–80 conditions
10 mL whole blood	1 × 10^7 PBMCs = 0.1 × 10^7 $CD14^+$ cells = 5,000,000 MDMs	16 conditions	10 conditions

- Recombinant murine CSF-1 (Peprotech)
- MACS separation buffer, containing BSA (Miltenyi Biotec)
- PBS without Mg^{2+} and Ca^{2+} (Gibco)
- Non-tissue culture-treated 10 cm dishes (Corning, Sigma)
- Non-tissue culture-treated 24-well plates (Falcon)

2.1.2 Protocol for isolation of murine bone marrow and differentiation into macrophages

All steps are performed at room temperature under sterile conditions.

Day 0

1. After euthanization of the wild-type mouse, collect tibiae and femora by dissociating the femur head from the hip socket for both hind legs. Make sure the femur head stays attached to the femur to prevent loss of bone marrow. Remove the paw, skin and muscle tissue from the bones using a scalpel. Remove any residual tissue attached to the bones using clean paper towels soaked in ethanol (Note 1).
2. Detach the tibiae from the femora and cut a small piece of both ends of the bones. Use a syringe with a 27G needle to flush the bones with full RPMI medium and strain the bone marrow through a 70 μm cell strainer (Note 2 and 3).

3. After collection of bone marrow, spin the cells down at 250 × *g* for 5 min and resuspend the cell pellet in 50 mL of full RPMI medium supplemented with 20 ng/mL murine M-CSF (Peprotech). Divide the bone marrow suspension over 10 non-tissue culture-treated 10 cm dishes (5 mL per plate) and incubate at 37 °C at 5% CO_2 (Note 4, 5, 6 and 7).

Day 3

4. Add 5 mL full RPMI medium supplemented with 20 ng/mL recombinant murine M-CSF to every 10 cm dish and incubate at 37 °C until day 7 (Fig. 1B)

Day 4 or 5

5. The preparation of conditioned medium from cancer cells starts on day 4 or 5 and is described in Section 2.3

Day 7

6. Discard the medium from all dishes and wash the cells with sterile PBS. Add 3 mL of MACS separation buffer (Miltenyi Biotec) to all dishes and incubate for at least 15 min at 37 °C for cell detachment (Note 8 and 9).
7. Add 3 mL full RPMI medium to the plate with MACS separation buffer and mix to harvest the BMDMs. Collect the cells from all dishes in 50 mL tubes (Note 10). Spin the cells down at 250 × *g* for 5 min and resuspend the cell pellet in 1 mL RPMI medium. Combine cell suspensions from one donor if more than one 50 mL tube was used to collect the cells from all dishes.
8. Count the cells using a manual counting chamber or an automated cell counter and proceed to Section 2.4.2 (Table 1).

2.2 Isolation of human peripheral blood mononuclear cells and differentiation into monocyte-derived macrophages (MDMs)

To perform these experiments with human cells, peripheral blood from healthy donors can be used to isolate peripheral blood mononuclear cells (PBMCs) using density gradient centrifugation. By utilizing CD14 microbeads, CD14-expressing monocytes can be isolated from these PBMCs. These monocytes will be used to obtain monocyte-derived macrophages (MDMs) by stimulation of 7 days with recombinant human M-CSF (Fig. 1A). An overview of the average cell yield and the corresponding number of experiments that can be performed per buffy coat can be found in Table 1.

2.2.1 *Materials*

- Healthy donor blood (buffy coat or whole blood)
- PBS without Mg^{2+} and Ca^{2+} (Gibco)
- Ficoll-Paque Plus (GE Healthcare)
- 50 mL tubes (Falcon)
- NH_4Cl erythrocyte lysis buffer
- MACS separation buffer, containing BSA (Miltenyi Biotec)
- CD14 microbeads, human (Miltenyi Biotec)
- MACS Columns for magnetic cell isolation: columns for positive selection (usually MS or LS columns, see manufacturer's recommendations, Miltenyi Biotec)
- MACS magnetic separator (Miltenyi Biotec)
- 5 mL polystyrene round-bottom tube with cell strainer cap (Falcon)
- Recombinant human CSF1 (Peprotech)
- Full RPMI medium: RPMI medium 1640 (Gibco) supplemented with 10% Fetal Bovine Serum (Sigma), 100 IU mL^{-1} penicillin and 100 μg mL^{-1} streptomycin (Roche)

2.2.2 *Protocol for isolation of human monocytes and differentiation into macrophages*

All steps are performed at room temperature under sterile conditions, unless stated otherwise.

Day 0

1. Per 25 mL of a buffy coat, one 50 mL tube is needed. In a 50 mL tube, add 25 mL of a buffy coat to 25 mL PBS. Mix gently. Repeat this step until the entire buffy coat is diluted (Note 11).
2. Prepare 50 mL tubes with 25 mL Ficoll-Paque (GE Healthcare). For every 25 mL diluted blood, one tube with Ficoll is needed. Gently add 25 mL diluted blood to the tube containing the Ficoll: slowly pipet the blood via the side of the tube on top of the Ficoll without breaking the surface.
3. Centrifuge at room temperature at 350 × *g* for 20 min without brakes.
4. Using a pipette, carefully remove the plasma layer and collect the PBMCs from the interphase layer into a new 50 mL collection tube (Fig. 1A).
5. Fill tubes containing PBMCs with ice cold PBS, spin down 200 × *g* for 8 min (with brakes). Carefully pipet off the supernatant and resuspend the pellet in 1 mL PBS. Fill tubes with PBS, and repeat this wash step two times. After the wash steps, combine the separate tubes of one blood sample together, spin down 200 × *g* for 8 min and discard supernatant.

6. Add 5 mL of sterile NH_4Cl erythrocyte lysis buffer to the washed PBMCs, incubate for 5 min, and centrifuge at 200 × *g* for 8 min and discard supernatant.
7. Add 1 mL of sterile NH_4Cl erythrocyte lysis buffer to washed PBMCs and incubate for 5 min. Spin down 200 × *g* for 8 min and discard supernatant.
8. Resuspend pellet in 5 mL MACS buffer (Miltenyi Biotec) and count cells. Take a 30 μL sample from the cell suspension for later purity check by flow cytometry (step 19). Spin down 200 × *g* for 8 min and discard supernatant.
9. Resuspend pellet in 80 μL MACS buffer per 10^7 cells.
10. Add 20 μL MACS CD14 MicroBeads per 10^7 cells, mix and incubate on ice for 20 min.
11. Add 10 mL of MACS buffer to wash, spin down 250 × *g* for 8 min and discard supernatant. Resuspend cells in 500 μL of MACS buffer per 10^8 cells. To remove clumps, pass cell suspension through a 70 μM cell strainer. To ensure all cells came through, pass another 500 μL MACS buffer through the filter.
12. Depending on the number of cells, choose the appropriate Miltenyi magnetic column (see manufacturer's recommendations). Place the column in the magnetic separator above a 15 mL tube.
13. Wash column with 3 mL MACS buffer. Allow liquid to flow through.
14. When all liquid has passed the filter, add the cell suspension to the column.
15. Wash filter three times with 3 mL MACS buffer. The flow-through contains the CD14-negative fraction. Keep these cells for the purity check in step 19.
16. After washing, remove the column from the magnet and place it into a new 15 mL tube. Add 5 mL MACS buffer to the column and plunge cells through the column. This tube contains the CD14-positive fraction (Note 12). Count the cells.
17. Take 30 μL from the $CD14^-$ and $CD14^+$ fractions (from step 15 and 16, respectively) for later purity check (step 19).
18. To generate monocyte-derived macrophages, culture the $CD14^+$ fraction in 5 mL full RPMI medium supplemented with 20 ng/mL recombinant human M-CSF (Peprotech) in non-tissue culture-treated plates at a density of 5×10^6 cells per 10 cm plate (Note 6 and 7).
19. Check the purity of your isolation by flow cytometry using the samples from the $CD14^-$ and $CD14^+$ fractions (from step 15 and 16, respectively) together with the total PBMC sample (from step 8) (see Section 2.4.3.3).

Day 3

20. Supplement culture with 5 mL fresh full RPMI medium containing 20 ng/mL recombinant human M-CSF and incubate at 37 °C until day 7

Day 4 or 5

21. The preparation of conditioned medium of cancer cells starts on day 4 or 5 and is described in Section 2.3

Day 7

22. Discard the medium from all dishes and wash the cells with sterile PBS. Add 3 mL of MACS buffer and incubate at 37 °C for 15 min (Note 9).

23. Collect the cells in a 50 mL tube by resuspending them in 3 mL of full RPMI (Note 10). Spin cells down at 250 × *g* for 5 min and resuspend the cell pellet in 1 mL RPMI medium. Combine cell suspensions if more than one 50 mL tube was used to collect the cells from the dishes.

24. Count the cells and proceed to Section 2.4.2 (Table 1).

2.3 Preparation of conditioned medium from mouse/human cancer cell lines

2.3.1 Materials

- Cancer cell lines of interest
- Culture medium with FCS and additional supplements
- Culture medium without FCS (serum-free)
- PBS, without Mg^{2+} and Ca^{2+} (Gibco)
- 0.05% Trypsin-EDTA (Gibco)
- Tissue culture-treated 6-well plates (Greiner CELLSTAR)
- Inhibitory or activating agents of interest to target cancer cells or cancer cell-derived mediators

2.3.2 Protocol for preparation of conditioned medium

The conditioned medium is prepared on day 5–7 of the protocol that is described in Sections 2.1 and 2.2.

Optional. To study the effect of cancer cell-derived proteins on macrophages, cancer cells can be treated with targeted therapies to block or activate these specific molecules prior to preparing conditioned medium from these cells. In case targeted therapies will be used in this assay, conditioned medium is prepared on day 4–7 of the protocol described in Sections 2.1 and 2.2. In this setting, cancer cells are seeded on day 4 (see step 1 of this section) and the extra day (day 5) is used to perform the treatment of cancer cells for a pre-determined time (for example, 24 h) to be followed by serum starvation at the end of day 6 (see step 2 of this section). Alternatively, cancer cells can be genetically manipulated, e.g., using CRISPR/Cas9 technology, to target specific genes of interest to assess how the proteins encoded by these genes are involved in paracrine activation of macrophages.

Day 5

1. Make sure the cancer cell lines are in good conditions for the experiment. Count the cancer cells and seed 200,000 cells per well in a 6-well plate (Note 13)

Day 6 (end of afternoon)

2. Discard the cell culture medium, add 1 mL serum-free medium to every well and incubate overnight for 16 h (Note 14). Here, 1 mL of medium is used instead of the usual working volume of 2 mL for 6-well plates to concentrate the conditioned medium (Note 15)

Day 7 (morning)

3. Harvest the conditioned medium and spin down at 250 × *g* for 5 min to remove detached cancer cells. The supernatant is the conditioned medium to be used for macrophage activation in the subsequent experiment (see Section 2.4, Note 16)

2.4 Readout assays to assess macrophage phenotype and activation status

A variety of methods can be used to assess macrophage phenotype and activation after exposure to cancer cell conditioned medium. Below, several assays are described to determine RNA or protein expression induced on *in vitro* generated macrophages.

2.4.1 Materials

Exposure of macrophages:

- Non-tissue culture-treated 24-well plates (Falcon)
- Full RPMI medium: RPMI medium 1640 (Gibco) supplemented with 10% Fetal Bovine Serum (Sigma), 100 IU mL^{-1} penicillin and 100 μg mL^{-1} streptomycin (Roche)
- Freshly prepared 2 × concentrated conditioned medium (serum-free, see Section 2.3.2)
- PBS without Mg^{2+} and Ca^{2+} (Gibco)
- Inhibitory or activating agents of interest targeting macrophages

RT-qPCR:

- 1.5 mL tubes (Eppendorf)
- RNA isolation kits, such as RNeasy Mini Kit (Qiagen) or ISOLATE II RNA Mini Kit (Bioline)

- Invitrogen AMV Reverse Transcriptase (Thermo Fischer) with Oligo(dT) primers
- SensiFAST™ SYBR® No-ROX Kit (Bioline)
- LightCycler 480 Thermocycler (Roche)
- Forward and reverse primers for genes of interest including reference gene (Table 2)

Western blotting:

- RIPA buffer (20 mM Tris-HCl (pH7.5), 1 mM EDTA, 1% Triton X-100, 0.1% sodium deoxycholate, 0.1% SDS, 150 mM NaCl.)
- cOmplete™ Protease Inhibitor Cocktail (Sigma)
- Halt™ Phosphatase Inhibitor Cocktail (Thermo Fischer)
- Pierce BCA Protein Assay Kit (Thermo Fischer)
- LDS NuPAGE sample buffer (4 ×) (Thermo Fischer)
- 100 mM dithiothreitol (DTT)
- NuPAGE Bis-Tris gel (Thermo Fischer)
- Transblot Turbo nitrocellulose or PVDF blots (BioRad)
- Transblot Turbo Transfer system (BioRad)
- Ponceau S staining (Sigma)
- Western blot antibodies (Table 3)
- 1 × TBS-Tween
- 10% and 5% Western blot blocking reagent (Roche) in TBS-T
- Optional: 3% Bovine Serum Albumin Fraction V (Roche) in TBS-T
- Optional: 5% non-fat milk in TBS-T

Table 2 Primers for RT-qPCR.

Gene	Forward (5′–3′)	Reverse (5′–3′)
Gapdh	AGGTCGGTGTGAACGGATTTG	TGTAGACCATGTAGTTGAGGTCA
Ccl2	GTTGGCTCAGCCAGATGCA	AGCCTACTCATTGGGATCATCTTG

Table 3 Western blot antibodies.

Antigen	Clone	Company	Dilution
β-Actin	D6A8	Cell Signaling	1:5000
Total S6	5G10	Cell Signaling	1:1000
Phospho-S6$^{Ser235/236}$	D57.2.2E	Cell Signaling	1:1000
Total 4EBP1	53H11	Cell Signaling	1:1000
Phospho-4EBP1$^{Thr37/46}$	Thr37/46	Cell Signaling	1:1000

Flow cytometry:

- Clean paper towels
- 96-well cell culture plate, U bottom (Greiner CELLSTAR)
- FACS buffer: PBS supplemented with 0.5% BSA and 2mM EDTA
- Fc Block (BD Biosciences)
- Fluorochrome-conjugated antibodies (Table 4)
- Cell viability staining (Table 4)
- Optional: Cytofix/Cytoperm Fixation/Permeabilization Solution Kit (BD Biosciences)
- Optional: Fluorochrome-conjugated antibodies for intracellular proteins

2.4.2 Exposure of macrophages to cancer cell conditioned medium

1. In a non-tissue culture-treated 24-wells plate, seed 300,000 differentiated BMDMs or MDMs per well in 250 μL full RPMI medium (Note 17 and 18). Keep 300,000 cells for purity check of the differentiated cells in Section 2.4.3.3, step 2 (Fig. 1C and D).
2. Add 250 μL of freshly prepared conditioned medium to the macrophage culture. Add non-conditioned medium (i.e., medium that has not

Table 4 Flow cytometry reagents.

	Marker	Fluorochrome	Clone	Target species	Company	Dilution
Cell surface antigens	CD45	BUV395	30-F11	Mouse	BD Biosciences	1:400
	CD11b	BV650	M1/70	Mouse/human	eBioscience	1:400
	F4/80	FITC	BM8	Mouse	eBioscience	1:400
	MHC-II	APC-eFluor 780	M5/114.15.2	Mouse	eBioscience	1:200
	CD206	AF488	MR5D3	Human	AbD Serotec	1:200
	CD68	FITC	Y1/82A	Human	BD Biosciences	1:200
	CD14	BV421	M5E2	Human	BD Biosciences	1:400
Viability stains	7-AAD				eBioscience	1:20
	Fixable Viability Dye eFluor-780				eBioscience	1:1000

been cultured on cancer cells) to one well as a control. Exposure of BMDMs or MDMs to conditioned medium is performed for a pre-determined time. In our experience, an exposure time of 24 h works well (Note 19 and 20).

3. *Optional.* Desired inhibitory or activating molecules targeting macrophages can be added to the culture at this point.
4. Proceed with Section 2.4.3.1 for quantification of gene expression, with Section 2.4.3.2 for western blotting of intracellular signaling pathways or with Section 2.4.3.3, step 3 for assessment of macrophage phenotype using flow cytometry.

2.4.3 Readout of macrophage polarization

Upon exposure of macrophages with cancer cell conditioned medium, macrophage polarization can be analyzed using several techniques, examples of which are described below. These techniques include analysis of cancer cell-induced gene expression, intracellular protein signaling and regulation of cell surface markers.

2.4.3.1 Quantitative RT-PCR to detect gene expression in macrophages

This method is used to analyze cancer cell-induced gene expression in macrophages. Here, we provide one example of differential chemokine expression upon exposure to cancer cell conditioned medium (Fig. 2A). Other genes that can be of interest to be studied in this context are cytokines (genes encoding IL-1β, IL-6, IL-10, TGF-β, TNF-α), chemokines (genes encoding CCL2, CCL8, CCL17, CCL22), or enzymes (genes encoding Arginase-1, iNOS). In addition, expression of targets for checkpoint inhibition, such as PD-L1 or SIRP-α, can be determined.

1. After 24 h of exposure to the conditioned medium, remove medium and wash cells once with PBS. Detach cells using 500 μL full MACS buffer per well and incubate at 37 °C for 15 min (Note 9). Collect the cells in a 1.5 mL tube. Use a new tube for every well of plated cells. Spin down at 250 × *g* for 5 min and discard the supernatant.
2. RNA can subsequently be isolated from cell pellets either using Trizol-chloroform isolation, followed by DNase I (Invitrogen) treatment, or using RNA isolation kits (for example, the Qiagen RNeasy kit, Note 21).
3. Determine RNA concentration and quality using NanoDrop 1000 UV–Vis Spectrophotometer (Thermo Fisher) or a similar device (Note 22). Quality is considered satisfactory when the A260/280 > 2.0.
4. Generate cDNA from a set input of RNA using AMV reverse transcriptase and Oligo(dT) primers (Invitrogen) or Maxima First Strand cDNA

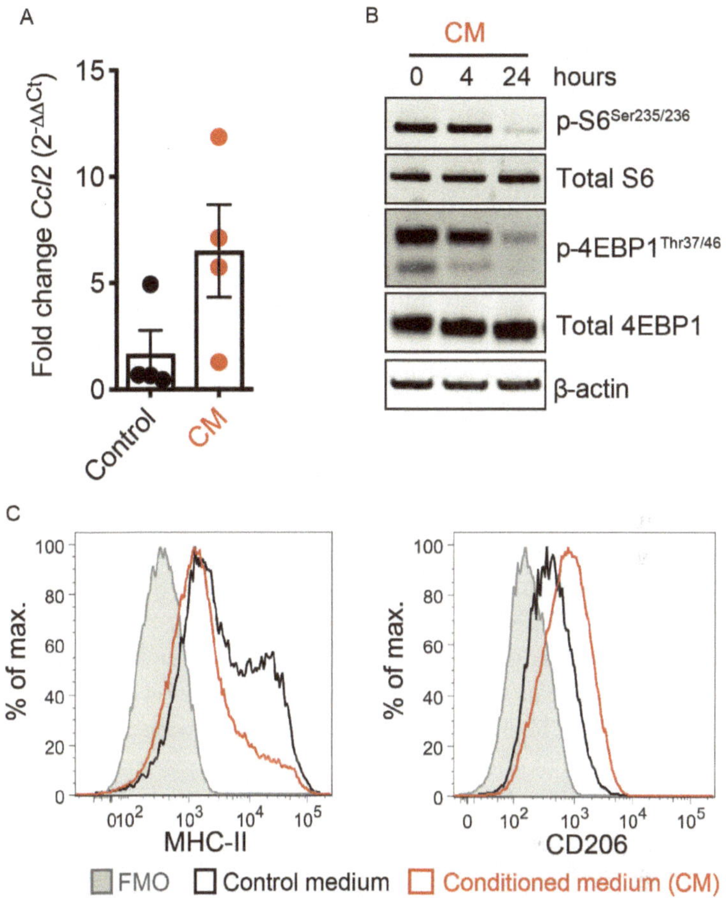

Fig. 2 Examples of assays used to assess macrophage activation after exposure to cancer cell conditioned medium. (A) Example of RT-qPCR analysis of BMDMs showing induction of *Ccl2* expression after exposure to cancer cell conditioned medium (CM), compared to control medium. (B) Western blot showing down-regulation of mTOR signaling after 4- and 24-h exposure of BMDMs to CM, as indicated by a decrease in phosphorylated S6 and phosphorylated 4EBP1. Beta-actin serves as a loading control. (C) Flow cytometry histograms showing down-regulation of MHC-II and up-regulation of CD206 expression on BMDMs after exposure to CM (red), compared to control medium (black). An unstained sample is shown in gray. All data represent mean with S.E.M.; FMO = fluorescence minus one (unstained for the indicated protein).

Synthesis Kit (Thermo Fisher). Usually, we use 20 ng of RNA input per gene that is tested. Analyze cDNA using SYBR-Green PCR with 500 nM primers for the gene of interest using a LightCycler 480 thermocycler (Roche). Include housekeeping gene for reference and perform cDNA analysis in technical duplicate (Table 2).

2.4.3.2 Western blotting to determine intracellular protein levels in macrophages

Western blotting is used to study intracellular signaling pathways in macrophages by antibody-based detection of phosphorylated proteins (Fig. 2B). A general overview of Western blotting is given, but similar protocols for this method can be used here.

1. After 24h of exposure to the conditioned medium, remove medium and wash macrophages once with PBS.
2. Detach cells from the plate using 500 μL full MACS buffer per well, incubate at 37 °C for 15 min (Note 9), and collect the cells in a 1.5 mL tube. Spin down at 250 × *g* for 5 min and discard the supernatant.
3. Lyse cells by resuspending cell pellets in 50 μL RIPA buffer or equivalent containing protease and/or phosphatase inhibitors. Incubate for 45 min on ice.
4. Spin down 16,000 × *g* for 10 min at 4 °C.
5. Harvest supernatant containing cell lysate and transfer to new tube. Measure protein concentration using the Pierce BCA Protein Assay Kit (Thermo Fisher).
6. Use equal amounts of protein per lysate (usually 20–40 μg) and add H_2O up to a total volume of 15 μL. Add 5 μL of 4 × LDS NuPAGE sample buffer (Thermo Fisher) supplemented with 100 mM DTT to obtain a final volume of 20 μL per sample.
7. Denature the samples by boiling at 100 °C for 5 min and spin down at 10,000 × *g* for 10 s.
8. Together with a molecular weight marker, load samples on a NuPAGE 4–12% Bis-Tris gel (Invitrogen) or any desired gel. Perform electrophoresis for approximately 45 min at 200 V. Depending on molecular weight of the protein to be analyzed, choose the appropriate gel to run your samples.
9. Transfer proteins from gel to membrane (nitrocellulose or PVDF). PVDF requires activation using 100% methanol. Transfer can be performed using the Trans-Blot Turbo system (Bio-Rad) according to the manufacturer's instructions.
10. *Optional.* After transfer of proteins, transfer efficiency can be checked using Ponceau S staining (Sigma).
11. Block membranes in 10% Western blot blocking reagent (Roche) in TBS-T for 1 h at room temperature.
12. Incubate membranes overnight at 4 °C with primary antibodies in 5% Western blot blocking reagent (Roche) in TBS-T (Note 23 and 24).

13. The next day, remove primary antibody mix and wash membranes three times for 10 min using TBS-T (Table 3).
14. Incubate membranes with the appropriate IRdye-conjugated secondary antibodies in 5% Western blot blocking reagent (Roche) in TBS-T for 1 h at room temperature. Wash membranes three times for 10 min in TBS-T prior to visualization.
15. Infrared signals can be visualized using the Odyssey Imaging System (LI-COR, Note 25).

2.4.3.3 Flow cytometry to assess macrophage purity and phenotype

Flow cytometry is used to verify macrophage purity and to assess macrophage phenotype upon cancer cell-induced activation (Fig. 2C). For general macrophage identification, we use $CD11b^{+}F4/80^{+}$ for mouse and $CD11b^{+}$ $CD14^{+}CD68^{+}$ for human. There are several markers used for looking at activation states, depending on the research questions. Often used markers include, but are not limited to, CD206, MHC-II/HLA-DR, PD-L1, Tie2, CD163 (human).

Verification of CD14 separation of human PBMCs

The three cell fractions from Section 2.2.2 (step 19) are used to verify the purity of the isolated $CD14^{+}$ cells of human PBMCs. In this step, CD11b and CD14 are used to discriminate the different cell populations.

1. Plate the 30 μL samples of the PBMC, $CD14^{-}$ and $CD14^{+}$ fractions in a U-bottom 96-well cell culture plate. Proceed with step 5 of this section

Verification of BMDM or MDM purity

After differentiation of murine bone marrow or human monocytes, macrophage purity is determined using flow cytometry. For BMDMs, the following markers are used: CD45, CD11b, F4/80 and the viability stain 7-AAD. For MDMs, antibodies against human CD11b, CD14, CD68 and a viability stain are used (Fig. 1C and D).

2. Plate 300,000 cells in a U-bottom 96-well cell culture plate and proceed with step 5 of this section

Assessment of macrophage phenotype upon conditioned medium exposure

For functional assessment of macrophages, cells are stained for general macrophage markers as described for verification of macrophage purity. In addition, expression of activation markers can be detected after gating on macrophage population to determine their phenotype. In Fig. 2C, we show an example of two activation markers, MHC-II and CD206, that can be assessed upon exposure of BMDMs and MDMs, respectively.

3. After exposure with cancer cell conditioned medium, remove medium and wash cells once with PBS. Detach cells from the plates using 500 μL MACS buffer by incubating at 37 °C for 15 min followed by scraping off the cells (Note 9).
4. Plate the cells in a U-bottom 96-well cell culture plate and proceed with step 5 of this section.

Antibody staining

5. Incubate cells 15 min on ice with Fc Block (1:100 in FACS buffer, BD Biosciences, Note 26) followed by surface antibody staining in FACS buffer for 30 min on ice (see Table 4 for a selection of commonly used markers to analyze macrophages).
6. Add 100 μL of FACS buffer to the samples and wash the cells by resuspending three times. Spin down at 380 × *g* for 2 min and discard supernatant by inverting the plate and blotting the plate against clean paper towels.
7. Stain the cells with a viability dye to exclude dead cells. When only surface staining is applied to the cells, a non-fixable viability staining, such as 7-AAD (BD Biosciences) is used.
8. *Optional.* Intracellular staining of proteins of interest can be performed using an intracellular staining kit, such as the Cytofix/Cytoperm Fixation/Permeabilization Solution Kit (BD Biosciences, Note 27). In this case, a fixable viability staining has to be used followed by fixation of the cells according to manufacturer's instructions (Table 4). After fixation and permeabilization of the cells, intracellular staining of cytokines is performed.

3. Notes

1. Be careful with removing the tissue from the bones, as the bones could break when a high force is applied.
2. Flushing of the bones can also be performed using sterile PBS (without Mg^{2+} and Ca^{2+}). In that case, start using full RPMI medium when seeding the cells in culture plates.
3. At this point, bone marrow can be frozen down to be used for differentiation into BMDMs in future experiments. However, differentiation from fresh bone marrow and direct use of cells gives highest cell

yield (Table 1). The use of frozen material usually gives only half of the amount of differentiated BMDMs.

4. As macrophages adhere very strongly to plastic wear, make sure to use non-tissue culture-treated plates in order to make the harvesting process less difficult.
5. When bone marrow is isolated from the two hind legs of the mouse, the average cell number that is reached is 50 million cells. We recommend to use ±5 million cells per 10 cm plate, giving 10 plates per mouse. When fewer cells are needed for the experiment (Table 1), freeze the remainder of the bone marrow for future use.
6. In some studies, GM-CSF is used to induce macrophage differentiation. However, we recommend to only use M-CSF, because GM-CSF can induce different gene and protein expression profiles or even a mixed population of dendritic cells and macrophages, compared to M-CSF (Helft et al., 2015; Lacey et al., 2012).
7. Culture in low volume could give issues such as liquid evaporation. In our protocol, low volume of medium (5 mL) is only used shortly (until day 3) and did not give any problems. Although, it is important to add 5 mL of supplemented medium to the culture at day 3 to prevent any issues during the second part of the culture.
8. Here, MACS separation buffer is used, but any other separation buffer containing EDTA should work as well.
9. Macrophages are hard to detach from the culture plates. In order to speed up the process, the dishes are incubated at 37 °C for at least 15 min.
10. For the cells that are difficult to detach, it is possible to scrape them off the plate using the pipet or a cell scraper (optional). This process may increase cell yield, but also induces more cell death.
11. If whole blood is being used to isolate human monocytes, it needs to be diluted 1:3 or 1:4 in Section 2.2.2, step 1.
12. At this stage, $CD14^+$ monocytes can be frozen down for future use. Note that freeze-thaw cycles can reduce total amount of cells.
13. Other culture plates can be used to prepare conditioned medium, depending on the amount of conditioned medium that is needed to expose the macrophages. For a six-well plate, we seed 200,000 cells per well from which 1 mL of conditioned medium per well will be obtained. With 1 mL of conditioned medium, four wells with macrophages can be exposed (250 μL per well, see Section 2.4.2).
14. Here, serum-free medium is used to specifically study the impact of cancer cell-derived molecules on macrophages. Researchers are free

to use media containing serum, but keep in mind that the impact on macrophages can also be influenced by the factors present in serum (e.g., growth factors and cytokines).

15. We recommend to produce 2 × concentrated conditioned medium. During the exposure step, conditioned medium will be added to macrophage culture in a 1:1 ratio (i.e., 250 μL conditioned medium to 250 μL RPMI medium). By utilizing 2 × conditioned medium, the final concentration of the conditioned medium during the exposure of macrophages will be 1 ×.

16. At this point, conditioned medium can be frozen down for future use. From our experience, fresh preparation and direct use of conditioned medium gives the best results.

17. Biological replicates for these experiments are BMDMs or MDMs from additional donors.

18. If higher RNA or protein levels are desired for the readout assays in Section 2.4.3, more macrophages can be seeded at this stage. Table 1 shows the corresponding number of experiments that can be performed when 500,000 cells per condition are seeded.

19. We recommend to expose the macrophages up to a maximum of 48 h starting at day 7, because viability of macrophages is limited. The optimal duration of the exposure assay is 1–2 days.

20. To study whether the exposure of tumor-derived factors impacts differentiation into macrophages, cancer cell conditioned medium can be added at start of bone marrow- or monocyte differentiation. In this setting, we recommend to start with equal cell numbers per plate and to use 50% of conditioned medium diluted in full RPMI medium supplemented with M-CSF as described before. As a readout, the proportion of cells that have differentiated into macrophages can be assessed using both microscopy and flow cytometry by comparing cells that were exposed to cancer cell conditioned medium to cells cultured in control medium.

21. When using RNA isolation kits, the supplied lysis buffer can be added directly to the PBS-washed plate. This way, macrophages are lysed directly in the plate.

22. When seeding 300,000 macrophages, the total yield varies between 1600 and 3200 ng RNA.

23. For blocking of membranes and incubation of antibodies (next step), 3% BSA in TBS-T or 5% non-fat milk in TBS-T can also be used as reagent.

24. Phosphorylated proteins are known to be detected more easily when blocking buffers without phosphatases are used, such as BSA.

25. Other methods to visualize proteins of interest on the membrane, such as the chemiluminescent method, can also be applied here.

26. Macrophages can very potently aspecifically bind antibodies, so the use of Fc Block is essential here.

27. Depending on the research question, it can be relevant to perform intracellular staining of the cell, for instance to measure cytokines, but also other (intracellular) proteins can be visualized using flow cytometry.

References

Brown, J. M., Recht, L., & Strober, S. (2017). The promise of targeting macrophages in cancer therapy. *Clinical Cancer Research*, *23*(13), 3241–3250. https://doi.org/10.1158/1078-0432.Ccr-16-3122.

DeNardo, D. G., Brennan, D. J., Rexhepaj, E., Ruffell, B., Shiao, S. L., Madden, S. F., et al. (2011). Leukocyte complexity predicts breast cancer survival and functionally regulates response to chemotherapy. *Cancer Discovery*, *1*(1), 54–67. https://doi.org/10.1158/2159-8274.Cd-10-0028.

DeNardo, D. G., & Ruffell, B. (2019). Macrophages as regulators of tumour immunity and immunotherapy. *Nature Reviews. Immunology*, *19*, 369–382. https://doi.org/10.1038/s41577-019-0127-6.

Gentles, A. J., Newman, A. M., Liu, C. L., Bratman, S. V., Feng, W., Kim, D., et al. (2015). The prognostic landscape of genes and infiltrating immune cells across human cancers. *Nature Medicine*, *21*(8), 938–945. https://doi.org/10.1038/nm.3909.

Helft, J., Bottcher, J., Chakravarty, P., Zelenay, S., Huotari, J., Schraml, B. U., et al. (2015). GM-CSF mouse bone marrow cultures comprise a heterogeneous population of CD11c(+)MHCII(+) macrophages and dendritic cells. *Immunity*, *42*(6), 1197–1211. https://doi.org/10.1016/j.immuni.2015.05.018.

Lacey, D. C., Achuthan, A., Fleetwood, A. J., Dinh, H., Roiniotis, J., Scholz, G. M., et al. (2012). Defining GM-CSF- and macrophage-CSF-dependent macrophage responses by in vitro models. *Journal of Immunology*, *188*(11), 5752–5765. https://doi.org/10.4049/jimmunol.1103426.

Lewis, C. E., & Pollard, J. W. (2006). Distinct role of macrophages in different tumor microenvironments. *Cancer Research*, *66*(2), 605–612. https://doi.org/10.1158/0008-5472.can-05-4005.

Mahmoud, S. M., Lee, A. H., Paish, E. C., Macmillan, R. D., Ellis, I. O., & Green, A. R. (2012). Tumour-infiltrating macrophages and clinical outcome in breast cancer. *Journal of Clinical Pathology*, *65*(2), 159–163. https://doi.org/10.1136/jclinpath-2011-200355.

Mantovani, A., Marchesi, F., Malesci, A., Laghi, L., & Allavena, P. (2017). Tumour-associated macrophages as treatment targets in oncology. *Nature Reviews. Clinical Oncology*, *14*(7), 399–416. https://doi.org/10.1038/nrclinonc.2016.217.

Medrek, C., Ponten, F., Jirstrom, K., & Leandersson, K. (2012). The presence of tumor associated macrophages in tumor stroma as a prognostic marker for breast cancer patients. *BMC Cancer*, *12*, 306. https://doi.org/10.1186/1471-2407-12-306.

Quail, D. F., & Joyce, J. A. (2013). Microenvironmental regulation of tumor progression and metastasis. *Nature Medicine*, *19*(11), 1423–1437. https://doi.org/10.1038/nm.3394.

Salvagno, C., Ciampricotti, M., Tuit, S., Hau, C. S., van Weverwijk, A., Coffelt, S. B., et al. (2019). Therapeutic targeting of macrophages enhances chemotherapy efficacy by unleashing type I interferon response. *Nature Cell Biology*, *21*, 511–521. https://doi.org/10.1038/s41556-019-0298-1.

Singhal, S., Stadanlick, J., Annunziata, M. J., Rao, A. S., Bhojnagarwala, P. S., O'Brien, S., et al. (2019). Human tumor-associated monocytes/macrophages and their regulation of T cell responses in early-stage lung cancer. *Science Translational Medicine*, *11*(479), eaat1500. https://doi.org/10.1126/scitranslmed.aat1500.

Steidl, C., Lee, T., Shah, S. P., Farinha, P., Han, G., Nayar, T., et al. (2010). Tumor-associated macrophages and survival in classic Hodgkin's lymphoma. *The New England Journal of Medicine*, *362*(10), 875–885. https://doi.org/10.1056/NEJMoa0905680.

Tan, K. L., Scott, D. W., Hong, F., Kahl, B. S., Fisher, R. I., Bartlett, N. L., et al. (2012). Tumor-associated macrophages predict inferior outcomes in classic Hodgkin lymphoma: A correlative study from the E2496 intergroup trial. *Blood*, *120*(16), 3280–3287. https://doi.org/10.1182/blood-2012-04-421057.

Thorsson, V., Gibbs, D. L., Brown, S. D., Wolf, D., Bortone, D. S., Ou Yang, T. H., et al. (2018). The immune landscape of cancer. *Immunity*, *48*(4), 812–830.e814. https://doi.org/10.1016/j.immuni.2018.03.023.

Zhang, Q. W., Liu, L., Gong, C. Y., Shi, H. S., Zeng, Y. H., Wang, X. Z., et al. (2012). Prognostic significance of tumor-associated macrophages in solid tumor: A meta-analysis of the literature. *PLoS One*, 7(12), e50946. https://doi.org/10.1371/journal.pone.0050946.

CHAPTER TEN

Protocol to assess the suppression of T-cell proliferation by human MDSC

Annika M. Bruger[a], Christophe Vanhaver[a], Kirsten Bruderek[b], Giada Amodio[c], Ece Tavukçuoğlu[d], Günes Esendagli[d], Silvia Gregori[c], Sven Brandau[b,†], Pierre van der Bruggen[a,e,*,†]

[a]de Duve Institute, Université catholique de Louvain, Brussels, Belgium
[b]Research Division, Department of Otorhinolaryngology, University Hospital Essen, West German Cancer Center, Essen, Germany
[c]San Raffaele Telethon Institute for Gene Therapy (SR-Tiget), San Raffaele Scientific Institute IRCCS, Milan, Italy
[d]Department of Basic Oncology, Hacettepe University Cancer Institute, Ankara, Turkey
[e]WELBIO, Brussels, Belgium
*Corresponding author: e-mail address: pierre.vanderbruggen@uclouvain.be

Contents

† Share senior authorship.

Methods in Enzymology, Volume 632
ISSN 0076-6879
https://doi.org/10.1016/bs.mie.2019.05.046

Abstract

Inhibition of T-cell proliferation is the most common approach to assess human myeloid-derived suppressor cell (MDSC) functions. However, diverse methodologies hinder the comparison of results obtained in different laboratories. In this chapter, we present a T-cell proliferation assay procedure based on allogeneic MDSC and T-cells that is potentially suitable to multi-center studies. The T-cells are isolated from non-cancerous donors and frozen for later use in different research groups. We observed that pure thawed T-cells showed poor proliferative capacities. To retain proliferation, T-cell-autologous mature dendritic cells are supplemented after thawing. MDSC are isolated from clinical samples and represent the sole variant between assays. Flow cytometry is used to assess T-cell proliferation by the dilution of a tracking dye.

Abbreviations

CFSE carboxyfluorescein succinimidyl ester
COST European Cooperation in Science and Technology
DC dendritic cell
EU European Union
FBS fetal bovine serum
HS human serum
MDSC myeloid-derived suppressor cell(s)
NK natural killer
PBMC peripheral blood mononuclear cells
PBS phosphate buffered saline
RT room temperature

1. Introduction

Myeloid-derived suppressor cells (MDSC) are heterogeneous populations of cells of the myeloid lineage with strong immunosuppressive capacities. Increased numbers of MDSC have been measured in the blood and tissue of patients with diverse conditions including cancer, arthritis and infectious diseases. In cancer, the increased presence of MDSC has been correlated with poor prognoses for disease progression and survival (Safarzadeh, Orangi, Mohammadi, Babaie, & Baradaran, 2018). Targeting the MDSC and their immunosuppressive capacities could be an attractive target for future combination immunotherapies for cancers. The characteristics of the different human MDSC and their role in diseases have been reviewed elsewhere (Amodio et al., 2019; Cassetta et al., 2019; Dorhoi et al., 2019; Umansky et al., 2019). Briefly, MDSC are broadly categorized based on their lineage of origin into monocytic, granulocytic and early stage MDSC. A defining character of all is their immunosuppressive function (Haile, Greten, & Korangy, 2012). However, both isolation strategies and functional assay procedures differ widely in the literature and among groups (Bruger et al., 2019; Cassetta et al., 2019). This complicates the comparability of MDSC sourced from different pathologies, tissues and in different groups. We here propose a protocol using one common allogeneic T-cell responder to test the suppressive function of MDSC sourced at different locations comparably.

Compared to murine MDSC, the study of human MDSC is complicated by various challenges: (i) The heterogeneity of MDSC populations and the lack of a specific cellular marker results in complex isolation strategies with high levels of variation among laboratories. While Lox-1 has been proposed as a specific MDSC marker, it does not seem to be expressed by all MDSC and is not yet widely adopted in isolation protocols (Condamine et al., 2016). (ii) Granulocytic MDSC (also known as PMN-MDSC) in particular possess a short ex vivo life-span and do not tolerate freezing, restricting the breadth of experiments that can be performed on any one sample (Trellakis et al., 2013). (iii) Patients' samples are restricted in the type and amount of tissue available for MDSC extraction. Usually, human MDSC are isolated from the peripheral blood of cancer patients at low yields. This limits the number and types of experiments and manipulations possible. Genetic, depletion or adoptive transfer models, such as in mice (Clavijo et al., 2017; Corzo et al., 2010; Pan et al., 2010), cannot be performed in humans.

Considering these restrictions, T-cell proliferation assays are the most common test used to assess MDSC-mediated suppression of T-cell functions. The principle of T-cell proliferation assays is convenient as they are fast to set up and easy to adapt according to MDSC yields: T-cells are stained with radioactive tritium or with a fluorescent tracker dye that halves in its intensity with every round of multiplication. The T-cells are co-cultured with MDSC and stimulated through CD3 and the co-stimulatory CD28 molecule. The results are read after 3–4 days of incubation. However, similarly to the isolation strategies for MDSC, many variables exist among laboratories that complicate the comparison of results obtained at different locations. These variations include the type and origin of T-cells used, e.g., $CD3^+$ or $CD8^+$ T-cells, autologous or allogeneic to the MDSC (Bruger et al., 2019).

Particularly the fragile nature of the MDSC, their low yields, and the large variation in T-cell proliferation protocols mean that MDSC suppression data from different laboratories are difficult to compare to each other. Since the MDSC themselves cannot be exchanged, a common suppression assay protocol is important to assess if MDSC share functional properties regardless of the origin of tissue or lineage. Are there distinct MDSC populations or is there a functional spectrum?

To address these questions, we established a T-cell proliferation assay protocol and reagents to share among different laboratories with in the pan-European COST Action Mye-EUNITER. Ideally, the MDSC would remain the only variant between experiments and laboratories. This approach is based on isolating and freezing T-cells from non-cancerous donors that can be shared. The following protocol has been developed and tested in the van der Bruggen laboratory in Brussels (Fig. 1). We are still validating the protocol with exchanged cells in COST Action Mye-EUNITER partner laboratories.

2. T-cell isolation and mature dendritic cell generation

T-cell proliferation is the most commonly used assays to assess the suppressive capacity of human MDSC. However, due to the large numbers of parameters involved, we observed procedural variations among laboratories (Bruger et al., 2019). Since none of the procedures is objectively superior, we agreed to adopt a common protocol in a multicenter study.

To achieve this, we advise to isolate allogeneic T-cells from non-cancerous donors (Sections 2.1–4). Specifically, in the van der Bruggen group (Brussels) we obtain up to 500 mL peripheral blood from hemochromatosis patients who have to be bled regularly as part of their routine treatment. This allows us to isolate T-cells in numbers large enough to be shared easily with other laboratories. The negative selection process, ensuring that CD3-engagement is not induced during the isolation process, is performed by separating $CD2^+$ and $CD2^-$ cells through rosetting with sheep

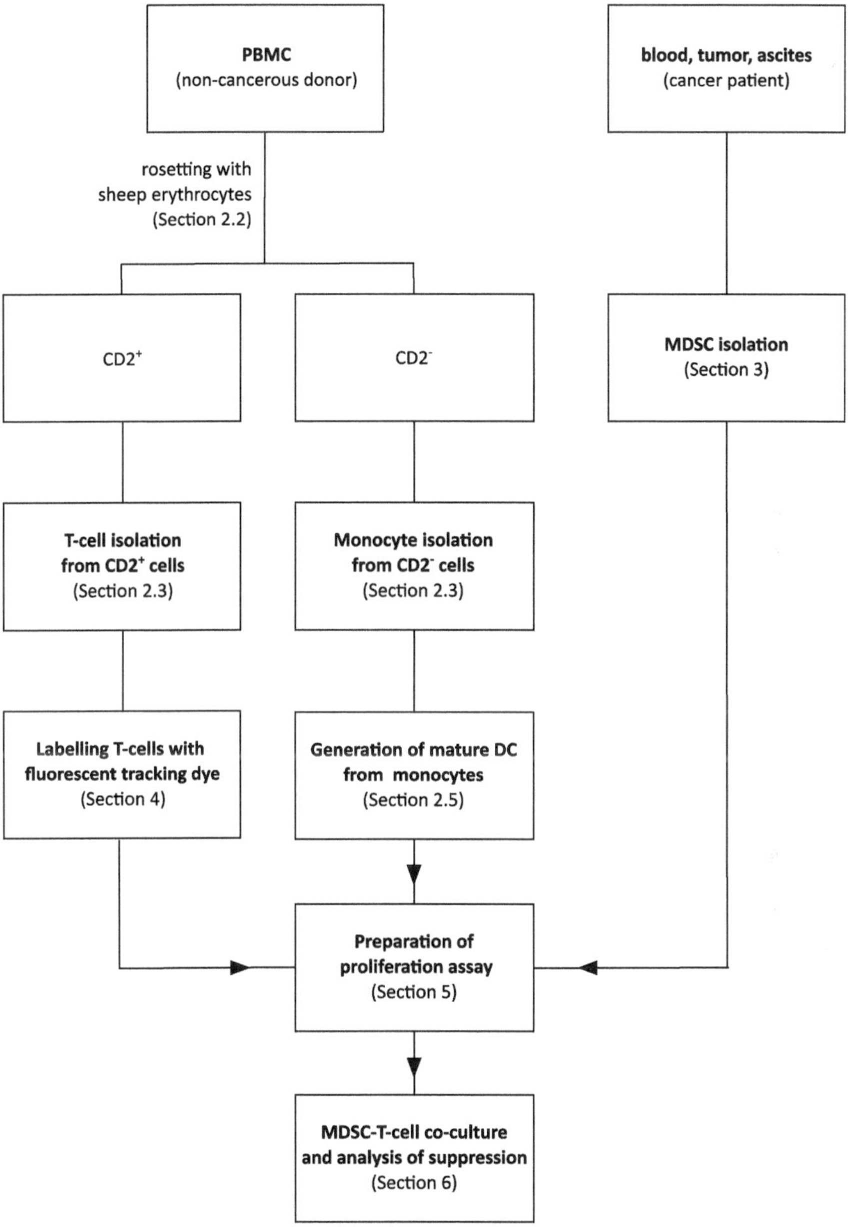

Fig. 1 Overview of the process. T-cells and monocytes are isolated from non-cancerous (here: hemochromatosis) donors. DCs are generated from the monocytes. T-cells and monocytes are frozen for future use in T-cell proliferation assays at different research groups. MDSC are isolated from cancer patients and used directly in T-cell proliferation assays.

erythrocytes, followed by the depletion of $CD56^+$ NK cells from the $CD2^+$ fraction (Sections 2.2–3). The T-cells are frozen at −80 °C and can be shared among laboratories. However, remaining myeloid cell impurities did not survive the thawing process, and, as a consequence, thawed T-cells barely proliferated. T-cell proliferation can be rescued by the addition of low numbers of autologous mature dendritic cells (DC), differentiated from the $CD2^-$ fraction (Sections 2.4–7).

2.1 PBMC isolation from whole blood

Sharing PBMC from the same donor within multi-center studies requires the isolation of allogeneic T-cells from non-cancerous donors. We use blood obtained from hemochromatosis patients for this purpose. We first isolate the PBMC, and then the T-cells and monocytes in subsequent steps.

a. Dilute whole blood 1:3 with cold PBS + 1 mM EDTA.

b. Separate the PBMC from the erythrocytes, neutrophils and platelets by density gradient centrifugation. Add 15 mL Lymphoprep™ (Alere Technologies) or Ficoll (Sigma) into a 50 mL Falcon tube at room temperature (RT) and carefully layer the blood onto the Lymphoprep without mixing the two solutions by holding the tube at a low angle. To separate the cells by density, centrifuge the cells for 20 min at 900*g* without breaks or acceleration to avoid disrupting the density gradient established during centrifugation.

c. Following the density gradient separation the PBMC settle in a thin layer below the plasma, and above the Lymphoprep solution and erythrocyte pellet (Fig. 2). Carefully remove the PBMC layer with a pipette and place it into a fetal bovine serum (FBS)-coated 50 mL Falcon tube. The Falcon is pre-coated with FBS for 24 h at 37 °C to prevent cells from adhering to the plastic.

d. Wash the cells with cold PBS + 1 mM EDTA and centrifuge at 400*g* for 10 min. EDTA is a Ca^{2+} chelating agent that prevents platelet aggregation.

e. The PBMC settle in the pellet. Carefully remove the supernatant and as much of residual platelets that settle in a white ring above the PBMC as possible. Remaining platelets could interfere in later steps with the DC generation through CD40L and with flow cytometry analyses by interacting with antibodies. Inspect the cell solution for platelets underneath a microscope if doubtful that all platelets have been successfully

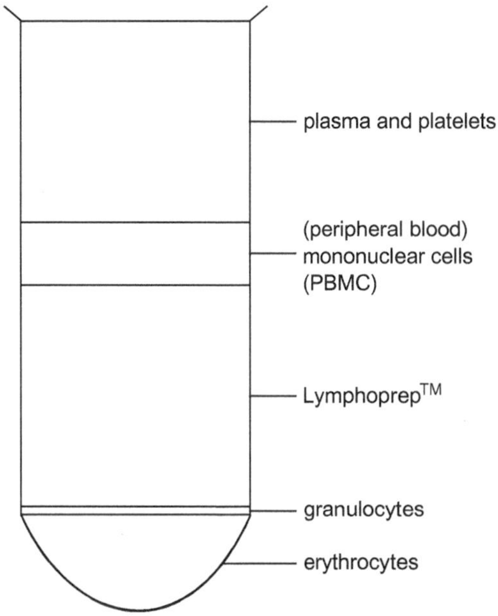

Fig. 2 Density gradient separation of peripheral blood. Whole blood diluted with cold PBS-EDTA is carefully layered onto Lymphoprep™ solution. Density separation is achieved after centrifugation at 900*g* for 20 min without acceleration or brakes. Shown is a sketch of how the blood components separate.

removed by aspiration. Should platelets remain, resuspend the PBMC in PBS + 1 mM EDTA and centrifuge them at 200*g* for 3 min and aspirate the supernatant. Platelets are too light to pellet at this speed.

f. Repeat the washing steps with 50 mL cold PBS + 1 mM EDTA twice.

g. Resuspend the PBMC in warm RPMI 1640 (Gibco) and count the cells. Retain 5×10^5 cells for flow cytometry assessment (Section 2.5).

2.2 Separation of CD2$^+$ and CD2$^-$ cells by rosetting with sheep erythrocytes

T-cells are isolated by negative selection to avoid any adverse effects caused by CD3-binding beads. Erythrocyte rosetting allows separating CD2$^+$ and CD2$^-$ cells from PBMC. CD2 is an adhesion molecule present on T-cells and natural killer (NK) cells, which spontaneously binds to the LFA3-homolog on sheep erythrocytes (Ocklind, 1988). CD2/LFA-3 interaction causes an aggregation of NK and T-cells around the

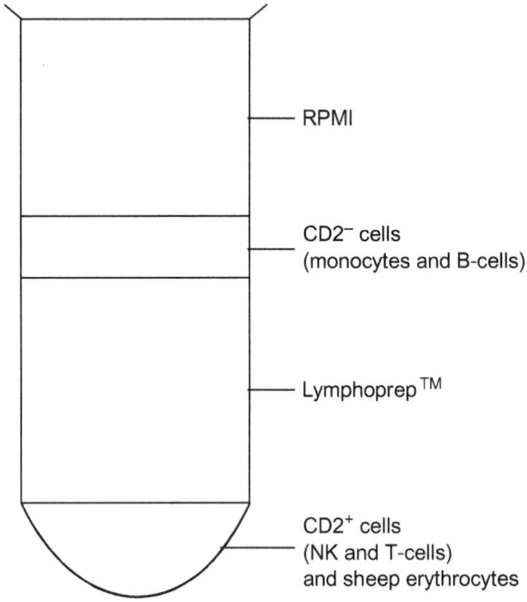

Fig. 3 Density gradient separation of $CD2^+$ and $CD2^-$ cells. Sheep erythrocytes express an LFA3-homolog that binds to human CD2 and thus causes the spontaneous aggregation of human NK and T-cells (Ocklind, 1988). The sketch indicates how $CD2^+$ and $CD2^-$ cells separate by density gradient.

erythrocytes, allowing $CD2^+$ NK and T-cells to be elegantly separated from $CD2^-$ B-cells and monocytes by a density gradient. The $CD2^+$ cells pellet together with the erythrocytes, while the $CD2^-$ cells settle in the mononuclear layer (Fig. 3). In the following sections, we first describe the preparation of sheep erythrocytes. Second, we describe the erythrocyte rosetting protocol.

2.2.1 Preparation of sheep erythrocytes

a. Use reagents equilibrated to room temperature. We use sheep erythrocytes in 50% Alsever's suspension provided by Labor Dr. Merck (E-400). 2-aminoethylisothiouronium bromide (AET) is provided by Sigma (A54601).

b. Check the hematocrit (% erythrocytes) of the sheep erythrocytes first, using the provider's website: http://www.labormerk.com/manufacturing/blood_products/

Table 1 Reference table for the resuspension of sheep erythrocytes in AET and RPMI 1640 according to the hematocrit content.

% Hematocrit	X: A.E.T. (mg)	Y: H_20 volume for AET (mL)	Z: RPMI 1640 (mL)
10	625	13.5	15.6
11	688	14.7	17.2
12	750	16.2	18.7
13	813	17.3	20.3
14	875	18.9	21.9
15	938	20.0	23.4
16	1000	21.6	25.0
17	1063	22.7	26.6
18	1125	24.3	28.0
19	1188	25.7	29.7
20	1250	27.0	31.2

c. Prepare the AET solution by dissolving **X** mg of AET in **Y** mL of milli-Q water (see Table 1). Add 1 M NaOH to bring the solution to pH 9. Note that the correct pH of the solution is of particular importance.

d. Filter the AET suspension with a 0.22 μm Steriflip (Millipore, catalog ♯ SCGP00525) into a 50 mL filter Falcon tube.

e. Resuspend sheep erythrocytes by shaking and/or pipetting, and collect 12.5 mL of the erythrocyte suspension. Wash the suspension by adding 37.5 mL of PBS and centrifuge at 910*g* for 8 min.

f. Aspirate supernatant. Leave some PBS to avoid aspirating any erythrocytes. Add the AET solution and resuspend the erythrocytes. Incubate for 15 min at 37 °C.

g. Pellet the erythrocyte suspension by centrifugation at 910*g* for 8 min. Remove AET supernatant. Resuspend in 50 mL PBS. The erythrocyte pellet should be of dense consistency and require repeated gentle agitation for complete resuspension.

h. Wash with 50 mL PBS by centrifugation at 910*g* for 8 min three times. The pellet's consistency should become less dense with each washing step. Perform an additional washing step if the supernatant is quite red which indicates erythrocyte lysis.

i. Resuspend the cells in **Z** mL of RPMI 1640 (Table 1). Store the cells at 4 °C for up to 15 days. Test the sheep erythrocytes on donor blood before use, and control the efficiency of the rosetting by flow cytometry as described in Section 2.5.

2.2.2 Separation of $CD2^+$ and $CD2^-$ cells from sheep erythrocytes

a. Prepare three solutions in advance to lyse the erythrocytes in later steps:
  i. 5 M NaCl stock solution, in distilled water, sterilized
  ii. 0.2% NaCl lysis solution (in sterile water, from stock)
  iii. 1.2% NaCl reconstitution solution (in sterile water, from stock)

b. Resuspend the PBMC in warm RPMI 1640 at 10^7 cells/mL or in at least 10 mL.

c. Add 1 mL sheep erythrocytes (preparation described in Section 2.2.1) for each 10^7 PBMC. Mix the cells well and incubate them at room temperature for 10 min.

d. The Lymphoprep density separation step is repeated with the PBMC and sheep erythrocyte suspension (Section 2.1, steps b and c).

e. The $CD2^-$ cells will settle in the PBMC layer (Fig. 3). Remove the $CD2^-$ cells, wash and count them as before (Section 2.1, steps e–g). $CD14^+$ cells are obtained from the $CD2^-$ cells (described in Section 2.4) and used to generate autologous mature dendritic cells (Section 2.6).

f. Remove the Lymphoprep solution carefully and retain the pellet. The pellet contains the sheep erythrocytes and the $CD2^+$ cells that attached to them.

g. Lyse the erythrocytes with 0.2% NaCl (prepared in Section 2.2.1) by resuspending the cells carefully but thoroughly. After 50–60 s reconstitute osmolarity with 1.2% NaCl (Section 2.2.1). Refer to Table 2 for the required volumes.

Table 2 Reference table for the use of lysis buffers according to the original volumes of blood used.

Blood volume before density gradient (mL)	0.2% NaCl (mL)	1.2% NaCl (mL)
50	7.5	17.5
40	6.0	14.0
30	4.5	10.5
20	3.0	7.0
10	1.5	3.5

h. Centrifuge for 5 min at 300*g* RT. Resuspend the pellet in warm RPMI 1640 and count the cells. Retain 5×10^5 $CD2^+$ and $CD2^-$ cells for flow cytometry assessment (Section 2.5).

2.3 Isolation of T-cells from $CD2^+$ cells by negative selection

Only T and NK cells express CD2. T-cells are further isolated from $CD2^+$ cells by depleting NK cells using anti-CD56 magnetic beads. CD56 is a cell adhesion glycoprotein that is prototypic marker for NK cells. The purity of the isolated cells and the success of the separation are assessed by flow cytometry in Section 2.5.

a. Use $CD2^+$ cells isolated from whole blood (Sections 2.2).

b. Count the cells and resuspend 10^7 cells in 60 μL PBS + 2% human serum (HS).

c. Per 10^7 cells add 20 μL Fc receptor (FcR) blocking reagent (Miltenyi, catalog ♯ 130-059-901) and 20 μL magnetic anti-CD56 beads (Miltenyi, catalog ♯ 130-050-401).

d. Mix well and incubate for 20 min at 4 °C.

e. Separate T-cells and NK cells using either manual or automatic magnetic columns. Keep the samples cool at all times using racks pre-cooled to 4 °C (e.g., Chill 15 Rack, Miltenyi, catalog ♯ 130-092-952). Follow the manufacturer's instruction if using manual separation. If the automatic AutoMACS separator is used, use the program possel_s.

f. The $CD56^+$ NK cells will be in the positive fraction, the T-cells in the negative eluate fraction. Retain both fractions, wash them by centrifugation at 300*g* for 8 min with RPMI 1640 and count the cells.

g. Retain 5×10^5 cells to assess the purity of the cells by flow cytometry (Section 2.5).

h. Freeze the T-cells at −80 °C as described in Section 2.8.1. These T-cells will be used in the proliferation assay in Section 5.

2.4 Isolation of $CD14^+$ cells from $CD2^-$ cells by positive selection

In our experience, frozen T-cells lack myeloid cells to sustain reliable proliferation in vitro. To supplement isolated T-cells (Section 2.4) with antigen presenting cells, autologous mature DC are differentiated from monocytes. These are isolated from the $CD2^-$ fraction (Section 2.2) by anti-CD14 magnetic bead selection. The purity of the isolated cells and the success of the separation are assessed by flow cytometry in Section 2.5.

a. Use $CD2^-$ cells isolated from whole blood (Section 2.2).
b. Count the cells and resuspend 10^7 cells in 60 μL PBS + 2% HS.
c. Per 10^7 cells add 20 μL FcR blocking reagent (Miltenyi, catalog ♯ 130-059-901) and 20 μL magnetic anti-CD14 beads (Miltenyi, catalog ♯ 130-050-201).
d. Mix well and incubate for 20 min at 4 °C.
e. Isolate monocytes by either manual or automatic magnetic columns. Keep the samples cool at all times using racks pre-cooled to 4 °C (e.g., Chill 15 Rack, Miltenyi, catalog ♯ 130-092-952). Follow the manufacturer's instruction if using manual separation. If the automatic AutoMACS separator is used, use the program possel_s.
f. The $CD14^+$ monocytes will be in the positive fraction. Wash the isolated cells by centrifugation at 300*g* for 8 min with RPMI 1640 and count them.
g. Retain 5×10^5 cells to assess the purity of the cells by flow cytometry (Section 2.5).
h. Use the $CD14^+$ monocytes to generate mature dendritic cells (Section 2.6).

2.5 Control of cell purity by flow cytometry

The purity of T-cells and monocytes is controlled by flow cytometry and compared to the individual steps of the isolation process (Fig. 4). Use own flow cytometry systems, colors and compensation controls. This laboratory uses a BD Fortessa and Diva software.

T-cells should be $CD2^+$, $CD3^+$ and (in part) $CD8^+$, and $CD14^-$ and $CD56^-$. NK cells should be $CD2^+$, $CD8\alpha^+$ and $CD56^+$, and $CD3^-$ and $CD14^-$. Monocytes should be $CD14^+$, and $CD2^-$, $CD3^-$, $CD8^-$ and $CD56^-$.

Samples

i. Unstained control
ii. PBMC
iii. $CD2^+$ cells
iv. $CD56^+$ NK cells
v. $CD56^-$ T-cells
vi. $CD2^-$ cells
vii. $CD14^+$ monocytes
viii. $CD14^-$ cells

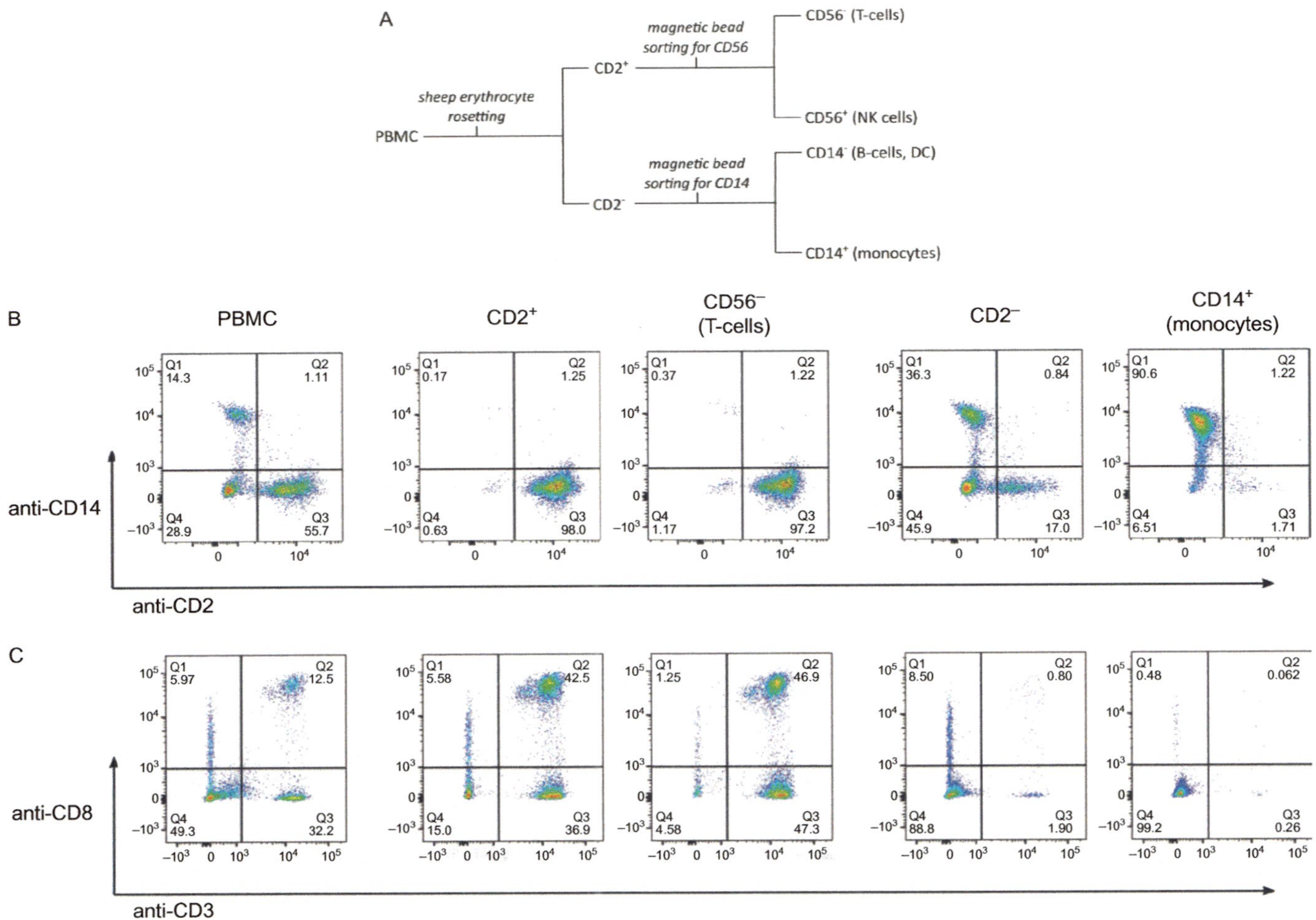

Fig. 4 See legend on next page.

Antibody-panel (BC = Beckman Coulter, BD = Becton Dickinson, BL = BioLegend):

1.	CD2	Alexa Fluor 647	clone LT2	Serotec, MCA1194A647
2.	CD3	Alexa Fluor 488	clone HIT3a	BL, 300319
3.	CD8α	PE	clone RPA-T8	BD, 555367
4.	CD14	PerCP	clone MoP9	BD, 345786

Procedure

a. The staining buffer contains PBS + 2% HS. Alternatively, the Brilliant Stain Buffer (BD, catalog # 563794) can be used.

b. Pellet the cells by centrifugation at 300*g* for 8 min. Aspirate the supernatant and resuspend the cells at 5×10^5 cells/mL (200 μL) in staining buffer. Seed 100 μL per sample into a 96-well plate or flow cytometry tubes. Add 100 μL of either staining buffer or antibody mix according to manufacturer's instructions to the cells. Incubate at RT for 20 min. Wash twice with 200 μL staining buffer by centrifugation at 300*g* for 5 min. Resuspend in 200 μL of staining buffer and proceed with flow cytometry analysis.

2.6 Generation of mature DC

Pure T-cells by themselves showed poor proliferative capacity after thawing as remaining myeloid cells do not survive the process. To overcome this problem, we generate and freeze autologous mature DC. These are thawed on the day of the proliferation assay (Section 2.7). This section describes the generation of mature DCs from monocytes over 7 days.

a. The complete medium for this section is RPMI 1640 + 10% FBS + 2 mM L-glutamine + 30 μg/mL penicillin + 100 μg/mL streptomycin. The staining buffer contains PBS + 2% HS.

Fig. 4 Assessment of T-cell and monocyte purity by flow cytometry after isolation. (A) Diagram of the separation workflow to obtain autologous T-cells and monocytes from whole blood. (B) Analysis of the expression of CD2 (NK and T-cells) and CD14 (monocytes) by PBMC and subsequently isolated populations. (C) Expression of CD3 (T-cells) and CD8 (NK and T-cells) by PBMC and subsequently isolated populations.

b. Use fresh CD14^{+} monocytes (Section 2.4). Count the cells and resuspend them at 1×10^6 cells/mL in complete medium pre-warmed to 37 °C.

c. Gently seed 1 mL (=1×10^6 cells) per well into a 12-well tissue culture plate.

d. Gently add 1 mL of complete medium supplemented with 200 ng/mL GM-CSF (final concentration: 100 ng/mL) and 20 ng/mL IL-4 (final concentration: 10 ng/mL). Incubate the cells at 37 °C, 5% CO_2 for 2 days.

e. On the second day of incubation, gently add 2 mL complete medium supplemented with 100 ng/mL GM-CSF (final concentration: 50 ng/mL) and 20 ng/mL IL-4 (final concentration: 10 ng/mL). Incubate the cells at 37 °C, 5% CO_2 for 3 days.

f. On the fifth day of incubation, add 50 ng/mL GM-CSF, 10 ng/mL IL-4 and 1 μg/mL biomunyl (Zambon, G00132) or 1 μg/mL LPS (Sigma) directly. Incubate the cells at 37 °C, 5% CO_2 for 2 days.

g. On the seventh day of incubation, visually inspect the cells for morphology using light microscopy. The cells should be non-adherent to the plastic and display dendrites.

h. Gently harvest the cells in 50 mL Falcons using 5 mL pipettes. Wash each well with 1 mL RPMI 1640. Count the cells and wash by centrifugation at 300*g* for 8 min.

i. Resuspend the cells at 1×10^6 cells/mL in RPMI 1640. Remove 5×10^5 cells for phenotypic assessment by flow cytometry. Freeze the rest at −80 °C as described in Section 2.8.2.

j. To control the cells' phenotype, pellet the cells by centrifugation at 300*g* for 8 min. Aspirate the supernatant and resuspend the cells at 2.5×10^5 cells/mL (200 μL) in staining buffer. Seed 100 μL per sample into a 96-well plate or flow cytometry tubes. Add 100 μL of either staining buffer or antibody mix to the cells. Incubate on ice for 20 min. Wash twice with 200 μL staining buffer by centrifugation at 300*g* for 5 min. Resuspend in 200 μL staining buffer and proceed with flow cytometry analysis.

k. Use own flow cytometry systems, colors and compensation controls. This laboratory uses a BD Fortessa and Diva software.

l. The mature dendritic cells are CD3^{-}, CD19^{-} and CD14^{-}, and CD1a^{+}, CD11b/c^{+} and CD83/86^{+} (Fig. 5).

m. Samples
   i. Unstained control
   ii. Stained sample

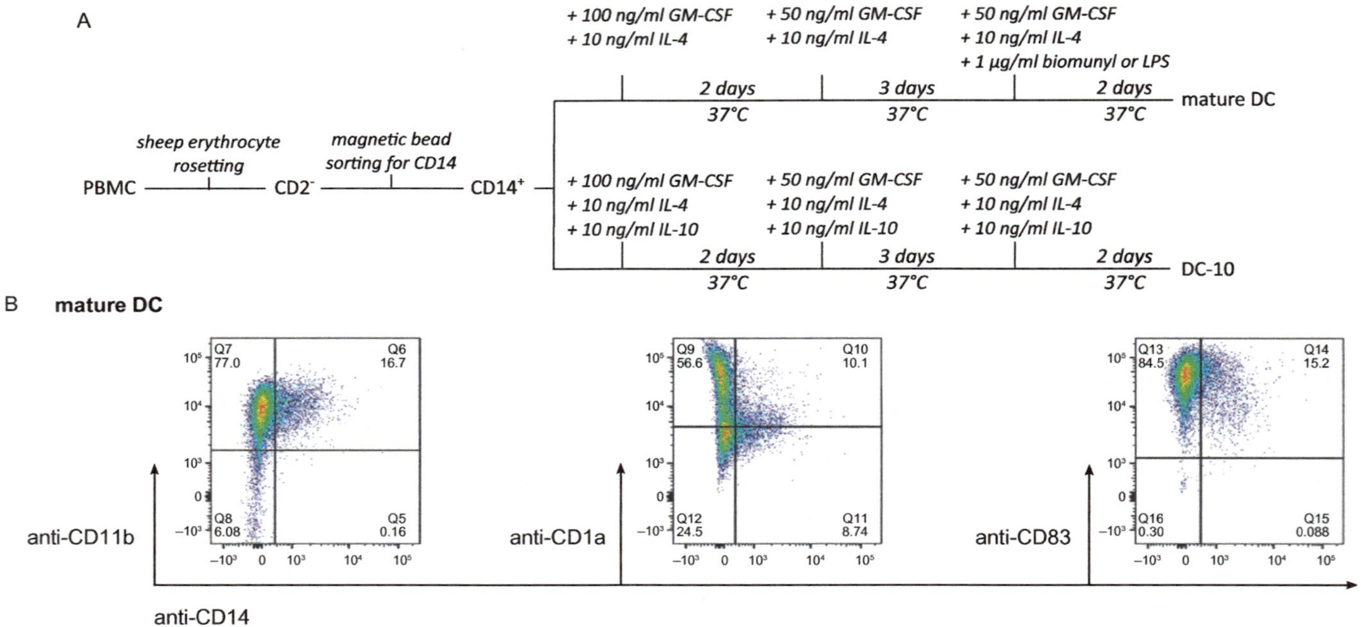

Fig. 5 Phenotypes of mature DC and DC-10s. (A) Workflow of DC differentiation protocol. Phenotypes of (B) mature DC and (C) DC-10 after 7 days of differentiation.

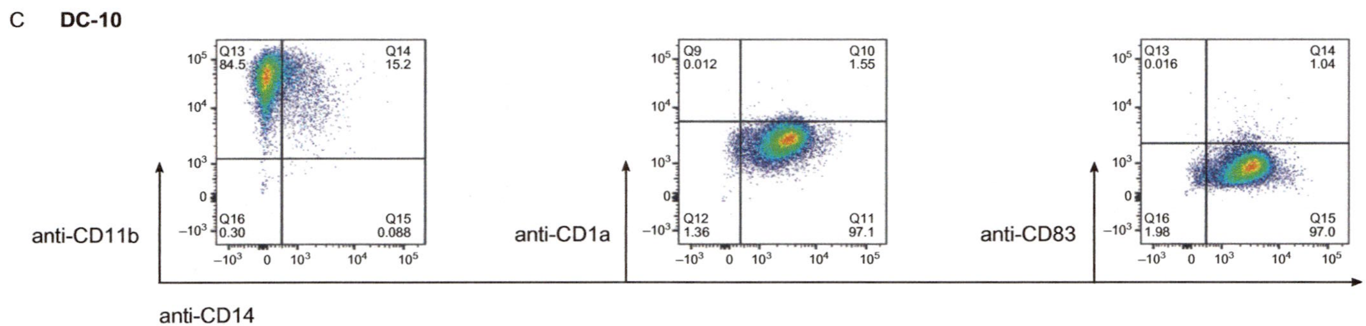

C
DC-10
anti-CD11b
anti-CD14
anti-CD1a
anti-CD83
Q13
84.5
Q14
15.2
Q16
0.30
Q15
0.088
Q9
0.012
Q10
1.55
Q12
1.36
Q11
97.1
Q13
0.016
Q14
1.04
Q16
1.98
Q15
97.0

Fig. 5—Cont'd

Antibody-panel (BC = Beckman Coulter, BD = Becton Dickinson, BL = BioLegend):

1.	CD3 (lineage control—T-cells)	Alexa Fluor 488	clone HIT3a	BL, 300319
2.	CD19 (lineage control—B-cells)	BV510	clone SJ25C1	BD, 562947
3.	CD1a	PE-Cy5	clone HI149	BD, 555808
4.	CD11b	PE	clone ICRF44	BD, 555388
5.	CD14	APC-Cy7	clone HCD14	BL, 325619
6.	CD83	APC	clone HB15a	BC, IM2218

2.7 Immunosuppressive DC-10

We recommend validating the proliferation assay with immunosuppressive DC-10 as a suppression control. Thus, if cells isolated as MDSC are not suppressive in the system, the procedure itself can be excluded as a cause. This section describes the generation of immunosuppressive DC-10 from monocytes (Gregori et al., 2010).

a. The complete medium for this section is RPMI 1640 + 10% FBS + 2 mM L-glutamine + 30 μg/mL penicillin + 100 μg/mL streptomycin. The staining buffer contains PBS + 2% HS.

b. Use the freshly $CD14^+$ monocytes (Section 2.4). Count the cells and resuspend them at 1×10^6 cells/mL in complete medium pre-warmed to 37 °C.

c. Gently seed 1 mL (=1×10^6 cells) per well into a 12-well tissue culture plate.

d. Gently add 1 mL complete medium supplemented with 200 ng/mL GM-CSF (final concentration: 100 ng/mL), 20 ng/mL IL-4 (final concentration: 10 ng/mL) and 20 ng/mL IL-10 (final concentration: 10 ng/mL, Invitrogen). Incubate the cells at 37 °C, 5% CO_2 for 2 days.

e. On the third day of culture, gently add 2 mL complete medium supplemented with 100 ng/mL GM-CSF (final concentration: 50 ng/mL), 20 ng/mL IL-4 (final concentration: 10 ng/mL) and 20 ng/mL IL-10 (final concentration: 10 ng/mL) without further manipulation to avoid cell activation. Incubate the cells at 37 °C, 5% CO_2 for additional 2 days.

f. On the fifth day of culture, add 50 ng/mL GM-CSF, 10 ng/mL IL-4 and 10 ng/mL IL-10 directly to the wells without further manipulation to avoid cell activation. Incubate the cells at 37 °C, 5% CO_2 for additional 2 days.
g. On the seventh day of culture, visually inspect the cells for morphology using light microscopy. The cells should be non-adherent to the plastic or other cells, and display dendrites.
h. Gently harvest the cells into 50 mL Falcons using 5 mL pipettes. Wash each well with 1 mL RPMI 1640. Count the cells and wash by centrifugation at 300*g* for 8 min.
i. Resuspend the cells at 1×10^6 cells/mL in RPMI 1640. Remove 5×10^5 cells for phenotypic assessment by flow cytometry. Freeze the rest at −80 °C as described in Section 2.8.2.
j. To control the DC-10 phenotype, pellet the cells by centrifugation at 300*g* for 8 min. Aspirate the supernatant and resuspend the cells at 2.5×10^5 cells/mL (200 μL) of staining buffer. Seed 100 μL per sample into well of a plate or flow cytometry tubes. Add 100 μL of either staining buffer or antibody mix to the cells. Incubate on ice for 20 min. Wash twice with 200 μL of staining buffer by centrifugation at 300*g* for 5 min. Resuspend in 200 μL of staining buffer and proceed with flow cytometry analysis.
k. Use own flow cytometry systems, colors and compensation controls. This laboratory uses a BD Fortessa and Diva software.
l. The DC-10 cells are $CD3^-$, $CD19^-$, $CD1a^-$, and $CD14^+$ $CD83^+$ $CD86^+$, $CD11b^+$ and $CD11c^+$ (Fig. 5).
m. Samples
  i. Unstained control
  ii. Stained sample

Antibody-panel (BC = Beckman Coulter, BD = Becton Dickinson, BL = BioLegend):

1.	CD3 (T cells)	Alexa Fluor 488	clone HIT3a	BL, 300319
2.	CD19 (B cells)	BV510	clone SJ25C1	BD, 562947
3.	CD1a	PE-Cy5	clone HI149	BD, 555808
4.	CD11b	PE	clone ICRF44	BD, 555388
5.	CD14	APC-Cy7	clone HCD14	BL, 325619
6.	CD83	APC	clone HB15a	BC, IM2218

2.8 Freezing and thawing of cells

In a multi-center study, the MDSC should ideally be the only variable parameter. To achieve this, the T-cells and mature DC isolated in Sections 2.3 and 2.6 are shared with each laboratory. Here, we describe the freezing and thawing protocols.

2.8.1 T-cells

a. Prepare freezing serum, which is composed of HS + 20% DMSO, and store it at 4 °C. Use the freezing serum at 4 °C.
b. Count T-cells (refer to Section 2.3) and centrifuge cells at 400*g* for 7 min.
c. Resuspend cells in cold RPMI without serum at 10^7 cells/mL.
d. Dilute the cell suspension 1:1 with freezing serum for a final concentration of 5×10^6 cells/mL.
e. Quickly transfer 1 mL to each cryovial.
f. Transfer the vials to −80 °C. We advise keeping the freezing process as temperature-controlled as possible. To achieve that, we recommend using Mr. Frosty™ Freezing Container (Thermo Scientific™ 5100-0001).
g. Thaw the T-cells the day before the planned proliferation assay. Rest the cells at 37 °C, 5% CO_2 overnight. Use FBS-coated 15 mL Falcon tubes and 48-well tissue culture plates.
h. For thawing the cells, use 12 mL pre-warmed RPMI 1640 + 5% HS + 30 μg/mL penicillin + 100 μg/mL streptomycin + 5 U/mL DNase I (Worthington) + 5 U/mL IL-2.
i. Thaw the cells half-way by holding the cryovial into a warm water-bath. Resuspend the half-thawed drop-wise cells in 10 mL warm medium.
j. Centrifuge the cells at 300*g* for 8 min. Aspirate the supernatant and resuspend the cells in 1 mL medium +5 U/mL IL-2. Transfer the cells into a 48-well plate and incubate at 37 °C, 5% CO_2 overnight.

2.8.2 Dendritic cells

a. Prepare freezing serum, which is composed of FBS + 20% DMSO and store it at 4 °C. Use freezing serum at 4 °C.
b. Count DC (refer to Sections 2.6 and 2.7) and centrifuge cells at 400*g* for 7 min.
c. Resuspend cells in cold RPMI without serum at 1×10^6 cells/mL.
d. Dilute the cell suspension 1:1 with freezing serum for a final concentration of 5×10^5 cells/mL.
e. Quickly transfer 1 mL to each cryovial.

f. Transfer the vials to −80 °C. We advise keeping the freezing process as temperature-controlled as possible. To achieve that, we recommend using Mr. Frosty™ Freezing Container (Thermo Scientific™ 5100-0001).

g. Transfer cells into liquid nitrogen after 24 h.

h. Thaw the mature dendritic cells on the day of the planned proliferation assay. Rest the cells at 37 °C, 5% CO_2 for at least 4 h before use. Use FBS-coated 15 mL Falcon tubes and 48-well tissue culture plates.

i. For thawing the cells, use 12 mL pre-warmed RPMI 1640 + 10% FBS + 30 μg/mL penicillin + 100 μg/mL streptomycin + 5 U/mL DNase I.

j. Thaw the cells half-way by holding the cryovial into a warm water-bath. Resuspend the half-thawed cells drop-wise in 10 mL warm medium.

k. Centrifuge the cells at 300*g* for 8 min. Aspirate the supernatant and resuspend the cells in 1 mL medium supplemented with 50 ng/mL GM-CSF and 10 ng/mL IL-4 for mature DC, and 50 ng/mL GM-CSF, 10 ng/mL IL4 and 10 ng/mL IL-10 for DC-10, respectively. Transfer the cells into a 48-well plate and incubate at 37 °C, 5% CO_2 for at least 4 h.

3. MDSC isolation

The aim of the protocol presented is to compare the suppressive function of MDSC isolated from different tissues and pathologies and in different research groups. Since there is no single marker that distinguishes MDSC, the isolation strategies reported vary greatly among laboratories. Here, we describe the methods used by Annika Bruger and Pierre van der Bruggen (Brussels) to isolate MDSCs from the blood, ascites and tumor tissue of cancer patients. Each of these tissues requires slightly different preparation steps, e.g., tumors or tumor biopsies specifically need to be dissociated. All isolation methods require density gradient separation steps to isolate low-density granulocytic MDSC from high-density granulocytes.

The MDSC are sorted from the tissue samples by flow cytometry (Section 3.4). The same panel of markers and gating strategy for all tissue samples should be used. The anti-CD206 marker for macrophages can be replaced with an additional granulocytic marker, anti-CD66b, in blood samples. This serves as an extra check on the identity of granulocytic MDSC.

3.1 MDSC preparation from blood

Human MDSC are most commonly isolated from the blood of cancer patients. The MDSC are isolated from the mononuclear layer following a density gradient separation. Furthermore, no CD45 stain is necessary to isolate hematopoietic cells. Additional cell types such as dendritic cells or early MDSC could be isolated instead. However, the sample sizes obtainable with blood are usually small, ranging from 20 to 100 mL, and MDSC presence in blood is rare (0.01–0.1% of $CD45^+$ cells) (own observations). This limits the yields of MDSC and thus the range of viable subsequent experiments.

a. Dilute whole blood 1:3 with PBS + 1 mM EDTA at room temperature. Do not use cold PBS + 1 mM EDTA as this might activate granulocytes and gMDSC.
b. Perform density gradient separation on the PBMC suspension as previously described in Section 2.1, steps b and c.
c. Following the density gradient separation, the PBMC settle in a thin layer below the serum, and above the Lymphoprep solution and erythrocyte pellet (Fig. 2). Carefully remove only the PBMC layer with a pipette and place it into an FBS-coated 50 mL Falcon tube.
d. Wash the cells with PBS + 1 mM EDTA and centrifuge at 300*g* for 10 min.
e. The PBMC settle in the pellet. Carefully remove the supernatant and as much of the white platelet ring covering the PBMC pellet as possible.
f. Repeat the washing steps with 50 mL PBS + 1 mM EDTA twice. Resuspend the PBMC a third time and centrifuge them at 200*g* for 3 min if platelets still remain.
g. Refer to MDSC staining protocol in Section 3.4.

3.2 MDSC preparation from ascites

Some cancers, in particular ovarian cancers, lead to an accumulation of fluid, called ascites, in the abdomen of patients in later stages of disease. In the clinic, ascites are usually collected and discarded. Ascites can be obtained in large volumes (up to 10 L, own experience) and can contain up to one trillion cells (including tumor cells). This enhances the MDSC yields obtainable. Cells from ascites are harvested by centrifugation and the MDSC are isolated from the mononuclear layer following density separation. A stain for CD45 (hematopoietic cells) is required, as epithelial or cancer cells could be present. Furthermore, due to the advanced stage of disease in most patients, patient consistency is difficult to maintain.

a. Pour the ascitic fluid into 200 mL flasks. Pellet the cells by centrifugation at 400*g* for 25 min. Aspirate the supernatant and repeat the harvesting step until no ascitic fluid remains.

b. Pool the cells, collect them in an FBS-coated 50 mL Falcon tube and wash with PBS + 1 mM EDTA and centrifuge at 400*g* for 8 min. Do not use cold PBS + 1 mM EDTA, as this might activate granulocytes and gMDSC.

c. Should the pellet contain many erythrocytes, lyse them with 3 mL 0.2% NaCl by resuspending the cells in the solution carefully and thoroughly. After 50–60 s of lysis reconstitute osmolarity with 7 mL 1.2% NaCl.

d. Resuspend the cells in warm RPMI 1640 to a total volume of 35 mL.

e. Perform density gradient separation on the cell suspension as previously described in Section 2.1, steps b and c.

f. Following the density gradient separation, the mononuclear cells settle in a thin layer below the medium, and above the Lymphoprep solution and erythrocyte pellet (in ascites this can vary in size, depending on how many erythrocytes are suspended in the ascitic fluid). Carefully remove only the PBMC layer with a pipette and place it into an FBS-coated 50 mL Falcon tube.

g. Wash the cells with warm PBS + 1 mM EDTA and centrifuge at 300*g* for 10 min.

h. The PBMC settle in the pellet. Carefully remove the supernatant.

i. Repeat the washing steps with 50 mL PBS + 1 mM EDTA twice.

j. Refer to staining protocol in Section 3.4.

3.3 MDSC preparation from solid tissue

Samples from tumors or tumor biopsies are rare, difficult to obtain and usually small, which limits the expected MDSC yield. Moreover, solid tissue samples require the most manipulation for MDSC isolation, as the tissue has to be gently dissociated. MDSC are isolated from the mononuclear fraction following density gradient separation. A stain for CD45 is required, as epithelial or cancer cells could be present.

a. We work with a hospital in very close proximity, and we receive tumor samples within 30 min after extraction. The specimens are placed in a 50 mL flask in PBS and transported on ice.

b. Place the tumor and any accompanying PBS into a sterile Petri dish on ice underneath a flow hood. Cut the tumor into very small pieces (1–2 mm) using a sterile scalpel. The pieces should be small enough to not block a 10 mL pipette.

c. Place the tumor pieces and fluid into a MACS dissociator C tube (Miltenyi, catalog ♯ 130-093-237). Add DNase I at 5 U/mL and Liberase™ TL and DL (thermolysin-dispase-low, both Sigma and Roche, catalog ♯ 05401020001 and LIBDL-RO) at 0.5 and 0.26 U/mL, respectively.
d. Dissociate the tumor cells using the pre-set program "h_tumor_01" on the gentleMACS™ dissociator (Miltenyi, catalog ♯ 130–093-235) for 45 s.
e. Incubate the tumor cells in the dissociator tube under agitation for 45 min at 37 °C.
f. Dissociate the tumor cells using the dissociator's pre-set program "h_tumor_01" for 45 s.
g. Transfer the tumor cells into an FBS-coated 50 mL Falcon tube. Rinse the dissociation tube with equal volume of human serum and add this to the tumor cells in the Falcon tube. The human serum stops the digestion by the liberase. Pellet the cells by centrifugation at 400*g* for 10 min.
h. Resuspend the tumor cells with 50 mL PBS + 1 mM EDTA per Falcon tube and pass the cells successively through a 70 and 40 μm filters, washing the cells by centrifugation at 400*g* for 8 min in between. Count and collect the cells through centrifugation at 400*g* for 8 min.
i. Resuspend the cells in warm RPMI 1640 to a total volume of 35 mL.
j. Perform density gradient separation on the cell suspension as previously described in Section 2.1, steps b and c.
k. Following the density gradient separation the mononuclear cells settle in a thin layer below the medium, and above the Lymphoprep solution and erythrocyte pellet (which is expected to be small in tumors). Carefully remove only the PBMC layer with a pipette and place it into an FBS-coated 50 mL Falcon tube.
l. Wash the cells with warm PBS + 1 mM EDTA and centrifuge at 300*g* for 10 min.
m. The PBMC settle in the pellet. Carefully remove the supernatant.
n. Repeat the washing steps with 50 mL PBS + 1 mM EDTA twice.
o. Refer to MDSC staining protocol in Section 3.4.

3.4 MDSC isolation by flow cytometry sorting

No single marker specific for MDSC has yet been discovered. Gating and isolation strategies for human MDSC thus vary widely within the literature. Bronte et al. (2016) proposed general and minimal guidelines for the detection and isolation of human MDSC. Working along these guidelines, we present here the gating and sorting strategy used in the van der Bruggen group as an example. Up to four populations can be sorted simultaneously by flow cytometry with a FACS Aria III (Fig. 6).

a. The staining buffer contains PBS + 2% HS. Alternatively, the Brilliant Stain Buffer (BD, catalog# 563794) can be used.

b. Count the cells and resuspend in staining buffer +20% FcR blocking reagent (Miltenyi, catalog ♯ 130-059-901) in 50 μL per 10^7 cells.

c. While the cells are incubating with the FcR blocking reagent prepare the staining mix for the MDSC isolation in 50 μL of staining buffer per 10^7 cells.

d. Adapt the panel to in-house requirements for machine settings and compensation matrices. Shown below is the panel used in the van der Bruggen group (Brussels) as an example.

e. Samples

  i. Unstained control

  ii. FMO controls

  iii. Full stain

  Antibody-panel (BC = Beckman Coulter, BD = Becton Dickinson, BL = BioLegend):

1.	Live cell stain	efluor 780		Affimetrix
2.	Lineage stain:			
	CD3 (T-cells)	BV510	clone HIT3a	BD 564713
	CD19 (B-cells)	BV510	clone SJ25C1	BD 562947
	CD56 (NK cells)	BV510	clone NCAM16.2	BD 563041
3.	HLA-DR	PE-Cy7	clone L243	BL 307616
4.	CD11b	APC	clone ICRF44	BD 550019
5.	CD14	PerCP	clone MΦP9	BD 345786
6.	CD15	PE	clone HI98	BD 555402
7.	CD16	BV421	clone 3G8	BD 562878
8.	CD33	BV711	clone WM53	BD 563171
9.	CD45	Alexa-Fluor 700	clone HI30	BD 304024
10.	CD66b (blood)	FITC	clone 80H3	BC IM0531U
	or CD206 (tissues)	FITC	clone 19.2	BD 551135

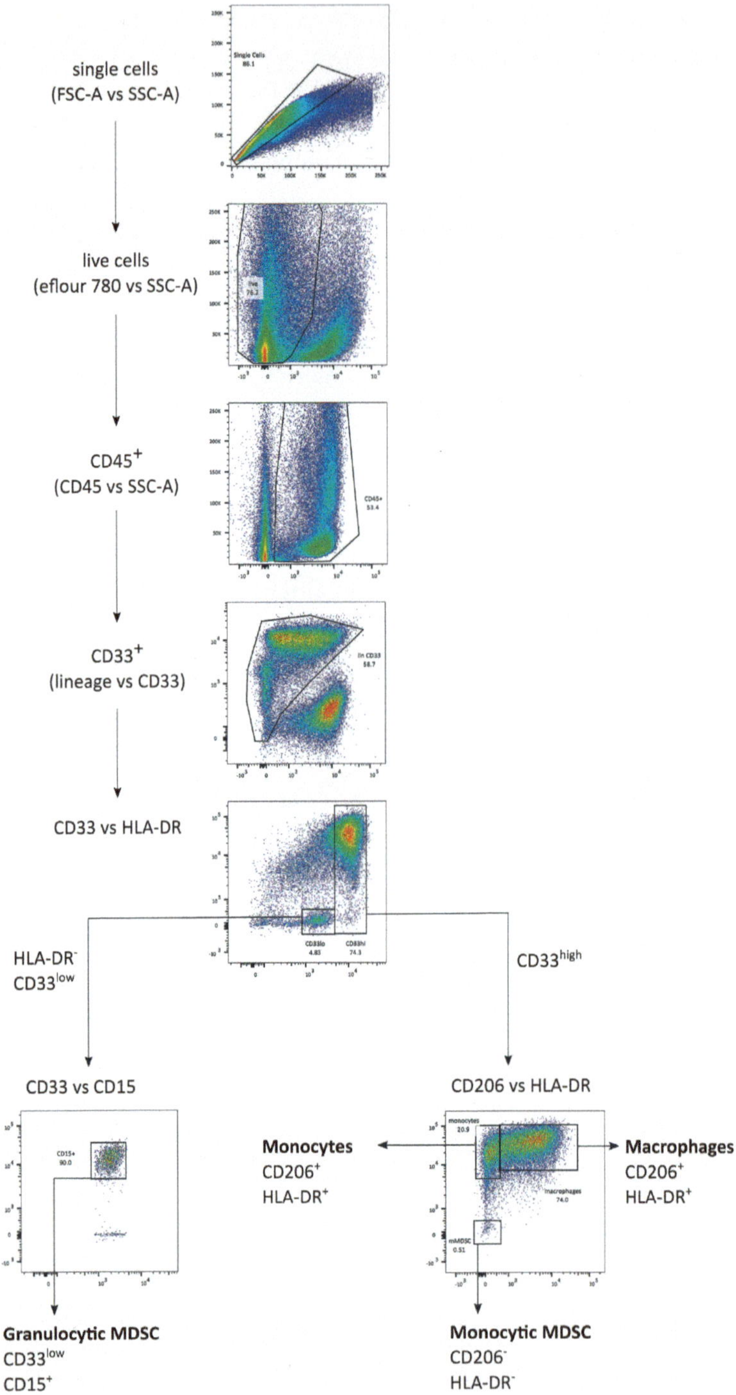

Fig. 6 MDSC gating and sorting strategy. Gating strategy employed in van der Bruggen laboratory in Brussels to isolate MDSC, monocytes and macrophages from blood, tumor and ascites. Further markers, such as CD11b, CD14, CD16 and CD66b, are used but not shown.

f. Incubate cells with 50 μL antibody mix per 10^7 cells (total: 100 μL) at RT in the dark for 20 min.
g. Wash the cells twice by centrifugation at 300*g* for 7 min with staining buffer.
h. Resuspend the cells at 500 μL per 10^7 cells or according to the sorter's cell density specifications in staining buffer.
i. This laboratory uses a BD FACS Aria III on the lowest speed setting with the largest nozzle, and Diva software.
j. For the gating strategy refer to Fig. 6.
k. Collect the MDSC, monocytes and macrophages in FACS tubes pre-coated with 1 mL FBS. Refer to Table 3 for an example of the yields obtainable from all three tissues from one ovarian cancer patient. Mind factors that could influence cell purity after the sorting procedure, including isolating multiple cell types simultaneously, debris, and cancer cells.
l. Use MDSC immediately. The cells, particularly gMDSC, cannot be frozen and have a short life span ex vivo.

4. Labeling of T-cells with fluorescent tracking dye

T-cell proliferation can be assessed in various ways. The most popular methods are labeling T-cells with radioactive thymidine or with fluorescent tracking dyes. For multi-center studies, we propose to use the latter. The T-cells are labeled with tracking dyes such as CFSE or CMFDA at the

Table 3 Example of monocytic cell yields from one ovarian cancer patient.

		Blood	**Tumor**	**Ascites**
Cells sorted	*(input)*	18 million	50 million	15 million
Monocytes	Yield	72,900	12,300	40,100
	Purity (% CD45$^+$)	*91%*	*92%*	*91%*
Macrophages	Yield	–	377,000	24,100
	Purity (% CD45$^+$)	–	*96%*	*93%*
Monocytic MDSC	Yield	487	6400	1200
	Purity (% CD45$^+$)	–	*38%*	–
Granulocytic MDSC	Yield	39,700	60,500	24,800
	Purity (% CD45$^+$)	*72%*	*88%*	*85%*

beginning of the experiment. Similar chemical properties mean these dyes are highly suitable to track cell proliferation. First, acetate groups on the dyes allow their entrance into the cells. Within the cytosol, esterases cleave off acetate groups on the dyes, rendering them cell impermeable. Second, the dyes covalently bond to several cytosolic proteins, including some with low turnover rates. Both events result in a stable high fluorescence within the cells. With each division, the dyes' intensities are halved between the two daughter cells. Thus, progressive rounds of division can be monitored by flow cytometry through the dilution of the tracking dye. With optimal labeling, up to eight peaks (or seven rounds of division) can be observed with CFSE (Baker, Sullam, & Levin, 1997; Quah & Parish, 2012; Quah, Warren, & Parish, 2007). T-cells can divide once every 24h (own observation). In our experience, the best results are obtained after 4 days of incubation, or about three rounds of divisions. At this point, the dye dilution results in nicely separated peaks and proliferating cells are not depleted of nutrients yet.

a. Use T-cells isolated in Section 2.3 and thawed 1 day in advance as described in Section 2.8.

b. Complete medium is RPMI 1640 + 5% HS + 30 μg/mL penicillin + 100 μg/mL streptomycin + 5 U/mL IL-2.

c. Count the T-cells. Collect the cells in a 15 mL Falcon tube and pellet the cells by centrifugation at 300*g* for 8 min. Discard the supernatant and resuspend the cells in warm RPMI 1640 at a concentration of 2×10^6 cells/mL.

d. Dilute the tracking dye (we use CMFDA, Life Technologies, at 0.5 μM/mL final concentration) in warm RPMI 1640. Optimize to own specific flow cytometry settings.

e. Add diluted tracking dye to the T-cells at a ratio of 1:1. Mix gently and incubate for 20 min at 37 °C in the dark. Mix gently occasionally.

f. Fill the Falcon tube with complete medium and wash the labeled T-cells once by centrifugation at 300*g* for 8 min. Resuspend the T-cells at 2×10^6 cells/mL in complete medium. Rest the cells at 37 °C, 5% CO_2 until seeding for the proliferation assay (proceed to Section 5.2).

5. Preparation of assay

T-cell proliferation assays contain many variables that diverge widely among different laboratories. The possible variables include the type of assay plate, the type of medium, the amount of IL-2 supplemented, the type of stimulation used, and the type of T-cell chosen (Bruger et al., 2019). We present our protocol set-up for multi-center use next.

5.1 Preparation of the plate for T-cell stimulation

a. Complete medium consists of RPMI 1640 + 5% HS + 30 μg/mL penicillin + 100 μg/mL streptomycin + 5 U/mL IL-2. Filter medium and warm to 37 °C before use. No supplementation with arginine, asparagine and tryptophan is required since RPMI 1640 already contains concentrations of the amino acids exceeding natural conditions. Low amounts of IL-2 are added to prevent the T-cells from starving of the cytokine, and to avoid an IL-2 sink effect.

b. The proliferation assay is assembled from components prepared at different times. The wells of the assay contain in 200 μL total (Fig. 7):

  i. *T-cells*: labeled, 30,000 or 50,000 per well (Section 4)

  ii. *Mature DC*: 1000 per well (Sections 2.6 and 2.8)

  iii. *anti-CD28 antibodies*: soluble, 2 μg/mL final concentration (Section 5.1, step h)

  iv. *MDSC/myeloid cells/DC-10*: variable concentrations (Section 3)

c. The T-cells are stimulated through anti-human CD3 (clone OKT3, BioLegend, catalog # number 317315) and anti-CD28 (clone CD28.2, BD Pharmingen, catalog # 555725) antibodies. These antibodies target the TCR and the co-stimulatory molecule CD28, respectively. The anti-CD3 antibody is immobilized to the tissue culture plate as T-cells require the stable cross-linking of TCRs and prolonged dwell times for activation. The anti-CD28 antibody is supplied in solution.

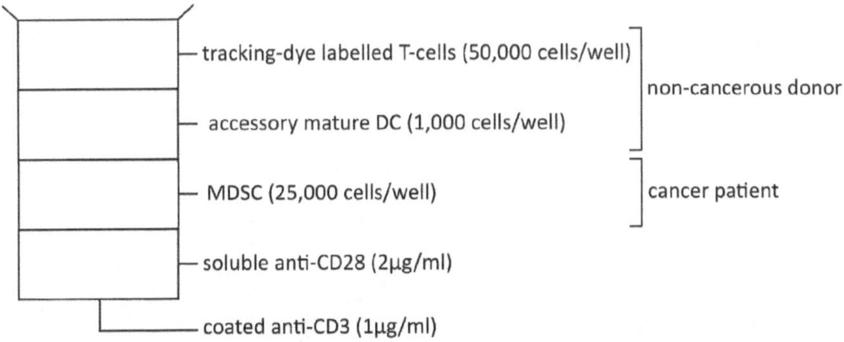

Fig. 7 Proliferation assay well design.

Controls

d. Control wells consist of the following:
  i. No stimulation, no MDSC, no accessory mature DC
  ii. No stimulation, no MDSC, plus accessory mature DC
  iii. Anti-CD3-anti-CD28 stimulation, no MDSC, no accessory mature DC
  iv. Anti-CD3-anti-CD28 stimulation, no MDSC, plus accessory mature DC

e. Add complete medium to control wells instead of stimulation or myeloid cells.

Anti-CD3 stimulation

f. Coat the wells of the flat-bottomed 96-well tissue culture plate by seeding 1 μg/mL anti-CD3 (clone: OKT3) diluted in sterile PBS at 100 μL/well. Use sterile PBS only for control wells without stimulation. Incubate the plate at 37 °C for at least 4 h. Keep sterile.

g. Gently remove all solution from the wells before seeding cells or anti-CD28 antibodies. Do not wash wells.

Anti-CD28 stimulation

h. Dilute anti-CD28 in pre-warmed complete medium to a concentration of 8 μg/mL (4 × final concentration = 2 μg/mL) and seed 50 μL/well. Use complete medium without anti-CD28 antibody in control wells.

5.2 Cells

T-cells

a. Count tracking dye-labeled T-cells (Section 4) and resuspend at 1 × or 0.6×10^6 cells/mL in complete medium. Seed at 30,000 or 50,000 cells/well.

Mature dendritic cells

b. Count the mature DC thawed and rested in Section 2.8. Harvest 2×10^5 mature DC into a FBS-coated 15 mL Falcon tube and collect by centrifugation at 300*g* for 8 min. Resuspend the cells in 10 mL complete medium (20,000 cells/mL). Seed the cells at 1000 cells/well. Use complete medium instead of myeloid cells in control wells.

Isolated MDSC, monocytes and macrophages

c. Use count provided by sorting machine.

d. Pellet the MDSC in the FACS tubes by centrifugation at 300*g* for 8 min. Transfer, handle and centrifuge the cells as little as possible to avoid activating or losing the cells.

e. Resuspend the cells in complete medium as previously calculated. Seed the cells at desired concentration (e.g., 25,000 cells/well). Use complete medium instead of MDSC in control wells.

6. MDSC-T-cell co-culture and analysis of suppression

6.1 Incubation

a. The cells are co-incubated for 4 days at 37 °C, 5% CO_2. After 4 days about three cycles of divisions have been concluded.

b. The tracking dye is diluted by 50% with each cycle of division but remains constant in non-proliferating cells. Since flow cytometry analysis is usually performed on log scales, 2–3 division cycles result in a dilution of the tracking dye that provides a good separation between signals from proliferating and non-proliferating cells. Too few division cycles result in higher concentrations of the tracking dye in proliferating cells, which could make difficult separating proliferating from non-proliferating peaks. Too many division cycles result in a larger separation of the peaks, but also in higher cell concentrations per well and nutrient depletion from the medium. This could result in increased cell death.

c. Proceed to reading the results by flow cytometry (Section 6.2).

6.2 Flow cytometry analysis

a. Inspect wells visually by light microscopy.

b. The following procedures are not performed under sterile conditions.

c. Dilute FcR blocking reagent 1:10 in PBS + 2% HS. Add 20 μL per well and incubate on the bench for 5 min.

d. In the meantime, prepare the antibody master mix in PBS + 2% HS. Calculate adding a total volume of 20 μL per 200 μL well. Previously test optimal antibody dilutions or refer to manufacturer's instructions.

1.	Live cell stain	e.g., efluor 780		Affimetrix
2.	CD3	PE	clone UCHT1	BD 555333
3.	CD8	APC	clone RPA-T8	BD 555369
4.	Sphero™ Rainbow Calibration Particles			BD 556288

e. The Rainbow beads are added as normalization constant at 5000 beads/well.

f. Add 20 μL of antibody master mix per well. Mix by transferring cells to a 96-well plate adequate for flow cytometry acquisition. Incubate at room temperature in the dark for 15 min.

g. Wash twice by centrifugation at 300*g* for 5 min and resuspend cells in 200 μL PBS + 2% HS.

h. Proceed to flow cytometry acquisition according to in-house machine specifications and settings. This laboratory uses a Fortessa flow cytometer and Diva software.

6.3 Data analysis

The van der Bruggen laboratory (Brussels) uses FlowJo software v10.0.7 (Treestar, Ashland, OR, USA) for flow cytometry data analysis and Excel (Microsoft) for further analysis and presentation of numeric results. Alternatively, FlowJo's incorporated proliferation modeling tool and the associated proliferation index are commonly used for analysis.

Flow cytometry analysis (on FlowJo)

a. Use compensation controls acquired to create compensation matrix.

b. Gate successively for the lymphocytes (FSC/SSC), single cells (FSC-A/FSC-H), live T-cells (viability dye/anti-CD3), and the proliferating cells (tracking dye/histogram). Finally, gate on proliferating (tracking dye^{low}) and non-proliferating (tracking dye^{high}) populations (Fig. 8). An additional step for the $CD8^+$ T-cells can be inserted (anti-CD8/anti-CD3), followed by the assessment of the tracking dye dilution.

c. Create a gate for the Rainbow beads.

d. Apply the gates to all samples. Control all samples and the imposed gating. Correct if necessary.

e. Extract numerical data for the following parameters for each sample (e.g., as an Excel file):

- **i.** Number of Rainbow beads.
- **ii.** Number of live T-cells.
- **iii.** Number of non-proliferating cells (tracking dye high).
- **iv.** Median of non-proliferation cells (tracking dye high).
- **v.** Number of proliferating cells (tracking dye low).
- **vi.** Median of proliferation cells (tracking dye low).

Proliferation assay data analysis

a. Import FlowJo data as described in previous section:

- **i.** Number of Rainbow beads
- **ii.** Number of live T-cells

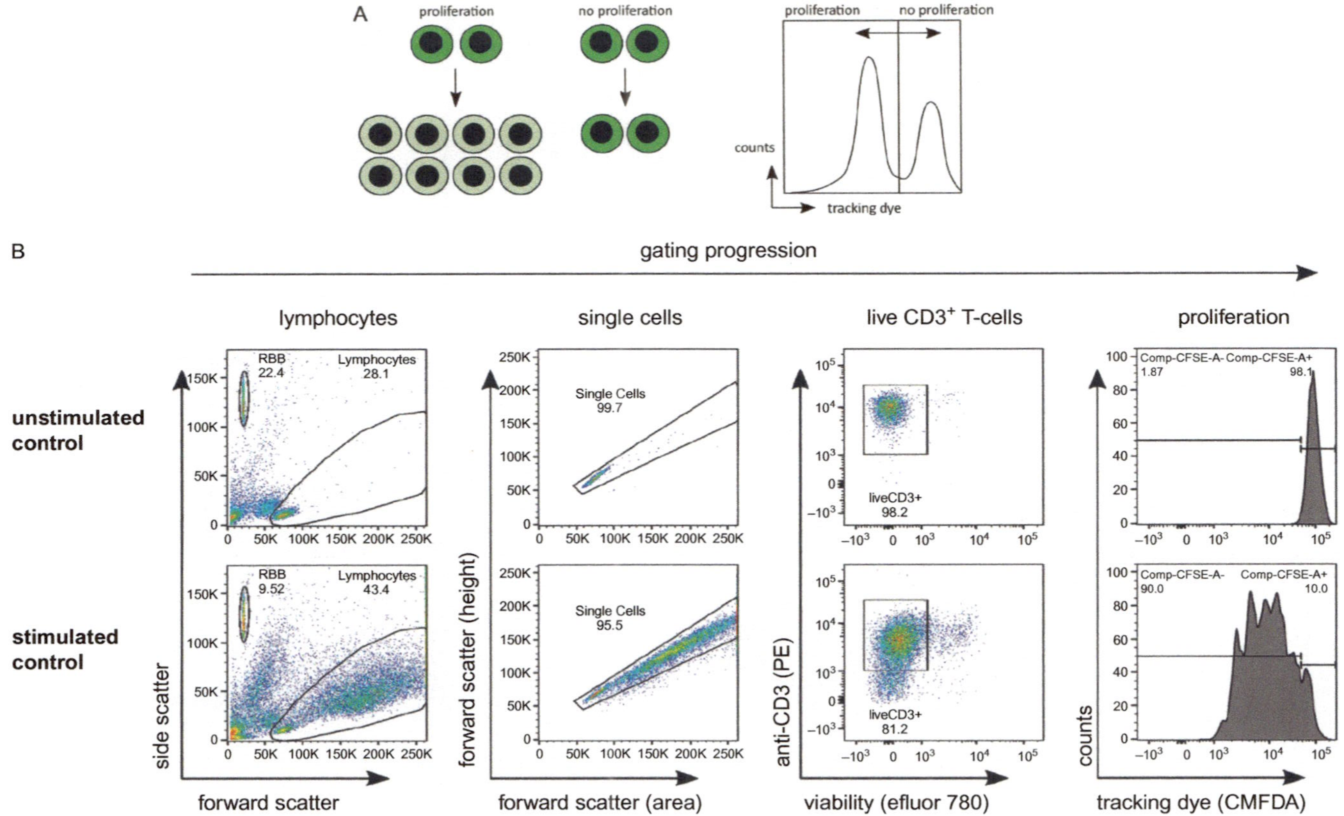

Fig. 8 Example of flow cytometry analysis of proliferation assay results. (A) Principle of proliferation assay. T-cells are labeled with a tracking dye (e.g., CFSE or CMFDA). The intensity of the dye decreases by a factor of two with each division within proliferating cells. Non-proliferating cells retain high concentrations of the dye. The results are acquired using flow cytometry. (B) Flow cytometry gating strategy. After 4 days the T-cells are stained with a viability dye, and anti-CD3 and anti-CD8 antibodies. First, lymphocytes are selected according to their cell size. Next, single cells and then live $CD3^+$ cells are selected. Finally, proliferation is assessed by tracking dye dilution. Extra steps can be inserted to focus on $CD8^+$ or $CD8^-$ cells specifically. (C) Example of the flow cytometry analysis of a MDSC-suppression assay.

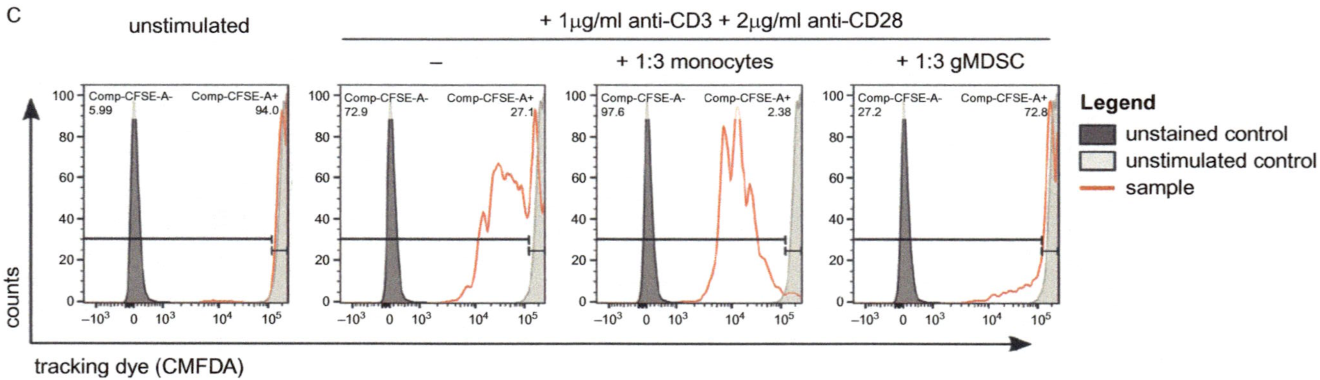
C
unstimulated
+ 1μg/ml anti-CD3 + 2μg/ml anti-CD28
–
+ 1:3 monocytes
+ 1:3 gMDSC
Comp-CFSE-A-
5.99
Comp-CFSE-A+
94.0
Comp-CFSE-A-
72.9
Comp-CFSE-A+
27.1
Comp-CFSE-A-
97.6
Comp-CFSE-A+
2.38
Comp-CFSE-A-
27.2
Comp-CFSE-A+
72.8
counts
tracking dye (CMFDA)
Legend
unstained control
unstimulated control
sample

Fig. 8—Cont'd

   iii. Number of non-proliferating cells (tracking dye^{high}).
   iv. Median of non-proliferation cells (tracking dye^{high}).
   v. Number of proliferating cells (tracking dye^{low}).
   vi. Median of proliferation cells (tracking dye^{low}).

b. Normalize the total live cell numbers, and the proliferating and non-proliferating counts to the 5000 Rainbow beads added per sample.
   i.

$$cell\ numbers = \frac{total\ lymphocyte\ count}{bead\ count} \times 10{,}000\ (\text{ref : total bead number})$$

c. Calculate:
   i.

$$number\ of\ divisions = log2\left(\frac{median\ non-proliferating\ cells}{median\ proliferating\ cells}\right)$$

   ii.

$$\#of\ initial\ cells\ that\ have\ undergone\ division = \frac{count\ proliferating\ cells}{2^{number\ of\ divisions}}$$

   iii.

$$\%\ of\ initial\ cells\ that\ have\ divided = \frac{count\ proliferating\ cells}{2^{number\ of\ divisions}} \bigg/ \left(count\ of\ non-proliferating\ cells + \frac{count\ proliferating\ cells}{2^{number\ of\ divisions}}\right) \times 100$$

(explanation: *initial cells that have divided/total number of cells at outset × 100)*

d. Present data as (Fig. 9):
   a. change in absolute cell numbers.
   b. number of divisions over time.
   c. percentage of dividing cells.

Comparing different experiments

a. To compare results from separate experiments set the proliferation of the stimulation controls (T-cells, + stimulation, no myeloid cells) as 1.
b. Normalize all other results to the control.

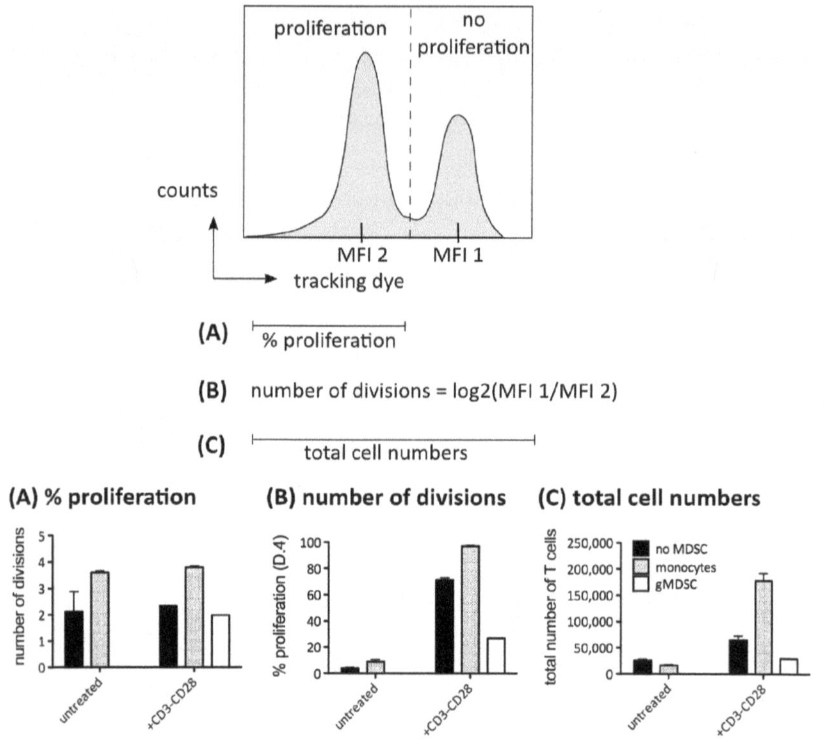

Fig. 9 Example of further analysis of proliferation assay results. Flow cytometry data is exported into excel files or similar. MDSC-mediated suppression of T-cell proliferation can then be assessed using different parameters. (A) The percentage of T-cells that proliferated. This can be calculated for the cells acquired at the end of the assay (Day 4, shown here) or calculated back to represent the fraction of the original cells in the assay (Day 0). (B) Fewer divisions can indicate that the cell cycle progression in inhibited. (C) The total number of cells present within wells after 4 days can indicate unexpected levels of apoptosis, either by strong levels of proliferation that depletes nutrients from the medium, or by interactions with myeloid cells.

7. Concluding remarks

The method presented here offers an approach to assess the suppressive functions of human MDSC through T-cell proliferation assays that can be applied to multi-center studies. At present, comparing studies of human MDSC functions is difficult due to the variation in the procedures of both MDSC isolation and function (Bruger et al., 2019). The reliable comparison of the function of MDSC isolated from different sources and in different research groups requires the MDSC themselves to be the only variable within the assay. We achieve this by isolating T-cells from allogeneic non-cancerous donors in one laboratory, freezing the cells and shipping them to participating groups. However, the isolated T-cells are poor proliferators after thawing. We solved this issue by the addition of low numbers of allogeneic in vitro generated mature DC that restore T-cell proliferation. Moreover, mature DC can be aliquoted, frozen, shared and thawed. Further aspects including the type of consumables and antibodies are standardized within the group. However, best shipping conditions for the cells still need to be determined to successfully validate the method in partner laboratories.

Acknowledgments

This chapter follows work supported by grant #7571 from the Région Wallonne Program BIOWIN: Convention IT-TARGETS (Belgium) and grant #1610119 from the Région Wallonne Program WALINNOV: Convention IMMUCAN (Belgium). Pierre van der Bruggen was supported by research program PDR T.0232.19 from the "Fonds de la Recherche Scientifique" (Belgium). Christophe Vanhaver was supported by a PhD fellowship grant from FRIA/FNRS (Belgium).

References

Amodio, G., Cichy, J., Conde, P., Matteoli, G., Moreau, A., Ochando, J., et al. (2019). Role of myeloid regulatory cells (MRCs) in maintaining tissue homeostasis and promoting tolerance in autoimmunity, inflammatory disease and transplantation. *Cancer Immunology, Immunotherapy*, *68*(4), 661–672. (Epub Oct 2018). https://doi.org/10.1007/s00262-018-2264-3.

Baker, G. R., Sullam, P. M., & Levin, J. (1997). A simple, fluorescent method to internally label platelets suitable for physiological measurements. *American Journal of Hematology*, *56*(1), 17–25. Retrieved from http://www.ncbi.nlm.nih.gov/pubmed/9298862.

Bronte, V., Brandau, S., Chen, S.-H., Colombo, M. P., Frey, A. B., Greten, T. F., et al. (2016). Recommendations for myeloid-derived suppressor cell nomenclature and characterization standards. *Nature Communications*, 7, 12150. https://doi.org/10.1038/ncomms12150.

Bruger, A. M., Dorhoi, A., Esendagli, G., Barczyk-Kahlert, K., van der Bruggen, P., Lipoldova, M., et al. (2019). How to measure the immunosuppressive activity of MDSC: Assays, problems and potential solutions. *Cancer Immunology, Immunotherapy*, *68*(4), 631–644. (Epub May 2018). https://doi.org/10.1007/s00262-018-2170-8.

Cassetta, L., Baekkevold, E. S., Brandau, S., Bujko, A., Cassatella, M. A., Dorhoi, A., et al. (2019). Deciphering myeloid-derived suppressor cells: Isolation and markers in humans, mice and non-human primates. *Cancer Immunology, Immunotherapy*, *68*(4), 687–697. https://doi.org/10.1007/s00262-019-02302-2.

Clavijo, P. E., Moore, E. C., Chen, J., Davis, R. J., Friedman, J., Kim, Y., et al. (2017). Resistance to CTLA-4 checkpoint inhibition reversed through selective elimination of granulocytic myeloid cells. *Oncotarget*, *8*(34), 55804–55820. https://doi.org/10.18632/oncotarget.18437.

Condamine, T., Dominguez, G. A., Youn, J.-I., Kossenkov, A. V., Mony, S., Alicea-Torres, K., et al. (2016). Lectin-type oxidized LDL receptor-1 distinguishes population of human polymorphonuclear myeloid-derived suppressor cells in cancer patients. *Science Immunology*, *1*(2), aaf8943. https://doi.org/10.1126/sciimmunol.aaf8943.

Corzo, C. A., Condamine, T., Lu, L., Cotter, M. J., Youn, J.-I., Cheng, P., et al. (2010). HIF-1α regulates function and differentiation of myeloid-derived suppressor cells in the tumor microenvironment. *The Journal of Experimental Medicine*, *207*(11), 2439–2453. https://doi.org/10.1084/jem.20100587.

Dorhoi, A., Glaría, E., Garcia-Tellez, T., Nieuwenhuizen, N. E., Zelinskyy, G., Favier, B., et al. (2019). MDSCs in infectious diseases: Regulation, roles, and readjustment. *Cancer Immunology, Immunotherapy*, *68*(4), 673–685. (Epub Dec 2018). https://doi.org/10.1007/s00262-018-2277-y.

Gregori, S., Tomasoni, D., Pacciani, V., Scirpoli, M., Battaglia, M., Magnani, C. F., et al. (2010). Differentiation of type 1 T regulatory cells (Tr1) by tolerogenic DC-10 requires the IL-10-dependent ILT4/HLA-G pathway. *Blood*, *116*(6), 935–944. https://doi.org/10.1182/blood-2009-07-234872.

Haile, L. A., Greten, T. F., & Korangy, F. (2012). Immune suppression: The hallmark of myeloid derived suppressor cells. *Immunological Investigations*, *41*(6–7), 581–594. https://doi.org/10.3109/08820139.2012.680635.

Ocklind, G. (1988). All CD2-positive human lymphocytes rosette with sheep erythrocytes in the presence of polyethylene glycol. *Journal of Immunological Methods*, *112*(2), 169–172. Retrieved from http://www.ncbi.nlm.nih.gov/pubmed/2901445.

Pan, P.-Y., Ma, G., Weber, K. J., Ozao-Choy, J., Wang, G., Yin, B., et al. (2010). Immune stimulatory receptor CD40 is required for T-cell suppression and T regulatory cell activation mediated by myeloid-derived suppressor cells in cancer. *Cancer Research*, *70*(1), 99–108. https://doi.org/10.1158/0008-5472.CAN-09-1882.

Quah, B. J. C., & Parish, C. R. (2012). New and improved methods for measuring lymphocyte proliferation in vitro and in vivo using CFSE-like fluorescent dyes. *Journal of Immunological Methods*, *379*(1–2), 1–14. https://doi.org/10.1016/j.jim.2012.02.012.

Quah, B. J. C., Warren, H. S., & Parish, C. R. (2007). Monitoring lymphocyte proliferation in vitro and in vivo with the intracellular fluorescent dye carboxyfluorescein diacetate succinimidyl ester. *Nature Protocols*, *2*(9), 2049–2056. https://doi.org/10.1038/nprot.2007.296.

Safarzadeh, E., Orangi, M., Mohammadi, H., Babaie, F., & Baradaran, B. (2018). Myeloid-derived suppressor cells: Important contributors to tumor progression and metastasis. *Journal of Cellular Physiology*, *233*(4), 3024–3036. https://doi.org/10.1002/jcp.26075.

Trellakis, S., Bruderek, K., Hütte, J., Elian, M., Hoffmann, T. K., Lang, S., et al. (2013). Granulocytic myeloid-derived suppressor cells are cryosensitive and their frequency does not correlate with serum concentrations of colony-stimulating factors in head and neck cancer. *Innate Immunity*, *19*(3), 328–336. https://doi.org/10.1177/1753425912463618.

Umansky, V., Adema, G. J., Baran, J., Brandau, S., Van Ginderachter, J. A., & Fridlender, G. (2019). Interactions among myeloid regulatory cells in cancer. *Cancer Immunology, Immunotherapy*, *68*(4), 645–660. (Epub Jul 2018). https://doi.org/10.1007/s00262-018-2200-6.

CHAPTER ELEVEN

Detection and functional evaluation of arginase-1 isolated from human PMNs and murine MDSC

Stefania Canè, Vincenzo Bronte*

Department of Medicine, Section of Immunology, University of Verona, Verona, Italy
*Corresponding author: e-mail address: vincenzo.bronte@univr.it

Contents

Methods in Enzymology, Volume 632
ISSN 0076-6879
https://doi.org/10.1016/bs.mie.2019.07.022

Abstract

Immunotherapy has emerged as a potent alternative for cancer treatment, unfortunately, the clinical benefit remains limited to few patients and immunotherapy resistance due to immunosuppressive tumor microenvironment represents the major reason of such a failure. Arginase-1 is one of the enzymes contributing to the establishment of such immunosuppression. Among the human immune cells, polymorphonuclear cells (PMNs) represent the major source of arginase-1, while myeloid-derived suppressor cells (MDSCs) are the main arginase-1 producing cells in mice. Due to arginase-1 potential impact in dampening the immune response, there is a growing interest in assaying arginase-1 levels and functions. Thus, in this chapter we propose how to evaluate the expression and activity of arginase in human peripheral blood-derived PMNs and in MDSCs isolated from tumor-bearing mice.

1. Introduction

Arginase (ARG) is a manganese metalloenzyme hydrolyzing L-arginine to L-ornithine and urea. It is found throughout the living organisms, in bacteria, yeasts, plants, invertebrates, and vertebrates, and is thought to have appeared first in bacteria (Dzik, 2014). The subsequent transfer of ARG to an eukaryotic cell likely occurred through mitochondria. In fact, most invertebrates, plants, bacteria, and yeasts have only one form of ARG (ARG2) that localizes in mitochondria, while the ureotelic organisms, which metabolize excess nitrogen as urea, express a second cytosolic isoform (ARG1). In humans, up to now, three different isoforms of ARG1 have been identified and produced by mRNA alternative splicing. Isoform 1, *alias* liver ARG1, consists of 322 amino acids (aa) (Dizikes, Grody, Kern, & Cederbaum, 1986) and is localized in the cytosol of hepatocytes where it contributes to the urea cycle and nitrogen metabolism; 330 aa isoform 2, identified in erythroid cells and PMNs (Munder et al., 2005), carrying eight additional amino acids beginning from aa 43, is important for L-arginine homeostasis via the competition with nitric oxide synthase (NOS) for the available L-arginine substrate; 236 aa isoform 3, missing amino acids 204–289, has still an undefined function and tissue distribution (UniProt-P05089).

Crystal structure resolution indicates that active human ARG1 is a 105 kDa homotrimer and each of 35 kDa subunit contains a binuclear Mn(II) located at the bottom of the catalytic cleft. In particular, the metal ions establish hydrogen bonds with histidine and aspartate residues, ensuring protein stability and ARG activation (Ash, 2004). The overall fold of each subunit belongs to the α/β family, consisting of a parallel, 8 stranded β-sheet flanked on both sides by numerous α-helices. Similarly to rat and mouse arginases, which have alkaline pH optimal activity, human ARG1 has a maximal enzyme velocity in the range of pH 9.0–9.5 with a pK value of 7.8 (Di Costanzo et al., 2005). In non-hepatic tissues, ARG1 reaction is thought to provide a source of ornithine, the biosynthetic precursor of proline and the polyamines required for collagen deposition and cellular proliferation, respectively (Casero, Murray Stewart, & Pegg, 2018), even though these conclusions were never experimentally confirmed. While this is true in general, in immune cells, like neutrophils, a direct role of ARG1 in decreasing L-arginine levels cannot be underestimated. In human neutrophils, ARG1 is stored in tertiary granules and, upon different stimuli (e.g., inteleukin-8 (IL-8), phorbol-12 myristate-13 acetate (PMA), Ca^{2+} ionophore A23187, ionomycin; Barrientos et al., 2013; Jacobsen, Theilgaard-Monch, Christensen, & Borregaard, 2007; Rotondo et al., 2009, 2011), is secreted and participate to the homeostasis of L-arginine in the extracellular milieu. L-arginine is a semi-essential amino acid that is generally provided by protein turnover but, in special cases (e.g., inflammation), needs to be supplemented with the diet. While the release of ARG1 from activated PMNs was shown to dampen T cell proliferation in vitro (Rotondo et al., 2011), increased ARG1 activity is required to sustain microbial clearance in neutrophils, through the depletion of L-arginine in the phagolysosome (Munder et al., 2005). Moreover, alteration of L-arginine levels was shown to contribute to the pathogenesis of Alzheimer's disease (Liu et al., 2014) and diabetes (Romero et al., 2012). Even though administration of L-arginine was though to prevent and or delay the onset of diabetes-associated vascular alterations, several studies, both in mice and humans, found no real benefit on clinical outcomes after prolonged supplementation of L-arginine (Lucas, Fulton, Caldwell, & Romero, 2014). These negative effects could be related to the ability of L-arginine to induce either the expression or activity of ARG1, consequently reducing L-arginine plasma levels, independently from ornithine production.

In mice ARG1 is constitutively expressed in the cytosol of hepatocytes while its expression is induced in myeloid cells (e.g., MDSC and alternatively activated macrophages) by cytokines like interleukin 4 and interleukin 13 (IL-4 and IL-13, respectively) (Munder et al., 1999; Munder, Eichmann, & Modolell, 1998), as well as by transforming growth factor-β (TGF-β) (Boutard et al., 1995) and granulocyte-macrophage colony-stimulating factor (GM-CSF) (Jost et al., 2003). The promoter region of mouse *ARG1* contains elements that control its transcription in response to IL-4, cyclic adenosine monophosphate (c-AMP), TGF-β and lipopolysaccharide (LPS) (Morris, 2002). In alternatively activated macrophages (M2 macrophages), ARG1 expression contributes to wound healing and tissue regeneration processes and in immune defense against multicellular pathogens, parasites and immune suppression and allergic inflammation (Barron et al., 2013; Campbell, Saville, Murray, Cruickshank, & Hardman, 2013; El Kasmi et al., 2008; Gray, Poljakovic, Kepka-Lenhart, & Morris, 2005; Herbert et al., 2010; Modolell, Corraliza, Link, Soler, & Eichmann, 1995; Pauleau et al., 2004; Pesce et al., 2009). In tumor-infiltrating dendritic cells (DCs) and myeloid-derived suppressor cells (MDSCs) ARG1 plays a role in suppression of T cell-mediated response (Liu et al., 2009; Narita et al., 2013) by downregulating the CD3ζ chain of the T cell receptor (TCR), even though a systematic analysis was never been done and controversial results keep the question open (Van de Velde et al., 2017). This mechanism seemed to be relevant for tumor escape in vivo, because injection of the ARG inhibitor N-hydroxy-nor-L-arginine (nor-NOHA) was shown to slow the growth of lung carcinoma in a dose-dependent manner (Rodriguez et al., 2004). Therefore, L-arginine degradation by ARG1 appears to exert a major negative impact on anti-tumor T cell function and, indeed, L-arginine directly modulates T cell metabolism by enhancing cell survival and sustaining their anti-tumor activity (Geiger et al., 2016). In this manuscript, however, very high concentrations of L-arginine were used in the in vitro assays. To dissect the contribution of myeloid-derived ARG1 to the immune response, conditional myeloid-specific *ARG1* knockout mice were developed (Arg1 flox/flox; Tie2cre) in which myeloid cells are nearly devoid of ARG1 activity, as a measure of urea production (El Kasmi et al., 2008). In these mice, the absence of ARG1 in myeloid compartment, unleashed the therapeutic activity of adoptively transferred, tumor-specific T cells by a mechanism requiring the release of NO by monocyte-derived DCs (Marigo et al., 2016).

On the light of this information, it appears that profound dissimilarities exist in the ARG1 biology between rodents and humans. While in mice ARG1 is a cytosolic enzyme induced in MDSC/M2 macrophages, in humans ARG1 is stored in PMN granules whose exocytosis is induced by different pro-inflammatory stimuli (Rotondo et al., 2011). Moreover, human ARG1 is constitutively expressed by PMNs and its expression is not upregulated by stimuli that affect the mouse analogue (Munder et al., 2005). Thus, in this chapter, we will describe assays optimized to determine the ARG1 level and activity (Corraliza, Campo, Soler, & Modolell, 1994), in human PMNs and mouse MDSCs.

2. Arginase inhibitors

In view of findings suggesting that altered L-arginine homeostasis can contribute to the pathogenesis and exacerbation of several conditions (e.g., cancer, autoimmune diseases, asthma), controlling intracellular and extracellular arginase activity is attractive for disease modulation, both in humans and mice. Targeting ARG is not an easy task. The few known ARG inhibitors suffer from poor structural diversity. Two types of inhibitors are known to bind to ARG (Di Costanzo, Pique, & Christianson, 2007). The first group comprises product analogues to the ligand. Such inhibitors prevent the natural hydrolysis taking place with the natural substrate. The second group comprises transition state analogues like hydroxyarginines, boronic acids, or sulfonamides (Boucher et al., 1994; Di Costanzo, Ilies, Thorn, & Christianson, 2010). Boronic acids, like 2-S-amino-6-boronohexanoic acid (ABH) and S-(2-boronoethyl)-L-cysteine (BEC), together with hydroxyarginine, like nor-N(ω)-hydroxy-L-arginine (nor-NOHA), react to form transition state analogs complexing the Mn^{2+} ions and sequestering them from the catalysis. Unfortunately, none of these inhibitors exhibit sufficient differences in affinity to provide isoform-selective inhibition, both in vitro and in vivo (Morris, 2009; Pudlo, Demougeot, & Girard-Thernier, 2017), beside been toxic in vivo to some extent. Recently, renewed interested in targeting ARG has brought scientists to design novel non-aminoacid-based inhibitors (Mortier et al., 2017) that selectively interact with ARG1, which is the prerequisite for the development of novel ARG1 modulators with therapeutic relevance. Furthermore, a novel inhibitor, CB-1158, was develop and the initial encouraging results in preclinical cancer models (Steggerda et al., 2017) have paved the way to start a phase I clinical trial for treatment of solid tumors. For the matter of this chapter we will evaluate ARG1 dependent effects using nor-NOHA, which has been broadly used in vivo and in vitro (Rotondo et al., 2011).

3. Evaluating ARG1 in activated and resting PMN: Immunofluorescence (IF) analysis

An immune-based method for the detection of ARG1 in purified PMNs, isolated from peripheral blood, is described below. Combining the specificity of antibody's binding to ARG1 with the high resolution capacity of confocal microscopy, it is possible to reveal the cellular distribution of ARG1 and changes occurring during PMN priming.

3.1 Equipment

1. BD vacutainer®, K_2EDTA
2. 15 and 50 mL polypropylene conical tubes (BD, Falcon)
3. Centrifuges
4. FACS Canto flow cytometer (BD Biosciences)
5. FlowJo Software (TreeStar)
6. 12 × 75–mm polypropylene tubes (BD, Falcon)
7. Cover glass, 13 mm in diameter, 1 mm thick (VWR)
8. SuperFrost Plus, glass slides, Thermo Fisher
9. 24-well plate (BD, Falcon)
10. Transfer pipette, 3.5 mL (Sarstedt)
11. Serological pipettes (5, 10, 25 mL, Sarstedt)
12. Leica TCS SP5 confocal laser scanning microscope (Leica Microsystems)

3.2 Material

1. Arginine-free RPMI (Biological Industries, Kibbutz Beit Haemek, Israel), supplemented with 10% FBS (Fetal bovine serum; Gibco), 10 U/mL penicillin and streptomycin and 0.01 M Hepes
2. Ficoll-Paque, 1.077 g/mL (GE Healthcare)
3. 4% Dextran (Mr 450,000–650,000; 31392, Sigma-Aldrich).
4. Dulbecco's phosphate-buffered saline (DPBS without Ca^{2+} and Mg^{2+}; Lonza BioWhittaker, BE17-515Q)
5. 0.2% NaCl solution, 0.22 μm filtered
6. 1.2% NaCl solution, 0.22 μm filtered
7. Ionomycin (I0634, Sigma-Aldrich)
8. IL-8 (130-108-979, Miltenyi Biotec).
9. 36.5-38% formaldehyde solution (F8775, Sigma-Aldrich)
10. Tris Buffered Saline (TBS)
11. Tween-20 (Sigma-Aldrich)
12. Triton X-100 (Sigma-Aldrich)
13. Normal goat serum (S-1000, Vector Laboratories)
14. Primary antibodies: Goat anti-human MPO (R&D, AF3667), rabbit anti-human H3 R2 + R8 + R17 citrullinated (Abcam, ab5103), mouse anti-human ARG1 (monoclonal antibody from homemade hybridoma, conjugated to Alexa Fluor 647)
15. Secondary antibodies: donkey anti-goat Alexa Fluor 488 (A-11055, Thermo Fisher) and goat anti-rabbit Alexa Fluor 546 (A-11035, Thermo Fisher)

16. 4′,6-diamidino-2-phenylindole dihydrochlorid (DAPI) (10236276001, Sigma-Aldrich)

Note: all experiments must be performed in accordance with local institutional ethical guidelines.

Note: collect blood within 10–15 min of the experiment to ensure optimal yield and resting PMNs. Avoid to purify PMNs from buffy coat. Cells tend to have an activated and already degranulate phenotype.

Note: avoid cold temperature, PMNs tends to become activated. Try to work at RT (20–22 °C) and to maintain the reagents at 37 °C before use.

3.3 Protocol

3.3.1 Isolation of PMNs from peripheral blood of healthy donors

1. Collect blood in K_2EDTA vacutainer tubes, dilute 1/2 with DPBS, and carefully layer onto a density gradient Ficoll-Paque. Centrifuge at 800 × *g* for 20 min at 24 °C, without brake.
2. Aspirate the top layer containing PBMC, keep momentarily aside, on ice. Erythrocytes (RBC) and PMNs sediment at the bottom of the tube, gently remove the excess of Ficoll-Paque and add DPBS and a volume of 4% dextran (both kept at 37 °C) to reach 1% final, gently invert the tube to mix and allow to sit at RT for 20 min.
3. Collect the upper PMN-rich layer avoiding the RBC, spin down and lyse the residual RBC by resuspending the pellet in 3 mL of 0.2% NaCl for 2½ min and then in 7 mL of warm 1.2% NaCl (both kept at 37 °C) to restore isotonicity. Centrifuge at 450 × *g* at 24 °C for 5 min.
4. Discard supernatant and wash pellet with 15 mL warm of PBS (kept at 37 °C) and centrifuge again at 450 × *g* at 24 °C to get rid of any remaining lysing solution. The pellet obtained at this point is enriched in PMNs.
5. Check the cell purity by FACS analysis on forward/side scatter parameters, and viability by trypan blue dye exclusion. Generally, PMNs yielded purity and viability is >95%. Dilute PMNs in warm L-arginine free RPMI to a concentration of 5×10^6 PMNs/mL.

Note: higher concentration of PMNs can result in spontaneous activation and degranulation.

3.3.2 Immunofluorescence analysis of ARG1 in activated PMNs

1. Prepare a 24-well cell culture plate by inserting sterile 13 mm round glass cover slip in each well.
2. Plate 2 × 105 cells in 300 μL L-arginine free RPMI containing 10% FBS in each well and incubate at 37 °C for 1 h for adherence.

3. Meanwhile, prepare 20 μM ionomycin in L-arginine free RPMI containing 10% FBS and add 100 μL per well (final concentration, 5 μM). Alternatively, add 100 μL of 80 nM of IL-8 (final concentration, 20 nM). Incubate the cells from 30 min up to 4 h at 37 °C, 5% CO2.
4. Fix cells with 4% paraformaldehyde solution (final concentration in the well) for 15 min at RT.
5. Carefully remove the supernatant and wash three times with PBS. Permeabilize the cells by adding 300 μL of 0.5% Triton X-100 in TBS and incubate for 1 h at RT.
6. Remove the liquid and add blocking buffer (20% normal goat serum in 0.1% Triton X-100). Let stand for 2 h at RT.
7. Without washing add primary antibodies diluted in PBS. The following antibodies are routinely used: anti-MPO (1:200, stock 0.2 mg/mL), anti-ARG-1 (1:1000, stock 1.5 mg/mL), anti-H3 citrullinated R2 + R8 + R17 (1:500, stock 0.8 mg/mL). Incubate O/N at 4 °C.
8. The following day, remove the primary antibodies and wash three times with 0.05% Tween-20 in TBS.
9. Dilute the secondary antibodies in blocking buffer as follows: anti-rabbit AF546 (1:2000, stock 2 mg/mL), anti-goat AF488 (1:2000, stock 2 mg/mL). Incubate for 1 h at RT.
10. Wash with 0.05% Tween-20 as before. Wash once with TBS only.
11. Stain DNA with 4′,6-diamidino-2-phenylindole (DAPI, 1 μg/mL) for 10 min, RT then wash twice with distilled water.
12. Set a 10 μL drop of ProLong Antifade reagent (Thermo Fisher) onto a glass slide and mount the cover slip. For microscopic analysis with immersion lenses let the mounting media to dry for at least 1 h. Store at 4 °C.
13. Acquire images at the Leica TCS SP5 confocal microscope.

4. Intracellular determination of ARG1 levels in PMN by flow activated cell sorting (FACS)

PMNs are a heterogeneous population of cells expressing ARG1. FACS analysis, by combining antibodies directed to different cellular surface markers, represents a sensitive and semi-quantitative method to evaluate total intracellular levels of ARG1 in PMN subtypes.

4.1 Equipment

1. BD vacutainer®, K2EDTA
2. 15 and 50 mL polypropylene conical tubes (BD, Falcon)

3. Centrifuges
4. FACS Canto flow cytometer (BD Biosciences)
5. FlowJo Software (TreeStar)
6. 12 × 75–mm polypropylene tubes (BD, Falcon)
7. Transfer pipette, 3.5 mL (Sarstedt)
8. Serological pipettes (5, 10, 25 mL, Sarstedt)

4.2 Material

1. Cytofix/Cytoperm kit (BD Biosciences)
2. Human FcR blocking reagent (Miltenyi Biotec, 130-059-901)
3. Staining buffer: DPBS + 0.5% w/v bovine serum albumin (BSA, A4503, Sigma-Aldrich)
4. Anti-CD16 (clone 3G8)-FITC, anti-CD15 (clone HI98)-PE, anti-CD66b (G10F5)-BV421 are from BD Biosciences, anti-ARG1 (clone 1.10 from homemade hybridoma, conjugated with Alexa FLuor-647, Thermo Fisher)

4.3 Protocol

1. Isolate PMNs as in Section 3.3.1 and incubate the cells with Fc-receptor blocking reagent diluted 1/25 at 4 °C for 10 min.
2. Add the mixture of antibodies (CD16/CD66b or CD15) and incubate at 4 °C for 20 min. The dose of each antibody depends on the lot number and must be evaluated before.
3. Wash the cells with staining buffer and proceed to fixed and permeabilized using the BD Cytofix/Cytoperm kit according to the manufacturer's instructions.
4. At the end of the incubation time, wash the cells extensively with permeabilization buffer diluted following BD protocol.
5. Add the antibody clone 1.10 AF-647 to anti-ARG1 and incubate 1 h at 4 °C. The amount of antibody should be predetermined using PMNs and the PBMC fraction as negative control.
6. Wash three times with permeabilization buffer and finally resupsend the cells in staining buffer.
7. Acquire directly to the flow cytometry.

Note: we recommend performing the staining protocol in 96-round well plate to avoid losing cells.

Note: do not let the stained PMNs standing for longer time before flow cytometry.

5. Determination of ARG1 activity from PMN-derived supernatant

The enzyme activity is determined by an end-point colorimetric method to evaluate the ornithine production, using L-arginine as a substrate. Beside evaluating the overall level of ARG-1 by commercial ELISA it is imperative to determine the activity of the protein, since it is an enzyme with defined biochemical properties (e.g., Vmax, Kd).

5.1 Equipment

1. BD vacutainer®, K_2EDTA
2. 15 and 50 mL polypropylene conical tubes (BD, Falcon)
3. Centrifuges
4. Transfer pipette, 3.5 mL (Sarstedt)
5. Cell culture treated petri dishes (Corning)
6. Serological pipettes (5, 10, 25 mL, Sarstedt)
7. BCA assay kit (Thermo Fisher)
8. Spectra Max microplate reader (Molecular Device)
9. Heating block
10. 2 mL Eppendorf tubes
11. pH-meter

5.2 Materials

1. Reagents for PMN isolation and activation as indicted in Section 3.3.1.
2. L-Arginine (A5131, Sigma-Aldrich)
3. L-Ornithine (O2375, Sigma-Aldrich)
4. $MnCl_2$ (244589, Sigma-Aldrich)
5. Ninhydrin (N4876, Sigma-Aldrich)
6. Acetic acid (45754, Sigma-Aldrich)
7. Tris/HCl and NaCl (Sigma-Aldrich)
8. Carbonate-bicarbonate buffer (C3041, Sigma-Aldrich)

5.3 Protocol

1. Determine the protein content in the supernatant by BCA assay.
2. Take 100 μg of total protein and bring to 100 μL volume with a carbonate solution, pH 10.

3. Adjust the pH to 9.5 then add 100 μL of 75 mM $MnCl_2$. Incubated 20 min at 55 °C.
4. Add 50 μL of 140 mM L-arginine and incubate for 2 h up to O/N at 37 °C.
5. Stop the reaction by adding 850 μL of acetic acid and add 300 μL of ninhydrin solution.
6. Incubate at 100 °C for 30 min. Remove from the heating block and let stand for 10 min avoiding light exposure.
7. Spin the samples at 14000 rpm for 5 min, take 250 μL of the supernatant and plate in a 96 well plate (flat bottom). Read at 492 nm.
8. Calculate L-ornithine concentration using a standard curve with serial dilutions ranging from 2.0 to 0.5 mmol/L. Reagent stock solutions to use are: 100 mmol/L L-arginine (fresh every day), 2 mmol/L L-ornithine (stable 1 month at 4 °C), 25 g/L ninhydrin in 10 mol/L acetic acid/ 2.4 mol/L phosphoric acid (stable for several months), 10 mmol/L $MnCl_2$ (stable for 1 month).

6. Evaluation of human PMN-derived ARG1 dependent suppressions of T cell proliferation and determination of ARG1 activity

Immune suppression exerted by degranulating PMNs on activated T cells can be measured in terms of inhibition of T cell proliferation due to the ability to induce proliferative arrest of actively dividing cells. Even though, in the protocol we recommend employ PBMCs, the use of purified T cells might help in avoiding potential interference from other cell subsets present among PBMCs.

6.1 Equipment

1. BD vacutainer®, K_2EDTA
2. 15 and 50 mL polypropylene conical tubes (BD, Falcon)
3. Centrifuges
4. FACS Canto flow cytometer (BD Biosciences)
5. FlowJo 7.6.5 Software (TreeStar)
6. 12 × 75-mm polypropylene tubes (BD, Falcon)
7. TruCount tubes (BD Biosciences)
8. Transfer pipette, 3.5 mL (Sarstedt)
9. Cell culture treated petri dishes (Corning)
10. Serological pipettes (5, 10, 25 mL, Sarstedt)
11. 96-well flat-bottom plate, cell culture treated
12. Incubator 8% CO_2

6.2 Materials

1. Anti-CD3 (5 μg/mL), clone OKT3, obtained after expansion and purification of a commercially available hybridoma
2. Anti-CD28 (5 μg/mL), clone CD28.2 (BioLegend, 302923)
3. Dulbecco's phosphate-buffered saline (DPBS without Ca and Mg; LonzaBioWhittaker, BE17-515Q)
4. CellTrace TM Violet Cell Proliferation Kit (Thermo Fisher)
5. Fetal bovine serum (FBS; Gibco). Note: FBS used for functional assay should be tested to guarantee that immunosuppression is detectable and not overcome by an excessive T cell proliferation
6. Human FcR blocking reagent (Miltenyi Biotec, 130-059-901)
7. L-Arginine (A5131, Sigma-Aldrich)
8. IMDM medium (12440053, Gibco)
9. Arginine-free RPMI (Biological Industries, Kibbutz Beit Haemek, Israel), supplemented with 10% FBS (Fetal bovine serum; Gibco), 10 U/mL penicillin and streptomycin and 0.01 M Hepes.
10. Ficoll-Paque, 1.077 g/mL (GE Healthcare).
11. 4% Dextran (Mr 450,000-650,000; 31392, Sigma-Aldrich).
12. 0.2% NaCl solution, 0.22 μm filtered.
13. 1.2% NaCl solution, 0.22 μm filtered.
14. Ionomycin (I0634, Sigma-Aldrich).
15. IL-8 (130-108-979, Miltenyi Biotec).
16. nor-NOHA (acetate) (10006861, Cayman Chemical).
17. Human serum (H3667, Sigma-Aldrich).
18. Anti-CD3, clone UCHT1 (Beckman Coulter), anti-CD4, clone SK3 (BD Biosciences) and anti-CD8, clone SK1 (BD Biosciences)
19. Staining buffer: DPBS + 0.5% w/v bovine serum albumin (BSA, A4503, Sigma-Aldrich)

6.3 Protocol

1. Prepare coating buffer with anti-CD3 (5 μg/mL final concentration in DPBS) to fill 96 well flat bottom microtiter plates with 200 μL/well using a multichannel pipettor. Fill an equal number of wells with 200 μL/well of DPBS without antibodies, for background proliferation measurement.
2. Incubate the plate overnight at 4 °C or alternatively at 37 °C for 1 h.
3. Follow the detailed protocol highlighted in Section 3.3.1 to isolate PMNs and resuspend them at 5×10^6 cells/ml in L-arginine free media.

4. Transfer PMNs in a petri dish and stimulate them with ionomycin (5 μM) or IL-8 (20 nM) for 30 min at RT.
5. Gently aspirate the media with transfer pipette. Spin for 10 min at 450 × *g* at 24 °C. PMNs and any remaining cells/debris will be pelleted at the bottom of the tube, leaving a cell-free supernatant.
6. Meanwhile, take the PBMC layer isolated in Section 3.3.1, transfer into a 50 mL conical tube and wash three times with ice cold DPBS containing 1% human serum.
7. Dilute an aliquot of PBMC suspension in trypan blue dye solution and evaluate the viable cell concentration.
8. Adjust the concentration of PBMCs to 2×10^7 cells/ml in DPBS and the concentration of CellTrace to 2 × in DPBS (final concentration ranging from 0.5 to 2 μM); quickly mix equal volumes of PBMCs and CellTrace and incubate for 5 min at 37 °C. Quench the reaction by adding 1/5 of the total volume of FBS, centrifuge 5 min at 300 × *g*, 4 °C. Check the incorporation of CellTrace by flow cytometry.
9. Plate CellTrace$^+$ PBMCs in a 24-well plate at 3×10^6 cell/well for at least 1 h at 37 °C in IMDM medium containing 10% FBS.
10. Harvest PBMCs, count them by trypan blue dye assay, adjust their concentration to 1×10^6 cells/mL in either supernatant from PMNs treated or left untreated. Add 150 μM L-arginine to all the conditions and plate in triplicate, 200 μL/well, in 96-well plates coated with anti-CD3 antibody extensively washed with DPBS to remove unbound anti-CD3 antibody. In some condition add nor-NOHA, an arginase inhibitor, at the final concentration of either 100 or 300 μM. Add 1 μL/well of soluble anti-CD28 (5 μg/mL) and incubate at 37 °C, 8% CO_2 for 4 days.
11. At day 4, pool triplicates and wash samples once with sorting and staining buffer, centrifuge the suspension 6 min at 300 × *g*, 4 °C, and discard the supernatant.
12. Block nonspecific binding with 25 μL FcR blocking reagent for 10 min at 4 °C.
13. Stain cells with anti-CD3 clone UCHT1, anti-CD4 clone SK3 and anti-CD8 clone SK1 for 20 min at 4 °C. Adjust the volume of staining mix to 100 μL with staining buffer.
14. Wash samples once with staining buffer, centrifuge the suspension for 5 min at 300 × *g*, 4 °C, and discard the supernatant.
15. Resuspend samples in 250 μL DPBS, transfer the mix in TruCount tubes, gently vortex, and proceed with flow cytometric acquisition and analysis.

7. Summary

Several data indicate that ARG1, secreted by activated PMNs or expressed by murine MDSC, by altering L-arginine homeostasis can modulate the immune response toward cancer. At the same time a new reinvigorating interest on understanding the molecular mechanisms driving human PMNs and murine MDSC differentiation has recently emerged due to the contribution of these cells in the pathogenesis of diseases, like systemic lupus erythematosus, rheumatoid arthritis, diabetes and cancer, among others. With this chapter we describe protocols for the detection and functional evaluation of ARG1 in both human PMNs, isolated from peripheral blood of healthy donors, and mouse MDSCs with the aim of defining the basic experimental approaches necessary to evaluate immune-derived ARG1.

Acknowledgment

This work is supported by the Italian Association for Cancer Research (18603 and 12182 that refers to Special Program Molecular Clinical Oncology 5 per mille), Cariverona Foundation (Project call, 2017) and by the Cancer Research Institute (Clinic and Laboratory Integration Program, CLIP-2017) to Professor Vincenzo Bronte.

Appendix

Determination of ARG1 levels and activity in mouse-derived myeloid cells

Isolate and purify MDSCs from the spleen of tumor-bearing mice following the instructions present in Basic Protocol 1 recently published by our group (Solito et al., 2019). Alternatively, bone-marrow derived MDSCs (BM-MDSCs) can be generated using Basic Protocol 2 (Solito et al., 2019). As control for MDSC-derived ARG-dependent suppression, we recommend to use cells obtained from conditional myeloid *Arg1*-deficient mice (e.g., ARG1fl/fl;Tie-2Cre mice).

1. ARG-dependent suppressive activity of myeloid cells

   Once obtained, proceed to evaluate ARG-dependent suppressive ability in MDSCs and BM-MDSCs following the Basic Protocol 3 (Solito et al., 2019) with minor modifications:

   1. At point 5 of Basic Protocol 3, MDSCs should be co-cultured with T lymphocytes in the presence of ARG inhibitor, nor-NOHA, previously titrated.

2. In case TCR transgenic mice, developed to recognize a specific antigenic peptide, are not available, it is possible to perform the assay using Basic Protocol 2 (Dolcetti, Peranzoni, & Bronte, 2010).

2. Evaluation of ARG1 protein levels in the pellet of myeloid cells

2.1 Equipment

1. 1.5 mL tubes
2. Minicentrifuge
3. SDS-PAGE running system and wet transfer apparatus
4. Heating block
5. Microplate reader
6. Bio-Rad ChemiDoc

2.2 Material

1. lysis buffer: 50 mM HEPES, 150 mM NaCl, 5 mM EDTA, 0.5% Triton, cocktail protease
   inhibitors, 1 mM NaOV4, 2 mM PMSF
2. cOmplete™ Protease Inhibitor Cocktail (11836145001, Sigma-Aldrich)
3. Tris-buffer saline (TBS) and Tween-20 (Sigma-Aldrich)
4. Laemmli buffer (161-0747, Bio-Rad)
5. NuPAGE, 10% bis-tris protein gel (NP0302BOX, Thermo Fisher)
6. 20X MOPS buffer (Thermo Fisher)
7. PVDF membrane 0.2 μm (88520, Thermo Fisher)
8. Transfer buffer: 25 mM Tris-HCl (pH 7.6), 192 mM glycine, 20% methanol (Sigma-Aldrich)
9. Non-fat dry milk (1706404, Bio-Rad).
10. Bovine serum albumin (BSA), (A2153, Sigma-Aldrich)
11. Primary antibodies: goat anti-mouse ARG1 (SAB2500101, Sigma-Aldrich); rabbit anti-mouse β actin (ab8227, Abcam)
12. Secondary antibodies: sheep anti-mouse-HRP (NA931, GE Healthcare); donkey anti-rabbit-HRP (NA934, GE Healthcare)
13. SuperSignal West Pico Chemiluminescent substrate (34580, Thermo Fisher)
14. BCA protein assay kit (23225, Thermo Fisher)

2.3 Protocol

1. Collect the cells (MDSCs, BM-MDSCs or M2 macrophages), wash them twice in cold PBS.
2. Disrupt 5×10^5 pelleted cells on ice in lysis buffer. Incubate on ice for 15 min. Vortex every 5 min, maximum speed.

3. Spin at 14000 rpm for 10 min at 4 °C, transfer the supernatant containing proteins into a new tube and quantify the protein content. Take 20 μg add 1 volume of Laemmli's buffer and carry out the denaturation 5 min at 100 °C.
4. Proteins are separated on a 10% bis-tris acrylamide gel in MOPS buffer and then transferred on a PVDF membrane (Millipore).
5. The membrane was saturated for 1 h at RT in TBS/0.05% Tween-20 (TBST) supplemented with 5% non-fat milk (blocking solution).
6. Hybridizations of primary antibody is carried out as follows: goat anti-ARG1 (1:500, stock 0.5 mg/mL) ON 4 °C in blocking solution; rabbit anti-actin antibody (1:3000, stock 0.3 mg/mL) for 1 h at RT in TBS-T/3% BSA.
7. After washing with TBST, hybridizations with the HRP-conjugated secondary antibodies sheep anti-mouse IgG (1:5000) and donkey anti-rabbit IgG (1:5000) both diluted in blocking buffer are performed for 1 h at RT. Detection of the proteins is carry out using SuperSignal West Pico Chemiluminescent following manufacturer's instructions.

3. Determination of ARG1 levels in mouse plasma, in tissue culture supernatant and in cell homogenate

3.1 Equipment

1. Microplate reader
2. Multichannel pipettes
3. Eppendorf tubes
4. 37 °C incubator
5. Sonicator
6. Minicentrifuge

3.2 Material

1. Mouse ARG1 ELISA kit (LS-F6864, LSBio).
2. Deionized or distilled water.
3. Dulbecco's phosphate-buffered saline (DPBS without Ca^{2+} and Mg^{2+}; Lonza BioWhittaker, BE17-515Q).
4. Lysis buffer: 50 mM Tris-HCl, 150 mM NaCl. Add at the time of lysis cocktail of protease inhibitors and PMSF (final concentration 2 mM).
5. cOmplete™ Protease Inhibitor Cocktail (11836145001, Sigma-Aldrich).

3.3 Protocol

1. Collect and pellet the cells by centrifugation, 450 × *g* at 4 °C for 10 min and remove the supernatant. Wash the cells three times with DPBS then resuspend in lysis buffer. Lyse the cells by ultrasonication. Centrifuge at 1500 × *g* at 4 °C for 10 min to remove cellular debris. Collect the supernatant and following manufacture's instruction assay ARG1 levels.
2. Collect plasma in heparin-containing tubes. Centrifuge the samples for 15 min at 1000 × *g* at 4 °C. Collect the supernatant for assaying following manufacture's instruction.
3. Remove cell culture supernatants and centrifuge the samples for 20 min at 1000 × *g* to remove cells and debris. Transfer the supernatant into a new tube and proceed to assay following the manufacture's instruction.

4. Quantify ARG activity in cells

4.1 Equipment

1. 24-well tissue culture plates
2. Platform rocker
3. 96-well plate
4. Microplate reader
5. Heating block
6. ELISA plate (15041, Thermo Fisher)
7. 96-well plate flat (353072, Falcon)

4.2 Material

1. Cultured cells (e.g., MDSC)
2. Dulbecco's phosphate-buffered saline (DPBS without Ca^{2+} and Mg^{2+}; Lonza BioWhittaker, BE17-515Q)
3. Lysis buffer: 0.001% Triton X-100 (Sigma-Aldrich), protease inhibitor cocktail and 2 mM PMSF. Prepare fresh each time.
4. Arginase activation solution: prepare a 10 mM $MnCl_2$, 50 mM Tris-HCl, pH 7.5 solution. Store indefinitely at room temperature.
5. Arginase substrate solution: make up a 0.5 M L-arginine, pH9.7 solution. Store indefinitely at room temperature
6. Urea standard solution: prepare a 100 mg/mL urea solution. Store for 1 year at room temperature. Prepare the standard starting from a concentration of 100 mg/mL up to 5 mg/mL
7. Quantichrom urea assay kit (DIUR-500, Bioassay Systems)

4.3 Protocol

1. To 5×10^5 cells plated in 24-well plate add lysis buffer (100 μL) per well and gently rock the plate for 15 min at 4 °C. Transfer into a new tube. Spin at 14000 rpm 10 min at 4 °C to remove debris.
2. Transfer 50 μL of each lysate to a 96-well flat plate. Add 50 μL of arginase activation solution to each well and incubate 10 min at 55 °C in a heating block.
3. Transfer 25 μL of each activated lysate to a new 96-well flat plate. Add 25 μL arginase substrate solution to each well and incubate from 1 h up to 24 h at 37 °C.
4. Prepare the urea standard and a blank. Mix of 1 vol of lysis buffer, 1 vol of arginase activation solution, and 2 vol arginase substrate solution as a diluent to match the content of the experimental samples.
5. Add 5 μL of each reacted sample, the serially diluted urea standard, and the blank solution control to an ELISA plate. Mix together the Quantichrom urea assay kit reagents following manufacturer's instructions, then add 200 μL per well to the ELISA plate.
6. Incubate 2–20 min at room temperature in the dark and quantify urea by reading the absorbance at 520 nm.

References

Ash, D. E. (2004). Structure and function of arginases. *The Journal of Nutrition*, *134*(10 Suppl), 2760S–2764S. discussion 2765S–2767S. https://doi.org/10.1093/jn/134.10.2760S.

Barrientos, L., Marin-Esteban, V., de Chaisemartin, L., Le-Moal, V. L., Sandre, C., Bianchini, E., et al. (2013). An improved strategy to recover large fragments of functional human neutrophil extracellular traps. *Frontiers in Immunology*, *4* , 166. https://doi.org/10.3389/fimmu.2013.00166.

Barron, L., Smith, A. M., El Kasmi, K. C., Qualls, J. E., Huang, X., Cheever, A., et al. (2013). Role of arginase 1 from myeloid cells in th2-dominated lung inflammation. *PLoS One*, *8*(4), e61961. https://doi.org/10.1371/journal.pone.0061961.

Boucher, J. L., Custot, J., Vadon, S., Delaforge, M., Lepoivre, M., Tenu, J. P., et al. (1994). N omega-hydroxyl-L-arginine, an intermediate in the L-arginine to nitric oxide pathway, is a strong inhibitor of liver and macrophage arginase. *Biochemical and Biophysical Research Communications*, *203*(3), 1614–1621.

Boutard, V., Havouis, R., Fouqueray, B., Philippe, C., Moulinoux, J. P., & Baud, L. (1995). Transforming growth factor-beta stimulates arginase activity in macrophages. Implications for the regulation of macrophage cytotoxicity. *Journal of Immunology*, *155*(4), 2077–2084.

Campbell, L., Saville, C. R., Murray, P. J., Cruickshank, S. M., & Hardman, M. J. (2013). Local arginase 1 activity is required for cutaneous wound healing. *The Journal of Investigative Dermatology*, *133*(10), 2461–2470. https://doi.org/10.1038/jid.2013.164.

Casero, R. A., Jr., Murray Stewart, T., & Pegg, A. E. (2018). Polyamine metabolism and cancer: Treatments, challenges and opportunities. *Nature Reviews. Cancer, 18*(11), 681–695. https://doi.org/10.1038/s41568-018-0050-3.

Corraliza, I. M., Campo, M. L., Soler, G., & Modolell, M. (1994). Determination of arginase activity in macrophages: A micromethod. *Journal of Immunological Methods, 174*(1–2), 231–235.

Di Costanzo, L., Ilies, M., Thorn, K. J., & Christianson, D. W. (2010). Inhibition of human arginase I by substrate and product analogues. *Archives of Biochemistry and Biophysics, 496*(2), 101–108. https://doi.org/10.1016/j.abb.2010.02.004.

Di Costanzo, L., Pique, M. E., & Christianson, D. W. (2007). Crystal structure of human arginase I complexed with thiosemicarbazide reveals an unusual thiocarbonyl mu-sulfide ligand in the binuclear manganese cluster. *Journal of the American Chemical Society, 129*(20), 6388–6389. https://doi.org/10.1021/ja071567j.

Di Costanzo, L., Sabio, G., Mora, A., Rodriguez, P. C., Ochoa, A. C., Centeno, F., et al. (2005). Crystal structure of human arginase I at 1.29-A resolution and exploration of inhibition in the immune response. *Proceedings of the National Academy of Sciences of the United States of America, 102*(37), 13058–13063. https://doi.org/10.1073/pnas.0504027102.

Dizikes, G. J., Grody, W. W., Kern, R. M., & Cederbaum, S. D. (1986). Isolation of human liver arginase cDNA and demonstration of nonhomology between the two human arginase genes. *Biochemical and Biophysical Research Communications, 141*(1), 53–59.

Dolcetti, L., Peranzoni, E., & Bronte, V. (2010). Measurement of myeloid cell immune suppressive activity. *Current Protocols in Immunology, Chapter 14, Unit 14 17, 91*, 14.17.1–14.17.25. https://doi.org/10.1002/0471142735.im1417s91.

Dzik, J. M. (2014). Evolutionary roots of arginase expression and regulation. *Frontiers in Immunology, 5*, Article 544, eCollection 2014. https://doi.org/10.3389/fimmu.2014.00544.

El Kasmi, K. C., Qualls, J. E., Pesce, J. T., Smith, A. M., Thompson, R. W., Henao-Tamayo, M., et al. (2008). Toll-like receptor-induced arginase 1 in macrophages thwarts effective immunity against intracellular pathogens. *Nature Immunology, 9*(12), 1399–1406. https://doi.org/10.1038/ni.1671.

Geiger, R., Rieckmann, J. C., Wolf, T., Basso, C., Feng, Y., Fuhrer, T., et al. (2016). L-arginine modulates T cell metabolism and enhances survival and anti-tumor activity. *Cell, 167*(3), 829–842.e813. https://doi.org/10.1016/j.cell.2016.09.031.

Gray, M. J., Poljakovic, M., Kepka-Lenhart, D., & Morris, S. M., Jr. (2005). Induction of arginase I transcription by IL-4 requires a composite DNA response element for STAT6 and C/EBPbeta. *Gene, 353*(1), 98–106. https://doi.org/10.1016/j.gene.2005.04.004.

Herbert, D. R., Orekov, T., Roloson, A., Ilies, M., Perkins, C., O'Brien, W., et al. (2010). Arginase I suppresses IL-12/IL-23p40-driven intestinal inflammation during acute schistosomiasis. *Journal of Immunology, 184*(11), 6438–6446. https://doi.org/10.4049/jimmunol.0902009.

Jacobsen, L. C., Theilgaard-Monch, K., Christensen, E. I., & Borregaard, N. (2007). Arginase 1 is expressed in myelocytes/metamyelocytes and localized in gelatinase granules of human neutrophils. *Blood, 109*(7), 3084–3087. https://doi.org/10.1182/blood-2006-06-032599.

Jost, M. M., Ninci, E., Meder, B., Kempf, C., Van Royen, N., Hua, J., et al. (2003). Divergent effects of GM-CSF and TGFbeta1 on bone marrow-derived macrophage arginase-1 activity, MCP-1 expression, and matrix metalloproteinase-12: A potential role during arteriogenesis. *The FASEB Journal, 17*(15), 2281–2283. https://doi.org/10.1096/fj.03-0071fje.

Liu, P., Fleete, M. S., Jing, Y., Collie, N. D., Curtis, M. A., Waldvogel, H. J., et al. (2014). Altered arginine metabolism in Alzheimer's disease brains. *Neurobiology of Aging, 35*(9), 1992–2003. https://doi.org/10.1016/j.neurobiolaging.2014.03.013.

Liu, Q., Zhang, C., Sun, A., Zheng, Y., Wang, L., & Cao, X. (2009). Tumor-educated CD11bhighIalow regulatory dendritic cells suppress T cell response through arginase I. *Journal of Immunology*, *182*(10), 6207–6216. https://doi.org/10.4049/jimmunol.0803926.

Lucas, R., Fulton, D., Caldwell, R. W., & Romero, M. J. (2014). Arginase in the vascular endothelium: Friend or foe? *Frontiers in Immunology*, *5*, 589. https://doi.org/10.3389/fimmu.2014.00589.

Marigo, I., Zilio, S., Desantis, G., Mlecnik, B., Agnellini, A. H., Ugel, S., et al. (2016). T cell cancer therapy requires CD40-CD40L activation of tumor necrosis factor and inducible nitric-oxide-synthase-producing dendritic cells. *Cancer Cell*, *30*(4), 651. https://doi.org/10.1016/j.ccell.2016.09.009.

Modolell, M., Corraliza, I. M., Link, F., Soler, G., & Eichmann, K. (1995). Reciprocal regulation of the nitric oxide synthase/arginase balance in mouse bone marrow-derived macrophages by TH1 and TH2 cytokines. *European Journal of Immunology*, *25*(4), 1101–1104. https://doi.org/10.1002/eji.1830250436.

Morris, S. M., Jr. (2002). Regulation of enzymes of the urea cycle and arginine metabolism. *Annual Review of Nutrition*, *22*, 87–105. https://doi.org/10.1146/annurev.nutr.22.110801.140547.

Morris, S. M., Jr. (2009). Recent advances in arginine metabolism: Roles and regulation of the arginases. *British Journal of Pharmacology*, *157*(6), 922–930. https://doi.org/10.1111/j.1476-5381.2009.00278.x.

Mortier, J., Prevost, J. R. C., Sydow, D., Teuchert, S., Omieczynski, C., Bermudez, M., et al. (2017). Arginase structure and inhibition: Catalytic site plasticity reveals new modulation possibilities. *Scientific Reports*, 7(1), 13616. https://doi.org/10.1038/s41598-017-13366-4.

Munder, M., Eichmann, K., & Modolell, M. (1998). Alternative metabolic states in murine macrophages reflected by the nitric oxide synthase/arginase balance: Competitive regulation by CD4+ T cells correlates with Th1/Th2 phenotype. *Journal of Immunology*, *160*(11), 5347–5354.

Munder, M., Eichmann, K., Moran, J. M., Centeno, F., Soler, G., & Modolell, M. (1999). Th1/Th2-regulated expression of arginase isoforms in murine macrophages and dendritic cells. *Journal of Immunology*, *163*(7), 3771–3777.

Munder, M., Mollinedo, F., Calafat, J., Canchado, J., Gil-Lamaignere, C., Fuentes, J. M., et al. (2005). Arginase I is constitutively expressed in human granulocytes and participates in fungicidal activity. *Blood*, *105*(6), 2549–2556. https://doi.org/10.1182/blood-2004-07-2521.

Narita, Y., Kitamura, H., Wakita, D., Sumida, K., Masuko, K., Terada, S., et al. (2013). The key role of IL-6-arginase cascade for inducing dendritic cell-dependent CD4(+) T cell dysfunction in tumor-bearing mice. *Journal of Immunology*, *190*(2), 812–820. https://doi.org/10.4049/jimmunol.1103797.

Pauleau, A. L., Rutschman, R., Lang, R., Pernis, A., Watowich, S. S., & Murray, P. J. (2004). Enhancer-mediated control of macrophage-specific arginase I expression. *Journal of Immunology*, *172*(12), 7565–7573. https://doi.org/10.4049/jimmunol.172.12.7565.

Pesce, J. T., Ramalingam, T. R., Mentink-Kane, M. M., Wilson, M. S., El Kasmi, K. C., Smith, A. M., et al. (2009). Arginase-1-expressing macrophages suppress Th2 cytokine-driven inflammation and fibrosis. *PLoS Pathogens*, *5*(4), e1000371. https://doi.org/10.1371/journal.ppat.1000371.

Pudlo, M., Demougeot, C., & Girard-Thernier, C. (2017). Arginase inhibitors: A rational approach over one century. *Medicinal Research Reviews*, *37*(3), 475–513. https://doi.org/10.1002/med.21419.

Rodriguez, P. C., Quiceno, D. G., Zabaleta, J., Ortiz, B., Zea, A. H., Piazuelo, M. B., et al. (2004). Arginase I production in the tumor microenvironment by mature myeloid cells inhibits T-cell receptor expression and antigen-specific T-cell responses. *Cancer Research*, *64*(16), 5839–5849. https://doi.org/10.1158/0008-5472.CAN-04-0465.

Romero, M. J., Iddings, J. A., Platt, D. H., Ali, M. I., Cederbaum, S. D., Stepp, D. W., et al. (2012). Diabetes-induced vascular dysfunction involves arginase I. *American Journal of Physiology. Heart and Circulatory Physiology*, *302*(1), H159–H166. https://doi.org/10.1152/ajpheart.00774.2011.

Rotondo, R., Barisione, G., Mastracci, L., Grossi, F., Orengo, A. M., Costa, R., et al. (2009). IL-8 induces exocytosis of arginase 1 by neutrophil polymorphonuclears in non-small cell lung cancer. *International Journal of Cancer*, *125*(4), 887–893. https://doi.org/10.1002/ijc.24448.

Rotondo, R., Bertolotto, M., Barisione, G., Astigiano, S., Mandruzzato, S., Ottonello, L., et al. (2011). Exocytosis of azurophil and arginase 1-containing granules by activated polymorphonuclear neutrophils is required to inhibit T lymphocyte proliferation. *Journal of Leukocyte Biology*, *89*(5), 721–727. https://doi.org/10.1189/jlb.1109737.

Solito, S., Pinton, L., De Sanctis, F., Ugel, S., Bronte, V., Mandruzzato, S., et al. (2019). Methods to measure MDSC immune suppressive activity in vitro and in vivo. *Current Protocols in Immunology*, *124*(1), e61. https://doi.org/10.1002/cpim.61.

Steggerda, S. M., Bennett, M. K., Chen, J., Emberley, E., Huang, T., Janes, J. R., et al. (2017). Inhibition of arginase by CB-1158 blocks myeloid cell-mediated immune suppression in the tumor microenvironment. *Journal for Immunotherapy of Cancer*, *5*(1), 101. https://doi.org/10.1186/s40425-017-0308-4.

Van de Velde, L. A., Subramanian, C., Smith, A. M., Barron, L., Qualls, J. E., Neale, G., et al. (2017). T cells encountering myeloid cells programmed for amino acid-dependent immunosuppression use rictor/mTORC2 protein for proliferative checkpoint decisions. *The Journal of Biological Chemistry*, *292*(1), 15–30. https://doi.org/10.1074/jbc.M116.766238.

CHAPTER TWELVE

Functional assay to assess T-cell inhibitory properties of myeloid derived suppressor cells (MDSCs) isolated from the tumor microenvironment of murine glioma models

Mahmoud S. Alghamri[a,b], Neha Kamran[a,b], Padma Kadiyala[a,b], Pedro Ricardo Lowenstein[a,b], Maria Graciela Castro[a,b,*]
[a]Department of Neurosurgery, University of Michigan Medical School, MSRB II, Ann Arbor, MI, United States
[b]Department of Cell and Developmental Biology, University of Michigan Medical School, MSRB II, Ann Arbor, MI, United States
*Corresponding author: e-mail address: mariacas@umich.edu

Contents

Abstract

Despite advances in uncovering the molecular mechanisms that mediate glioma progression and the implementation of novel therapeutic modalities, patients' prognosis remains dismal. This is due to both systemic and local tumor induced immune suppression. We are particularly interested in the role played by infiltrating immunosuppressive myeloid derived suppressor cells (MDSCs) in the glioma tumor microenvironment (TME). This immunosuppressive TME also interferes with the effectiveness of immunotherapies against glioma. Development of multipronged treatment approaches is imperative when aiming to generate a robust anti-glioma immune response. Evaluating

Methods in Enzymology, Volume 632
ISSN 0076-6879
https://doi.org/10.1016/bs.mie.2019.05.047

the inhibitory potential of MDSCs within the TME is an important aspect for developing effective treatments for glioma. Herein, we discuss methodology to assess the inhibitory effects of MDSCs isolated from the TME using a mouse glioma model.

1. Introduction

Myeloid derived suppressor cells (MDSCs) are a heterogeneous population of myeloid cells that give rise to immature macrophages, dendritic cells, granulocytes, and monocytes (Khaled, Ammori, & Elkord, 2013; Marvel & Gabrilovich, 2015; Parker, Beury, & Ostrand-Rosenberg, 2015). They mainly originate in pathological conditions such as persistent infections, chronic inflammation, or cancer, which alter normal hematopoiesis and block the myeloid cells' differentiation pathway. Immature MDSCs are highly immune suppressive; large numbers of MDSCs are recruited to the tumor microenvironment (TME) which hinder immunosurveillance and aid tumor immune evasion (Angell et al., 2016; Arihara et al., 2013; Diaz-Montero et al., 2009; Huang et al., 2013; Jordan et al., 2013; Sun et al., 2012; Yang et al., 2017; Zhang et al., 2016). Such scenario could impact survival and adversely affect prognosis.

MDSCs are classified into two subtypes; granulocytic MDSCs (also known as PMN-MDSCs) and monocytic MDSCs (M-MDSCs) (Youn, Nagaraj, Collazo, & Gabrilovich, 2008). In mice PMN-MDSCs are identified as $CD11b^{+}$ $Ly6G^{+}$ $Ly6C^{low}$ (Gr-1^{high}) whereas, the M-MDSCs are identified as $CD11b^{+}$ $Ly6G^{-}$ $Ly6C^{+}$ (Gr-1^{low}). In humans, PMN-MDSCs are identified as $CD11b^{+}CD14^{-}CD15^{+}$ or $CD11b^{+}CD14^{-}CD66b^{+}$, and M-MDSCs are identified as $CD11b^{+}CD14^{+}$HLA-$DR^{-/low}$ $CD15^{-}$ (Youn et al., 2008). Both MDSCs subtypes express immune suppressive molecules such as arginase (ARG1), iNOS, TGFβ, IL-10, COX2, indoleamine 2,3-dioxygenase (IDO), and PD-L1 (Gabrilovich, 2017; Kamran, Chandran, Lowenstein, & Castro, 2016; Kamran et al., 2017). Many studies have shown that the mechanisms of suppression also appear to be specific to the MDSCs subtype. For example, M-MDSC suppress T-cell responses both in antigen-specific and non-specific manners utilizing mechanisms associated with production of nitric oxide (NO), whereas PMN-MDSCs exert their function in an antigen-specific manner and require production of reactive oxygen species (ROS) (Movahedi et al., 2008; Youn et al., 2008). Direct contact of MDSCs with T cells results in the nitration of the T cell receptor and the CD8 molecule, which renders T cells' unresponsive to specific stimuli (Nagaraj et al., 2007). MDSCs have also been shown to support the development of FoxP3+ regulatory T cells (Tregs) through several mechanisms (Huang et al., 2006).

Several reports have shown the relation between immunosuppressive myeloid cells and tumor progression in a variety of tumors (Almand et al., 2001; Filipazzi et al., 2007; Hoechst et al., 2008; Iclozan, Antonia, Chiappori, Chen, & Gabrilovich, 2013; Kusmartsev et al., 2008; Poschke, Mougiakakos, Hansson, Masucci, & Kiessling, 2010; Romano et al., 2015; Wang et al., 2013). MDSCs from cancer patients have been shown to express elevated levels of PDL1, iNOS, arginase, ROS and IDO (Dubinski et al., 2016; Zea et al., 2005). MDSCs isolated from melanoma, glioma, squamous cell carcinoma, non-small cell lung cancer, breast cancer, hepatocellular carcinoma, gastric cancer, colorectal and chronic myeloid leukemia cancer patients have been shown to suppress antigen-specific and CD3-ligation induced T cell proliferation (Filipazzi et al., 2007; Hoechst et al., 2008; Iclozan et al., 2013; Kusmartsev et al., 2008; Poschke et al., 2010; Romano et al., 2015; Wang et al., 2013). MDSC expansion has been negatively associated with tumor progression, metastatic burden, response to immunotherapies and progression-free survival (Almand et al., 2001; Kamran et al., 2017; Romano et al., 2015).

The exact phenotypic, and molecular characteristics of MDSCs have not been identified to date. Recently, an algorithm defining the true immunosuppressive MDSCs characteristics based on phenotypic, biochemical, and functional characteristic has been proposed (Bronte et al., 2016). The algorithm classifies MDSCs mainly based on their T-cell inhibitory potential. Therefore, it is crucial to perform reproducible functional assays to uncover the functional properties of any MDSCs population. Here, we detail an *in vitro* reproducible methodology to test the T-cell inhibitory potential of MDSCs isolated from mouse brain tumors.

2. Principle of the assay

The assay utilizes OT-1 Rag2 knockout mice that carry a transgene encoding for T-cell receptor specific for a chicken ovalbumin peptide antigenic fragment (257–264 (SIINFEKL)), which is presented by the MHC class I molecule H2-Kb. When splenocytes from OT-1 mice are cultured *in vitro* in the presence of SIINFEKL peptide, T-cells will proliferate. Splenocytes from OT-1 mice are labeled with carboxyfluorescein succinimidyl ester (CFSE) (cell permeable and stable intracellular dye) prior to *in vitro* culturing. As the OT-1T-cells proliferate in response to SIINFEKL stimulation, the resulting daughter cells incorporate half the number of CFSE-tagged molecules (Fig. 1). By measuring the intensity of CFSE in Naïve T-cells vs stimulated T-cells, we can quantify T cells

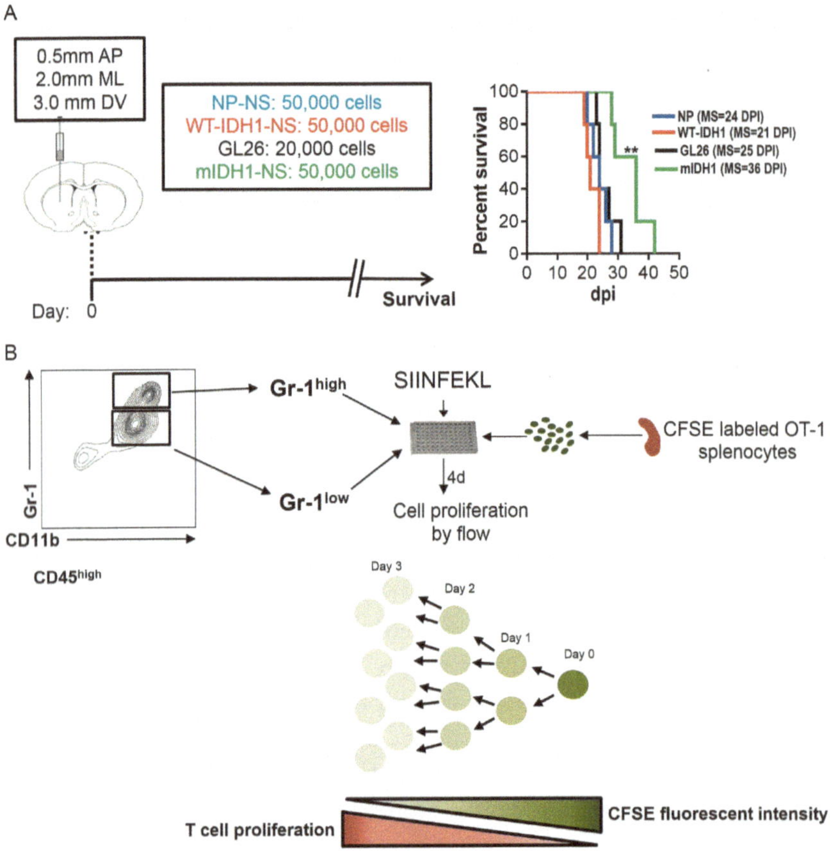

Fig. 1 Schematic illustrating NS implantation and the *in vitro* T-cell proliferation assay to analyze immune suppressive potential of MDSCs. (A) Schematic shows stereotactic coordinates for injection of syngeneic mouse glioma NS and GL26 cell line as well as the cell number, and expected median survival. AP: anteroposterior; ML: mediolateral; DV: dorsoventral; MS: median survival. (B) OT-1 splenocytes are labeled with the CFSE dye and co-cultured with Gr-1^{high} or Gr-1^{low} MDSCs sorted from tumor. When OT-1 splenocytes divide due to SIINFIKLE stimulation, the resulting daughter cells incorporate half the number of cytoplasmic carboxyfluorescein- (CFSE) fluorescent molecules. As cells T cells divide in response to SIINFEKL stimulation, fluorescence intensity decreases exponentially. T cell proliferation can be assessed by measuring the corresponding decrease in the dye intensity via flow cytometry.

proliferation. If SIINFEKL stimulated OT-1T-cells are cultured with inhibitory immune cells (such as MDSCs) T-cells' proliferation will diminish. Please note that this protocol has been optimized only in OT-1 Rag2 knockout mice. A list of reagents and the sources are listed in Table 1.

Table 1 Reagents/materials used in the protocol.

Reagent	Source	Catalog Number
Dulbecco's Modified Eagle Medium (DMEM)	Thermofisher	Cat #: 12430054
heat-inactivated fetal bovine serum (FBS)	Atlanta Biologicals	Cat #: S11150
Penstrep (10,000 U/mL)	Thermofisher	Cat #:15140122
G418	Thermofisher	Cat #:10131035
Trypsin-EDTA (0.05%), phenol red	Thermofisher	Cat #:25300054
DPBS, no calcium and no magnesium	Thermofisher	Cat #:14190144
SIINFEKL (OVA)	Anaspec	Cat #: 60193-1
RBC Lysis Buffer	Biolegend	Cat #: 420301
β-Mercaptoethanol	Sigma-Aldrich	Cat #: MS620-10ML
Percoll	Sigma-Aldrich	Cat #: P4937-500ML
Roswell Park Memorial Institute Medium (RPMI)	Thermofisher	Cat #: 12633012
10 × Hanks' Balanced Salt Solution (HBSS)	Thermofisher	Cat #: 14185052
1 × Hanks' Balanced Salt Solution (HBSS)	Thermofisher	Cat #: 14175095
Ethyl alcohol (Ethanol)	Sigma-Aldrich	Cat #: E7023-1 L
Corning 70 μm cell strainer	Sigma-Aldrich	Cat #: CLS431751-50EA
Ketamine	Zoetis	Cat #: 000680
Dexmedetomidine	Precedex	Cat #: 122692–5
Covidine Monoject 1 mL tuberculin Syringes	Thermofisher	Cat #: 22–257-154
26-Gauge precision glide needles	BD Bioscience	Cat #: 30511

3. Isolation of OT-1 spleenocytes from Rag 2 knockout mice

Note: Six to eight-week old female OT-1 Rag 2 knockout mice were purchased from Taconic Farm (Cat number 2334-F, Rensselaer, NY) and were housed in pathogen free conditions at the University of Michigan. All experimental studies were performed in compliance with Institutional Animal Care & Use Committee (IACUC).

1. Sacrifice an OT-1 Rag2 knockout mouse by injecting intraperitoneally (i.p.) 400 μL of anesthetics containing Ketamine (75 mg/Kg, Pfizer, NY) and dexmedetomidine (1 mg/kg) to induce deep anesthesia.
2. Place the anesthetized mouse on a clean dissection board and check for pedal and tail reflexes. When no responses are observed in both tests, the mouse is considered to be in a deep anesthetized state.
3. Pin the mouse down ventral side up using 26G needles and disinfect the abdomen with 70% ethanol.
4. Hold the skin above the peritoneal cavity using surgical forceps and with scissors make two slits on either side of the cavity.
5. Carefully extract the spleen that is located to the left side of the abdomen, inferior to the stomach.
6. Place the spleen in a 1.5 mL Eppendorf tube containing 1 mL of Dulbecco's modified eagle media (DMEM, Gibco, NH) supplemented with 10% fetal bovine serum (FBS, Gibco, NH) 100 unit/mL penicillin, and 0.3 mg/mL L-glutamine (Gibco, NH) (i.e, solution 1).
7. To prepare the splenocytes' cell suspension, place the spleen onto a 70 μm (Alkali Inc., FL) strainer attached to a 50 mL conical tube.
8. Using the plunger end of a syringe, carefully mash the spleen through the strainer.
9. Wash the cells through the strainer with 20 mL of solution 1.
10. Centrifuge the cell suspension at 500 × *g* for 5 min and aspirate the supernatant.
11. Resuspend the cell pellet in 2 mL of red blood cells (RBCs) lysis buffer (Biolegend) and incubate the cells at 4 °C for 2 min.
12. Add at least 20 mL of solution 1 and centrifuge the cells at 500 × *g* for 5 min at 4 °C.
13. Discard the supernatant and resuspend the cells in 2 mL of solution 1 and place the tube on ice.

14. To check cell viability, perform a cell count using trypan blue and a hemocytometer.
15. After counting, wash the splenocytes twice with Dulbeco's phosphate-buffered saline (DPBS, Gibco, NH) and resuspend them in DBPS at a concentration of 1.0×10^6 cells/mL.

4. Labeling OT-1 splenocytes with carboxyfluorescein succinimidyl ester

1. Prepare 10 mM stock solution of carboxyfluorescein succinimidyl ester (CFSE, Invitrogen, CA) by adding 9 μL DMSO to a vial of Cell Trace CFSE staining solution. Then dilute the stock solution in 20 mL of DPBS for a 5 μM staining solution.
2. Also, prepare Roswell Park Memorial Institute (RPMI, Gibco, NH) media supplemented with 10% FBS, 100 unit/mL penicillin, and 55 μM 2-mercaptoethanol (i.e., solution 2).
3. Resuspend the 1 mL of cells in DPBS from part 3 in 1 mL of 5 μM CSFE solution prepared in step 1 (final concentration of CFSE will be 2.5 μM).
4. Incubate the cells in the dark for 5 min at 37 °C with occasional shaking.
5. At the end of the incubation, add 2 mL of pure FBS to quench the staining reaction.
6. Pellet the splenocytes at 500 × *g* for 5 min at 4 °C (the cell pellet will appear yellow at this point).
7. Wash the pellet with solution 2 check cell viability by performing a cell count using trypan blue (Invitrogen) and a hemocytometer.
8. After counting, resuspend the splenocytes at a concentration of 100,000 cells/ 50 μL in solution 2.

5. MDSCS isolation from tumor

We have established protocol to isolate and maintain brain tumor neurospheres (NS) generated using the sleeping beauty (SB) transposase system (Calinescu et al., 2015; Koschmann et al., 2016; Núñez et al., 2019). Tumor implantation was preformed using the SB generated-NS as previously described (Baker, Castro, & Lowenstein, 2015). Briefly, 6–8 weeks old C57BL6 female mice were anesthetized using ketamine (75 mg/kg) and dexmedetomidine (1 mg/kg) prior to stereotactic implantation with

50,000 NS in the right striatum. The coordinates for NS implantation are 0.5 mm anterior and 2.0 mm lateral from the bregma and 3.0 mm ventral from the dura (Fig. 1A). NS were injected at a rate 1 μL/min (Baker et al., 2015). At moribund stage, tumor infiltrating immune cells were purified from the tumor tissue by percoll™ plus lymphocyte purification kit as described before (Kamran et al., 2017). A detailed procedure for percoll purification is detailed in the next section. After purification, immune cells were stained with Alexa fluor 700-CD45 (Biolegend), APC-CD11b (Biolegend), PE-Texas Red-Gr-1 (Biolegend). Granulocytic and monocytic MDSCs were sorted based on the expression of $CD45^{high}$ $Gr\text{-}1^{high}$ and $CD45^{high}$ $Gr\text{-}1^{low}$, respectively (Fig. 2). Purified $Gr\text{-}1^{high}$ and $Gr\text{-}1^{low}$ MDSCs were counted and diluted to 100,000 cells/50 μL solution 2 in preparation for co-culturing with OT-1 splenocytes.

6. Isolation of tumor infiltrating lymphocytes by density gradient

1. Prepare stock isotonic percoll™ (SIP, GE Healthcare life sciences), 4 mLl per brain, by mixing 9 parts of percoll with one part of 10 × HBSS (Gibco) without Ca^{++} and Mg^{++}.
2. Prepare SIP at 70% in 1 × HBSS (Gibco) without Ca^{++} and Mg^{++}, 3 mL per brain dispensed into a 15 mL polypropylene conical tube
3. Resuspend the pellet with 7 mL of DMEM 10% FCS media
4. Add 3 mL of SIP to the cell suspension pellet to make a final 30% SIP
5. Slowly layer the 10 mL cell suspension on top of the 1 mL of 70% SIP (pipet slowly and carefully). This is the most critical step of the procedure, use a pipette-aid set in the gravity mode avoiding mixing of the 70% and 30% solutions. A very clear flat line should be visible at the 70%–30% interface.
6. Centrifuge 20 min at 800 × *g* 18 °C. Make sure centrifuge will stop with minimal or no brake so that the interphase is not disturbed.
7. Using a transfer pipet, gently remove the layer of debris from the top of the tube and collect 2.0–3.0 mL of the 70%–30% interphase into a clean conical tube containing 8 mL of 1 × HBSS. Ensure that the percoll containing the interphase is diluted about threefold, mix a few times by inversion and centrifuge 7 min at 500 × *g* at 18 °C.
8. Resuspend pellet in 1 mL of blocking buffer and proceed to the cell staining procedure for MDSC cells sorting as described in Section 5.

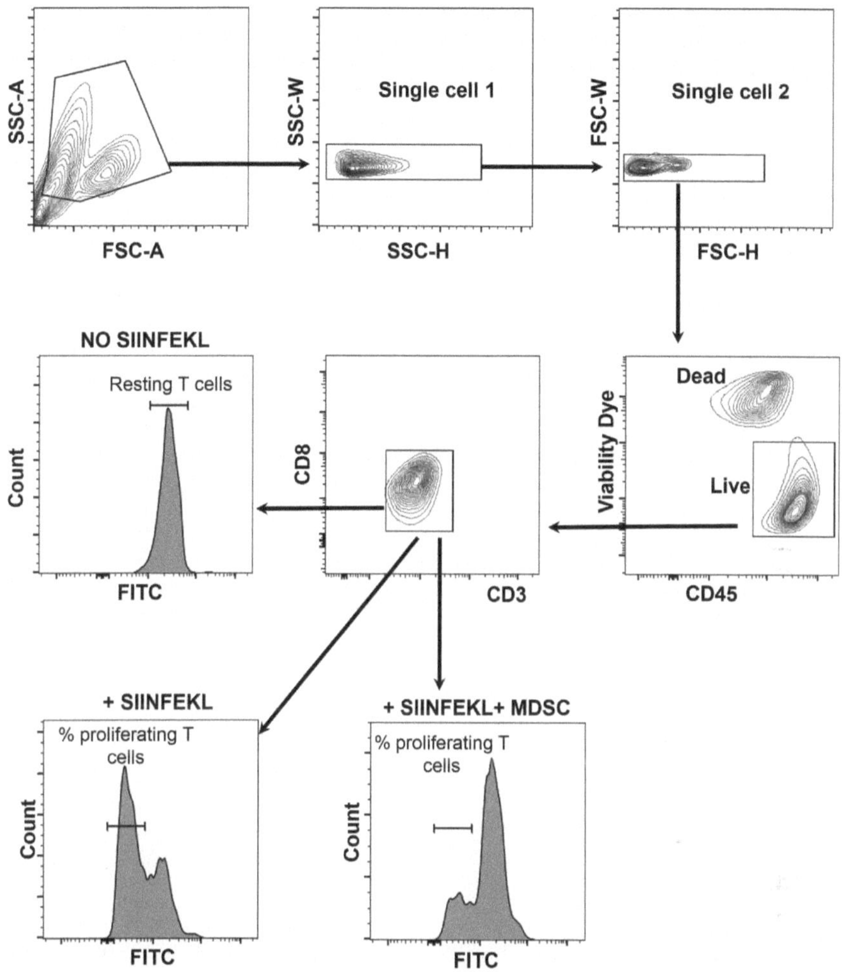

Fig. 2 Gating strategy to detect percentage of proliferating T cells. Splenocytes from OT-1 mice were extracted, labeled with CFSE dye, and stimulated *in vitro* with SIINFEKL peptide in the absence or presence of MDSCs for 4 days. Cells were stained with efluor 780 viability dye, Alexa fluor 700-CD45, Percp cy5.5-CD3, and Pacific Blue-CD8. Percentage of proliferating T cells were gated based on control condition (no SIINFEKL stimulation).

7. Coculturing MDSCS with OT-1 splenocytes

1. Plate the CFSE labeled splenocytes at a density of 100,000 cells in 50 μL in a 96-well round bottom plate (Thermo-fisher, MA) for the following conditions:

(a) Condition 1: Unstimulated splenocytes (No SIINFEKL)
(b) Condition 2: Stimulated OT-1 splenocytes (+ SIINFEKL)
(c) Condition 3: Stimulated OT-1 splenocytes + 50 μL Gr-1^{high} MDSCs
(d) Condition 4: Stimulated OT-1 splenocyte + 50 μL Gr-1^{low} MDSCs

2. Prepare 1 mL of 200 nM SIINFEKL peptide (AnaSpec Inc., CA) in solution 2 to stimulate the OT-1 splenocytes.
3. For condition 1, add 150 μL of solution 2 (NO SIINFEKL should be added).
4. For condition 2, add 100 μL of 200 nM SIINFEKL peptide to the wells from step 1 + 50 μL of solution 2 (final SIINFEKL concentration in each well will be 100 nM).
5. For condition 3 and 4, add 100 μL of 200 nM SIINFEKL peptide to the wells from step 1 + 50 μL (100,000 cells) of sorted Gr-1^{high} or Gr-1^{low} MDSCs, respectively (final SIINFEKL concentration in each well will be 100 nM).
6. Incubate the cells at 37 °C, 5% CO_2 humidified incubator for 4 days before flow staining for PE-Texas red-CD45 (Biolegend), Percp cy5.5-CD3 (Biolegend), PB-CD8 (Biolegend), and APC-Cy7 live/dead staining (ebioscience).
7. Proceed with Flow cytometry analysis to determine the percentage of T-cells proliferation based on CFSE expression as illustrated in Fig. 2.

8. Notes

1. It is important to stain the splenocytes with CFSE in FBS free media. FBS quenches the CFSE staining, therefore avoid using FBS containing media to label the OT-1 splenocytes (see Section 4, steps 1–4)
2. CFSE overlaps with the FITC channel, therefore do not use FITC conjugated antibody when staining for CD3, CD8, or when MDSCs are being sorted by flow cytometry.
3. We notice that using a round bottom 96-well plate for plating the cells has led to optimal T cell proliferation compared to plating the cells in flat bottom plate. This is probably due to the enhanced antigen presentation, which leads to a greater cell proliferation when the OT-1 splenocytes are stimulated by SIINFEKL.
4. When plating the splenocytes in a 96-well plate, avoid plating at the edge of the plate, instead, add PBS to all wells at the edges of the plate.

5. SIINFEKL is a peptide that can be degraded by freeze-thawing. Therefore, it is important to prepare small aliquots of the peptide upon receiving from the manufacture to prevent freeze-thawing
6. To get enough MDSCs for performing splenocyte: MDSCs co-culturing, you may need to pool together cells from more than one animal's tumor. In this protocol, we plated 100,000 MDSCs/ well.
7. As a positive control condition, we recommend to co-culture $CD11b^+$ $Gr\text{-}1^+$ cells isolated from BM of WT mice (tumor free animal) with OT-1 splenocytes. We have previously observed that $CD11b^+$ $Gr\text{-}1^+$ cells isolated from BM of WT mice do not inhibit T-cell proliferation.
8. We also recommend co-culturing CFSE labeled MDSCs splenocytes in a dose dependent manner; for example, the ratio of splenocyte: MDSCs could be 1:1, 1:2, or 1:4.
9. It is important to add the β-ME to the splenocytes culture media to reduce the oxidative stress on the splenocytes cultured in the 96-well plate.
10. Use appropriate antibody isotype to gate for positive population of CD45, CD3, CD8.

Acknowledgments

This work was supported by NIH/NINDS Grants, R37-NS094804, R01-NS105556 and 1R21NS107894 to M.G.C.; NIH/NINDS Grants R01-NS076991, and R01-NS096756 to P.R.L.; NIH/NIBIB: R01-EB022563; NIH/NCI U01CA224160; the Department of Neurosurgery, the Rogel Cancer Center, Program in Cancer Hematopoiesis and Immunology (CHI), the ChadTough Foundation and Leah's Happy Hearts Foundation to M.G.C. and P.R.L. RNA Biomedicine Grant F046166 to M.G.C.; NIH/NCI T32-CA009676 to M.S.A.

References

Almand, B., Clark, J. I., Nikitina, E., van Beynen, J., English, N. R., Knight, S. C., et al. (2001). Increased production of immature myeloid cells in cancer patients: A mechanism of immunosuppression in cancer. *Journal of Immunology*, *166*(1), 678–689.

Angell, T. E., Lechner, M. G., Smith, A. M., Martin, S. E., Groshen, S. G., Maceri, D. R., et al. (2016). Circulating myeloid-derived suppressor cells predict differentiated thyroid cancer diagnosis and extent. *Thyroid*, *26*(3), 381–389. https://doi.org/10.1089/thy.2015.0289.

Arihara, F., Mizukoshi, E., Kitahara, M., Takata, Y., Arai, K., Yamashita, T., et al. (2013). Increase in CD14+HLA-DR-/low myeloid-derived suppressor cells in hepatocellular carcinoma patients and its impact on prognosis. *Cancer Immunology, Immunotherapy: CII*, *62*(8), 1421–1430. https://doi.org/10.1007/s00262-013-1447-1.

Baker, G. J., Castro, M. G., & Lowenstein, P. R. (2015). Isolation and flow cytometric analysis of glioma-infiltrating peripheral blood mononuclear cells. *Journal of Visualized Experiments*, *105*, 53676. https://doi.org/10.3791/53676.

Bronte, V., Brandau, S., Chen, S. H., Colombo, M. P., Frey, A. B., Greten, T. F., et al. (2016). Recommendations for myeloid-derived suppressor cell nomenclature and characterization standards. *Nature Communications*, *7*, 12150. https://doi.org/10.1038/ncomms12150.

Calinescu, A.-A., Núñez, F. J., Koschmann, C., Kolb, B. L., Lowenstein, P. R., & Castro, M. G. (2015). Transposon mediated integration of plasmid DNA into the subventricular zone of neonatal mice to generate novel models of glioblastoma. *Journal of Visualized Experiments: JoVE*, *96*, 52443. https://doi.org/10.3791/52443.

Diaz-Montero, C. M., Salem, M. L., Nishimura, M. I., Garrett-Mayer, E., Cole, D. J., & Montero, A. J. (2009). Increased circulating myeloid-derived suppressor cells correlate with clinical cancer stage, metastatic tumor burden, and doxorubicin-cyclophosphamide chemotherapy. *Cancer Immunology, Immunotherapy*, *58*(1), 49–59. https://doi.org/10.1007/s00262-008-0523-4.

Dubinski, D., Wolfer, J., Hasselblatt, M., Schneider-Hohendorf, T., Bogdahn, U., Stummer, W., et al. (2016). CD4 + T effector memory cell dysfunction is associated with the accumulation of granulocytic myeloid-derived suppressor cells in glioblastoma patients. *Neuro-Oncology*, *18*(6), 807–818. https://doi.org/10.1093/neuonc/nov280.

Filipazzi, P., Valenti, R., Huber, V., Pilla, L., Canese, P., Iero, M., et al. (2007). Identification of a new subset of myeloid suppressor cells in peripheral blood of melanoma patients with modulation by a granulocyte-macrophage colony-stimulation factor-based antitumor vaccine. *Journal of Clinical Oncology*, *25*(18), 2546–2553. https://doi.org/10.1200/JCO.2006.08.5829.

Gabrilovich, D. I. (2017). Myeloid-derived suppressor cells. *Cancer Immunology Research*, *5*(1), 3–8. https://doi.org/10.1158/2326-6066.CIR-16-0297.

Hoechst, B., Ormandy, L. A., Ballmaier, M., Lehner, F., Kruger, C., Manns, M. P., et al. (2008). A new population of myeloid-derived suppressor cells in hepatocellular carcinoma patients induces CD4(+)CD25(+)Foxp3(+) T cells. *Gastroenterology*, *135*(1), 234–243. https://doi.org/10.1053/j.gastro.2008.03.020.

Huang, B., Pan, P. Y., Li, Q., Sato, A. I., Levy, D. E., Bromberg, J., et al. (2006). Gr-1+CD115+ immature myeloid suppressor cells mediate the development of tumor-induced T regulatory cells and T-cell anergy in tumor-bearing host. *Cancer Research*, *66*(2), 1123–1131. https://doi.org/10.1158/0008-5472.CAN-05-1299.

Huang, A., Zhang, B., Wang, B., Zhang, F., Fan, K. X., & Guo, Y. J. (2013). Increased CD14(+)HLA-DR (−/low) myeloid-derived suppressor cells correlate with extrathoracic metastasis and poor response to chemotherapy in non-small cell lung cancer patients. *Cancer Immunology, Immunotherapy*, *62*(9), 1439–1451. https://doi.org/10.1007/s00262-013-1450-6.

Iclozan, C., Antonia, S., Chiappori, A., Chen, D. T., & Gabrilovich, D. (2013). Therapeutic regulation of myeloid-derived suppressor cells and immune response to cancer vaccine in patients with extensive stage small cell lung cancer. *Cancer Immunology, Immunotherapy*, *62*(5), 909–918. https://doi.org/10.1007/s00262-013-1396-8.

Jordan, K. R., Amaria, R. N., Ramirez, O., Callihan, E. B., Gao, D., Borakove, M., et al. (2013). Myeloid-derived suppressor cells are associated with disease progression and decreased overall survival in advanced-stage melanoma patients. *Cancer Immunology, Immunotherapy*, *62*(11), 1711–1722. https://doi.org/10.1007/s00262-013-1475-x.

Kamran, N., Chandran, M., Lowenstein, P. R., & Castro, M. G. (2016). Immature myeloid cells in the tumor microenvironment: Implications for immunotherapy. *Clinical Immunology*, *189*, 34–42. https://doi.org/10.1016/j.clim.2016.10.008.

Kamran, N., Kadiyala, P., Saxena, M., Candolfi, M., Li, Y., Moreno-Ayala, M. A., et al. (2017). Immunosuppressive myeloid cells' blockade in the glioma microenvironment enhances the efficacy of immune-stimulatory gene therapy. *Molecular Therapy, 25*(1), 232–248. https://doi.org/10.1016/j.ymthe.2016.10.003.

Khaled, Y. S., Ammori, B. J., & Elkord, E. (2013). Myeloid-derived suppressor cells in cancer: Recent progress and prospects. *Immunology and Cell Biology, 91*(8), 493–502. https://doi.org/10.1038/icb.2013.29.

Koschmann, C., Calinescu, A.-A., Nunez, F. J., Mackay, A., Fazal-Salom, J., Thomas, D., et al. (2016). ATRX loss promotes tumor growth and impairs nonhomologous end joining DNA repair in glioma. *Science Translational Medicine, 8*(328), 328ra328. https://doi.org/10.1126/scitranslmed.aac8228.

Kusmartsev, S., Su, Z., Heiser, A., Dannull, J., Eruslanov, E., Kubler, H., et al. (2008). Reversal of myeloid cell-mediated immunosuppression in patients with metastatic renal cell carcinoma. *Clinical Cancer Research, 14*(24), 8270–8278. https://doi.org/10.1158/1078-0432.CCR-08-0165.

Marvel, D., & Gabrilovich, D. I. (2015). Myeloid-derived suppressor cells in the tumor microenvironment: Expect the unexpected. *The Journal of Clinical Investigation, 125*(9), 3356–3364. https://doi.org/10.1172/JCI80005.

Movahedi, K., Guilliams, M., Van den Bossche, J., Van den Bergh, R., Gysemans, C., Beschin, A., et al. (2008). Identification of discrete tumor-induced myeloid-derived suppressor cell subpopulations with distinct T cell-suppressive activity. *Blood, 111*(8), 4233–4244. https://doi.org/10.1182/blood-2007-07-099226.

Nagaraj, S., Gupta, K., Pisarev, V., Kinarsky, L., Sherman, S., Kang, L., et al. (2007). Altered recognition of antigen is a mechanism of CD8+ T cell tolerance in cancer. *Nature Medicine, 13*(7), 828–835. https://doi.org/10.1038/nm1609.

Núñez, F. J., Mendez, F. M., Kadiyala, P., Alghamri, M. S., Savelieff, M. G., Garcia-Fabiani, M. B., et al. (2019). IDH1-R132H acts as a tumor suppressor in glioma via epigenetic up-regulation of the DNA damage response. *Science Translational Medicine, 11*(479), pii: eaaq1427.

Parker, K. H., Beury, D. W., & Ostrand-Rosenberg, S. (2015). Myeloid-derived suppressor cells: Critical cells driving immune suppression in the tumor microenvironment. *Advances in Cancer Research, 128*, 95–139. https://doi.org/10.1016/bs.acr.2015.04.002.

Poschke, I., Mougiakakos, D., Hansson, J., Masucci, G. V., & Kiessling, R. (2010). Immature immunosuppressive CD14+HLA-DR−/low cells in melanoma patients are Stat3hi and overexpress CD80, CD83, and DC-sign. *Cancer Research, 70*(11), 4335–4345. https://doi.org/10.1158/0008-5472.CAN-09-3767.

Romano, A., Parrinello, N. L., Vetro, C., Forte, S., Chiarenza, A., Figuera, A., et al. (2015). Circulating myeloid-derived suppressor cells correlate with clinical outcome in Hodgkin Lymphoma patients treated up-front with a risk-adapted strategy. *British Journal of Haematology, 168*(5), 689–700. https://doi.org/10.1111/bjh.13198.

Sun, H. L., Zhou, X., Xue, Y. F., Wang, K., Shen, Y. F., Mao, J. J., et al. (2012). Increased frequency and clinical significance of myeloid-derived suppressor cells in human colorectal carcinoma. *World Journal of Gastroenterology, 18*(25), 3303–3309. https://doi.org/10.3748/wjg.v18.i25.3303.

Wang, L., Chang, E. W., Wong, S. C., Ong, S. M., Chong, D. Q., & Ling, K. L. (2013). Increased myeloid-derived suppressor cells in gastric cancer correlate with cancer stage and plasma S100A8/A9 proinflammatory proteins. *Journal of Immunology, 190*(2), 794–804. https://doi.org/10.4049/jimmunol.1202088.

Yang, G., Shen, W., Zhang, Y., Liu, M., Zhang, L., Liu, Q., et al. (2017). Accumulation of myeloid-derived suppressor cells (MDSCs) induced by low levels of IL-6 correlates with poor prognosis in bladder cancer. *Oncotarget, 8*(24), 38378–38388. https://doi.org/10.18632/oncotarget.16386.

Youn, J. I., Nagaraj, S., Collazo, M., & Gabrilovich, D. I. (2008). Subsets of myeloid-derived suppressor cells in tumor-bearing mice. *Journal of Immunology*, *181*(8), 5791–5802.

Zea, A. H., Rodriguez, P. C., Atkins, M. B., Hernandez, C., Signoretti, S., Zabaleta, J., et al. (2005). Arginase-producing myeloid suppressor cells in renal cell carcinoma patients: A mechanism of tumor evasion. *Cancer Research*, *65*(8), 3044–3048. https://doi.org/10.1158/0008-5472.CAN-04-4505.

Zhang, S., Ma, X., Zhu, C., Liu, L., Wang, G., & Yuan, X. (2016). The role of myeloid-derived suppressor cells in patients with solid tumors: A meta-analysis. *PLoS One*, *11*(10). e0164514https://doi.org/10.1371/journal.pone.0164514.

CHAPTER THIRTEEN

FACS isolation and analysis of human circulating and tumor neutrophils

Agnieszka Swierczak, Jeffrey W. Pollard*
MRC Centre for Reproductive Health, Queen's Medical Research Institute, The University of Edinburgh, Edinburgh, United Kingdom
*Corresponding author: e-mail address: jeff.pollard@ed.ac.uk

Contents

Methods in Enzymology, Volume 632
ISSN 0076-6879
https://doi.org/10.1016/bs.mie.2019.07.023

Abstract

The role of neutrophils in tumor growth and metastasis is still controversial. Studies in clinically relevant models of cancer have shown that neutrophils can promote tumor growth and development and metastasis, or inhibit it. Thus, further analysis is required to fully elucidate the role of neutrophils in cancer. A number of different methods are available for neutrophil isolation and characterization. However, Fluorescence-activated cell sorting (FACS) is particularly effective for isolating neutrophils and assessing their phenotype as it allows for the simultaneous use of multiple cell surface markers, can be used for isolation of both blood and tumor neutrophils and features a high purity and high yields.

1. Introduction

Whole tissues are comprised of complex environments with epithelial cells embedded in stromal elements. The stromal cell compartment typically contains cells of mesenchymal origin, including fibroblasts, endothelial cells and adipocytes as well as abundant extracellular matrix. In addition there is a hematopoietic cell component that is primarily made up of innate immune cells such as macrophages, mast cells, innate lymphoid cells and neutrophils, as well as some adaptive immune cells particularly T cells.

In the face of infection or damage, neutrophils are typically the first effector cells to arrive at inflammatory sites often recruited by chemokines secreted by sentinel mast cells or macrophages embedded within tissues. Neutrophils are the most abundant immune cell in the human circulation and are well poised for this early effector role. Their development, termed granulopoiesis, begins in the bone marrow where they are derived from myeloid precursors under the control of granulocyte colony-stimulating factor (G-CSF; CSF3) after which they are released into the blood stream. Circulating neutrophils were originally believed to have a short-life span of ~5 h which can be extended during inflammatory responses to several days. However, it now appears that the circulating neutrophil lifespan may be longer than originally thought (Bekkering, 2013).

Neutrophils contain a number of granules that enclose a battery of proteins, proteases, antibiotics and reactive oxygen species. These include antimicrobial proteins, such as lactoferrin and lysozyme, the matrix metalloproteinases, MMP-8, MMP-9 and MMP-25, the serine proteases, cathepsin G and neutrophil elastase (NE), and the enzyme, myeloperoxidase (MPO). These proteins contribute to the ability of neutrophils to migrate to sites of inflammation or tissue damage, destroy invading pathogens and infected/damaged cells and modulate inflammatory responses (Smith, 1994). Neutrophils can also contribute to tissue repair via phagocytosis of tissue debris, release of growth or pro-angiogenic factors, and neutrophil apoptosis and subsequent clearance by macrophages is a key step in resolution of inflammatory responses during wound healing (Soehnlein & Lindbom, 2010).

In contrast to the positive roles in host defense and repair neutrophils are often implicated in tissue injury and inflammatory diseases due to their release of toxic factors, such as reactive oxygen species (ROS), proteolytic enzymes, antimicrobial proteins and the release of neutrophil extracellular traps (NETs). Furthermore these toxic products can also contribute to the development of a number of inflammatory diseases including, systemic lupus erythematous (SLE), rheumatoid arthritis (RA), diabetes, atherosclerosis, vasculitis, thrombosis and cancer (Jorch & Kubes, 2017). However, as mentioned above, these neutrophil-derived factors do not always cause tissue damage and in fact can have a beneficial effect under certain inflammatory conditions (Gaut et al., 2001; Reber et al., 2017; Wilgus, Roy, & McDaniel, 2013). Thus, it is now recognized that neutrophils are complex cells that produce cytokines and chemokines (Mantovani, Cassatella, Costantini, & Jaillon, 2011), modulate the activities of neighboring cells, and contribute to both the commencement and resolution of inflammation as well as its persistence in several pathological conditions (Greenlee-Wacker, 2016).

The role of neutrophils in tumor growth and metastasis is still controversial. Early research indicated anti-tumoral roles for neutrophils (Di Carlo et al., 2001).

However, research using more clinically relevant models of cancer has shown that neutrophils can promote tumor growth and development (Coffelt et al., 2015; Huh, Liang, Sharma, Dong, & Robertson, 2010; Kowanetz et al., 2010; Swierczak et al., 2014). Furthermore, neutrophils and/or neutrophil-derived factors have also been implicated in promoting tumor metastasis (Jablonska, Lang, Sionov, & Granot, 2017). Neutrophils have also been shown to play key roles in immunomodulation during cancer progression (Coffelt et al., 2015; Kersten et al., 2017) and have been shown to form a key component of the pre-metastatic/metastatic niche (Kaplan et al., 2005; Wculek & Malanchi, 2015). In a recent study neutrophil-derived NETs were found to play a role in the awakening and proliferation of dormant cancer cells via integrin activation and FAK/ERK/MLCK/YAP signaling (Albrengues et al. 2018). While neutrophils in the circulation were found to form clusters with circulating tumor cells and enhance metastasis (Szczerba et al. 2019).

On the other hand, neutrophils have also been shown to inhibit metastatic seeding (Granot et al., 2011). These conflicting reports have somewhat been explained via the recent identification of a neutrophil phenotypic spectrum, and possible sub-populations of neutrophils in tumors (Michaeli et al., 2017; Mishalian et al., 2013; Sagiv et al., 2015). Thus, it has been proposed that neutrophils may carry out pro- or anti-tumor functions depending on cancer stage and microenvironment (Fridlender et al., 2009; Mishalian et al., 2013; Singhal et al., 2016).

There are a number of methods available to assess neutrophil function and phenotype including gene and protein expression analysis, functional studies, assessment of cytokine and chemokine production and analysis of cell surface markers. Nonetheless, in order to carry out these studies neutrophils must be first isolated. This isolation can be achieved from blood and/or tissue via a variety of methods such as; (1) density gradient separation, (2) magnetic cell isolation, (3) fluorescence-activated cell sorting (FACS), and (4) laser capture microdissection. However, it is important to remember that neutrophils by nature are very sensitive to changes in their environment, hence it is crucial to think carefully about which method is most suited to the analyses to be carried out.

Laser capture microdissection allows for isolation of cells according to spatial position and morphology. However, it is limited by the use of a maximum of three markers, the need to fix the cells, the time-consuming nature of the protocol, the restriction to tissues, and the low yield. Density gradient separation is a quick method for isolating neutrophils, although there can be issues with purity as it relies on differences in cell size and density to separate the cells. This can be overcome by the addition of a dextran sedimentation step. However, this can activate neutrophils (Quach & Ferrante, 2017). Furthermore, density gradient separation is rarely successful for isolation of tumor/tissue neutrophils. Magnetic cell isolation is also a quick method for neutrophil isolation, both from tissue and circulation, but there are some caveats. For example, positive selection, where beads are conjugated to antibodies against neutrophil cell surface markers, can also lead to neutrophil activation and/or changes in downstream signaling pathways. To minimize this negative selection can be used whereby all other cell populations are removed leaving only the neutrophils. However, many immune cells share markers with neutrophils and hence the purity can be an issue (Zhou et al., 2012). Furthermore, the yield is much lower using negative bead selection. A study assessing leukocyte gene expression after positive-, negative-selection, and FACS found that there were minimal changes in gene expression signatures after negative selection and FACS. Furthermore, stress response genes were more highly expressed in cells isolated by negative and positive selection but not FACS (Beliakova-Bethell et al., 2014). Although this study did not uniquely assess neutrophils it still indicates that FACS is a better method for neutrophil isolation given that it results in the highest purity and is less likely to induce a stress response. See Table 1 for a summary of the methods available for neutrophil isolation.

FACS is particularly effective for isolating neutrophils as it can be used to isolate neutrophils from both the circulation and tissue, and allows for the simultaneous use of multiple cell surface markers, hence resulting in higher purity but also assessment of phenotype. Commonly used markers for mouse

Table 1 Summary of the methods available for neutrophil isolation.

Method	Purity	Activation	Yield	Use	Markers
Laser capture microdissection	Good	Minimal	Low	Tumor	3
Density gradient separation	Moderate	Possible	Low	Blood	Multiple
Magnetic bead isolation—POS	Good	Possible	High	Tumor and Blood	1
Magnetic bead isolation—NEG	Moderate	Minimal	Moderate	Tumor and Blood	Multiple
Cell sorting	Good	Minimal	High	Tumor and Blood	Multiple

neutrophil isolation are CD11b and Ly6G (Rose, Misharin, & Perlman, 2012), while CD11b, CD15 and CD66b are typically used to identify human neutrophils (Condamine et al., 2016; Sagiv et al., 2015).

2. Before you start

2.1 Ethics regulations

2.1.1. All human and mouse clinical samples must be collected following local ethics regulations and only after the approval for such studies from the Institution.

2.2 Neutrophil maturity

2.2.1. The methods described here are for isolation of mature neutrophils.

2.3 Sample processing

2.3.1. Neutrophils are highly dynamic cells and can become easily activated. Hence we recommend keeping samples at 4 °C or on ice as much as possible. We also recommend not to carry out any steps to remove debris, or to enrich cells prior to FACS as this can lead to neutrophil activation.

2.3.2. Blood samples should be kept in EDTA tubes at 4 °C until time of processing.

2.3.3. Tissues should be kept in Dulbecco's modified minimal essential media (DMEM) without serum at 4 °C until time of processing.

2.3.4. Ideally all samples should be processed on the day of collection. However, if necessary they can be left overnight at 4 °C.

- **2.3.4.1.** All samples must be treated the same.
- **2.3.4.2.** Leaving the samples overnight may lead to cell death and changes in surface marker expression. It may also affect neutrophil activation state. In our experience cell death is minimal (<5%) and phenotype and activation state do not change significantly. However, it is recommended to initially compare fresh samples with samples left overnight to ensure that no substantial changes are occurring.

2.4 Sterile conditions

2.4.1. For all assays and in vivo experiments sterile conditions must be maintained by carrying out protocols in a Level 2 Biohazard hood. This is required for human material.

2.4.2. All buffers must be filtered using 0.22 µm filters.

2.5 Consumables/materials

2.5.1. Polypropylene tubes are recommended for FACS or flow cytometry as they reduce cell clumping and cells sticking to the sides of the tube.

2.5.2. Polystyrene tubes are not recommended as they often do not seal completely on a number of machines leading to inefficient event collection or no collection at all.

2.5.3. In some cases different antibody clones can give different results hence it is important to use the same clone for each experiment and to fully verify it.

2.5.4. Myeloid cells express high levels of Fc receptors hence it is important to use an effective blocking agent prior to antibody staining. Serum or CD16/CD32 antibodies are routinely used as blocking agents. If serum is chosen as a blocking agent we recommend getting serum from AB blood types.

2.6 Choice of fluorochromes

2.6.1. It is recommended that the brightest and most common fluorochromes are used for makers that are expected to have a low density, and to use the remaining fluorochromes for markers with a high density.

2.6.2. For markers that will be visualized on the same plot it is recommended to choose fluorochromes with emissions that are as far apart as possible on the spectrum.

2.6.3. Tandem dyes, including the brilliant violets, are more prone to degradation or non-specific binding if left in a mix or in light for a prolonged period of time. A staining buffer is available if using the brilliant violet dyes to minimize this.

2.6.4. PerCP is not recommended for cell sorting as this fades quickly.

2.6.5. See Table 2 for examples of multi-color panels.

Table 2 Examples of fluorochrome combinations for multi-color FACS panels.

6 color	8 color	10 color
	BV421	BV421
		BV510
	BV711	BV711
PerCP-Cy5.5	PerCP-Cy5.5	PerCP-Cy5.5
FITC	FITC	FITC
PE	PE	PE
PE-Cy7	PE-Cy7	PE-Cy7
APC	APC	APC
APC-Cy7	APC-Cy7	APC-Cy7
		AF700

2.6.6. A "Dump" channel/gate is often used which contains all antibodies targeting cells that are not of interest and/or need to be excluded from your downstream analysis on the same fluorochrome.

2.6.7. The Biologend online flow cytometry web tools are extremely helpful when designing multi-color FACS panels (https://www.biolegend.com/webtoolstab).

2.7 Viability dyes

2.7.1. Viability dyes are typically chosen to fit within the panel design. The most commonly used is DAPI, however, other options include propidium iodide (PI) and 7-AAD.

2.7.2. Regardless of which dye is chosen it is essential to include a viability dye as dead cells can take up antibodies non-specifically and also they can become autofluorescent thus leading to false positive staining. Furthermore, they can release DNA which is sticky and will cause cells to clump.

2.8 Setting up a template

2.8.1. All antibodies must be titrated prior to template set up.

2.8.1.1. It is important to use the same sample type that you will be using for your experiments as cell surface expression can vary between blood and tissue and also within tissue types.

2.8.1.2. A range of antibody concentrations are tested that range from above and below the amount recommended by the supplier. For example if the suggested volume is 2 μL a titration would include 0, 0.5, 1, 2, 3 and 5 μL. It is sometimes necessary to titrate even further based on the results of the initial titration.

2.8.1.3. The ideal antibody concentration would clearly separate the positive population from the negative population and have the positive population sitting above 10^3 for intensity. However, this will not be possible for all markers due to dim or continuous expression. In this case the best possible separation of positive and negative populations should be chosen.

2.8.2. A tube of unstained cells must be used to set up the experimental template prior to the beginning of the experiment in order to determine background and set voltages accordingly.

2.8.3. Beads and/or single stained positive controls need to be used to determine positive signals for each cell surface marker in order to set voltages accordingly.

2.8.4. Fluorescence Minus One (FMO) are required to verify the specificity of signals.

2.8.4.1. FMOs contain all of the antibodies used in the panel minus one.

2.8.4.2. FMOs are useful as they take into account the interaction of each antibody/fluorochrome and how this affects the spread of each marker in the panel. FMOs are therefore considered to show the true negative population.

2.8.4.3. FMOs should be used when assessing placement of gates for positive populations, especially if the positive population does not separate clearly from the negative population due to dim or continuous expression.

2.8.4.4. FMOs are particularly important to use for samples that are innately autofluorescent such as tumors in order to account for background.

2.8.5. Isotype controls

2.8.5.1. Isotype controls are antibodies raised against an antigen not found on the cell type or sample hence they account for any non-specific binding and/or artifacts.

2.8.5.2. The use of isotype controls is not recommended as they do not account for interaction between the panel antibodies/fluorochromes and hence often do not show the true negative population.

2.8.6. It is recommended to prepare an antibody master mix rather than adding single antibody aliquots to each samples in order to avoid mistakes and reduce experimental bias.

2.9 Compensation

2.9.1. For multi-color FACS staining fluorescence compensation is essential.

2.9.2. A compensation matrix needs to be set up prior to starting the experiment.

2.9.3. Each individual fluorochrome has to be compensated against all of the other fluorochromes.

2.9.3.1. Beads and/or single stained positive controls are used for this.

2.9.3.2. A final check on a sample stained with all the antibodies should be carried out after final adjustments are made using beads.

2.9.4. Compensation should be checked on a monthly basis for analysis and prior to each sort. If there is not enough sample for single stained controls beads can be used but all final settings must be checked on a fully stained sample.

2.10 Final checks

2.10.1. All final panels need to be checked by Flow facility staff to confirm that they are workable for compensation, isolation of positive populations and for the machines that are available within the facility.

2.11 Nozzle size for sorting

2.11.1. Generally the nozzle diameter should be no <5–7 times the diameter of the cells.

2.11.2. For sorting circulating neutrophils a nozzle size of 70 μm is recommended to ensure efficient sorting. This is because overall blood cells are relatively small and will pass through a 70 μm nozzle more easily and thus reduce sort times.

2.11.3. For sorting tissue neutrophils a nozzle size of 100 μm is recommended to ensure optimal sorting. This is because overall blood cells in tissue are larger and will pass through a 70 μm very slowly, leading to unnecessarily long sort times.

3. Blood sample preparation

3.1 Materials

3.1.1. Blood Collection Tubes + EDTA

3.1.1.1. For example we use S Monovette 9 mL K3E (SARSTEDT; 02.1066.001): 1.6 mg/mL EDTA

3.1.2. Red Blood Cell (RBC) Lysis Buffer (Biolegend)

3.1.3. Human serum from AB donors (Sigma-Aldrich)

3.1.4. DAPI (Biolegend)

3.1.5. DMEM

3.1.6. 5 ml Polypropylene FACS tubes

3.1.7. 5 mL 35 μm filter top FACS tubes

3.1.8. 50 mL conical tubes

3.1.9. Trypan blue

3.1.10. Sorting antibodies (see Table 3)

3.1.10.1. These are the minimum markers we recommend for sorting neutrophils. The panel set up will depend on the machine specifications available hence we are not providing any fluorochrome recommendations. All panel design must first be checked against the specifications of the machines available and verified as workable by flow facility staff.

3.1.11. Phenotyping antibodies (see Table 4)

3.1.11.1. These panels were set up for use on a BD 6-laser Fortessa flow cytometer and are only meant to be used as a guide. All panel design must first be checked against the specifications of the machines available and verified as workable by flow facility staff.

Table 3 Antibodies required for isolation of circulating neutrophils.

Antibody	Clone	Cell surface marker
CD3	UCHT1	T cells
CD11b	ICRF44	Myeloid cells
CD45	HI30	Leukocytes
HLA-DR	L243	Monocytes
CD15[a]	W6D3	Neutrophils
CD14	M5E2	Monocytes/Neutrophils

[a]It is recommended that a class IgG1 CD15 antibody is used as other classes can result in non-specific binding.

Table 4 Antibodies required for phenotyping circulating neutrophils.
GENERAL PANEL: Used for gating on neutrophils

Marker	Clone	Fluorochrome	Cell type	Concentration
CD45	HI30	AF700	Leukocytes	1:200
CD11b	ICRF44	APC-Cy7	Myeloid cells	1:200
HLA-DR	L243	FITC	Monocytes	1:100
CD294 (Dump)	BM16	PECF594	Eosinophils	1:200
CD3 (Dump)	UCHT1	PECF594	T cells	1:200
CD19 (Dump)	HIB19	PECF594	B cells	1:200
CD56 (Dump)	B159	PECF594	NK cells	1:50
CD15[a]	W6D3	BV510	Neutrophils	1:50
PANEL 1: Neutrophil phenotypic markers				
CD32	FUN-2	PE	↑ Upon degranulation	1:100
CD63	H5C6	BV421	↑ Upon degranulation	1:100
CD88	S5/1	PerCP-Cy5.5	↑ Upon activation	1:50
CD89	A59	APC	↑ Upon activation	1:200
CD66b	G10F5	PE-Cy7	↑ Upon activation	1:50
PANEL 2: Neutrophil phenotypic markers				
CD54	HA58	APC	↑ Upon activation	1:50
CD181	8F1/CXCR1	PE	↑ Upon activation	1:50
CD182	5E8/CXCR2	PerCP-Cy5.5	↑ Upon activation	1:50
CD43	CD43-10G7	PE-Cy7	↓ Upon activation	1:50
CD62L	DREG56	BV421	↓ Upon activation	1:100

[a]It is recommended that a class IgG1 CD15 antibody is used as other classes can result in non-specific binding.

3.2 Equipment

3.2.1. Vortex
3.2.2. Refrigerated centrifuge
3.2.3. Flow sorter/analysis machine

3.3 Solutions

3.3.1. Sort Buffer: 0.1% w/v bovine serum albumin (BSA): phosphate buffered saline (PBS)
3.3.2. FACS Buffer: 1% w/v BSA:PBS
3.3.3. 1X RBC Lysis Buffer: Dilute in distilled H2O
3.3.4. 10% v/v AB serum: Dilute in sort buffer

3.4 Processing of blood samples

3.4.1. ~20 mL of whole blood is to be collected into 2 × 10 mL blood EDTA collection tubes
3.4.2. Transfer each sample collection tube into a 50 mL conical tube
3.4.3. Wash the original collection tubes twice with 5 mL RBC lysis buffer and add the 5 mL to the 50 mL conical tube
3.4.4. Top up each conical tube to 50 mL with RBC lysis buffer
3.4.5. Do not combine the sample at this stage
3.4.6. Incubate for 15 min at 4 °C
3.4.7. Wash cells by centrifugation for 5 min at 400G and 4 °C
3.4.8. Discard the supernatant
3.4.9. Add 5 mL DMEM and re-suspend cells
3.4.10. Combine tubes 1 and 2 for each sample into 1 50 mL conical tube.
3.4.11. Top up to 50 mL DMEM
3.4.12. Wash cells by centrifugation for 5 min at 400G and 4 °C
3.4.13. Discard the supernatant
3.4.14. For isolation of neutrophils by sorting go to Section 5
3.4.15. For phenotypic analysis of neutrophils by flow cytometry go to Section 6

4. Tumor sample preparation

4.1 Tissue size

4.1.1. The tumor protocol has been optimized for 0.05–2.0 g tumors and therefore volume adjustments will be necessary for larger or smaller tissues/tumors.

4.2 Materials

4.2.1. DMEM
4.2.2. 60 mm petri dishes
4.2.3. Razor blades/scalpel blades
4.2.4. 15 mL conical tubes
4.2.5. 50 mL conical tubes
4.2.6. PBS
4.2.7. DNAse (Sigma-Aldrich)
4.2.8. Liberase TL and Liberase DL (Roche)
4.2.9. 100 μm strainers/filters
4.2.10. Human AB serum (Sigma-Aldrich)
4.2.11. DAPI (Biolegend)
4.2.12. 5 mL Polypropylene FACS tubes
4.2.13. 5 mL 35 μm filter top FACS tubes
4.2.14. Trypan blue
4.2.15. Sorting antibodies (see Table 5)

4.2.15.1. These are the minimum markers we recommend for sorting neutrophils. The panel set up will depend on the machine specifications available hence we are not providing any fulorochrome recommendations. All panel design must first be checked against the specifications of the machines available and verified as workable by flow facility staff.

Table 5 Antibodies required for isolation of tumor/tissue neutrophils.

Antibody	Clone	Cell surface marker
CD3	UCHT1	T cells
CD11b	ICRF44	Myeloid cells
CD45	HI30	Leukocytes
CD294	BM16	Eosinophils
CD15[a]	W6D3	Neutrophils
CD163	GHI/61	TAMs

[a]It is recommended that a class IgG1 CD15 antibody is used as other classes can result in non-specific binding.

4.2.16. Phenotyping antibodies (see Table 6)

4.2.16.1. These panels were set up for use on a BD 6-laser Fortessa flow cytometer and are only meant to be used as a guide. All panel design must first be checked against the specifications of the machines available and verified as workable by flow facility staff.

Table 6 Antibodies required for phenotyping tumor/tissue neutrophils.
GENERAL PANEL: Used for gating on neutrophils

Marker	Clone	Fluorochrome	Cell type	Concentration
CD45	HI30	AF700	Leukocytes	1:200
CD11b	ICRF44	APC-Cy7	Myeloid cells	1:200
HLA-DR	L243	FITC	Monocytes	1:100
CD294 (Dump)	BM16	PECF594	Eosinophils	1:200
CD3 (Dump)	UCHT1	PECF594	T cells	1:200
CD19 (Dump)	HIB19	PECF594	B cells	1:200
CD56 (Dump)	B159	PECF594	NK cells	1:100
CD15[a]	W6D3	BV510	Neutrophils	1:100
PANEL 1: Neutrophil phenotypic markers				
CD32	FUN-2	PE	↑ Upon degranulation	1:50
CD63	H5C6	BV421	↑ Upon degranulation	1:50
CD88	S5/1	PerCP-Cy5.5	↑ Upon activation	1:50
CD89	A59	APC	↑ Upon activation	1:100
CD66b	G10F5	PE-Cy7	↑ Upon activation	1:50
PANEL 2: Neutrophil phenotypic markers				
CD54	HA58	APC	↑ Upon activation	1:50
CD200R	OX-108	PE	Immunomodulatory	1:50
CD274	MIH2	APC	T cell co-stimulation	1:50
CD43	CD43-10G7	PE-Cy7	↓ Upon activation	1:50
CD62L	DREG56	BV421	↓ Upon activation	1:50

[a]It is recommended that a class IgG1 CD15 antibody is used as other classes can result in non-specific binding.

4.3 Equipment

4.3.1. Vortex
4.3.2. Refrigerated centrifuge
4.3.3. Flow sorter/analysis machines

4.4 Solutions

4.4.1. Sort Buffer: 0.1% v/v BSA:PBS
4.4.2. 10% v/v AB serum: Dilute in sort buffer

4.5 Processing of tissue samples

4.5.1. Collect tissue into ice-cold DMEM
4.5.2. If more than one tumor is being processed at a time it is recommended to leave the other tumors on ice until time of processing as leaving the tumors at room temperature can lead to cell death and changes in cell surface marker expression.
4.5.3. Transfer the tissue into a 60 mm petri dish
4.5.4. Remove all medium
    4.5.4.1. Left over medium can inactivate the digestion enzymes
4.5.5. Weigh, and record the tissue weight
4.5.6. Add 1 mL of serum free PBS
4.5.7. Mince the tissue using a razor blade/scalpel blade or tissue macerator (see Fig. 1A and B)
4.5.8. Ensure that the tissue is as finely minced as possible as large chunks of tissue will affect the digestion efficiency
4.5.9. Transfer the tumor lysates into a 15 mL conical tube by cutting the top of an Eppendorf P1000 tip

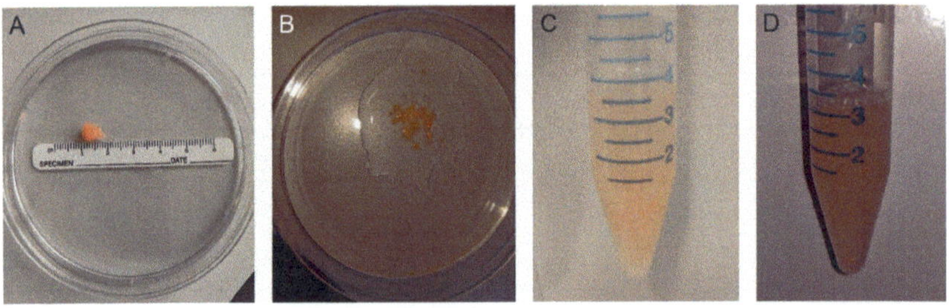

Fig. 1 Tumor/tissue processing and digestion. (A and B) Tissue/tumor samples were finely minced in 1 mL of PBS on ice. (C) Minced tumor samples were transferred into a 15 mL Falcon tube containing the digestion mix and incubated at 37 °C and 700–800 RPM until (D) fully digested.

4.5.10. Adjust to 3 mL with serum free PBS
4.5.11. Add
40 μL of Liberase DL stock solution 28 U/mL.
120 μL of Liberase TL stock solution 14 U/mL.
30 μL of DNAse I stock solution 15 mg/mL
4.5.12. Mix by vortexing and incubate for 30–90 min at 37 °C and 700–800 rpm
4.5.13. Initially digest for 30 min and check samples visually; if the suspension is smooth go to step 4.5.15 (See Fig. 1C and D)
  4.5.13.1. If samples are not smooth digest for a further 15 min; if after 15 min the suspension is smooth go to step 4.5.15 (see Fig. 1C and D)
  4.5.13.2. Repeat 15 min digestion until sample is smooth or the digestion time reaches 90 min (see Fig. 1C and D)
  4.5.13.3. Do not digest for >90 min as the sample will be over-digested and cell surface markers may be cleaved off
4.5.14. To end the enzymatic digestion adjust volume to 10 mL with sort buffer
4.5.15. Filter the samples though a 100 μm strainer and add 10 mL of sort buffer
4.5.16. Wash cells by centrifugation for 5 min at 400G and 4 °C
4.5.17. Discard the supernatant
4.5.18. For isolation of neutrophils by sorting go to Section 5
4.5.19. For phenotypic analysis of neutrophils by flow cytometry go to Section 6

5. FACS isolation of neutrophils

5.1. *Re*-suspend cells in 1 mL of 10% v/v AB serum in a fresh FACS tube
5.2. Count cells using trypan blue exclusion
5.3. Incubate cells on ice for a minimum of 30 and a maximum of 60 min
5.4. Wash cells in 3 mL sort buffer by centrifugation for 5 min at 400G and 4 °C
5.5. Discard supernatant
5.6. Re-suspend cells in 500 μL sort buffer
5.7. Label two fresh FACS tubes
UNSTAINED EXPERIMENT
5.8. Filter cells through a 35 μm filter top FACS tube
5.9. Add 480 μL of cells to the EXPERIMENT tube

5.10. Add 20 µL of cells to the UNSTAINED tube and make up to 500 µL with sort buffer

5.11. Make up antibody master mix and add to EXPERIMENT tube

- **5.11.1.** See Table 3 for blood neutrophil markers
- **5.11.2.** See Table 5 for tumor neutrophil markers
- **5.11.3.** In addition make up single color tubes (either using beads or cells) for compensation checks. It is recommended to carry out compensation checks prior to each sort.
- **5.11.4.** Make up master mix by adding 1:160 of each antibody/10^7 cells to master mix
  - **5.11.4.1.** It is important to first verify that these antibody concentrations are resulting in optimal staining
  - **5.11.4.2.** If staining is sub-optimal using these concentrations antibody titrations well need to be carried out

5.12. Mix cells by vortexing

5.13. Incubate antibodies on ice for a minimum of 30 and a maximum of 60 min protected from the light

5.14. Wash cells in 3 mL sort buffer by centrifugation for 5 min at 400G and 4 °C

5.15. Re-suspend cells in sort buffer so that the final cell concentration is $5–10 \times 10^6$ cells/mL

- **5.15.1.** Try to keep the cell concentration as low as possible to minimize clumping and overloading the sorter

5.16. Add 1 mL sort buffer to the UNSTAINED tube

5.17. Just prior to sorting add 3 µM of DAPI (or any viability dye) to the EXPERIMENT tube

5.18. Filter all samples through a 35 µm filter top FACS tube again just prior to sorting

- **5.18.1.** Clumping can clog the sorting machine but also affect sort purity and yield. Primary digested tissues are particularly prone to clumping. Hence it is strongly recommended that blood and especially tumor samples are filtered just prior to sorting, and filtered again during sorting if necessary.

5.19. Maintain cell temperature at 4 °C as much as possible during sorting to avoid neutrophil activation.

5.20. See Fig. 2A–H for blood neutrophil gating strategy

5.21. See Fig. 3 for tumor neutrophil gating strategy

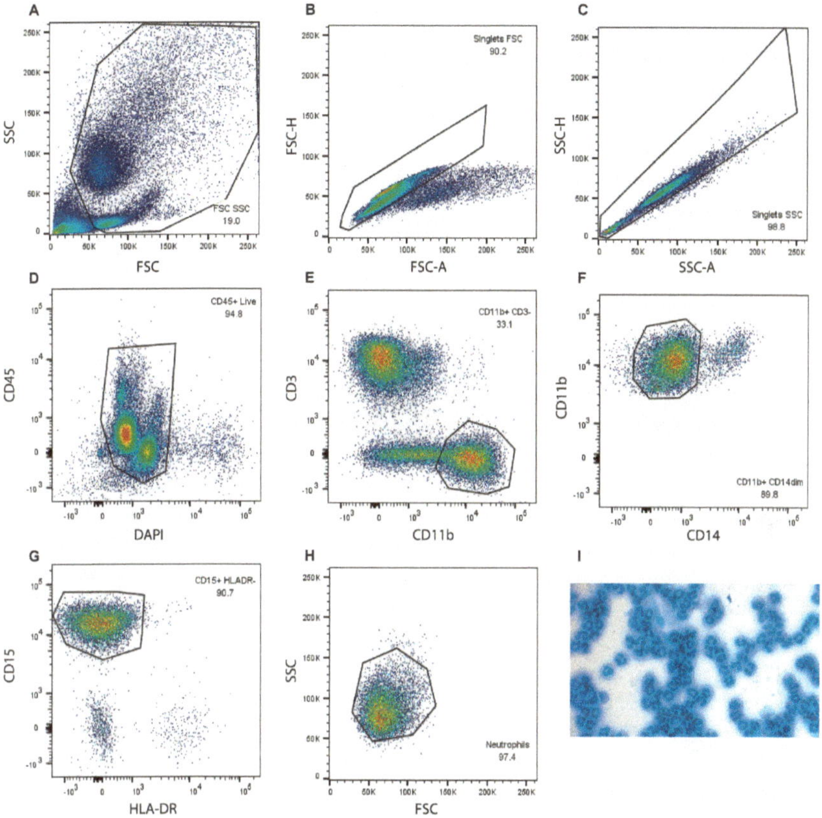

Fig. 2 Gating strategy for FACS isolation of human circualting neutrophils. Cells were first (A) gated on based on morphology to discard debris, after which (B and C) doublets were excluded. (D) Live leukocytes were then selected for as CD45+ DAPI- cells. Next (E) CD11b+ myeloid cells were selected and T lymphotyes excluded using CD3, (F) contaminating monocytes were then removed by gating for CD11b+CD14- cells. Finally (G) neutrophils were selected as CD15+HLA-DR- cells (HLA-DR was used to confirm monocyte exclusion) and (H) a final check was carried out as using a second morphology gate. (I) Sorted cells were spun down and stained using the Rapid Romanovsky staining procedure to confirm neutrophil purity.

6. Neutrophil phenotypic analysis by flow cytometry

6.1. Re-suspend cells in 1 mL of FACS buffer
6.2. Count cells using trypan blue exclusion
6.3. Aliquot 500,000 cells/panel into a fresh FACS tube
6.4. Make up to 1 mL 10% v/v AB serum
6.5. Incubate cells on ice for a minimum of 30 and a maximum of 60 min

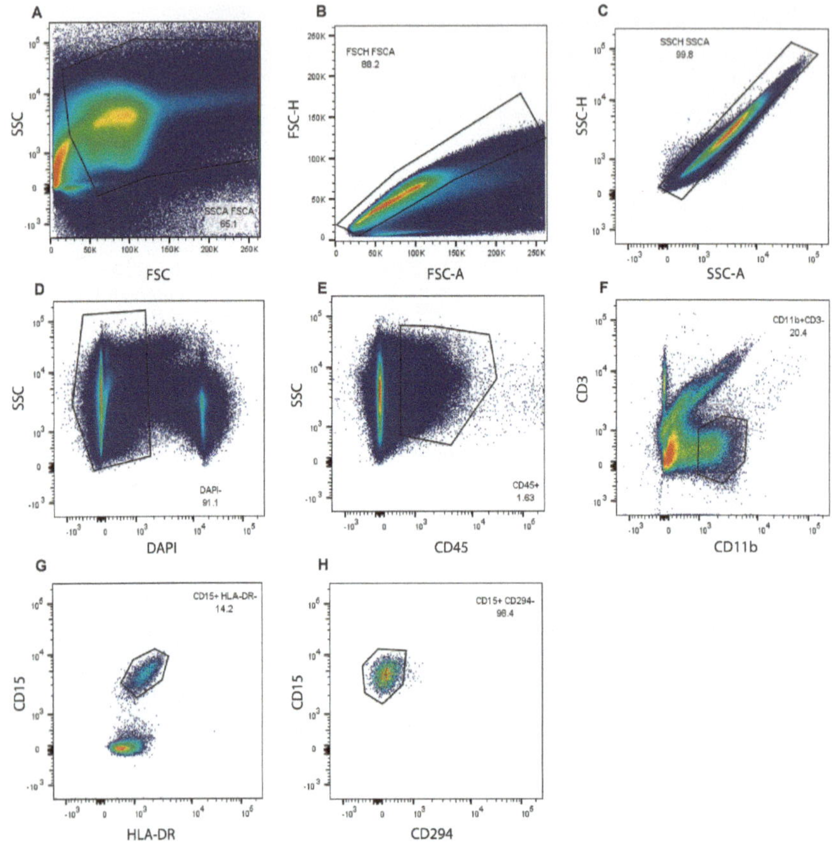

Fig. 3 Gating strategy for FACS isolation of human tumor/tissue neutrophils. (A) A morphology gate was used first to discard debris, after which (B and C) doublets were removed. (D) Live cells were then gated for as DAPI- cells, after which (E) leukocytes were gated as CD45+ cells. (F) Myeloid cells were then selected as CD11b+ cells and T lymphotyes excluded using CD3, finally (G) neutrophils were gated for as CD15+HLADR- (HLA-DR was used to exclude macrophages) and (H) CD15+CD294- cells (CD294 was used to exclude eosinophils.

6.6. Wash cells in 3 mL sort buffer by centrifugation for 5 min at 400G and 4 °C

6.7. Discard supernatant

6.8. Re-suspend cells in 300 μL FACS buffer

6.9. Label fresh FACS tubes

GENERAL PANEL

PANEL 1

PANEL 2

6.10. Make up master mixes (GEN, PANEL 1, PANEL 2)

6.10.1. See Table 4 for blood neutrophil markers

6.10.2. See Table 6 for tumor neutrophil markers

6.11. Filter cells through a 35 μm filter top FACS tube
6.12. Add 100 μL of cells per FACS tube
6.13. Mix cells by vortexing
6.14. Incubate antibodies on ice for a minimum of 30 and a maximum of 60 min protected from the light
6.15. Wash cells in 3 mL sort buffer by centrifugation for 5 min at 400G and 4 °C
6.16. Re-suspend cells in 300 μL FACS buffer
6.17. Just prior to analysis add 3 μM of DAPI to each tube
6.18. See Fig. 4 for blood neutrophil gating strategy
6.19. See Fig. 5 for tumor neutrophil gating strategy
6.20. Collect 5–10,000 events in the neutrophil gate

7. Expected cellular yield for sorting

7.1 Blood neutrophils

7.1.1. In our experiments a threshold of 500,000 neutrophils was set per sample. However, there was some variation between normal controls and cancer samples, for example; in normal controls, an average of 454,063 neutrophils were collected per 20 mL of blood, with a minimum of 103,024 and a maximum of 700,611. However, in cancer patients (in this case breast and ovarian) an average of 502,108 neutrophils were collected per 20 mL, with a minimum of 370,736 and a maximum of 836,389 (Fig. 6).

7.2 Tumor neutrophils

7.2.1. There were no differences between benign and cancer tissue samples in terms of total cell counts (Fig. 7A), nor was there any correlation between tumor weights and the number of neutrophils within the samples (Fig. 7B).

8. Analysis of FACS data

8.1 Quantifying cell populations

8.1.1. Flow percentages are often used to designate changes in population abundance. However, these values are relative to changes in the abundance of every other cell type within the sample and therefore are often not representative. Furthermore, they do not take into consideration the total numbers of cells within the sample.

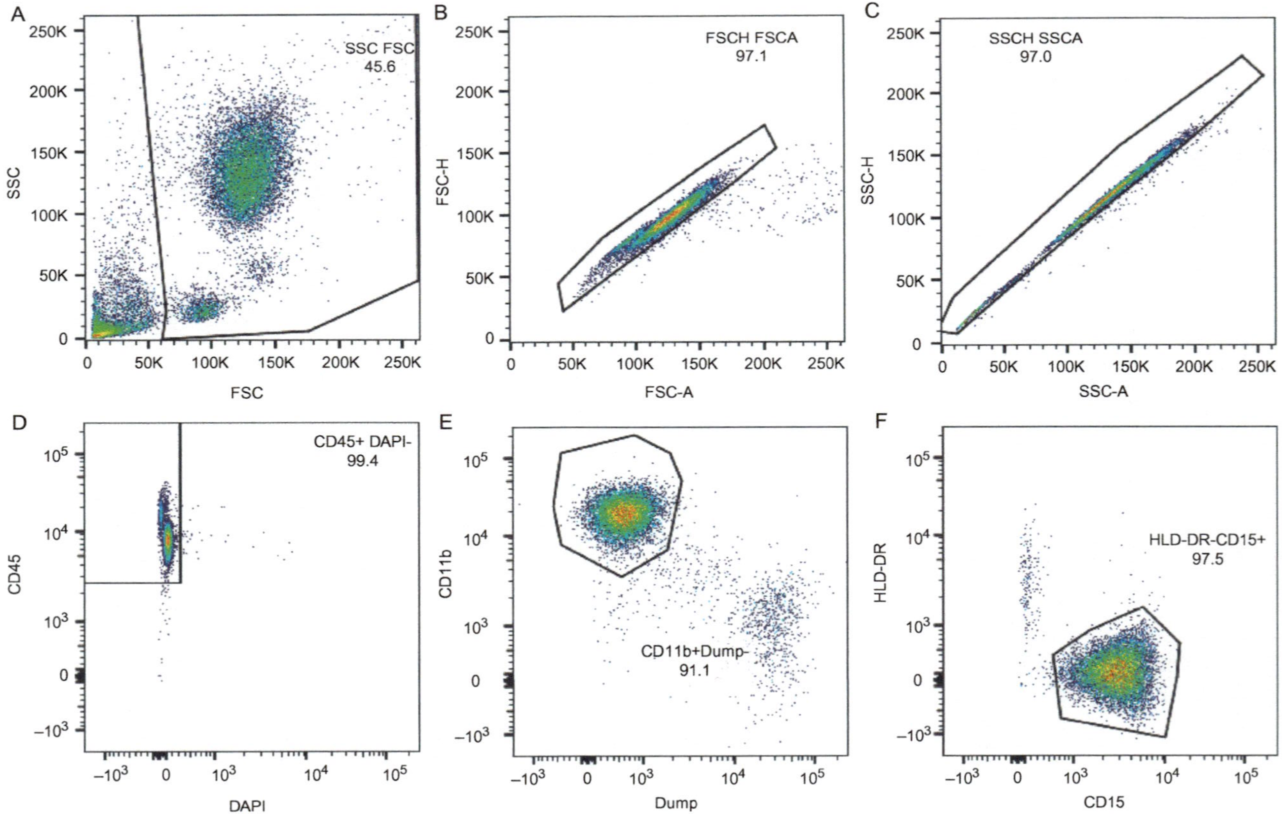

Fig. 4 Gating strategy for phenotypic analysis of circulating neutrophils. (A) A morphology gate was first used to remove debris, after which (B and C) doublets were removed. (D) Live leukocytes were then gated as CD45 + DAPI- cells and from this population (E) myeloid cells were selected as CD11b + DUMP- (the dump gate included markers for T cells, B cells, NK cells and eosinophils). (F) Neutrophils were gated as CD15 + HLA-DR- cells (HLA-DR was used to confirm exclusion of monocytes).

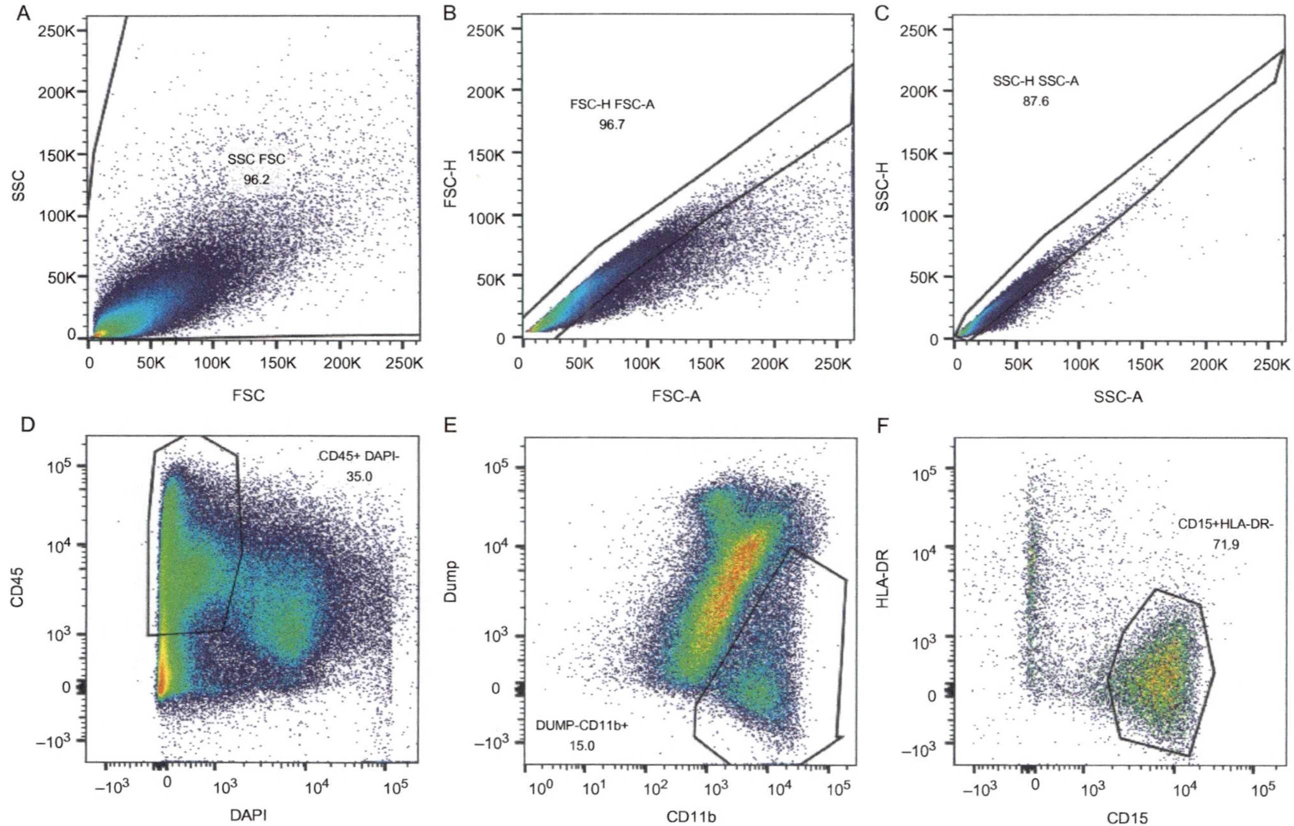

Fig. 5 Gating strategy for phenotypic analysis of tumor/tissue neutrophils. (A) A morphology gate was first used to remove debris, after which (B and C) doublets were removed. (D) Live leukocytes were next selected as CD45+DAPI- cells, after which (E) myeloid cells were gated as CD11b+DUMP- (the dump gate included markers for T cells, B cells, NK cells and eosinophils). Finally (F) neutrophils were isolated as CD15+HLA-DR- cells (HLA-DR was used to confirm exclusion of monocytes).

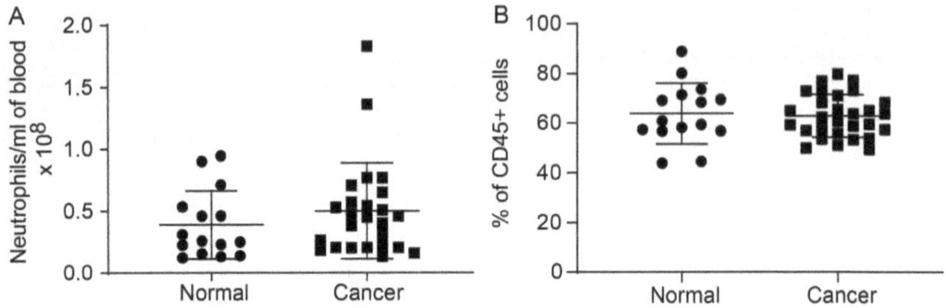

Fig. 6 Neutrophil yield from blood. On average ~500,000 cells were isolated from normal/control ($N=15$) and cancer ($N=25$) patient samples as assessed by (A) total cell counts and, (B) on average neutrophils made up ~60% of leukocytes in both normal/ control and cancer patient samples.

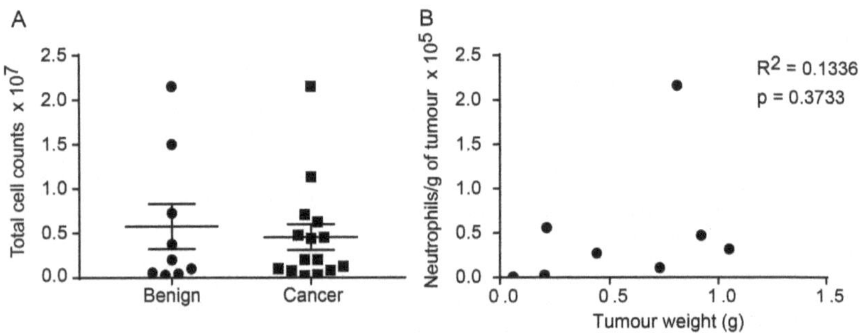

Fig. 7 Neutrophil yield from tumors/tissues. (A) There was no difference between benign ($N=9$) and tumor ($N=15$) samples regarding the total number of cells within the tissue, and (B) there was no correlation between dry tumor weight at the number of neutrophils per gram of tumor/tissue. Pearson correlation coefficient was calculated using Graphpad prism software. Each dot represents one sample. ($N=8$, $R^2=$correlation coefficient, $P=$statistical value).

8.2 For analyzing changes in population abundance two strategies are recommended

8.2.1. Total cell counts are calculated and then the percentage of cells are multiplied by the total cell counts. Typically neutrophil percentages are calculated as a percentage of CD45+DAPI- (live) cells.

8.2.2. Flow cytometry cell counting beads can also be used to determine cell concentration and are often considered more accurate as they are run at the same time as the sample on the flow cytometer. One example are the 123eCount beads™ from ThermoFisher (catalog number: 01–1234-42).

8.3 Analyzing expression of cell surface markers

8.3.1. Mean Fluoresence Intensity (MFI) calculates the mean fluorescence of the marker/antibody. However, MFI is only useful if the data is normally distributed, which is rare in flow cytometry. The more the data is skewed the further the mean drifts, becoming less and less representative. Furthermore, mean FI does not take into consideration cell size, and hence a large cell, with more membrane and consequently more surface protein, can appear brighter than a smaller cell. It is not recommended that mean FI be used for analysis of marker expression.

8.3.2. Median Fluoresence Intensity, is also often referred to as MFI but instead of calculating the mean, calculates the median, which is insensitive to outliers, and hence is more accurate than mean fluorescence intensity when the data is skewed. However, median can often be too insensitive to outliers resulting in inaccuracies.

8.3.3. Geometric Mean (GeoMean) calculates a mean, or average, by using the product of their values. Geometric mean is recommended for analyzing flow data as it takes outliers, or skewness of data, into consideration but is not significantly affected by a large difference in one, or a few, data points.

8.3.4. To assess expression of cell surface markers:

$$Expression = GeoMean\ Panel\ 1\ or\ 2 - GeoMean\ General\ Panel$$

9. Useful tips/troubleshooting

9.1 RBC lysis

9.1.1. Efficient RBC lysis is required for blood samples as large numbers of erythrocytes can lead to autofluoresence and a low acquisition rate, thus compromising the analysis, as well as the purity and yield of sorts.

9.2 Collection tubes

9.2.1. Any size polypropylene collection tubes can be used for sorting as long as they comply with the sorting machine. However, an adequate volume of serum containing medium or buffer is necessary to ensure cell viability. The sides of the collection tube must be coated to avoid drying out of the sorted cells. This is achieved by vortexing the collection tube just prior to sorting.

9.3 Tissue digestion

9.3.1. This protocol utilizes Liberase enzyme blends (HPLC purified), other compounds such as collagenase mixtures are available, although these may contain additional proteases that can cleave extracellular markers, or lead to artifacts. Beware of changing batches from the producers as these can vary and thus effect survival of cell surface markers.

9.4 Staining

9.4.1. IgG2a class antibodies can lead to non-specific binding hence if the data is unclear it is recommended to try an IgG1 class antibody.

9.5 Sort protocol

9.5.1. Built in sorting modes are available to optimize various aspects of collection:

9.5.2. PURITY: This mode is typically used if high purity is important for downstream experiments, such as -omics studies. In this mode any target events that fall close to non-target events are excluded, resulting in a lower yield but higher purity.

9.5.3. YIELD: This mode is typically used when it is more important to recover as many target events as possible from the sort and purity is not as much of a concern. In this mode all cells that fall into the sort gate will be collected, resulting in a higher yield but lower purity.

9.5.4. SINGLE: This mode is typically used when extremely accurate counting of the output cells is critical for downstream experiments.

9.6 Sort purity checks

9.6.1. It is crucial to carry out a sorting purity during set up of the sorting protocol settings; this is done by taking an aliquot of cells after the run and re-acquiring them to verify purity.

9.6.2. These protocols result in a purity of ~99.5% (Fig. 2I) with a minor eosinophil contamination (~0.5%). However, if eosinophil contamination is an issue, an anti-CD294 antibody (clone: BM16) can be added to gate out these cells.

9.7 Sort yields for tumors

9.7.1. Neutrophil abundance varies greatly between tumor types and in some cases, it may be difficult to sort a sufficient number of neutrophils from one tumor. In this case samples can be pooled if available.

9.8 Phenotypic analysis markers

9.8.1. Keep in mind that these panels are recommended for assessment of neutrophil activation. There are a number of other markers available that can be used for further phenotypic analysis of neutrophils.

Acknowledgments

The authors would like to thank Dr. Shonna Johnston, Dr. William Ramsay and Dr. Mari Pattison from the QMRI Flow Cytometry and Cell Sorting Facility for their assistance in the development of these methods. Research reported in this methods article was supported by the Wellcome Trust (101067/Z/13/Z) and MRC Centre grant MR/N022556/1.

References

Albrengues, J., Shields, M. A., Ng, D., Park, C. G., Ambrico, A., Poindexter, M. E., et al. (2018). Neutrophil extracellular traps produced during inflammation awaken dormant cancer cells in mice. *Science, 361*, eaao4227. https://doi.org/10.1126/science.aao4227.

Bekkering, S. (2013). Another look at the life of a neutrophil. *World Journal of Hematology, 2*, 44–58. https://doi.org/10.5315/wjh.v2.i2.44.

Beliakova-Bethell, N., Massanella, M., White, C., Lada, S. M., Du, P., Vaida, F., et al. (2014). The effect of cell subset isolation method on gene expression in leukocytes. *Cytometry. Part A, 85*, 94–104. https://doi.org/10.1002/cyto.a.22352.

Coffelt, S. B., Kersten, K., Doornebal, C. W., Weiden, J., Vrijland, K., Hau, C. S., et al. (2015). IL-17-producing γδ T cells and neutrophils conspire to promote breast cancer metastasis. *Nature, 522*, 345–348. https://doi.org/10.1038/nature14282.

Condamine, T., Dominguez, G. A., Youn, J.-I., Kossenkov, A. V., Mony, S., Alicea-Torres, K., et al. (2016). Lectin-type oxidized LDL receptor-1 distinguishes population of human polymorphonuclear myeloid-derived suppressor cells in cancer patients. *Science Immunology, 1*, aaf8943. https://doi.org/10.1126/sciimmunol.aaf8943.

Di Carlo, E., Forni, G., Lollini, P. L., Colombo, M. P., Modesti, A., & Musiani, P. (2001). The intriguing role of polymorphonuclear neutrophils in antitumor reactions. *Blood, 97*, 339–345. https://doi.org/10.1182/blood.V97.2.339.

Fridlender, Z. G., Sun, J., Kim, S., Kapoor, V., Cheng, G., Ling, L., et al. (2009). Polarization of tumor-associated neutrophil phenotype by TGF-β: "N1" versus "N2" TAN. *Cancer Cell, 16*, 183–194. https://doi.org/10.1016/j.ccr.2009.06.017.

Gaut, J. P., Yeh, G. C., Tran, H. D., Byun, J., Henderson, J. P., Richter, G. M., et al. (2001). Neutrophils employ the myeloperoxidase system to generate antimicrobial brominating and chlorinating oxidants during sepsis. *Proceedings of the National Academy of Sciences of the United States of America, 98*, 11961–11966. https://doi.org/10.1073/pnas.211190298.

Granot, Z., Henke, E., Comen, E. A., King, T. A., Norton, L., & Benezra, R. (2011). Tumor entrained neutrophils inhibit seeding in the premetastatic lung. *Cancer Cell, 20*, 300–314. https://doi.org/10.1016/j.ccr.2011.08.012.

Greenlee-Wacker, M. C. (2016). Clearance of apoptotic neutrophils and resolution of inflammation. *Immunological Reviews, 273*, 357–370. https://doi.org/10.1111/imr.12453.

Huh, S. J., Liang, S., Sharma, A., Dong, C., & Robertson, G. P. (2010). Transiently entrapped circulating tumor cells interact with neutrophils to facilitate lung metastasis development. *Cancer Research, 70*, 6071–6082. https://doi.org/10.1158/0008-5472.CAN-09-4442.

Jablonska, J., Lang, S., Sionov, R. V., & Granot, Z. (2017). The regulation of pre-metastatic niche formation by neutrophils. *Oncotarget*, *8*, 112132–112144. https://doi.org/10.1111/jcpt.12372.

Jorch, S. K., & Kubes, P. (2017). An emerging role for neutrophil extracellular traps in noninfectious disease. *Nature Medicine*, *23*, 279–287. https://doi.org/10.1038/nm.4294.

Kaplan, R. N., Riba, R. D., Zacharoulis, S., Bramley, A. H., Vincent, L., Costa, C., et al. (2005). VEGFR1-positive haematopoietic bone marrow progenitors initiate the premetastatic niche. *Nature*, *438*, 820–827. https://doi.org/10.1038/nature04186.

Kersten, K., Coffelt, S. B., Hoogstraat, M., Verstegen, N. J. M., Vrijland, K., Ciampricotti, M., et al. (2017). Mammary tumor-derived CCL2 enhances pro-metastatic systemic inflammation through upregulation of IL1β in tumor-associated macrophages. *OncoImmunology*, *6*, e1334744. https://doi.org/10.1080/2162402X.2017.1334744.

Kowanetz, M., Wu, X., Lee, J., Tan, M., Hagenbeek, T., Qu, X., et al. (2010). Granulocyte-colony stimulating factor promotes lung metastasis through mobilization of Ly6G+Ly6C+ granulocytes. *Proceedings of the National Academy of Sciences of the United States of America*, *107*, 21248–21255. https://doi.org/10.1073/pnas.1015855107.

Mantovani, A., Cassatella, M. A., Costantini, C., & Jaillon, S. (2011). Neutrophils in the activation and regulation of innate and adaptive immunity. *Nature Reviews. Immunology*, *11*, 519–531. https://doi.org/10.1038/nri3024.

Michaeli, J., Shaul, M. E., Mishalian, I., Hovav, A. H., Levy, L., Zolotriov, L., et al. (2017). Tumor-associated neutrophils induce apoptosis of non-activated CD8 T-cells in a TNFα and NO-dependent mechanism, promoting a tumor-supportive environment. *OncoImmunology*, *6*, e1356965. https://doi.org/10.1080/2162402X.2017.1356965.

Mishalian, I., Bayuh, R., Levy, L., Zolotarov, L., Michaeli, J., & Fridlender, Z. G. (2013). Tumor-associated neutrophils (TAN) develop pro-tumorigenic properties during tumor progression. *Cancer Immunology, Immunotherapy*, *62*, 1745–1756. https://doi.org/10.1007/s00262-013-1476-9.

Quach, A., & Ferrante, A. (2017). The application of dextran sedimentation as an initial step in neutrophil purification promotes their stimulation, due to the presence of monocytes. *Journal of Immunology Research*, *2017*, 1–10. https://doi.org/10.1155/2017/1254792.

Reber, L. L., Gillis, C. M., Starkl, P., Jönsson, F., Sibilano, R., Marichal, T., et al. (2017). Neutrophil myeloperoxidase diminishes the toxic effects and mortality induced by lipopolysaccharide. *The Journal of Experimental Medicine*, *214*, 1249–1258. https://doi.org/10.1084/jem.20161238.

Rose, S., Misharin, A., & Perlman, H. (2012). A novel Ly6C/Ly6G-based strategy to analyze the mouse splenic myeloid compartment. *Cytometry. Part A*, *81*, 343–350. https://doi.org/10.1002/cyto.a.22012.

Sagiv, J. Y., Michaeli, J., Assi, S., Mishalian, I., Kisos, H., Levy, L., et al. (2015). Phenotypic diversity and plasticity in circulating neutrophil subpopulations in cancer. *Cell Reports*, *10*, 562–573. https://doi.org/10.1016/j.celrep.2014.12.039.

Singhal, S., Bhojnagarwala, P. S., O'Brien, S., Moon, E. K., Garfall, A. L., Rao, A. S., et al. (2016). Origin and role of a subset of tumor-associated neutrophils with antigen-presenting cell features in early-stage human lung cancer. *Cancer Cell*, *30*, 120–135. https://doi.org/10.1016/j.ccell.2016.06.001.

Smith, J. A. (1994). Neutrophils, host defense, and inflammation: A double-edged sword. *Journal of Leukocyte Biology*, *56*, 672–686. https://doi.org/10.1002/jlb.56.6.672.

Soehnlein, O., & Lindbom, L. (2010). Phagocyte partnership during the onset and resolution of inflammation. *Nature Reviews. Immunology*, *10*, 427–439. https://doi.org/10.1038/nri2779.

Swierczak, A., Cook, A. D., Lenzo, J. C., Restall, C. M., Doherty, J. P., Anderson, R. L., et al. (2014). The promotion of breast cancer metastasis caused by inhibition of CSF-1R/CSF-1 signaling is blocked by targeting the G-CSF receptor. *Cancer Immunology Research*, *2*, 765–776. https://doi.org/10.1158/2326-6066.CIR-13-0190.

Szczerba, B. M., Castro-Giner, F., Vetter, M., Krol, I., Gkountela, S., et al. (2019). Neutrophils escort circulating tumour cells to enable cell cycle progression. *Nature*, *566*, 553–557.

Wculek, S. K., & Malanchi, I. (2015). Neutrophils support lung colonization of metastasis-initiating breast cancer cells. *Nature*, *528*, 413–417. https://doi.org/10.1038/nature16140.

Wilgus, T. A., Roy, S., & McDaniel, J. C. (2013). Neutrophils and wound repair: Positive actions and negative reactions. *Advances in Wound Care*, *2*, 379–388. https://doi.org/10.1089/wound.2012.0383.

Zhou, L., Somasundaram, R., Nederhof, R. F., Dijkstra, G., Faber, K. N., Peppelenbosch, M. P., et al. (2012). Impact of human granulocyte and monocyte isolation procedures on functional studies. *Clinical and Vaccine Immunology*, *19*, 1065–1074. https://doi.org/10.1128/CVI.05715-11.

Further reading

Deryugina, E. I., Zajac, E., Juncker-Jensen, A., Kupriyanova, T. A., Welter, L., & Quigley, J. P. (2014). Tissue-infiltrating neutrophils constitute the major in vivo source of angiogenesis-inducing MMP-9 in the tumor microenvironment. *Neoplasia*, *16*, 771–788. https://doi.org/10.1016/j.neo.2014.08.013.

Nozawa, H., Chiu, C., & Hanahan, D. (2006). Infiltrating neutrophils mediate the initial angiogenic switch in a mouse model of multistage carcinogenesis. *Proceedings of the National Academy of Sciences*, *103*, 12493–12498. https://doi.org/10.1073/pnas.0601807103.

Spicer, J. D., McDonald, B., Cools-Lartigue, J. J., Chow, S. C., Giannias, B., Kubes, P., et al. (2012). Neutrophils promote liver metastasis via Mac-1-mediated interactions with circulating tumor cells. *Cancer Research*, *72*, 3919–3927. https://doi.org/10.1158/0008-5472.CAN-11-2393.

CHAPTER FOURTEEN

Flow cytometry-based isolation of tumor-associated regulatory T cells and assessment of their suppressive potential

Kevin Kos[a], Martijn van Baalen[b], Denize A. Meijer[a], Karin E. de Visser[a,*]

[a]Division of Tumor Biology & Immunology, Oncode Institute, Netherlands Cancer Institute, Amsterdam, The Netherlands
[b]Flow Cytometry Facility, Netherlands Cancer Institute, Amsterdam, The Netherlands
*Corresponding author: e-mail address: k.d.visser@nki.nl

Contents

Abstract

Regulatory T cells (Tregs) play a major role in establishing an immunosuppressive tumor microenvironment. In order to fully uncover their role and molecular regulation in tumor-bearing hosts, it is critical to combine phenotypical characterization with functional analyses. A standard method to determine the suppressive potential of Tregs is with an in vitro suppression assay, in which the impact of freshly isolated Tregs on T cell

Methods in Enzymology, Volume 632
ISSN 0076-6879
https://doi.org/10.1016/bs.mie.2019.07.035

proliferation is assessed. The assay requires the isolation of substantial numbers of Tregs from tissues and tumors, which can be challenging due to low yield or cell damage during sample preparation. In this chapter, we discuss a flexible suppression assay which can be used to assess the suppressive potential of low numbers of murine Tregs, directly isolated from tumors. We describe methods for tissue preparation, flow cytometry-based sorting of Tregs and optimal conditions to perform a suppression assay, to obtain reliable and reproducible results.

1. Introduction

There is a growing appreciation for the influential role of the tumor microenvironment on cancer biology (Binnewies et al., 2018). This has led to the realization that the immune system is not only involved in tumor clearance, but can also support tumor growth and metastasis via different mechanisms, including the induction of intra-tumoral and systemic immunosuppression (Kitamura, Qian, & Pollard, 2015). One of the key cell types that is involved in exerting immunosuppression in cancer patients is the Foxp3$^+$ CD4$^+$ regulatory T cell (Treg) (Togashi, Shitara, & Nishikawa, 2019).

In mice, regulatory T cells are characterized by high expression of the transcription factor Foxp3, which arms these cells with immunosuppressive properties (Fontenot, Gavin, & Rudensky, 2003; Li, Li, Tsun, & Li, 2015). Other proteins that are highly expressed by Tregs include CD25, CTLA-4 and GITR (Vignali, Collison, & Workman, 2008). Tregs play a central role in many aspects of immune homeostasis, and are, for example, critical in resolving inflammation, dampening immunity toward food- and commensal microbial antigens and controlling adipose inflammation (Josefowicz et al., 2012; Littman & Rudensky, 2010; Sharma & Rudra, 2018). Importantly, Tregs are the gate keepers of peripheral tolerance and thus responsible for clearing auto-reactive T cells from the periphery (Chang, Zheng, & Liu, 2009). To perform this role, Tregs have a wide arsenal of immunosuppressive abilities, including the release of cytokines, inhibition of T cell priming and direct effector cell killing (Vignali et al., 2008). In a cancer setting, many of these immunosuppressive mechanisms can be directly employed to prevent anti-tumor immunity and promote tumor growth (Togashi et al., 2019). However, the type of suppressive effector mechanism that is used in a certain setting depends on many factors, such as signals from the local environment, and the activation status of the Treg or the target cell.

Therefore, when investigating Tregs, it is important to combine a comprehensive phenotypical analysis with functional assays that assess their suppressive potential.

The suppressive potential of Tregs is best assessed with a suppression assay. For this assay, Tregs are isolated from a source of interest and subsequently co-cultured with activated conventional $CD4^+$ and $CD8^+$ T cells (responder cells) in several ratios. These responder cells are labeled with a fluorescent proliferation dye prior to the start of the co-culture. By analyzing the intensity of the proliferation dye after several days, one can assess to what extent Tregs from a source of interest can suppress responder cell proliferation (Fig. 1). Although the focus of this 6 day protocol is on testing the suppressive potential of tumor-derived Tregs, the assay is applicable to murine Tregs obtained from virtually any tissue, disease model, or genetically manipulated context. In addition, the protocol allows modifications to fit the user's research question. This includes changing the type of responder cells, or supplementing the co-culture with blocking/stimulating agents, e.g., antibodies or cytokines of choice.

Satisfactory execution of a suppression assay can be quite tedious due to many different variables that have to be taken into account. For example, results can be influenced by the chosen isolation method, choice of stimulatory signals and responder cells, and culture conditions. This chapter describes a protocol for the isolation of Tregs from murine mammary tumors, spleens and lymph nodes by flow cytometry-based cell sorting and analysis of their suppressive potential on splenic $CD4^+$ and $CD8^+$ T lymphocytes. We provide detailed descriptions of optimal sample

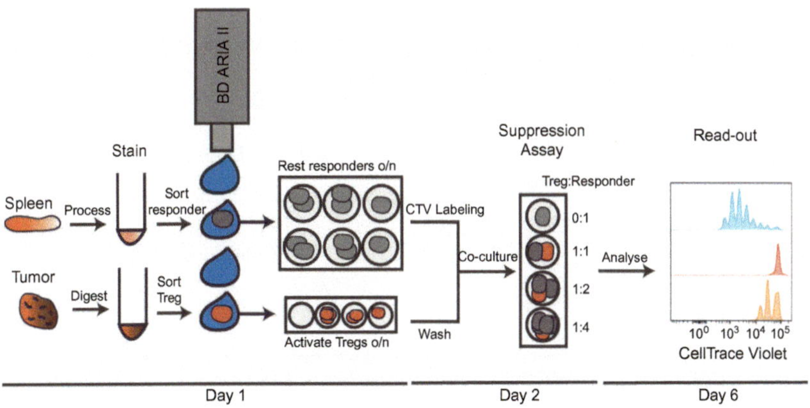

Fig. 1 Schematic overview of the 6-day suppression assay described in this protocol.

preparation, sorting strategies, co-culture conditions, execution and analysis of the suppression assay.

2. Equipment and materials

2.1 Preparation of single-cell suspensions from freshly isolated murine tissues

1. Freshly isolated murine tissues of interest in PBS on ice
2. McIlwain Tissue Chopper (Ted Pella Inc.) including chopping discs, blades and scalpel
3. Plunger of a 2mL syringe
4. Cell strainer, mesh size 70μm (Falcon)
5. Shaking waterbath or shaker
6. Red Blood Cell (RBC) lysis buffer: 155mM NH_4Cl, 10mM $KHCO_3$, 0.1mM EDTA in H_2O, pH 7.2–7.4 (Note 1).
7. Freshly prepared tumor digestion mix: 3mg/mL Collagenase A (Roche) in serum free DMEM (Dulbecco), Deoxyribonuclease I from bovine pancreas (DNAse) 25μg/mL (Sigma-Aldrich)
8. Inactivation buffer: DMEM medium supplemented with 10% vol/vol FCS
9. Sorting buffer: 1 × PBS, 0.5% BSA, 2mM EDTA

2.2 Antibody staining and cell sorting

1. Freshly prepared single cell suspensions from tissues of interest on ice
2. Fluorescently labeled anti-mouse antibodies (Table 1)
3. LIVE/DEAD™ Fixable Aqua Dead Cell Stain Kit (Thermofisher)

Table 1 Recommended fluorochrome-conjugated antibody panel for optimized Treg/responder cell sorting from murine tumor tissues.

Antigen	Fluorochrome	Clone	Concentration	Marker	Manufacturer
CD3	APC-eF780	145-2C11	1:400	T lymphocytes	BioLegend
CD4	PE	GK1.5	1:200	$CD4^+$ T lymphocytes	Thermofisher
CD25	APC	PC61	1:400	Regulatory T cells	Thermofisher
CD8	FITC	53-6.7	1:400	$CD8^+$ T lymphocytes	BD Biosciences

4. 7-AAD Viability Staining Solution (Thermofisher)
5. Fc block: purified anti-mouse CD16/32 antibody (BD Biosciences)
6. 1.4 mL Push Cap U-bottom tubes (Micronic)
7. 5 mL polypropylene round-bottom tubes, 12 × 75 mm (Falcon)
8. 5 mL polypropylene tubes with 35 μm cell strainer snap cap (Falcon)
9. Collection/Culture buffer: IMDM (Dulbecco), 9% FCS, 100 IU/mL penicillin, 100 μg/mL streptomycin, β-mercapto-ethanol 0.2% (Merck), GlutaMAX 1% (Thermofisher)
10. A 3-laser flow cytometry-based cell sorter, with the ability to detect at least five different fluorescent labels

2.3 Suppression assay

1. Freshly sorted Tregs and responder cells on ice
2. Recombinant murine IL-2 (Peprotech)
3. Dynabeads™ Mouse T-Activator CD3/CD28 (Thermofisher)
4. CellTrace™ Violet Cell Proliferation dye (Thermofisher)
5. A 3-laser analytical flow cytometer, with the ability to detect at least five different fluorescent labels

2.4 Essential laboratory materials

1. 1 × sterile phosphate buffered saline (PBS)
2. 12-multichannel pipette
3. Tissue culture-treated 96-well U-bottom plates (Greiner bio one)
4. Tissue culture-treated 6-well plates (Greiner bio one)
5. Sterile filter tips
6. 15 and 50 mL polypropylene tubes (Falcon)
7. 1.5 mL Eppendorf tubes (Eppendorf)
8. Refrigerated (plate) centrifuge
9. Cell counting equipment (manual/automated)
10. Ice bucket

3. Preparation of single-cell suspensions from freshly isolated murine tissues

The suppression assay can be performed with Tregs obtained from any tissue, dependent on the research question of the experiment. It is important to realize that the number of Tregs that can be obtained will greatly vary between different tissues or stages of disease. Tregs can be quite easily isolated from spleen and lymph nodes and are therefore commonly used for standard assays. In contrast, murine tumors may harbor few intra-tumoral T cells, and accordingly yield a low number of Tregs. Isolating sufficient numbers of Tregs from tumors can therefore be challenging, but is essential for high quality results. Tregs and responder cells are isolated from single cell suspensions of murine tissues via flow cytometry-based sorting. Typically, each murine tissue requires a specific approach for optimal preparation of single cell suspensions. In this protocol, we will describe the isolation of Tregs from murine spleen, lymph nodes and mammary tumors.

Keep cells sterile, on ice and protected from light, unless otherwise stated throughout the protocol. Before starting processing of tissues, cool PBS and sorting buffer on ice and warm the tumor digestion mix to 37°C (10mL in a 15mL tube per tumor).

3.1 Spleen and lymph nodes

1. Remove fat attached to the spleen/lymph nodes and place organ of interest on a (separate) 70 μm cell strainer, inserted into a 50 mL tube.
2. Gently mash the tissue with the plunger through the filter, while adding ~15 mL cold sterile PBS until cells have passed through the cell strainer.
3. Centrifuge the cells at 300 *g*, 4 °C for 5 min and aspirate supernatant.
4. Incubate the spleen cell pellet (skip this step for the lymph node cells) in 1 mL of sterile RBC lysis buffer for 2 min at room temperature. Stop RBC lysis with 10 mL of cold sorting buffer and subsequently pellet the cells.
5. Resuspend the cells in 1 mL (spleen) or 500 μL (lymph nodes) of sorting buffer on ice and count (manually/automated) the number of cells in suspension.

3.2 Mouse mammary tumors

The following protocol applies to tumors that measure up to a maximum of 1500 mm^3 (Note 2).

1. Place the tumor on a chopping disc cleaned with 70% ethanol. Remove non-tumor tissue such as fat tissue. Make sure the tumor is free of lymph nodes, to prevent contamination with lymph node-derived Tregs. Pre-cut the tumor in small pieces using a clean scalpel.
2. Place the disc under the tissue chopper and chop the tumor at least 3 ×, to a homogenous sample. Use maximum cutting speed and blade force, 5–10 μm blade travel. Alternatively, other cutting instruments can be used to fragment the tumor.

3. Scrape the chopped tumor from the disc with a scalpel and carefully insert into a tube with 10 mL of warmed digestion mix. Incubate at 37 °C for 1 h with gentle agitation. This can be done in a shaking water bath, or with a shaker placed in an incubator.
4. After incubation, pass the cells using a plunger through a 70 μm cell strainer capped into a 50 mL tube and add 15 mL of cooled inactivation buffer to stop the enzymatic digestion.
5. Centrifuge the cells at 300 *g*, 4 °C for 5 min and aspirate supernatant.

Resuspend the cells in 1 mL of sorting buffer on ice and count the cells. The suppression assay requires the co-culture of Treg and responder cells. Here, the responder cells are of a combination of splenic $CD4^+$ $CD25^-$ and $CD8^+$ T cells, which are co-cultured with Tregs in five different ratios for 96 h.

In order to estimate the portion of the single cell suspensions that should be stained and sorted to obtain sufficient cells to perform the assay, it is important to calculate the required number of cells first. Testing a single condition in five Treg:responder ratios requires a total input of 125,000 responder cells and 50,000 Tregs. Per ratio, 25,000 responder cells are plated. To obtain a Treg:responder ratio of 1:1, 25,000 Tregs are co-cultured with 25,000 responder cells. The remaining 25,000 Tregs are serially diluted, with linearly fewer Tregs for each ratio. When testing Tregs from multiple conditions, multiply the number of required responder cells accordingly. This suppression assay has been optimized for an input of 50,000 Tregs but can be downscaled to an input of 10,000 Treg cells.

As each tissue yields a different number of Tregs, we have summarized our typical sorting yields here. From a spleen of a naïve adult wild-type FVB mouse, $0.8 * 10^6$ $CD4^+$ $CD25^+$ Tregs, $3.0 * 10^6$ $CD8^+$ T cells and $10 * 10^6$ $CD4^+$ $CD25^-$ T cells can be recovered. The number of Tregs that can be obtained from a tumor is size and model dependent. From a tumor sized 1500 mm^3, we generally recover between 50,000 and 100,000 Tregs.

Estimate the portion of single cell suspension that should be fluorescently labeled and sorted accordingly, and proceed with the antibody staining procedure (Note 3).

4. Fluorescent antibody staining and cell sorting

After the preparation of single cell suspensions, cells are fluorescently labeled in order to identify and sort Tregs and responder cells from the samples. Here, a thoroughly tested and optimized approach for the fluorescent staining and sorting of Tregs is presented. The choice of antibody-fluorochrome conjugate combination dictates the purity of the isolated Treg population. $CD4^+$ Tregs in mice are characterized by high expression of the transcription factor Foxp3. However, unless Foxp3-reporter mice are used, it is not possible to sort live Tregs based on this marker, because detection of intracellular Foxp3 requires fixation and permeabilization. To circumvent this, it is widely accepted to sort Tregs based on surface expression of CD4 and CD25 (IL2-receptor α-chain) (Thornton & Shevach, 1998). Since CD25 can also be transiently expressed on conventional T cells upon TCR stimulation, it is important to confirm Treg identity of sorted $CD4^+$ $CD25^{high}$ cells by intracellular Foxp3 staining, as shown in Fig. 2A (Note 4). Additionally, tumor tissue-derived Tregs express relatively lower levels of CD4 as compared to Tregs derived from other tissues due to enzymatic treatment. Therefore, a relatively bright anti-CD4 fluorochrome-antibody conjugate must be used to discriminate these cells from the CD4 negative fraction, as illustrated in Fig. 2B. Using the following protocol, Tregs can be directly sorted from the single cell suspensions without the need for additional purification steps.

Cell suspensions are first incubated with Fc receptor block solution (Note 5), followed by incubation with fluorescently labeled antibodies (Table 1). Finally, cells are stained with a viability dye to distinguish viable from dead cells. To compensate for spectral overlap, single fluorochrome stained samples should be prepared for each fluorochrome (Note 6). To determine positive populations for CD4-PE and CD25-APC, fluorescence minus one (FMO) controls should be included for each tissue that is sorted.

1. Plate the previously calculated number of cells from each single cell suspension in a 96-well plate. Plate a maximum of $6 * 10^6$ cells per well. Keep the plate on ice.
2. In the same 96-well plate, plate cells in 6 additional wells for each single fluorochrome sample and an unstained sample. For splenocytes, plate $\sim 1 * 10^6$ cells/well. When sorting tumor samples, plate compensation beads instead of cells to save sample for sorting.
3. Plate 2 extra wells for each tissue sample for CD4-PE and CD25-APC FMO controls. Plate same number of cells/well as determined in Section 4, step 1. The plate now contains cells for sorting, single fluorochrome controls, and FMO controls.
4. Prepare the following solutions in cold sterile sorting buffer (viability dye in PBS), 50 μL per well.

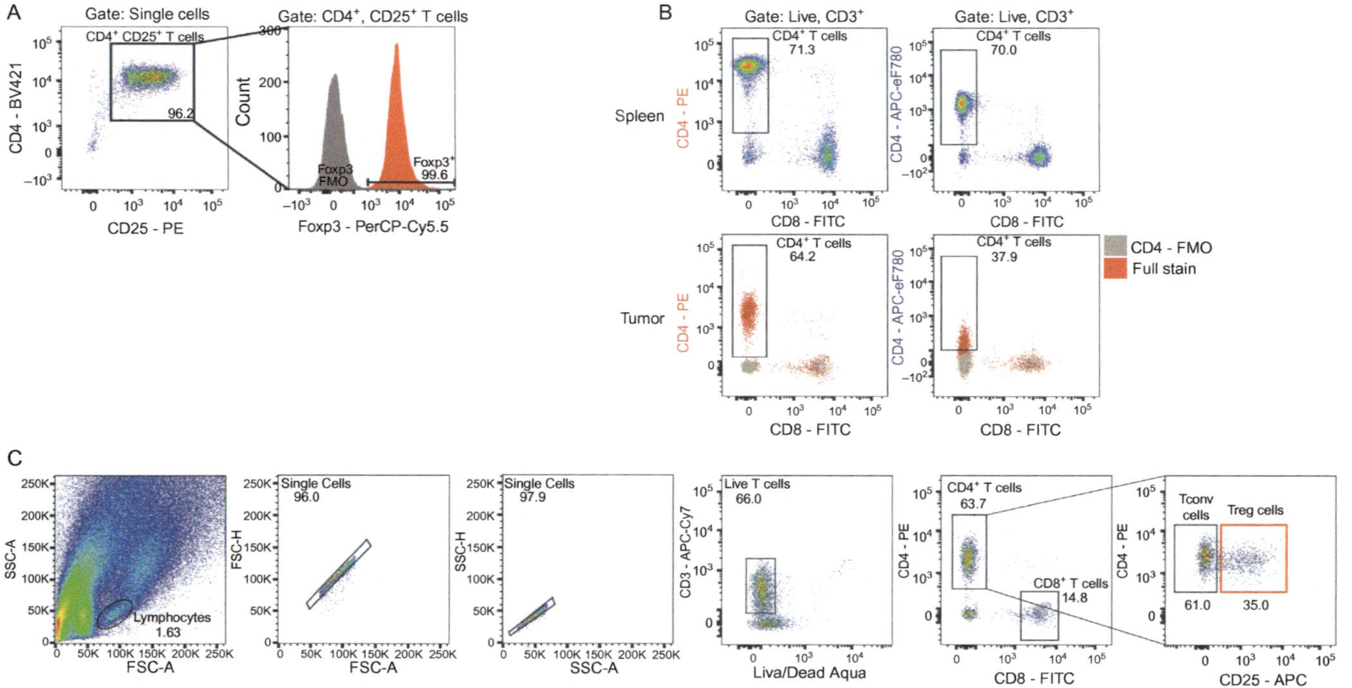

Fig. 2 Flow cytometry-based sorting of regulatory T cells. (A) Splenic CD4$^+$ CD25$^+$ cells were sorted as described and intracellularly stained for Foxp3 to confirm Treg identity. (B) Single cell suspensions from spleen and tumor were fluorescently labeled according to Table 1, or an alternative panel using a relatively dim antibody for CD4. CD4 expression on live CD3$^+$ T cells is shown. (C) Recommended gating strategy to sort Tregs from fluorescently labeled single cell suspensions. Tumor tissue is shown.

Solutions for fluorescent labeling of single cell suspensions

a. Antibody mix (Note 7)

i. Prepare the mix using the antibodies and concentrations described in Table 1. Prepare 50 μL antibody mix per well. For example, if 60 * 10^6 splenocytes are sorted, calculate for 10 wells. Prepare 50 μL extra (10 wells * 50 μL) + 50 μL = 550 μL antibody mix.

b. Viability dye

i. Fixable LIVE/DEAD Aqua cell stain kit (1:1000 in PBS)

c. Fc receptor block solution.

i. Dilute purified anti-mouse CD16/32 antibody (1:100)

Control solutions for flow-based sorting

d. CD4-PE and CD25-APC FMO controls.

Prepare FMO controls for each tissue that will be sorted.

i. Prepare the antibody mix as in a. but without the CD4-PE antibody.

ii. Prepare the antibody mix as in a. but without the CD25-APC antibody.

e. Single fluorochrome controls (Note 6).

i. Prepare single fluorochrome controls by diluting each fluorochrome (including the viability dye) from Table 1 separately. Use the indicated concentration.

5. Centrifuge the 96-well plate for 2 min at 380 *g* after preparation of the solutions. Flick off the supernatant in a waste bin in the flow hood and gently press the plate upside down on a tissue.
6. Resuspend cell samples (all, except single fluorochrome controls) in 50 μL of Fc receptor block solution and incubate 5 min on ice. After incubation, centrifuge the plate and discard supernatant.
7. During this incubation, add 50 μL of single fluorochrome controls to designated wells.
8. Resuspend the fluorescent antibody mix and FMO controls and add 50 μL to the designated wells. Gently mix with a pipette and incubate the plate for 20 min protected from light at 4 °C.
9. After incubation, wash the plate 1 × with PBS (Note 8).
10. Stain cells with viability dye according to manufacturer's instructions. After incubation, wash the cells 1 × with sorting buffer.
11. Resuspend samples for single fluorochrome- and FMO controls in 200 μL sorting buffer and collect in 1.4 mL U-bottom tubes.
12. Resuspend each sample for sorting in 200 μL sorting buffer. Filter the sample through a 35 μm cell strainer into a 5 mL polypropylene tube. Combine all samples from matching tissue in a single tube. Replace

the cell strainer cap if clogged to enhance cell recovery. Tregs/ responders will be sorted directly from this tube. Keep tubes sterile, on ice and protected from light.

13. Adjust sample volume to an appropriate cell concentration with sorting buffer, to increase sorting efficiency (Note 9). Guidelines:
  a. 70 μm nozzle: $14 * 10^6$–$21 * 10^6$ cells/mL
  b. 85 μm nozzle: $10 * 10^6$–$15 * 10^6$ cells/mL
  c. 100 μm nozzle $4 * 10^6$–$6 * 10^6$ cells/mL

14. Prepare 5 mL polypropylene collection tubes with 1 mL collection medium for collection of Treg/responders (Notes 10 and 11).

15. Sort cells in an aseptic fashion. Keep samples cooled at all times during the sorting process. Keep the collected cells on ice.

16. Apply the following gating strategy on single cells that are negative for the viability dye. See Fig. 2C for gating Tregs in tumor tissue (Note 12).
  a. Regulatory T cells:
    i. $CD3^+$, $CD8^-$, $CD4^+$, $CD25^+$
  b. SPLEEN RESPONDERS—Conventional $CD4^+$ T cells:
    i. $CD3^+$, $CD8^-$, $CD4^+$, $CD25^-$
  c. SPLEEN RESPONDERS—$CD8^+$ T cells:
    i. $CD3^+$, $CD4^-$, $CD8^+$, $CD25^{+/-}$

17. Sort both $CD4^+$ and $CD8^+$ responder cells simultaneously in the same collection tube (Note 13).

18. Sort Tregs from each tissue in a separate collection tube (Note 14).

19. Perform a purity check on collected cells after sorting (Note 15):
  a. Resuspend 20 μL of the collected cell suspension in 180 μL PBS
    i. Run ddH_2O at high differential pressure until sample line is clean.
    ii. Record 500–1000 events in the gate of the sorted subset.
  b. Mix 20 μL of collection buffer with 180 μL PBS.
    i. Run ddH_2O at high differential pressure until sample line is clean.
    ii. Record for the same amount of time to determine background signal.
  c. Determine the percentage of the population of interest from total events (Note 16).

20. Immediately proceed with the suppression assay (Note 17).

5. Suppression assay

A schematic representation of the assay is shown in Fig. 3. Freshly sorted Tregs and responder cells are first cultured separately overnight to allow for activation and resting, respectively. The following morning,

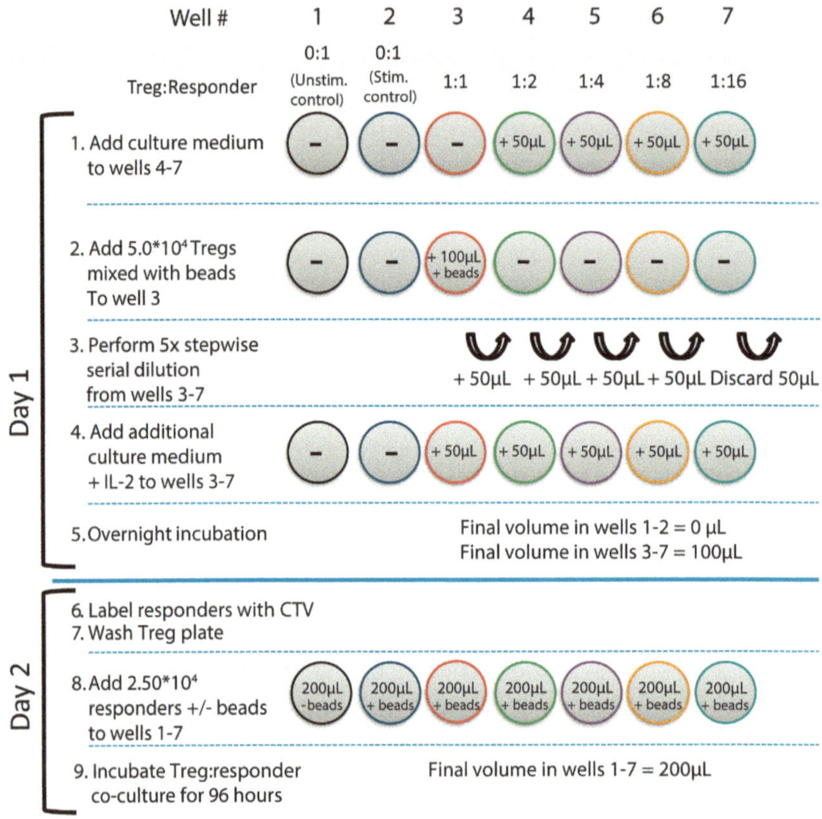

Fig. 3 Schematic plate layout for the suppression assay. Follow stepwise instructions to co-culture Tregs and responder cells in five Treg:Responder ratios in seven wells, per Treg condition. Repeat for other conditions if necessary.

responder cells are labeled with CellTrace™ Violet Proliferation (CTV) Dye (Note 18) and co-cultured with Tregs at different ratios for 96h. Finally, responder cell proliferation is analyzed by flow cytometry to determine the suppressive potential of the Tregs. This protocol is optimized for the analysis of low numbers of Tregs, and has been validated for an input as low as 10,000 Tregs per condition. However, if Treg yield is sufficient, for robustness we strongly recommend performing the assay with 50,000 Tregs, and technical replicates for test conditions and controls.

5.1 Controls

In order to adequately determine how Tregs affect responder cell proliferation, it is essential to include appropriate controls during the co-culture of Tregs and responder cells.

1. To determine the maximum proliferative potential of responder cells in a given time period, responder cells should be cultured with stimulatory signals in the absence of Tregs. This is essential to verify that responder cells have the capacity to proliferate, allowing for a window of Treg mediated inhibition of proliferation.
2. To determine the background proliferative activity of unstimulated responder cells in a given time period, responder cells should be cultured in the absence of both Tregs and stimulatory signals. This is an essential control to verify that responder cells do not proliferate in the absence of stimulation. Additionally, the cells from this control are used to set up CTV signal for data acquisition.

Additional considerations have to be taken into account in order to draw reliable conclusions from the assay. Although cell isolation by flow sorting allows for near 100% purity of the target populations, this method of cell sorting is performed under relatively high pressure conditions. As a result, shear stress may damage cells and alterations may occur in the redox state and metabolic profile (Llufrio, Wang, Naser, & Patti, 2018). This procedure may therefore directly impact the performance of cells in this assay. It is essential that after sorting, cells are first cultured separately overnight, to allow cells to recover from sorting-induced stress as shown in Fig. 4A (Note 19).

Another consideration is the method of responder cell stimulation, as there is a fine balance between excessive and inadequate stimulation. In this assay, responder T cells are stimulated with anti-CD3 and anti-CD28 antibodies covalently bound to magnetic beads, leading to both TCR activation and co-stimulation. Although beads are simple and extremely reproducible in use, these beads do not fully mimic true antigen-presenting cells (APCs) and bypass any inhibitory effect on proliferation and priming mediated by Treg signaling to APCs. If APCs are required, the user can exchange beads with APCs of choice.

It is important to titrate bead:responder ratios because excessive stimulation of responder cells can lead to loss of Treg mediated suppression, whereas inadequate activation may prevent responder cell proliferation. We have found that a ratio of bead:responder of 1:5 provides a window for suppression while simultaneously activating responder cells.

Additional to responder cell activation, overnight activation of Tregs prior to responder cell co-culture greatly enhances suppressive potential. Treg pre-activation may be essential when sorting from homeostatic conditions (Fig. 4B) (Note 20).

Finally, T cell proliferation protocols typically include supplementation of IL-2 to culture medium. However, supplementing the co-culture

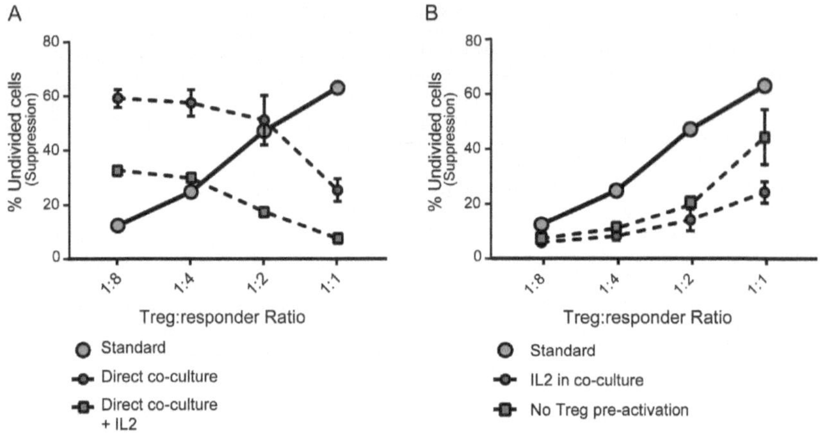

Fig. 4 Impact of culture conditions on responder cell proliferation. The percentage of undivided responder cells is plotted for Treg:Responder ratios following suppression assays with various conditions. Standard data: Assay was performed as recommended, including overnight resting, resulting in ratio-dependent proliferation. (A) When a resting period is omitted and cells are directly co-cultured, ratio-dependent proliferation is lost. (B) Adding IL-2 (300 U/mL) to the co-culture, or omitting Treg pre-activation reduces assay sensitivity. Two technical replicates were used per condition, mean ± SEM is shown from two independent biological replicates.

medium with IL-2 highly increases responder cell proliferation in the presence of Tregs, thus severely reducing the window of Treg suppression. It is therefore strongly recommended to perform the assay in the absence of IL-2. (Note 21) (Fig. 4B).

5.2 Day 1: Treg pre-activation and responder resting: See Fig. 3

5.2.1 Responder cells

1. Determine the number of responder T cells and centrifuge the cells for 5 min, 250 *g* at 4 °C. Discard supernatant and resuspend the cells in culture medium at a concentration of $1.0 * 10^6$ cells/mL. Do not add beads or recombinant IL-2.
2. Plate 2 mL of the cell suspension per well in a tissue culture-treated 6-wells plate.
3. Incubate the cells overnight at 37 °C, 5% CO_2.

5.2.2 Tregs

4. Accurately determine the number of Tregs and subsequently centrifuge the cells for 5 min, 250 *g* at 4 °C. Discard supernatant and resuspend the cells in culture medium at a concentration of $5.0 * 10^5$ cells/mL.

5. Treg and responder cells will be co-cultured in the Treg:responder ratios 1:1, 1:2, 1:4, 1:8 and 1:16. The following protocol applies to an input of 50,000 Tregs for each condition, which are all seeded into the "1:1" well.
6. Add CD3/CD28 coated Dynabeads™ in a 1:5 bead:cell ratio to the Treg cell suspension and mix (Note 22).
7. In a tissue culture-treated round-bottom 96-wells plate, add 50 μL of culture medium to wells 4–7; add 100 μL of the Treg cell suspension to well 3 (Fig. 3).
8. Transfer 50 μL of Treg cell suspension to the well designated for the 1:2 ratio. Mix thoroughly, but avoid bubble formation. This will result in a twofold dilution of both Tregs and beads.
9. Repeat step 8 in consecutive wells to serially dilute 1:4, 1:8 and 1:16 Treg:responder ratios. Discard the leftover 50 μL of Treg suspension.
10. Finally, add 50 μL of culture buffer + murine recombinant IL-2 (600 U/mL) to all wells that now contain Tregs. Final volume per well is 100 μL.
11. Incubate the 96-well plate overnight at 37 °C, 5% CO_2.

5.3 Day 2: Treg:Responder co-culture

1. Harvest responder cells from the 6-well plate by thorough resuspension. Collect the cells in a 15 mL tube and centrifuge the tube 5 min, 250 *g*. Keep an aliquot of cells apart for Section 5.4, step 2 (~25 K cells)
2. Label responder cells with CellTrace™ Violet according to manufacturer's instructions.
3. After labeling, accurately determine the cell number. Recovery rate after sorting, overnight culture and labeling is typically ~40%.
4. Discard the supernatant and resuspend the responder cells at $1.25 * 10^5$ cells/mL in culture medium.
5. Remove at least 200 μL from the suspension ($2.5 * 10^4$ cells) and reserve for unstimulated controls.
6. Add CD3/CD28 coated Dynabeads™ in a 1:5 bead cell ratio to the responder cell suspension and mix.
7. Wash the 96-well plate containing the activated Tregs with culture medium.
8. Start Treg:responder co-culture by adding 200 μL of the responder cell suspension to all ratios and mix by pipette. Also add 200 μL of responder cells to the stimulation control condition in well 2. Add cells from Section 5.3, step 5 (unstimulated control condition) to well 1.
9. Incubate the cells at 37 °C, 5% CO_2 for 96 h (Note 23).

5.4 Day 6: Analysis

The suppressive potential of Tregs will be determined by measuring the proliferation of responder cells on day 6 (Note 24). As both CD4 and CD8 T cells are used as responder cells, proliferation can be assessed for each cell type separately. Cells will again be fluorescently labeled according to Table 1, with the exception of LIVE/DEAD Aqua as this channel is now reserved for CTV. Instead of LIVE/DEAD Aqua, we recommend using 7-AAD viability staining solution to detect dead cells (Note 25).

1. After incubation, wash cells 1 × in sorting buffer.
2. Re-stain cells according to Section 4, steps 2–10, omit FMO controls. Use beads for single fluorochrome controls.
   a. A small sample of the stimulated control cells can be used as a single fluorochrome control for CTV.
   b. A small sample of unstained stimulated control cells are spiked into the CTV sample to obtain a negative population.
   c. A small sample of unstained stimulated control cells are stained with 7-AAD.
3. Resuspend cells in 100 μL sorting buffer +7-AAD (1:100) and proceed with cell analysis on a flow cytometer.
4. Prior to compensation, acquire an unstimulated responder cell sample and adjust detector gain of the channel used for CTV. Set the signal of the undivided peak to an intensity of 10^5.
5. Run compensation controls and set up gates. Gate Live, single, $CD3^+$ CTV^+ cells to exclude Tregs from the analysis (Note 26).

Record sufficient CTV^+ cells to perform descriptive and analytical statistics with the predetermined level of confidence. Ideally, a minimum number of 2000 CTV^+ events is recorded.

5.5 Data analysis

To assess the suppressive potential of Tregs, perform fluorescence compensation on recorded samples and gate: Single, Live, $CD3^+$, CTV^+ cells. The first peak (intensity 10^5) indicates the undivided population. Each subsequent peak indicates consecutive cell division. The CTV^- fraction represents the Tregs. At this point, it is not possible to discriminate Tregs from responder cells based on CD25 expression, as responder cells will have upregulated CD25 in response to strong TCR stimulation. Confirm stimulation-dependent proliferation in controls. Draw gates to determine the percentage of divided/undivided fraction of the cells (Fig. 5). For each sample, determine the percentage of the undivided population, and plot this according to Treg:responder ratios (see graph type in Fig. 4). The impact of Tregs from different conditions on responder cell proliferation can now be assessed (Note 27). An appropriate statistical test can be applied using the mean of the variable "% of undivided cells."

6. Notes

1. Commercially available RBC lysis buffers can contain fixation solutions. These should be avoided to maintain viability and cell function.
2. Mechanical disruption and enzymatic digestion is used to dissociate tumor tissue into a single cell suspension. Both treatments can result in cell damage and subsequent cell death. Additionally, enzymatic digestion can lead to a reduction of the available recognizable epitopes of cell surface markers. Preparation of single cell suspensions from tumor tissue should be optimized, i.e., duration and force of mechanical disruption, enzyme choice, concentration and incubation time should be determined empirically per tumor type (Abuzakouk, Feighery, & O'Farrelly, 1996; Autengruber, Gereke, Hansen, Hennig, & Bruder, 2012). The described digestion method is optimized for mammary tumors (Salvagno & de Visser, 2016).
3. The abundance of Tregs varies per tissue type. To sort sufficient cells to perform the assay, it is important to fluorescently label an adequate proportion of single cell suspensions. For tumor samples, we recommend to use 100% of the available tissue for sorting. For responder cells, sort at least 2–3 × the required number of cells to compensate for cells lost during washing and labeling steps.
4. Contamination of activated T-helper cells within the $CD25^{high}$ Treg population may depend on mouse strain, genotype, tissue and homeostatic conditions of mice within the animal housing facility.
5. Murine myeloid- and B cells have high expression of Fc receptors, that specifically bind the Fc domain of fluorescently labeled antibodies. This may lead to false interpretation of positive signal, and could lead to sorting of a contaminated population. Therefore, incubate single cell suspensions in αCD16/32 Fc blocking reagent.
6. Guidelines for single fluorochrome controls
   a. The compensation matrix is determined with an algorithm that is embedded in the cell sorter software. To calculate the compensation values, the algorithm requires an accurate determination of the

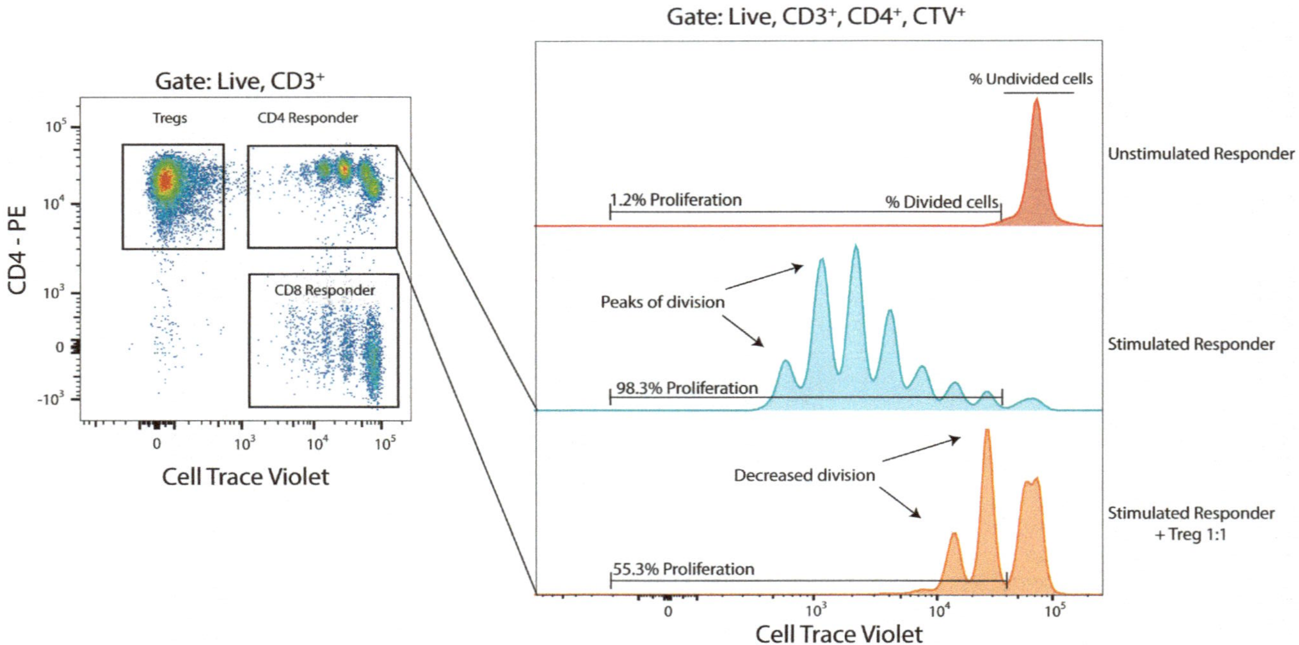

Fig. 5 Recommended analysis of the suppression assay. A suppression assay was performed as described with splenic Tregs and analyzed 96 h after co-culture (day 6 of the assay). Responder cells do not proliferate in the absence of stimulation, but strongly divide when stimulated. Tregs potently suppress responder cell proliferation. Treg:Responder ratio of 1:1 is shown.

median fluorescent intensities (MFI) from the negative population and positive fluorochrome carrier population for each channel. An accurate estimate of the median is dependent on the variance, and the minimal number of negative and positive events required is between 2000 and 5000 for each.

b. The median fluorescent intensity of the positive population must be at least as high as the highest median fluorescent intensity of an experimental sample. It is recommended to perform compensation on cells obtained from tissue that will be used for sorting Tregs. When sample is limited, the use of antibody-binding beads is recommended.

c. The median fluorescent intensities of the negative and positive populations must be within the linear range of the detectors that are used.

d. The antibody-fluorochrome conjugates used to prepare the single stains must be the same as those used in the experimental samples.

7. When the recommended antibody panel is modified, we advise to titrate antibody concentrations to determine the stain index and effect on data spread in secondary channels. The goal is to maintain optimal discrimination between negative and positive cell populations.

8. Washing entails the process of centrifuging the plate for 2 min at 380 *g*, subsequently flicking off the supernatant and resuspending all samples in PBS followed by centrifugation. Lastly, flick of the supernatant and continue.

9. Nozzle diameter is determined as follows: average diameter of cells of interest * 6. For lymphocytes, the 70 μm nozzle is acceptable. It also allows for acceptable duration of the sort. Using larger nozzles results in higher recovery and increased cell function, but samples may require pre-enrichment to ensure acceptable sort times.

10. Sorting charged droplets that contain cells of interest can lead to accumulation of electrostatic charge inside a tube made of insulating polymers. Therefore, polypropylene tubes should be used to avoid build-up of electrostatic charge. Additionally, sorting into polypropylene tubes results in lower cell adherence to the tube walls, as compared to polystyrene tubes.

11. Addition of at least 1–2% FCS to the collection buffer has been shown to strongly reduce negative effects of cell sorting on cell viability, redox and metabolic perturbations (Brennan, Mansky, Roberts, & Lichtman, 1975; Llufrio et al., 2018).

12. $CD8^+$ and $CD4^+$ T cell proliferation can be mitigated by multiple cell types, especially in the tumor microenvironment. To investigate the specific role of regulatory T cells on the suppression of cytotoxic T cells, the isolated Treg fraction must be free from contamination of other cell types. In this specific assay, CD3 is used to uniquely identify cells of interest, and exclude other cell types that may bias the results. For example, DCs and B lymphocytes.
13. Sorting both CD4 and CD8 responder cells allows for investigating T cell-specific suppression. In this assay, cells are sorted in physiological ratios (CD4:CD8 = ~3:1 in spleen), but this can be adjusted to the user's preference, including sorting CD4 and CD8 T cells separately. Investigating suppression of CD4 and CD8 T cells provides a general method of assessing suppression of Tregs. Alternatively, it is possible to use virtually any other cell type of interest as responders.
14. Sorting Tregs and responder cells from spleen samples can be performed relatively fast, and takes approximately ~20–30 min per condition in our experience. In contrast, sorting tumor samples can take up to multiple hours due to high cell concentration and a low Treg fraction. In case of insufficient Treg yield from tumors, an optional pre-enrichment of tumor samples may be performed. Sterile isolation of $CD45^+$ cells from the tumor sample via positive magnetic selection may increase yield and does not interfere with surface staining. Alternatively, percoll gradient separation may increase yield, although we have not tested this in this assay.
15. Doublets with $CD3^+$ cells and incorrect determination of charge delay timing (drop delay) may result in contamination or insufficient isolation of the cells of interest. A post-sort (purity) check can be performed to evaluate the collected samples.
16. The median fluorescent intensity may be slightly lower after sorting. Adjustment of the gates may be required to estimate to percentage correctly.
17. In our experience, storing cells on ice for up to 1 h after sorting does not impact cell viability.
18. CellTrace™ Violet is a fluorescent dye which covalently binds free amines on the surface and the inside of cells. When cells divide, the dye is equally divided over daughter cells, resulting in a 50% reduction of signal intensity per division. This allows for the visualization of proliferation in the responder cell pool. Alternatively, responder cells can be labeled with fluorescent proliferation dyes such as CFSE. This does require modification of recommended antibody panels.
19. For cell sorting, we used a FACSAria II (BD Biosciences) with a 70 μm nozzle and 70 psi. Likely, the observed negative effect on proliferation

when cells are co-cultured directly after cell sorting is due to shear stress from the cuvette-based cell sorter. If a resting period is undesired, this effect may be minimized by using a jet-in-air based cell sorter at lower sheath pressure, e.g., with a 100 μm nozzle and 20 psi. When omitting sorting-induced stress is not required (If using magnetic-based cell isolation, for example), alternative assays exist as discussed here (Collison & Vignali, 2011).

20. In our experience, Treg pre-activation does not impact pre-existing differences in suppressive potential between Treg populations.

21. Conventional CD4 T cells produce high levels of IL-2, whereas Tregs do not, although essential for survival of both. Accordingly, a proposed mechanism of Treg suppression is CD25-mediated depletion of IL-2 from the local microenvironment (Chen & Oppenheim, 2011).

22. Beads added for Treg activation on day 0 will not induce responder cell proliferation during co-culture, for which additional beads are added.

23. Co-culture time may be adjusted as preferred, but should be minimally 72 h.

24. Further in-depth analysis may include flow-cytometric analysis of responder cells and Tregs focused on the expression of cell surface receptors, intracellular cytokine production and cell-death. Culture medium can be collected after co-culture for cytokine analysis.

25. For analysis, an LSR II SORP (BD Biosciences) was used. The configurations are provided in Table 2.

Table 2 LSR II configurations for suppression assay analysis.

Antigen	Fluorochrome	Configuration FACSAria II	Configuration LSR II SORP
CD3	APC-Cy7	633 nm laser (20 mW); 750LP, 780/60	638 nm laser (40 mW); 750LP, 780/60
CD4	PE	488 nm laser (20 mW); 565LP, 585/42	561 nm laser (40 mW); 565LP, 585/42
CD8	FITC	488 nm laser (20 mW); 502LP, 530/30	488 nm laser (50 mW); 505LP, 525/50
CD25	APC	633 nm laser (20 mW); 660/20	638 nm laser (40 mW); 670/14
Free amines	L/D Aqua	405 nm laser (25 mW); 502LP, 530/30	
Free amines	CTV		405 nm laser (40 mW); 450/50
dsDNA	7-AAD		561 nm laser (40 mW); 600LP, 610/20

26. Divided $CD4^+$ responder cells may lose CTV to such an extent that these cells become indistinguishable from Tregs during analysis. In case of prolonged co-culture, it is advisable to additionally label the Treg cells. This could be done by using alternative fluorescent proliferation labels, congenic markers, or including Foxp3 in the analysis panel.
27. Additional methods for analyzing T cell proliferation that may provide higher sensitivity are reviewed here (McMurchy & Levings, 2012). Alternatively, specialized software to analyze proliferation data can be used (ModFit LT™, Flowjo LLC).

References

Abuzakouk, M., Feighery, C., & O'Farrelly, C. (1996). Collagenase and dispase enzymes disrupt lymphocyte surface molecules. *Journal of Immunological Methods*, *194*(2), 211–216.

Autengruber, A., Gereke, M., Hansen, G., Hennig, C., & Bruder, D. (2012). Impact of enzymatic tissue disintegration on the level of surface molecule expression and immune cell function. *European Journal of Microbiology & Immunology*, *2*(2), 112–120.

Binnewies, M., Roberts, E. W., Kersten, K., Chan, V., Fearon, D. F., Merad, M., et al. (2018). Understanding the tumor immune microenvironment (TIME) for effective therapy. *Nature Medicine*, *24*(5), 541–550.

Brennan, J. K., Mansky, J., Roberts, G., & Lichtman, M. A. (1975). Improved methods for reducing calcium and magnesium concentrations in tissue culture medium: Application to studies of lymphoblast proliferation in vitro. *In Vitro*, *11*(6), 354–360.

Chang, X., Zheng, P., & Liu, Y. (2009). Selective elimination of autoreactive T cells in vivo by the regulatory T cells. *Clinical Immunology (Orlando, Fla.)*, *130*(1), 61–73.

Chen, X., & Oppenheim, J. J. (2011). Resolving the identity myth: Key markers of functional CD4+FoxP3+ regulatory T cells. *International Immunopharmacology*, *11*(10), 1489–1496.

Collison, L. W., & Vignali, D. A. A. (2011). In vitro Treg suppression assays. *Methods in Molecular Biology (Clifton, N.J.)*, *707*, 21–37.

Fontenot, J. D., Gavin, M. A., & Rudensky, A. Y. (2003). Foxp3 programs the development and function of CD4+CD25+ regulatory T cells. *Nature Immunology*, *4*, 330.

Josefowicz, S. Z., Niec, R. E., Kim, H. Y., Treuting, P., Chinen, T., Zheng, Y., et al. (2012). Extrathymically generated regulatory T cells control mucosal T H 2 inflammation. *Nature*, *482*(7385), 395–399.

Kitamura, T., Qian, B. Z., & Pollard, J. W. (2015). Immune cell promotion of metastasis. *Nature Reviews Immunology*, *15*(2), 73–86.

Li, Z., Li, D., Tsun, A., & Li, B. (2015). FOXP3+ regulatory T cells and their functional regulation. *Cellular and molecular immunology*, *12*(5), 558–565.

Littman, D. R., & Rudensky, A. Y. (2010). Th17 and regulatory T cells in mediating and restraining inflammation. *Cell*, *140*(6), 845–858.

Llufrio, E. M., Wang, L., Naser, F. J., & Patti, G. J. (2018). Sorting cells alters their redox state and cellular metabolome. *Redox Biology*, *16*, 381–387.

McMurchy, A. N., & Levings, M. K. (2012). Suppression assays with human T regulatory cells: A technical guide. *European Journal of Immunology*, *42*(1), 27–34.

Salvagno, C., & de Visser, K. E. (2016). In J. Ursini-Siegel & N. Beauchemin (Eds.), *Purification of immune cell populations from freshly isolated murine tumors and organs by consecutive magnetic cell sorting and multi-parameter flow cytometry-based sorting BT—The tumor microenvironment: Methods and protocols* (pp. 125–135). New York, NY: Springer New York.

Sharma, A., & Rudra, D. (2018). Emerging functions of regulatory T cells in tissue homeostasis. *Frontiers in Immunology*, *9*, 883.

Thornton, A. M., & Shevach, E. M. (1998). CD4 + CD25 + immunoregulatory T cells suppress polyclonal T cell activation in vitro by inhibiting interleukin 2 production. *The Journal of Experimental Medicine*, *188*(2), 287 LP–296.

Togashi, Y., Shitara, K., & Nishikawa, H. (2019). Regulatory T cells in cancer immunosuppression—Implications for anticancer therapy. *Nature Reviews Clinical Oncology*, *16*, 356–371.

Vignali, D. A. A., Collison, L. W., & Workman, C. J. (2008). How regulatory T cells work. *Nature Reviews Immunology*, *8*(7), 523–532.

CHAPTER FIFTEEN

Assessment of lipid load in tumor-infiltrating Tregs by flow cytometry

Ilenia Pacella[a], Alfonso Grimaldi[b], Silvia Piconese[a,c,*]
[a]Dipartimento di Medicina Interna e Specialità Mediche, Sapienza Università di Roma, Rome, Italy
[b]Center for Life Nano Science @ Sapienza, Istituto Italiano di Tecnologia, Rome, Italy
[c]Istituto Pasteur Italia—Fondazione Cenci Bolognetti, Rome, Italy
*Corresponding author: e-mail address: silvia.piconese@uniroma1.it

Contents

Abstract

Regulatory T cells (Tregs), expressing the transcription factor Foxp3, are defined as immunosuppressive cells able to modulate a variety of immune cells in order to avoid unwanted and excessive immune responses; however, in the tumor context, Treg function contribute to inhibit immune surveillance, thus promoting cancer progression.

In tumor microenvironment, where the availability of metabolic resources is strongly limited, Tregs are expanded and may exploit different metabolic routes to achieve a metabolic advantage, prevailing over effector cells. In this context an important role of lipid metabolism has been described thanks to the possibility to evaluate the intracellular lipid content selectively in tumor-infiltrating Tregs (TUM-Tregs). Taking into account the heterogeneous and complex build of tumor mass, we set-up a combined protocol that optimizes tumor-infiltrating lymphocytes (TIL) extraction from the tissue through a Percoll density gradient, and assesses ex vivo the lipid load in whole TUM-Treg population, evaluating by flow cytometry the incorporation of an intensely fluorescent lipophilic fluorophore able to specifically stain neutral lipids. This method provides an important advantage compared to the traditional technique based on microscopy, whose lipid level evaluation is limited to a tissue section, and hence may not be representative of the entire population.

Methods in Enzymology, Volume 632
ISSN 0076-6879
https://doi.org/10.1016/bs.mie.2019.07.024

1. Introduction

Regulatory T cells (Tregs) represent a specialized CD4$^+$ T cell subset phenotypically characterized by the constitutive surface expression of high level of interleukin-2 receptor chain α (IL-2Rα), also known as CD25 (the high affinity subunit of IL-2R) and especially by the expression of the lineage-determining forkhead box protein 3 (FOXP3) transcription factor. Both molecules are essential for human and murine Treg development and function, and CD25 is important also for Treg survival (Sakaguchi, Vignali, Rudensky, Niec, & Waldmann, 2013).

Tregs play critical role in maintaining immunological self-tolerance and immune homeostasis by preventing autoimmune responses against self-antigens and excessive inflammatory responses (Levings, Allan, d'Hennezel, & Piccirillo, 2006). However, in cancer, Tregs can contribute to tumor progression by suppressing anti-tumor immunity (Zou, 2006). Indeed, Treg infiltration is associated to poor prognosis in most cancer types (Fridman, Zitvogel, Sautes-Fridman, & Kroemer, 2017).

In the last decades, it has become clear that one of the main limitations to the ability of immune cells to eliminate cancer cells is represented by the metabolic rewiring and the deregulation of cellular energetics in tumor cells, altering the balance and availability of nutrients within the tumor microenvironment, and thus impairing immune cell effector functions (Buck, Sowell, Kaech, & Pearce, 2017). Despite the local metabolite restriction, a huge expansion of activated Tregs has been described in tumor tissue in mice and humans, indicating that the metabolic requirements for Treg survival and function in the tumor milieu may be peculiar and has been demonstrated to involve lipid metabolism (Pacella et al., 2018). Intracellular lipid accumulation and an increased lipid metabolism signature, driven by the adipogenic factor peroxisome proliferator-activated receptor gamma (PPARγ), were found to be associated to Treg expansion and accentuation of their phenotype and function in visceral adipose tissue (VAT) (Cipolletta et al., 2012). Also in tumor tissue, highly proliferative and activated tumor-infiltrating Tregs (TUM-Tregs) display a high lipid content that correlates with an increased anabolic and catabolic lipid metabolism (Pacella et al., 2018). The possibility to measure these metabolic aspects of TUM-Tregs and hence, to highlight their metabolic peculiarity, may be exploited to understand the metabolic routes that may support Tregs in the unfavorable tumor microenvironment.

Traditionally, lipid levels are assessed biochemically in the whole tissue or in specific cell types by microscopy using fluorophore-conjugated lipids, lipid-binding probes or lipid analogs and lipid soluble dyes (Daemen, van Zandvoort, Parekh, & Hesselink, 2016), but this technique has significant limitations. In particular, it is subjected to observer bias and usually it analyzes a tiny part of a heterogeneous tissue.

Tumor tissue has a complex structure consisting of a variety of different cell populations, signaling molecules, cellular interaction and extracellular matrix (Whiteside, 2008), which may make difficult the analysis of a single specific cell population.

Here, we have developed a strategy to efficiently evaluate the intracellular neutral lipid content in the whole TUM-Treg population by flow cytometry, combining a protocol for tumor-infiltrating lymphocytes (TIL) enrichment from tumor tissue, intracellular lipid labeling and Foxp3 staining.

In particular, to stain lipids, we have used the intrinsically lipophilic fluorophore BODIPY™ 505/515, 4,4-Difluoro-1,3,5,7-Tetramethyl-4-Bora-3a,4a-Diaza-s-Indacene (Thermo Fisher Scientific, Waltham, MA USA, hereafter BODIPY), a cell membrane-permeable dye characterized by a combination of non polar structure and long-wavelength absorption and fluorescence, which stains intracellular neutral lipid stores in a specific way and may be detected by flow cytometry.

Flow cytometry has been already used to assess lipid levels in order to sort specific cells in a mixed population (Steinbring, Graja, Jank, & Schulz, 2017): the combined method described here demonstrates the feasibility of estimating lipid mass in specific cell types within a complex tissue such as tumor.

2. Tumor cell line culture and tumor model

1. Upon thawing from liquid nitrogen, mouse colon adenocarcinoma MCA38 cell line is cultured in conventional 75-cm^2 cell culture flasks in complete Dulbecco's modified eagle medium (DMEM) containing high D-glucose (4500 mg/L) and 1 mM sodium pyruvate and supplemented with 10% fetal bovine serum (FBS), 2 mM L-glutamine, 1% penicillin/streptomycin, 1% non-essential aminoacids, 10 mM 2-[4-(2-hydroxyethyl)piperazin-1-yl]ethanesulfonic acid (HEPES) buffer and 50 μM 2-mercaptoethanol (see Notes 1–2). Cells are routinely maintained at 37 °C in humidified 5% CO_2 atmosphere.

2. Before full confluence is attained (i.e., 90–95% confluent flasks, obtained after 2 days from cell splitting), medium is discarded and MCA38 cells are gently washed with room temperature (RT) PBS, detached with 2 mL of trypsin/EDTA (see Notes 3–4), diluted with culture medium, and split at a 1:2 subcultivation ratio to allow cell propagation for maintenance cultures and experimental evaluations (see Note 5). MCA38 cells are subcultivated at least two to three times after thawing to ensure a proper growth capacity in vivo.
3. To obtain transplanted tumors and to assure a tumor growth suitable for subsequent experimental determinations, MCA38 cells are collected from flasks in an exponential phase of growth, washed two times with abundant volume of PBS to completely remove medium by centrifugation 1500 RPM for 5 min, carefully resuspended in a homogeneous single cell suspension and then subcutaneously injected at 5×10^5 cells per mice into the right flank of C57BL/6NCrl wild-type male mice in 200 μL of sterile saline solution (see Notes 6–7). In this setting, solid tumors become palpable after 5–7 days.

3. Tumor processing and TIL extraction

1. After 14 days from MCA38 cells in vivo injection, mice bearing tumors of 8 × 8 mm in size are sacrificed by cervical dislocation and tumors are collected separately in PBS.
2. Tumor tissues are cut in small pieces (see Note 8), transferred on 70 μm cell strainer placed onto a 50 mL falcon tube, and mechanically dissociated with a syringe plunger, by adding PBS 2% FBS buffer pre-warmed at RT to rinse the filter and to favor cell release and collection from the tissues.
3. Cell suspensions are harvested by centrifugation at 1600 RPM for 10 min, and mononuclear cells infiltrating tumor samples are enriched through 40/80 Percoll density gradient centrifugation, performed as follows: the cells are resuspended in 5 mL of 40% Percoll solution and gently overlaid onto 5 mL of 80% Percoll solution in a 15 mL tube, which is then centrifuged at $325 \times g$ at RT for 23 min without brake. After centrifugation, the white opaque layer at the interface between the two Percoll solutions, mostly containing leukocytes, is collected, transferred into a new 15 ml tubes, diluted by filling up the collection tubes with PBS, mixed thoroughly by inverting, and washed twice by centrifugation at $425 \times g$ at RT for 10 min (see Note 9).

4. Combined staining of intracellular FOXP3 and lipid content

1. After isolation, TILs are counted, seeded at least 5×10^5 cells for each sample per well in a 96-well plate round bottom (see Note 10), and washed by adding PBS and spinning at 1800 RPM for 5 min (see Note 11).
2. At first, cells are stained for 15 min at 37 °C in the dark with 100 μL/well of a mix of Fixable Viability Dye eFluor780 (which labels dead cells) and BODIPY, both used 1 μL/mL and diluted in azide-free and serum/protein-free PBS. In this step, to obtain a negative control for BODIPY fluorescence, named fluorescence-minus-one (FMO) control, an aliquot of a sample stained with Viability Dye only is prepared.
3. After incubation, samples are washed with PBS by centrifugation at 1800 RPM for 5 min; then the staining for surface antigens is performed by incubating cells for 20 min at 4 °C in the dark with 50 μL/well of a solution of PBS 2%FBS containing the following fluorochrome-conjugated antibodies necessary for TUM-Tregs identification: anti-CD4 clone RM4-5 Brilliant Violet 605, anti-CD25 clone PC61 Brilliant Violet 510 (see Note 12).
4. Thereafter, cells are washed with PBS 2%FBS as above to remove antibody excess, and are treated for 30 min at 4 °C with 100 μL/well of a fixation and permeabilization solution (Fix/Perm) optimized for transcription factors and nuclear proteins staining (see Note 13).
5. Once cells are fixed and permeabilized, the Fix/Perm solution is discarded after centrifugation at 1800 RPM for 5 min, and intracellular staining is performed by incubating samples for 30 min at RT in the dark with 50 μL/well of the antibody mix containing anti-Foxp3 antibody clone FJK-16s PE-eFluor610, diluted in permeabilization buffer (see Note 12–13).
6. Finally, samples are washed, resuspended in 200 μL/well of PBS, and transferred into FACS tubes for flow cytometry.

5. Flow cytometry acquisition and analysis

1. Samples are acquired with BD LSRFortessa™ analyzer that is configured with the following lasers: 100 mW 488-nm blue, 40 mW 640-nm red and 50 mW 405-nm violet, and operated by BD FACSDiva software™.

2. For a thorough run of the experimental samples, fluorescence compensation setting for the multicolor flow cytometry is established by acquisition of compensation beads prepared as follows:
   - Unstained control compensation beads (blank);
   - Single fluorescent compensation beads stained with the same fluorochrome-labeled anti-mouse antibodies used in the experiment: Foxp3 PE-eFluor610, CD25 Brilliant Violet 510, CD4 Brilliant Violet 605;
   - For BODIPY dye compensation control, beads may be stained with antibodies conjugated to fluorochromes that have similar emission spectrum of the dye.
   - For compensation control of Viability dye, a small aliquot of the same experimental samples is treated with repeated (three times) warm-cold cycles: 1 min at 65 °C, then immediately 1 min on ice. After this treatment, the heat-killed cells are combined 1:1 with live cells and then stained with Viability Dye (see Note 14).

   Once cytometer settings are set up, a compensation matrix is automatically calculated from BD FACSDiva software™ and applied to the experiment.
3. To complete the experimental setup, at first the FMO control of the BODIPY staining (and possibly the other FMO controls) is acquired to check the autofluorescence of the sample and the compensation setup (see Note 15).
4. All cells in the sample are acquired to allow a suitable analysis of TUM-Treg population that is performed by FlowJo software (Tree Star Inc., version 10.0.8rl). FlowJo is employed to manage multiparametric flow cytometry data in order to select the cell population of interest through a specific gating strategy (Fig. 1) and to evaluate within it the intracellular neutral lipid content in terms of geometric mean fluorescence intensity (gMFI) of BODIPY in the FITC channel (Fig. 2) (see Note 16). In details:
   - For gating strategy, at first dead/unwanted cells and doublets are excluded, then $CD4^+$ cells are gated and within this gate TUM-Tregs are identified as $CD25^+Foxp3^+$ cells;
   - BODIPY gMFI is calculated in selected TUM-Treg population through the specific option available as follows: Workspace, then Add statistics, Geometric mean and select BODIPY in Parameter section.

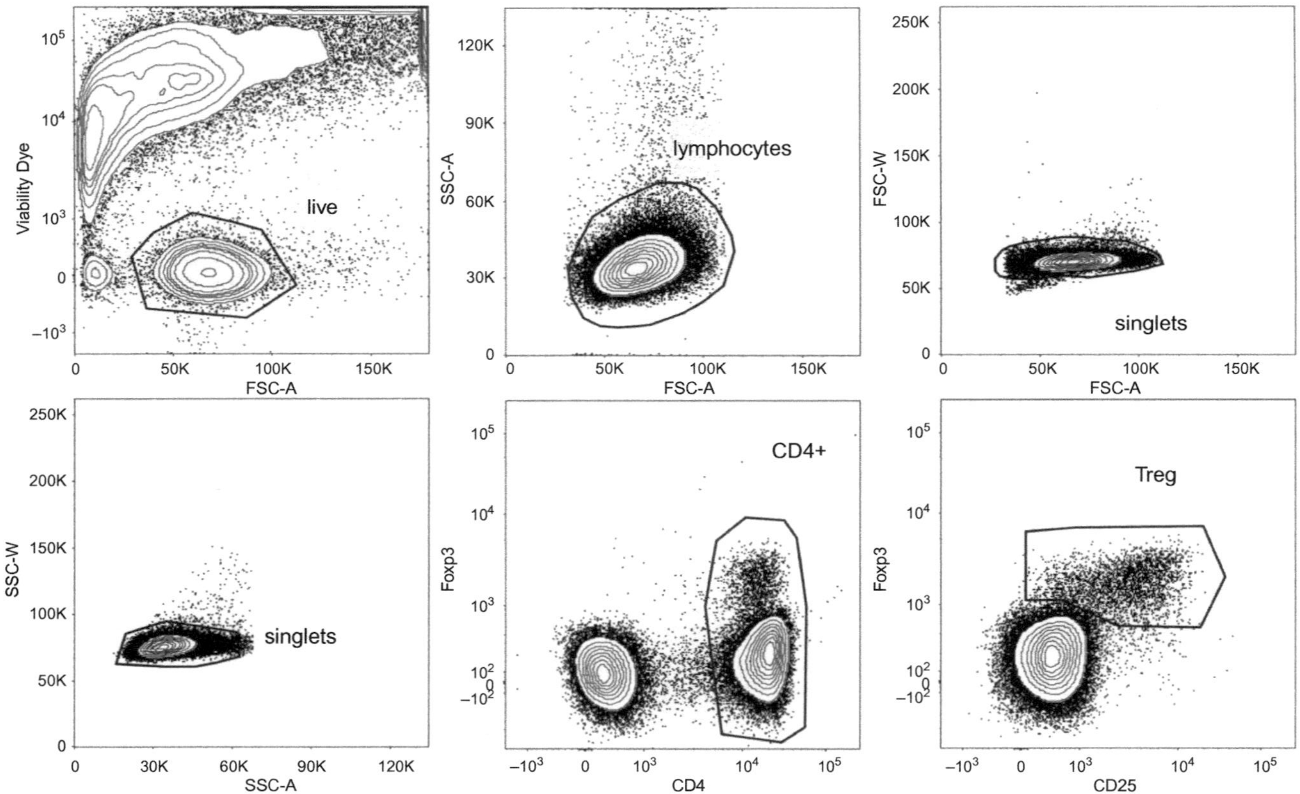

Fig. 1 Gating strategy for Treg identification in murine tumor tissue in flow cytometry. TILs extracted from murine tumor samples, as described in Section 3, are seeded in a round-bottomed 96-well plate and stained ex vivo as detailed in Section 4. Thereafter, the multiparametric flow cytometry data are analyzed as follows: viability dye negative area and forward and side scatter areas (FSC-A/SSC-A) are used to select living lymphocytes among total TIL samples; then, cell doublets are excluded in FSC-A/FSC-W and SSC-A/SSC-W panels, $CD4^+$ cells are selected, and in this gate Tregs are identified as $CD25^+Foxp3^+$ cells.

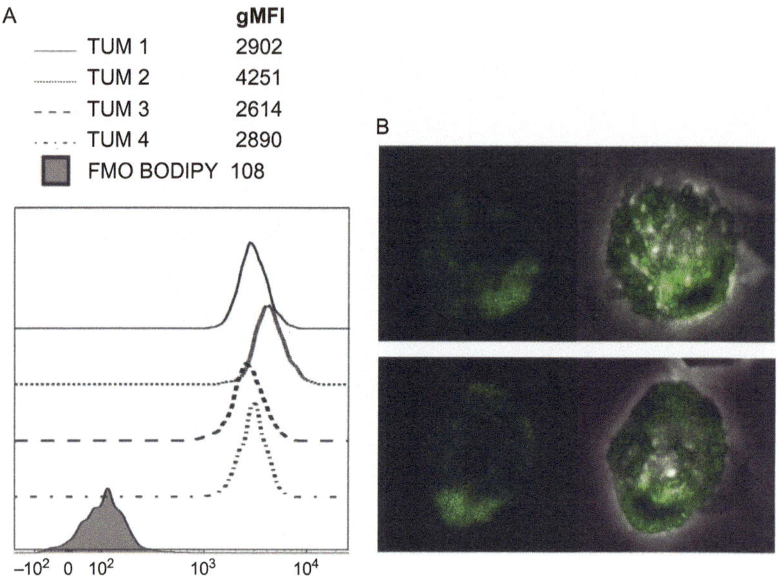

Fig. 2 Detection of intracellular neutral lipid by BODIPY cell incorporation. (A) TUM-Treg from 4 single-processed tumor samples, gated as in Fig. 1, are analyzed for intracellular BODIPY levels by estimating the geometric mean fluorescence intensity (gMFI) of BODIPY fluorescence in the FITC channel (empty peaks) after cell fixation and permeabilization. As negative control, gMFI of BODIPY in TUM-Treg FMO control is evaluated (filled peak in gray). (B) The localization of BODIPY dye mainly in the intracellular compartment is confirmed by confocal microscopy analysis of BODIPY-stained splenocytes. BODIPY fluorescence alone or merged with phase-contrast is shown. Magnification: 60 ×.

6. Concluding remarks

In this chapter we describe a combined protocol that allows a quantitatively and qualitatively efficient isolation of TILs from murine tumor tissue and provides a straightforward strategy to estimate intracellular lipid levels in specific cell types within a mixed population, taking advantage of flow cytometry as detection method. In this protocol, the propensity of the lipophilic fluorochrome BODIPY to partition into lipids and its fluorescence are used to assess intracellular lipid load by flow cytometric measure of its incorporation in total TUM-Treg, without any requirement for cell sorting and organic extraction, and allowing to preserve a representative analysis of the entire population. Indeed, the possibility to enrich TIL fraction from tissues through the above described extraction procedure, and the

BODIPY capability to work well also after cell fixation and permeabilization (Fig. 2A), allow combining lipid labeling with Foxp3 intranuclear staining, being the latter essential for Treg discrimination. Moreover, a huge variety of antibody panels for flow cytometry staining that are compatible with BODIPY may be developed and hence, many other cellular parameters may be contemporary assessed. Since flow cytometry measures many signals in the same cell and is generally performed on live cells, the relationship between additional cell features and lipid content may be evaluated in single cell types from a mixed cell population, not only ex vivo but also in assays in vitro, where lipid metabolism is modulated. This approach may be useful to define the specific phenotype and/or functional characteristics of TUM-Tregs that are associated to lipid load. Finally, taking into account the flow cytometry potentiality, the described strategy may be easily exploited to compare the lipid profile of different populations in the same tumor milieu. Because BODIPY partitions into all hydrophobic compartments, a weakness of this method is the detection of total lipid mass with little regard to the specific type of neutral lipids, but it is an efficient compromise for the lipid analysis possibly in any rare population within complex tissues.

7. Notes

1. Please note that optimal culture conditions (i.e., culture medium, supplements) for other cell lines may differ quite considerably.
2. After thawing, cells are seeded approximately $5.0–6.0 \times 10^6$ per flasks in 75-cm^2 flasks.
3. Washing cells with PBS before trypsinization is necessary to remove serum leftovers that are known to inactivate the enzyme and therefore reduce detachment.
4. The optimal time of incubation with trypsin/EDTA for a monolayer cell detachment may vary considerably with cell type and degree of confluence; 2–4 min at 37 °C could be sufficient to detach most malignant cell lines of human and murine origin and this procedure could be facilitated by gently hitting the side of the flask with the hands (mechanical strokes).
5. To ensure a proper cell line growth and the reproducibility of results over time, cells should be constantly kept under exponential growth conditions and subcultured for a relatively low, predetermined number of passages to avoid also the cell line exhaustion.
6. Since malignant cell lines may have a different growth capability in vivo, depending also on mouse strain, cell number and volume per injection

should be finely tuned by preliminary dose-scaling experiment to set up the specific experimental conditions.

7. For cell lines that tend to easily form clumps, it is necessary to bring the cells into a single cell suspension by pipetting in a small volume before counting, in order to ensure the injection of the same cell number and hence, a similar tumor growth in all biological replicates.
8. The tissue processing may be facilitated by using the optimized and ready-to-use program "m_impTumor_02_01" of gentleMACS™ Octo Dissociator (from Miltenyi Biotech, Bergisch Gladbach, Germany) that allows a semi-automated and standardized tissue disaggregation.
9. In some solid tumor models, especially if highly vascularized, after cell isolation with Percoll density gradient, it may be necessary to deplete red blood cells (that stratify within leukocytes layer) by incubation with a lysis buffer for 3 min at 4–8 °C, to enhance sample quality and hence, to improve its analysis.
10. The number of cells that it would be necessary to stain for a suitable flow cytometry analysis and Treg characterization is 5×10^5 cells per sample at least; however, since a huge Treg expansion has been described in murine tumor tissues (Pacella et al., 2018; Piconese, Valzasina, & Colombo, 2008; Valzasina, Piconese, Guiducci, & Colombo, 2006), also samples with lower TIL counts may be sufficient for Treg analysis by flow cytometry. The yield of TILs from MCA38 subcutaneous tumor model of 8×8 mm in size may vary between 3×10^5 and 10^6 cells/g of tumor tissue, but this count may change in different solid tumor models. A preliminary setting up experiment to evaluate TIL recovery from tumor tissue may be useful.
11. Each flow cytometry procedure, and especially the use of fluorochrome-conjugated antibodies, should be performed in the dark to preserve fluorochromes from incidental excitation.
12. In these steps, if necessary and if the TIL number is enough, FMO controls for all the other antibodies may be prepared, by adding to the staining mix all-but-one antibody for each control. Alternatively, splenocytes may be used as a replacement for these controls.
13. The staining of Foxp3 is sensitive to the fixation method. Cell fixation with paraformaldehyde is not recommended for Foxp3 detection. To perform these steps, some reagents have been formulated and are commercially available to optimize the staining with antibodies to transcription factors and nuclear proteins, including Foxp3 Transcription Factor Staining Buffer Set (from Thermo Fisher Scientific,

Waltham, MA USA). In such kit, Fix/Perm solution is obtained by mixing one part of concentrate buffer with three parts of diluent, whereas permeabilization buffer is prepared by diluting 10 × stock solution (Permeabilization buffer) at 1:10 in sterile deionized water.

14. Since TIL recovery is often not enough, splenocytes may be used as surrogate for this compensation control.
15. The availability of a BODIPY FMO control in each experiment allows normalizing autofluorescence, and to compare BODIPY fluorescence not only between different cell populations in the same sample, but also between data obtained in separate experiments.
16. To check BODIPY localization and to confirm its specificity for intracellular neutral lipid staining, TILs or splenocytes as alternative, stained with BODIPY dye, may be analyzed by confocal microscopy (Fig. 2B).

Acknowledgments

SP and IP are supported by Associazione Italiana Ricerca sul Cancro (AIRC) IG-19784 and International Network Institut Pasteur PTR n. 20-16. The authors wish to thank the imaging facility at Center for Life Nano Science @ Sapienza, Istituto Italiano di Tecnologia, for support and technical advice.

References

Buck, M. D., Sowell, R. T., Kaech, S. M., & Pearce, E. L. (2017). Metabolic instruction of immunity. *Cell, 169*, 570–586.

Cipolletta, D., Feuerer, M., Li, A., Kamei, N., Lee, J., Shoelson, S. E., et al. (2012). PPAR-gamma is a major driver of the accumulation and phenotype of adipose tissue Treg cells. *Nature, 486*, 549–553.

Daemen, S., van Zandvoort, M., Parekh, S. H., & Hesselink, M. K. C. (2016). Microscopy tools for the investigation of intracellular lipid storage and dynamics. *Molecular Metabolism, 5*, 153–163.

Fridman, W. H., Zitvogel, L., Sautes-Fridman, C., & Kroemer, G. (2017). The immune contexture in cancer prognosis and treatment. *Nature Reviews. Clinical Oncology, 14*, 717–734.

Levings, M. K., Allan, S., d'Hennezel, E., & Piccirillo, C. A. (2006). Functional dynamics of naturally occurring regulatory T cells in health and autoimmunity. *Advances in Immunology, 92*, 119–155.

Pacella, I., Procaccini, C., Focaccetti, C., Miacci, S., Timperi, E., Faicchia, D., et al. (2018). Fatty acid metabolism complements glycolysis in the selective regulatory T cell expansion during tumor growth. *Proceedings of the National Academy of Sciences of the United States of America, 115*, E6546–E6555.

Piconese, S., Valzasina, B., & Colombo, M. P. (2008). OX40 triggering blocks suppression by regulatory T cells and facilitates tumor rejection. *The Journal of Experimental Medicine, 205*, 825–839.

Sakaguchi, S., Vignali, D. A., Rudensky, A. Y., Niec, R. E., & Waldmann, H. (2013). The plasticity and stability of regulatory T cells. *Nature Reviews. Immunology, 13*, 461–467.

Steinbring, J., Graja, A., Jank, A. M., & Schulz, T. J. (2017). Flow cytometric isolation and differentiation of adipogenic progenitor cells into brown and brite/beige adipocytes. *Methods in Molecular Biology*, *1566*, 25–36.

Valzasina, B., Piconese, S., Guiducci, C., & Colombo, M. P. (2006). Tumor-induced expansion of regulatory T cells by conversion of CD4+CD25- lymphocytes is thymus and proliferation independent. *Cancer Research*, *66*, 4488–4495.

Whiteside, T. L. (2008). The tumor microenvironment and its role in promoting tumor growth. *Oncogene*, *27*, 5904–5912.

Zou, W. (2006). Regulatory T cells, tumour immunity and immunotherapy. *Nature Reviews. Immunology*, *6*, 295–307.

CHAPTER SIXTEEN

Preparation of single cells from tumors for single-cell RNA sequencing

Yonina Bykov[a], Sarah H. Kim[b], Dmitriy Zamarin[a,c,d,e,*]

[a]Ludwig Collaborative Laboratory, Memorial Sloan Kettering Cancer Center, New York, NY, United States
[b]Department of Surgery, Memorial Sloan Kettering Cancer Center, New York, NY, United States
[c]Department of Medicine, Memorial Sloan Kettering Cancer Center, New York, NY, United States
[d]Department of Medicine, Weill Cornell Medical College, New York, NY, United States
[e]Parker Institute for Cancer Immunotherapy, Memorial Sloan Kettering Cancer Center, New York, NY, United States
*Corresponding author: e-mail address: zamarind@mskcc.org

Contents

Abstract

Intratumoral heterogeneity of cancer cells and tumor-infiltrating immune cells is increasingly being viewed as a key factor driving tumor progression and response to therapy. Over the past several years, technological advances have created powerful tools to analyze the tumor microenvironment on a single-cell level, including mass cytometry and single-cell RNA sequencing, which is particularly pertinent to tumor immunology and cancer immunotherapy. The integrity and reliability of the data generated from these single-cell technologies, however, are highly influenced by the process and quality of sample preparation, which, if carried out inappropriately, has a potential to produce misleading results. In this chapter, we describe a protocol for the generation of single cell suspensions from human tumor samples that has been optimized for single-cell RNA sequencing. This protocol can be easily adapted for other single cell applications such as mass and flow cytometry. Throughout the entire workflow, we aim to maximize viability and minimize factors contributing to cellular stress that could affect downstream analyses.

Methods in Enzymology, Volume 632
ISSN 0076-6879
https://doi.org/10.1016/bs.mie.2019.05.057

1. Introduction

As analyses of cellular heterogeneity of tumors move to the forefront of cancer immunology, techniques allowing the examination of individual cells are becoming increasingly important. Single-cell RNA sequencing provides a snapshot of transcriptomes of individual cells, making it easier to understand regulation and function of diverse populations of cancer cells, immune cells, and other stromal cells in tumor microenvironment (See, Lum, Chen, & Ginhoux, 2018). Furthermore, understanding single-cell transcriptomes can help unravel differentiation and activation status of various immune infiltrates (See et al., 2018). This knowledge is instrumental in understanding tumor biology and mechanisms underlying differences in disease progression and responses to treatments among patients. In fact, multiple groups have successfully utilized single-cell RNA sequencing technology to profile populations of tumor-infiltrating immune cells (Andor et al., 2019; Azizi et al., 2018; Lawson, Kessenbrock, Davis, Pervolarakis, & Werb, 2018; Müller & Diaz, 2017; Tirosh et al., 2016). Other technologies have also been developed to study immune infiltrate on a single-cell basis (Arakaki et al., 2009). In order to accurately define and characterize populations, however, it is imperative to use high-quality samples—i.e. single cell suspensions of maximum-achievable viability, low clumping, and minimal stress induction by the dissociation process. In this chapter, we will discuss important considerations for the generation of single cell suspensions of a reliable quality for use in single-cell RNA sequencing and other single-cell techniques. We include a protocol that has been optimized for high viability and thorough dissociation that is sufficiently gentle as to avoid significant effects on the transcriptome. The protocol described below has been optimized for transcriptomic analysis using the 10 × Genomics platform, however, it can be adapted for a number of other single-cell applications, including other single-cell RNA and DNA sequencing platforms and flow and mass cytometry.

2. A discussion of the extracellular matrix

When generating single cell suspensions, it is necessary to directly target the extracellular matrix (ECM) to break the tissue apart. Held together by a complex web of diverse proteins and glycoproteins, individual cells form complex structures organized into tissues and organs. This sticky extracellular web, or the ECM, is key in cell-cell adhesion, and many of its components also play a role in communication between cells. The ECM plays a central role in cellular attachment. A typical cell is studded with integral proteins that span the membrane and attach to a variety of ECM proteins including collagen, laminin, thrombospondins, vitronectin and fibronectin (Maziveyi & Alahari, 2017).

These proteins are common targets for enzymatic means of dissociating tissue into single cells. Collagen is one of the most common ECM proteins, and is composed of both collagenous and noncollagenous repeating domains. Up to 28 different subunits of collagen have been detected in vertebrates and higher invertebrates (Halper & Kjaer, 2014). Same or different repeating subunits are organized into chains, three of which twist around each other (Halper & Kjaer, 2014) to form long strings of collagen resistant to stress. Different combinations of subunits can form a variety of collagen subtypes. The collagen can then form networks between cells, or provide attachment to basement membranes (Halper & Kjaer, 2014).

Fibronectin is also frequently found in the ECM, where it forms a matrix that can bind a multitude of other ECM components including proteoglycans, integrins, and collagen (Halper & Kjaer, 2014; Zollinger & Smith, 2017). Due to its broad binding capabilities and its ability to initiate the formation of the ECM's web-like structure, it is considered the master organizer of the ECM (Halper & Kjaer, 2014; Sabatier et al., 2009). While fibronectin is also commonly associated with morphogenesis, cell migration and inflammation, its levels tend to be reduced in most tumors (Halper & Kjaer, 2014). Fibrillin also contributes to the stability of the ECM, while thrombospondins and vitronectin are mainly involved in regulation of ECM structure (Halper & Kjaer, 2014; Mosher & Adams, 2012; Schvartz & Dalia Seger, 1999). The elasticity of the ECM that contributes to its resistance to mechanical degradation is largely accounted for by elastin (Halper & Kjaer, 2014).

Another common ECM protein is laminin, which serves to stabilize the structure of the matrix through its ability to adhere to a variety of molecules including collagen, integrins and the basal membrane. It is most frequently found in the basement membrane, although up to 16 distinctly-localizing isoforms have been discovered in human and mouse tissue (Halper & Kjaer, 2014).

In addition to the multitude of adhesion protein families described above, there is a number of integral protein families that provide sites for attachment of these proteins. The four most common groups are integrins, selectins, the immunoglobulin superfamily (IgSF) and cadherins (Samanta & Almo, 2015). Integrins are generally known for their ability to bind large extracellular proteins such as collagen and fibronectin, while selectins primarily adhere to carbohydrates on plasma membranes of cells (Maziveyi & Alahari, 2017; Samanta & Almo, 2015). Both IgSF and cadherins recognize proteins known as adaptor molecules on plasma membranes of neighboring cells, and are found in most tissues (Samanta & Almo, 2015).

Many of the above-mentioned ECM components are ubiquitously expressed—however, the ECM of cancer tissue, including primary lesions and metastases, undergoes several key changes during cancer evolution. During the course of hematogenous metastasis, expression of ECM proteins is altered several times. Extracellular connections are first modified to allow for the rolling of tumor cells on endothelial surfaces, following which the tumor cells must attach to ligands on the endothelium to end their movement. Cancer cells can also undergo diapedesis, during which they squeeze through the endothelium to attach to underlying tissue (Milde-Langosch et al., 2017). Most of these processes are mediated by selectins, but adhesion molecules such as intercellular adhesion molecule-1, CD44, dystroglycans, cell surface glycoprotein, oncofetal fibronectin and tenascin C are also upregulated in many cancers (Burdick et al., 2014; Halper & Kjaer, 2014; Inamori et al., 2009; Xu et al., 2016). These molecules allow cancer cells to form stronger attachments to each other and to surrounding tissue. As a result, some tumors can be particularly difficult to dissociate in comparison to normal tissue. Breast, pancreatic, hepatocellular and head and neck squamous cell tumors are known to be especially resistant to dissociation.

3. Evolution of single cell suspension preparation

The simplest and earliest method of dissociating tissue into single cells involved mechanical dissociation alone. While relatively easy to perform, this method produced poor-quality suspensions containing undissociated clumps of cells, debris, and low viability (Madden & Burk, 1961). However, as early as 1961, various groups began incorporating enzymatic dissociation into their protocols. Madden et al., for instance, published a method for dissociating solid tumors using trypsin and deoxyribonuclease (Madden & Burk, 1961). They performed the dissociations at room temperature, and tested both human and mouse tumors with some success. While they did demonstrate that the use of enzymes significantly improves yield and viability compared to mechanical dissociation alone, trypsin had relatively low activity against collagen and elastin (Madden & Burk, 1961). Their results have been recapitulated by a number of labs since, with some improvement over the years. Trypsin is still commonly used; however, collagenase and DNase have gained popularity in generating single cell suspensions. Nonetheless, mechanical dissociation assisted by incubation with digestive enzymes still frequently produces samples with plenty of clumps and low viability. A protocol optimizing the procurement of single, viable cells from tumors is indispensable when considering highly sensitive downstream applications.

4. Issues relevant to single cell suspensions

There are several relevant considerations when evaluating the quality of fresh single cell suspensions. The first step in processing tissue is the selection of the sample itself, whereby tumor type and location may play a role. The current protocol has been optimized for the dissociation of ovarian tumors, and while it has also been successfully used for breast and melanoma tumors, it may be necessary to make adjustments for specific cancer types due to the inherent diversity in features of the ECM. Optimal suspensions with high viability and low amounts of debris can be generated best from biopsies of soft tumor nodules, as can be seen in Table 1. Omental tissue is also a good option, and it is best to trim away excess fat and select areas with dispersed disease (as opposed to tough nodules). Excess fat can clog the filter during the first filtration and stick to cells, causing clumps in the processed sample. Poor selections for processing include grossly necrotic and/or burnt (from cautery) tissue and calcified or fluid-filled masses. These tissue types are generally not highly cellular, and cell suspensions tend to have an abundance of debris despite filtration.

Viability of a sample indicates the proportion of live cells out of the total number of cells. With all other factors held constant, efficient processing plays a key role in ensuring the highest possible viability of the sample. Once the sample is removed from the patient, it is important to quickly trim grossly necrotic or burnt areas and submerge the sample in cold serum-free media. The sample is to be kept on ice for the entirety of the processing time, with the only exception being the incubation with digestive enzymes and red blood cell lysis. It is also necessary to optimize the duration of the enzymatic dissociation to the shortest possible period resulting in dissociation. The combination of a high temperature and stress induced by prolonged

Table 1 A variety of conditions and tissue types were used to optimize this dissociation protocol.

Collection attempt	Sample site	Dissociation method	% Viable cells pre-FACS	% Viable cells post-FACS
1	Retroperitoneal nodule	GentleMACS + liberase	52	N/A
1	Retroperitoneal nodule	GentleMACS + HRA	86	N/A
1	Omental nodule	GentleMACS + liberase	30	N/A
1	Omental nodule	GentleMACS + HRA	50	N/A
2	Abdominal wall nodule biopsy	GentleMACS + HRA	86	>90
3	Omentum tumor	GentleMACS + HRA	75	>90
4	Omentum tumor	GentleMACS + HRA	75	>90
5	Omentum tumor	GentleMACS + HRA	75	>90
5	Paracolic gutter mass (calcified)	GentleMACS + HRA	50	>90

Each collection attempt corresponds to a surgery performed on a patient diagnosed with ovarian cancer. When matched samples were dissociated using either liberase or Miltenyi HRA enzymes, samples dissociated with HRA enzymes consistently yielded suspensions of superior viability and reduced undissociated clumps (not pictured). Starting with collection attempt 2, dissociated cells were always centrifuged at 4 °C, resulting in more consistent viability. Samples from collection attempt 1 were not sorted, as they were excluded from downstream analysis. Post-FACS viability reflects viability of both CD45+ and CD45− sorted populations. Viability is also partially dependent on sample type—biopsies typically yield samples with fewer dead cells, as they are not harvested using a cautery. Calcified masses should not be used due to low viability and debris.

exposure to dissociative enzymes, induces cellular stress responses culminating in cell death or upregulation of transcriptional signatures that can have a profound impact on the downstream data processing and interpretation.

While maximizing the percentage of viable cells within a processed sample, it is also important to avoid clumps of cells (Zilionis et al., 2017) and debris in the suspension. In order to minimize the presence of cellular aggregates, it is imperative to increase the surface-area-to-volume ratio prior to enzymatic dissociation by cutting the sample into small pieces 3–4 mm^3 in volume. Cutting the tumors into smaller pieces results in reduced viability and is not recommended. The enzymes used for the dissociation step are also very important. While there has been success in using liberase/collagenase and DNase to generate single cell suspensions, in our experience, the H-R-A enzymes supplied by Miltenyi generally yield cleaner suspensions with fewer clumps of cells. Although the exact identity of these enzymes is proprietary, these enzymes are specifically targeted to the extracellular matrix of human tumors and are non-reactive with cell surface markers (according to the manufacturer). Performance of the red blood cell lysis step after dissociation allows for further sample clean-up. Addition of a second filtration step after red blood cell lysis serves to catch most of the debris and cellular aggregates that escaped the filter during the first filtration. Finally, dead cells should be sorted out from the cell suspension once processing is complete, which can be accomplished by column-based or flow cytometry-based sorting.

Flow cytometry-based sorting is often considered to be the gold standard in separating single cells into populations of interest. Flow cytometers have the capability to detect scattered light and fluorescence emitted by fluorophore-linked antibodies on individual cells passing through a laser in fluorescence-activated cell sorting (FACS) (Cossarizza et al., 2017). Cytometers with sorting capabilities have additional functionality that allows the user to adjust the liquid stream to a series of miniscule, uniform droplets. Each droplet containing a cell of interest can then be charged and deflected into a designated collection tube using an electric field, effectively sorting a sample cell by cell (Cossarizza et al., 2017).

Flow cytometry-based sorting is highly advantageous for the purposes of sorting populations for single-cell RNAseq and other single-cell analysis methods. The sorting is highly accurate, as it is dependent on the specific and sensitive binding of a fluorophore-linked antibody to the cellular marker of interest. Overall, the level of purity of populations sorted via FACS tends to be higher than that of populations sorted using magnetic beads (Yan, Ding, Tian, Ge, & Jin, 2009). It is also quite easy to remove dead cells from the original sample by using a viability dye. Furthermore, loss of sample during sorting is minimal in our experience, when compared to column sorting using magnetic beads. This is immensely important when working with precious tissue and small biopsies.

There are, however, several pitfalls associated with this sorting technique. As with any ex vivo manipulation of human-derived cells, the process of staining and sorting cells by FACS involves a drastic change from physiological conditions. Cells are stained and sorted in buffers with reduced amounts of nutrients and are subjected to changes in temperature and pressure as they move through the cytometer's fluidics system. In fact, it has been shown that FACS can increase cellular levels of reactive oxygen species, and widely alter the metabolome in astrocytes (Llufrio, Wang, Naser, & Patti, 2018). Sorting by FACS can also induce expression of genes associated with cellular stress in CD4+ and CD8+ T cells, although it is worthwhile to note that induction of cellular stress has also been observed in T cells sorted using beads (Beliakova-bethell, Massanella, White, Lada, & Du, 2014).

Once the sample is sorted according to the populations of interest (our workflow involves sorting into CD45− and CD45+ populations for separate downstream processing), it is of use to consider the effect of the population type on the expected downstream data quality and complexity. For instance, CD45− populations are quite heterogeneous, as they can include tumor cells (which often have diverse gene expression and genetic/epigenetic profiles arising from high mutation rates and genomic instability) as well as fibroblasts and epithelial cells (Bantikassegn, Song, & Politi, 2015). In contrast, CD45+ populations tend to cluster into distinct, highly specialized subtypes of cells (Chen, Ye, & Guo, 2019), with higher consistency in gene expression defining the specific clusters between different samples. These features may help distinguish whether certain observations generated from the data are reflective of the actual biologic differences versus a batch effect of sample processing.

5. Preparation of single cell suspensions from human tumor samples

5.1 Equipment

1. GentleMACS Octodissociator with heater jackets (Miltenyi, 130-096-427)
2. C-tubes (Miltenyi, 130-096-334)
3. 50-mL falcon tubes (Falcon, 32618059)
4. 70 μm filters (Falcon, 352350)
5. 100 μm filters (Falcon, 352360)
6. Sterile forceps (Unomedical, 84011182)
7. Sterile scalpel (Bard-Parker, 371610)
8. 60 mm Petri dish (Falcon, 351029)
9. 5 mL syringe (BD, 309646)
10. Hemocytometer (Reichert, 1492)

5.2 Chemicals

1. Tumor dissociation kit, human H, R, A enzymes (Miltenyi, 130-095-929)
2. Serum-free RPMI
3. RPMI supplemented with 10% fetal calf serum (FCS)
4. ACK lysis buffer
5. 4% Trypan blue (Gibco, 15250061)
6. Dimethyl sulfoxide (DMSO) (Sigma-Aldrich, D4540)

5.3 Protocol

Preparing reagents

1. Reconstitute enzyme H in 3 mL serum free RPMI 1640.
2. Reconstitute enzyme R in 2.7 mL serum free RPMI 1640.
3. Reconstitute enzyme A in 1 mL Buffer A.
4. Make aliquots of 200 μL enzyme H, 100 μL enzyme R and 25 μL enzyme A (all stored in separate tubes), and store at −20 °C until ready for use.

Preparing single cell suspensions from human tumor samples

1. Overall, to ensure maximal preservation of tumor biology in the sample ultimately used for analysis, target time from tumor resection to encapsulation should be 2 h or under. It is thus critical to ensure the timely availability of all reagents and downstream processing instruments (e.g. FACS, 10 × Genomics instrument).
2. Following tumor resection or biopsy, immediately place tissue sample in a 15-mL falcon tube with ≥5 mL serum-free RPMI on ice.
   a. Note: When receiving fresh sample, quickly trim away cauterized and/or grossly necrotic tissue from sample prior to placing in media.
3. Working inside a laminar flow hood, set up a sterile C tube on ice, and add 5 mL of serum free media.
4. Thaw an aliquot each of enzymes H, R and A at room temperature and immediately place on ice.
5. Place a clean 60 mm Petri dish on ice, and add sufficient serum-free RPMI to form a layer with a thickness of approximately 0.5 cm.
6. Transfer sample to the Petri dish and using a sterile scalpel (aiding with sterile forceps for stability) cut into 3–4 mm^3 pieces.
   a. Mincing the tissue too finely results in reduced viability, while failing to break up larger pieces produces suboptimally dissociated suspensions with a large number of clumps.

7. Transfer no >0.5 g of tissue to the pre-filled C tube, and add 1 aliquot each of enzymes H, R and A.
   a. Adding too much tissue to a single C tube exhausts the dissociative enzymes and produces suspensions with a large amount of clumps. It is preferable to use a smaller amount of tissue per C tube to achieve optimally dissociated suspensions.
   b. If a higher number of cells is desired, it is suggested to partition the sample into several parts and proceed with this protocol using multiple setups.
8. Cover the C tube with its cap and turn cap until a click is heard.
9. On the GentleMacs Octodissociator, select the 37C_h_TDK_3 program. Place the C tube (cap-side down) on the dissociator and place a heating jacket (provided with the Octodissociator) over the C tube.
10. Incubate for 30 min.
11. During the incubation, set up on ice a 50 mL Falcon tube with a 100 μm filter over the top, and pre-wet with 5 mL of RPMI supplemented with 10% FCS.
12. Once the incubation is complete, pour the sample over the filter, and use the back of a sterile 5 mL syringe to push cells through the filter.
13. Add 5 mL RPMI + 10% FCS 1–2 times to pass the rest of the cells through the filter and to inactivate the dissociative enzymes.
   a. To reduce loss of sample, when performing the washes for this step rinse C tube and rotator located on the cap with the 5 mL of RPMI + 10% FCS over the filter.
   b. For a sample that is difficult to pass through the filter, an additional wash with 5 mL RPMI + 10% FCS is recommended.
14. Centrifuge cells at 1500 rpm for 4 min at 4 °C, decant supernatant and resuspend in 2 mL of ACK lysis buffer.
15. Incubate sample in the lysis buffer for 3 min at room temperature.
   a. Incubation time can be reduced if pellet does not have a large amount of red blood cells.
   b. While the purpose of the ACK lysis buffer is to remove red blood cells from the sample, prolonged incubation in the ACK lysis buffer can induce stress responses in the rest of the cells and should be avoided, even at the expense of retaining some red blood cells in the sample.
16. Add 8 mL of cold RPMI + 10% FCS to the sample to reduce activity of the ACK lysis buffer.

17. Place a 70 μm filter on a clean 50 mL Falcon tube (on ice) and pour the sample over the filter. Wash with 5 mL RPMI + 10% FCS.
18. To ensure a homogenous suspension, pipet up and down with a serological pipet. Add 10 μL of the final sample to 90 μL of 4% Trypan blue in a microcentrifuge tube.
19. Pipet up and down to mix, and count using a hemocytometer. Alternative automated live/dead counting methods such as Countess™ II (ThermoFisher cat. AMQAX100) can be used instead, if available.
    a. If doing a manual count, the following formula can be used to calculate the number of cells in a sample:

$$T = \left(\frac{C}{4}\right) * 10{,}000 * D * V$$

    where T is total number of cells, C is cells counted, D is dilution factor, and V is sample volume.

6. Downstream applications

The resultant single-cell suspensions are now ready for long-term live cryopreservation or downstream processing for a variety of applications, including flow and mass cytometry, preparation of primary cell lines and organoids and single cell RNA sequencing. Below, we discuss the downstream sample processing for scRNAseq using 10 × Genomics platform. Although the previously-mentioned techniques can tolerate input of suboptimally-dissociated samples, 10 × scRNAseq yields best results when used for samples with a viability of at least 80%, with minimal to no clumps or doublets in the final sample (Zilionis et al., 2017).

Sample preparation for 10 × RNA sequencing

1. Centrifuge cells at 1500 RPM for 4 min at 4 °C and resuspend in cold RPMI +10% FCS to a concentration of approximately 1 million cells/mL.
2. Removal of dead cells. This can be accomplished using magnetic beads/column-based or FACS-based methods as outlined below. In our experience, we prefer FACS-based approach, as it allows for simultaneous sorting of cells into populations of interest and removal of cell clumps and doublets:
    a. Miltenyi (MACS® Dead Cell Removal Kit: Cat. No: 130-090-101), according to manufacturer's protocol.

   b. Stain cells using protocol of choice for flow cytometry assisted sorting to separate populations of interest. Our typical workflow involves sorting CD45 + and CD45 − cells for separate downstream processing for scRNAseq. The staining solution should include a viability dye to exclude dead cells. To maximize viability, we have found that standard 70 μm nozzle with sorting pressure of 70 psi gives optimal results.
3. Collect sorted live populations in 1 mL RPMI +0.04% BSA. Place samples immediately on ice.
4. Samples are ready to proceed to barcoding for single-cell RNA sequencing. While we prefer to target 5000 cells per run on the 10 × Genomics platform, it is in general recommended to submit more cells due to expected loss during encapsulation.

Sample preparation for storage

1. Prepare freezing solution with sterile 90% FCS and 10% DMSO.
   a. DMSO is light-sensitive; freezing media should be protected from light.
2. Resuspend sample in freezing solution at a concentration of approximately 5 million cells/mL and distribute aliquots of 1 mL in cryovials.
3. Immediately place on ice, then transfer to freezing containers and freeze ≥4 h at −80 °C.
4. Samples can then be transferred to a regular box in a −80 °C freezer for storage. If cells will be stored for more than 1 month before use, it is highly recommended to store at or below −160 °C in liquid nitrogen.

In summary, optimal conditions for dissociation of tumor samples are absolutely necessary to ensure biological relevance of the datasets generated from downstream processing such as single-cell RNA sequencing. Factors such as prolonged exposure to elevated temperature or prolonged dissociation can lead to cell death and induction of stress signatures that may cloud the interpretation of the results. On the contrary, inadequate dissociation may result in excessive cell clumping. While removal of cell clumps can be successfully achieved by filtration and FACS, it could inadvertently lead to biased results among the poorly-dissociated samples.

It is important to note that the protocol detailed in this chapter has been optimized for ovarian cancer samples. While these conditions may be applicable to several tumor types, the extracellular matrix can vary significantly among different tumors and from the healthy tissue, which may be present within the tumor sample. Different types of cancers have been characterized as either "tough," "medium," or "soft" tumors by Miltenyi, based on their resistance to dissociation. Specific types of cancers also tend to either upregulate or downregulate various components of their ECM relative to healthy tissue, and as such it may be necessary to optimize the generation of single cell suspensions for the tumor type of interest. The latter could be accomplished by altering duration of incubation with the dissociative enzymes, or selection of alternative GentleMACS programs that may be better suited for the particular tumor type.

References

Andor, N., Simonds, E., Czerwinski, D., Chen, J., Grimes, S., Wood-Bouwens, C., et al. (2019). Single-cell RNA-Seq of lymphoma cancers reveals malignant B cell types and co-expression of T cell immune checkpoints. *Blood*, *133*(10), 1119–1129. https://doi.org/10.1182/blood-2018-08-862292.

Arakaki, A., Ooya, K., Akiyama, Y., Hosokawa, M., Komiyama, M., Iizuka, A., et al. (2009). TCR-β repertoire analysis of antigen-specific single T cells using a high-density microcavity array. *Biotechnology and Engineering*, *106*(2), 311–318.

Azizi, E., Carr, A. J., Plitas, G., Cornish, A. E., Konopacki, C., Prabhakaran, S., et al. (2018). Single-cell map of diverse immune phenotypes in the breast tumor microenvironment. *Cell*, *174*(5), 1293–1308. e36. https://doi.org/10.1016/j.cell.2018.05.060.

Bantikassegn, A., Song, X., & Politi, K. (2015). Isolation of epithelial, endothelial, and immune cells from lungs of transgenic mice with oncogene-induced lung adenocarcinomas. *American Journal of Respiratory Cell and Molecular Biology*, *52*(4), 409–417. https://doi.org/10.1165/rcmb.2014-0312MA.

Beliakova-bethell, N., Massanella, M., White, C., Lada, S., & Du, P. (2014). The effect of cell subset isolation method on gene expression in leukocytes. *Cytometry*, *85*(1), 94–104. https://doi.org/10.1002/cyto.a.22352.The.

Burdick, M. M., Cuckler, C. M., Goetz, D. J., Tees, D. F. J., Liu, T., Delgadillo, L. F., et al. (2014). CD44 variant isoforms expressed by breast cancer cells are functional E-selectin ligands under flow conditions. *American Journal of Physiology. Cell Physiology*. https://doi.org/10.1152/ajpcell.00094.2014.

Chen, H., Ye, F., & Guo, G. (2019). Revolutionizing immunology with single-cell RNA sequencing. *Cellular & Molecular Immunology*, *16*, 242–249. https://doi.org/10.1038/s41423-019-0214-4.

Cossarizza, A., Chang, H. D., Radbruch, A., Akdis, M., Andrä, I., Annunziato, F., et al. (2017). Guidelines for the use of flow cytometry and cell sorting in immunological studies. *European Journal of Immunology*, *47*(10), 584–1797. https://doi.org/10.1002/eji.201646632.

Halper, J., & Kjaer, M. (2014). In J. Halper (Ed.), *Progress in heritable soft connective tissue diseases*. New York: Springer.

Inamori, K., Willer, T., Harper, H. A., Campbell, K. P., Yoshida-Moriguchi, T., de Bernabé, D. B.-V., et al. (2009). Loss of α-dystroglycan laminin binding in epithelium-derived cancers is caused by silencing of LARGE. *Journal of Biological Chemistry*, *284*(17), 11279–11284. https://doi.org/10.1074/jbc.c900007200.

Lawson, D. A., Kessenbrock, K., Davis, R. T., Pervolarakis, N., & Werb, Z. (2018). Tumour heterogeneity and metastasis at single-cell resolution. *Nature Cell Biology*, *20*(12), 1349–1360. https://doi.org/10.1038/s41556-018-0236-7.

Llufrio, E. M., Wang, L., Naser, F. J., & Patti, G. J. (2018). Sorting cells alters their redox state and cellular metabolome. *Redox Biology*, *16*, 381–387. https://doi.org/10.1016/j.redox.2018.03.004.

Madden, R. E., & Burk, D. (1961). Production of viable single cell suspensions from solid tumors. *Journal of the National Cancer Institute*, *27*(4), 841–861. https://doi.org/10.1093/jnci/27.4.841.

Maziveyi, M., & Alahari, S. K. (2017). Cell matrix adhesions in cancer: The proteins that form the glue. *Oncotarget*, *8*(29), 48471–48487.

Milde-Langosch, K., Stürken, C., Maltseva, D. V., Rodin, S., Oliveira-Ferrer, L., Schumacher, U., et al. (2017). Selectin-independent adhesion during ovarian cancer metastasis. *Biochimie*, *142*, 197–206. https://doi.org/10.1016/j.biochi.2017.09.009.

Mosher, D. F., & Adams, J. C. (2012). Adhesion-modulating/matricellular ECM protein families: A structural, functional and evolutionary appraisal. *Matrix Biology*, *31*(3), 155–161. https://doi.org/10.1016/j.matbio.2012.01.003.

Müller, S., & Diaz, A. (2017). Single-cell mRNA sequencing in cancer research: Integrating the genomic fingerprint. *Frontiers in Genetics*, *8*, 1–10. https://doi.org/10.3389/fgene.2017.00073.

Sabatier, L., Chen, D., Fagotto-Kaufmann, C., Hubmacher, D., McKee, M. D., Annis, D. S., et al. (2009). Fibrillin assembly requires fibronectin. *Molecular Biology of the Cell*, *20*, 846–858. https://doi.org/10.1091/mbc.E08.

Samanta, D., & Almo, S. C. (2015). Nectin family of cell-adhesion molecules: Structural and molecular aspects of function and specificity. *Cellular and Molecular Life Sciences*, *72*(4), 645–658. https://doi.org/10.1007/s00018-014-1763-4.

Schvartz, I., & Dalia Seger, S. S. (1999). Vitronectin. *The International Journal of Biochemistry & Cell Biology*, *31*, 539–544.

See, P., Lum, J., Chen, J., & Ginhoux, F. (2018). A single-cell sequencing guide for immunologists. *Frontiers in Immunology*, *9*(OCT), 1–13. https://doi.org/10.3389/fimmu.2018.02425.

Tirosh, I., Izar, B., Prakadan, S. M., Ii, M. H. W., Treacy, D., Trombetta, J. J., et al. (2016). Dissecting the multicellular ecosystem of metastatic melanoma by single-cell RNA-seq. *Science*, *352*(6282), 189–196. https://doi.org/10.1126/science.aad0501.Dissecting.

Xu, Z., Sun, P., Ju, J., Wang, L., Di, D., Chen, L., et al. (2016). Downregulation of human intercellular adhesion molecule-1 attenuates the metastatic ability in human breast cancer cell lines. *Oncology Reports*, *35*(3), 1541–1548. https://doi.org/10.3892/or.2016.4543.

Yan, H., Ding, C., Tian, P., Ge, G., & Jin, Z. (2009). Magnetic cell sorting and flow cytometry sorting methods for the isolation and function analysis of mouse CD4+ CD25+ Treg cells. *Journal of Zhejiang University Science*, *10*(12), 928–932. https://doi.org/10.1631/jzus.B0920205.

Zilionis, R., Nainys, J., Veres, A., Savova, V., Zemmour, D., Klein, A. M., et al. (2017). Single-cell barcoding and sequencing using droplet microfluidics. *Nature Protocols*, *12*(1), 44–73. https://doi.org/10.1038/nprot.2016.154.

Zollinger, A. J., & Smith, M. L. (2017). Fibronectin, the extracellular glue. *Matrix Biology*, *60–61*, 27–37. https://doi.org/10.1016/j.matbio.2016.07.011.

CHAPTER SEVENTEEN

Characterization of the tumor immune infiltrate by multiparametric flow cytometry and unbiased high-dimensional data analysis

Romina E. Araya, Romina S. Goldszmid*
Inflammatory Cell Dynamics Section, Cancer and Inflammation Program, Center for Cancer Research, National Cancer Institute, NIH, Bethesda, MD, United States
*Corresponding author: e-mail address: rgoldszmid@mail.nih.gov

Contents

Abstract

The tumor microenvironment (TME) is a highly complex and dynamic ensemble of cells of which a variety of immune cells are a major component. The unparalleled results obtained with immunotherapeutic approaches have underscored the importance of examining the immune landscape of the TME. Recent technological advances have

Methods in Enzymology, Volume 632
ISSN 0076-6879
https://doi.org/10.1016/bs.mie.2019.11.012

incorporated high-throughput techniques at the single cell level, such as single cell RNA sequencing, mass cytometry, and multi-parametric flow cytometry to the characterization of the TME. Among them, flow cytometry is the most broadly used both in research and clinical settings and multi-color analysis is now routinely performed. The high dimensionality of the data makes the traditional manual gating strategy in 2D scatter plots very difficult. New unbiased visualization techniques provide a solution to this problem. Here we describe the steps to characterize the immune cell compartment in the TME in mouse tumor models by high-parametric flow cytometry, from the experimental setup to the analysis methodology with special emphasis on the use of unsupervised algorithms.

1. Introduction

Immunotherapy has transformed cancer treatment resulting in a paradigm shift: from targeting the tumor to targeting the patient's immune system. However, despite the unprecedented responses observed, only a subset of patients fully benefits of such approach (Kim & Chen, 2016; Pitt et al., 2016; Sharma, Hu-Lieskovan, Wargo, & Ribas, 2017). A major hurdle for successful anti-cancer therapy is overcoming the immunosuppressive tumor microenvironment (TME) (Hanahan & Coussens, 2012). This underscores the importance of dissecting the immunological landscape of the TME (Binnewies et al., 2018). Recent technological advances applied to the study of the TME have shed light on its complexity and diversity not only among tumor types but also within patients with similar tumor types (Aran et al., 2017; Bindea et al., 2013; Newman et al., 2015).

Both cells from the myeloid and lymphoid lineages can be found in the TME, and studies have shown that defining tumors based on the immune composition of the TME may have predictive value for therapy outcome and disease progression (Herbst et al., 2014; Spranger, 2016). These initial studies have mostly focused on the overall presence of T cells and were based on immuno-histochemistry analysis of tumor sections (Pagès et al., 2018). Due to the highly complex and dynamic nature of the TME, traditional histological and immuno-histochemistry analysis are now being replaced by high-throughput techniques such as multiparametric fluorescence-based flow cytometry (FACS), mass cytometry (CyTOF) and single cell RNA sequencing (scRNAseq). While these high-throughput techniques have increased our capabilities for a deeper understanding of the TME, they have also introduced an overwhelming amount of data, and with it, the need for unbiased high-dimensional data analysis methods.

Given its broad availability in research and diagnostic laboratories all around the world, FACS remains a very popular and versatile phenotyping technique. It allows the study of both surface and intracellular markers at a single cell level. Until not so long ago, flow cytometry data was analyzed in two dimensions by sequential gating performed with pairs of markers and visualized as two-axes plots (scatter, density, pseudocolor plots, etc.) (Bendall, Nolan, Roederer, & Chattopadhyay, 2012). This strategy is very limited when dealing with the rapidly growing number of parameters that can now be measured by FACS. In addition, visualization of pairwise comparisons when working with 20+ parameters is biased by the subjective determination of the gates, and excessively time consuming considering that it leads to 2^n number of scatter plots (being n = number of parameters). Most importantly, the segmented observation of the data in a pairwise manner can result in the loss of relevant information, particularly when studying rare populations or subtle changes in fluorescence intensity of certain markers due to a biologically meaningful cause (e.g., cell activation status).

In order to overcome these limitations, several algorithms were developed and adapted for the analysis of high-parametric datasets obtained from CyTOF, scRNAseq and multi-color flow cytometry (Amir et al., 2013; Becht et al., 2019; Chen et al., 2016; Levine et al., 2015; Newell, Sigal, Bendall, Nolan, & Davis, 2012; Van Gassen et al., 2015). These algorithms provide dimensionality reduction visualization methods, which reduce the high-dimensional data (multi-parameter) into a two- or three-dimensions plot, and also unbiased clustering methods that conserve the global information of the dataset (i.e., each parameter and their multivariate relationships). However, several critical steps must be considered for the successful use of these methods to analyze fluorescence-based flow cytometry data, in particular when working with complex cellular sources such as tumor samples.

In this chapter, we provide a detailed description of a method to characterize the immune landscape of the TME of mouse tumor models. This includes a general protocol for tissue processing and important considerations for the optimization of high-parametric flow cytometry panel design, data acquisition, and analysis that can be easily adapted to any tissue of interest. A summary of the steps described are shown in Fig. 1.

2. Cell culture and tumor models

(1) After thawing, mouse tumor cell lines are routinely maintained in standard culture conditions (37 °C, 5% CO_2) in medium (e.g., Dulbecco's

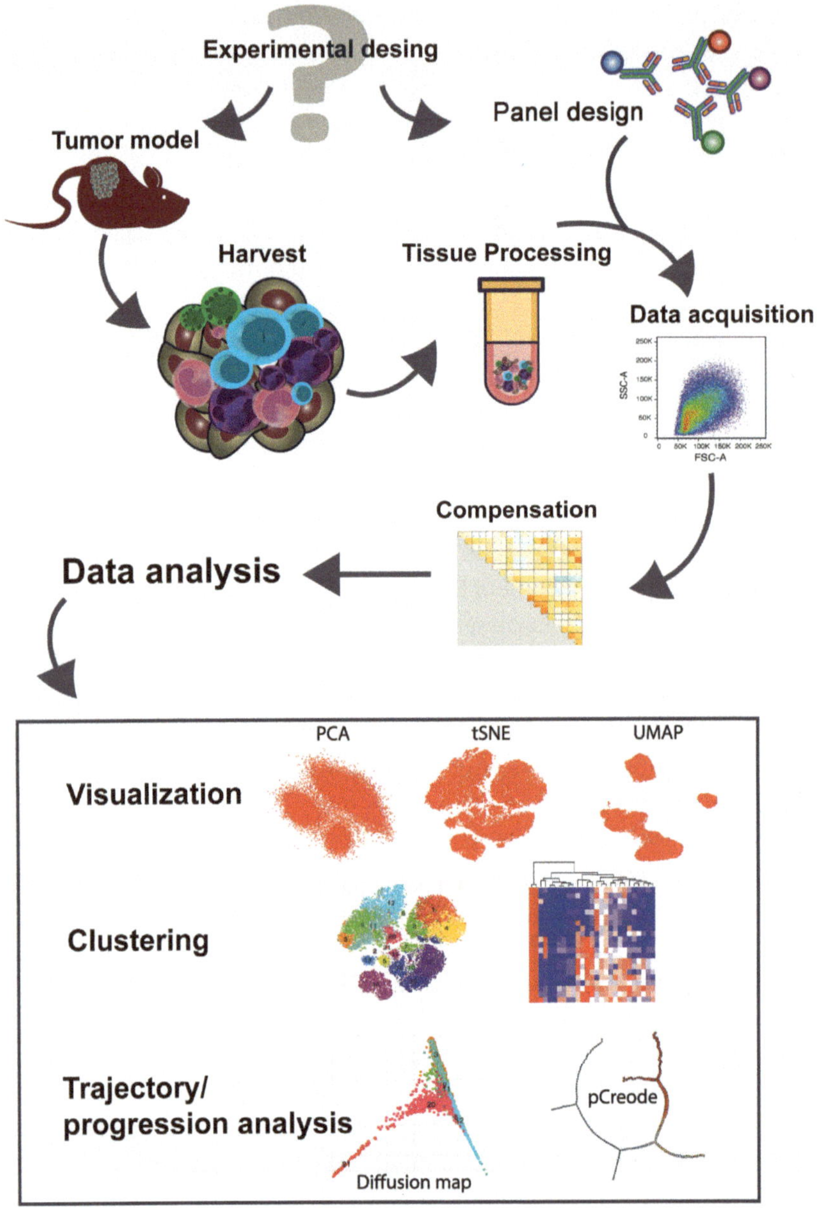

Fig. 1 Workflow of experimental design and data analysis of high-parametric flow cytometry by unsupervised algorithms applied to the study of the TME. Schematic representation of the workflow for the characterization of the tumor immune infiltrate by unsupervised high-dimensional flow cytometry.

Modified Eagle Medium (DMEM) or Roswell Park Memorial Institute (RPMI)-1640) supplemented with 10% fetal calf serum (FCS), L-glutamine, and additional supplements (e.g., non-essential amino acids, sodium pyruvate) depending on the cell line (see Note 1).

(2) When cells reach 75–80% confluence, the spent medium is discarded and the cells are gently washed with prewarmed (37 °C) Hank's balanced salt solution (HBSS) or phosphate buffered saline (PBS) to remove remaining FCS. The cell monolayer is then detached with cell dissociation enzymes (e.g., trypsin or TripLE Express) by adding enough volume to cover the monolayer and incubating at 37 °C, 5% CO_2 for a few minutes (see Notes 2–6). When the cells become loosely attached, they are collected, washed to remove any enzyme remnants, diluted, and reseeded to obtain cells in the exponential growth phase (50–60% confluence) for injection (see Note 7).

(3) For injection, the cells are collected as described above and washed at least twice with HBSS or PBS. Cells are then resuspended in HBSS or PBS at the desired concentration. The number of cells to be injected will depend on the tumor model selected for the study (see Note 8). For subcutaneous injection the volume should not exceed 100–200 μL.

(4) For subcutaneous injections, mice are shaved in the lower right flank leaving a square area of around 1 in. between the rib cage and the hip clear of hair.

(5) Before tumor injection the skin is cleaned with alcohol (e.g., with an alcohol pad) and time is allowed for the alcohol to evaporate.

(6) The cells should be injected in the center of the shaved and disinfected area using a 27G × 5/8 needle to minimize leakage.

(7) Ideally, harvest will be done when tumors reach 5–10 mm. However, the exact size and time point will be tumor model-dependent and should be determined by the investigator during the experimental optimization process.

3. Harvest

(1) Mice are euthanized using CO_2 followed by cervical dislocation as per ACUC guidelines (see Note 9).

(2) Mouse fur and skin are cleaned spraying ethanol 70%.

(3) The skin is opened from the center of the abdomen with scissors and then below until the pelvis to uncover the tumor area.

(4) Tumors are collected with a scalpel, making sure of removing the tumor attached to the inner side of the skin.
(5) Tumors are collected into RPMI medium containing 0.5% FCS and stored in wet ice or 4 °C until processing (see Note 10).

4. Tumor processing

(1) Keep everything on wet ice or 4 °C until digestion. To facilitate the digestion process, transfer the tumor into a petri dish and chop it into small pieces of about 1–2 mm. Depending on the tumor characteristics, the appropriate enzymatic digestion process should be setup (see Note 11).
(2) Transfer the digested tumor (single cell suspension) to a new tube passing through a 40 μm cell strainer (see Note 12). Quickly wash the cell strainer adding 10 mL of cold buffer or medium containing 2–10% of FCS to stop the enzymatic digestion (see Note 13).
(3) Spin the cells down and lyse red blood cells with 1 mL of ACK buffer, during 3–5 min. Wash with 10 mL of FACS buffer (see Note 14).
(4) Resuspend the cells in enough volume of FACS buffer (see Note 15). Separate an aliquot for cell counting if needed and transfer enough cells from the single cell suspension into a 96-well V-bottom plate for staining (see Note 16).

5. Design and validation of high parametric flow cytometry panels

5.1 Panel design

(a) *Define your goal*: Before designing a flow cytometry panel it is important to define the purpose of the study. If the goal is to have a general overview of the whole immune landscape of the TME, then it will be more appropriate to select markers or combination of markers that allow the unequivocally (as much as possible) definition of different immune cell subsets (e.g., lymphocytes, innate lymphoid cells, monocytes, macrophages, neutrophils, dendritic cells). The more lineage specific markers selected, the more comprehensive the characterization will be. However, the goal may be to evaluate the phenotype and/or activation status of a certain cell type, in which case a "dump" channel can be created to exclude cell populations that are not of interest in the analysis (see Note 17).

(b) *Selection of fluorochromes*: Once the desired list of markers is defined, the next step is the fluorochrome selection. This is determined based on fluorochrome brightness (staining index), marker expression level and cytometer configuration. The simplest formula is assigning the brightest fluorochromes to the dimmest markers or those expressed at a low density (Mahnke & Roederer, 2007) (see Notes 18 and 19). Staining index values can be obtained from the different antibody supplier companies, and it is highly recommended to obtain the values specific for the instrument that will be used in the study given that fluorescence intensity and resolution may vary among cytometers.

(c) *Spillover and spread error*: When several markers are needed to define a cell type, the spectral overlap among fluorochromes can be a problem. Certain channels are more prone to suffer from spill over among them. Although compensation normally corrects for this defect, there are cases in which the spectral overlap is so high that the signals result impaired and the resolution is lost (Ashhurst, Smith, & King, 2017). It is recommended to spread the markers that are co-expressed on the same cell along as many lasers as possible, this will minimize the spillover and preserve the characteristics of the signal (see Note 20). In addition to the specific markers, dyes that discriminate live/dead cells and Fc receptor blocking antibodies must be included in the panel.

(d) *Specificity controls*: Once the panel is designed, specificity controls are needed in order to set the positive/negative boundaries during analysis. These controls include:

  (i) *Isotype control*: traditionally considered control for "non-specific" antibody binding, which may result from Fc-mediated binding, cross-reactivity, or non-specific cell adhesion. In order to accurately account for the non-specific binding, the isotype control should be from the same species, same heavy and light chains, conjugated to the same fluorochrome, and have the same fluorophore/antibody ratio (i.e., number of fluorescent molecules bound per antibody). In addition, isotype control antibodies have different activities, which may result in varying levels of non-specific binding. Therefore, finding the ideal isotype control is highly unlikely. Furthermore, isotype controls do not discriminate signal from spectral overlap or cell autofluorescence. For these reasons, and the availability of Fc receptor blocking antibodies, the usefulness of isotype controls is debated and more and more they are being replaced by fluorescence minus one control.

(ii) *Fluorescence minus one (FMO)*: FMO controls consist of all the antibodies present in the mix, except one. The empty channel is the one the FMO controls for. This allows to determine any fluorescence signal that is detected either due to spillover from other channels or cell autofluorescence. This is particularly relevant when working with panels of >6 colors as the major source of background signal is generally due to spectral overlap. FMOs allow to determine the positive/negative boundaries and to maximize the resolution between channels. However, FMOs would not discriminate signal due to non-specific antibody binding.

(iii) *Biological control*: As it is the case in most experimental setups, proper biological controls should be included. For example, when staining for cytokines or activation markers after stimulation, an unstimulated control should be included. In some cases, these controls are the most appropriate for setting positive/negative boundaries.

5.2 Antibody titration

The optimal antibody concentration to be used in a given panel depends on the expression level of the antigen and the fluorochrome of choice and it needs to be determined experimentally doing a titration. For that purpose, serial dilutions of the antibody are tested and the staining index (SI) is obtained as follows (see Note 21):

$$SI = \frac{MFIpos - MFIneg}{2 * SDneg}$$

where MFIpos: mean fluorescence intensity positive population; MFIneg: mean fluorescence intensity negative population, and SDneg: standard deviation of MFIneg.

5.3 Compensation

Compensation corrects the effects of spillover among different channels and it should be done for each flow cytometry experimental run. Most instruments nowadays have a built-in software to do this automatically, please refer to the manufacturer manual for detailed procedures. Although compensation can be done by the instrument, it is highly recommended to save single-color compensation files to allow for post-acquisition compensation adjustments during analysis. Having adequate single-color compensation controls is key to obtaining a proper compensation matrix that will accurately correct the spillover. Below are basic rules to prepare compensation controls:

(a) Single-color controls need to be as bright or brighter than the experimental stain, and at least 10% of the population should have a positive signal.

(b) Background fluorescence should be similar for positive and negative controls (see Note 22).

(c) The fluorochrome used in the compensation control must match the one used in the experimental sample. This is particularly important when using tandem dyes or dyes detected with the same excitation laser and filters (e.g., FITC vs Alexa Fluor 488).

(d) Compensation controls must be treated the same way as the samples, e.g., fixation/permeabilization treatments may alter the fluorochrome.

(e) Antibodies used for single-color controls should be titrated both when using cells or synthetic compensation beads, so that the positive peak is set within the linear range of each detector and the PMT values do not need to be changed (see Note 23).

6. Application Settings

Setting up Application Settings ensures consistency and reproducibility over time across experiments and instruments. Optimal PMT voltages need to be determined for each particular tissue as a function of the tissue's autofluorescence. Once optimal PMT voltages are determined and saved as the Application Settings, the same setting can be applied to standardize experiments regardless of daily variability of the instrument's performance (Meinelt et al., 2012) (see Note 24).

(1) Obtain the robust standard deviation of the electronic noise (rSD_{EN}), and the maximum linear fluorescence (LinMax) for each parameter from the baseline report (see Note 25). The positive and negative signal should be out of these values. The optimal PMT voltages gating in the population of interest should be:

  (a) The PMT values that place the negative unstained population at least 2.5 times above the noise of the cytometer ($2.5 \times rSD_{EN}$).

  (b) The PMT values that keep the positive stained population within the linear range ($MFI_{pos} < LinMax$).

(2) Once optimal PMT voltages are determined for each parameter in the population of interest, the final settings are saved as Application Setting and used for further experiments (see Notes 26 and 27).

(3) Definition of target values: target values are defined with the established Application Settings using standardized bright beads, like rainbow calibration particles (Thermo Fisher Scientific) or CS&T beads (BD Biosciences) to create a template as shown in Fig. 2. Target values can be used to verify the Application Settings in use and to transfer the Application Setting to a different instrument (provided that the cytometer parameters are similar under the same or different optical configurations) by adjusting the positive gate to that shown in the template.

7. Data visualization, clustering and analysis

The traditional flow cytometry data analysis performed by consecutive bivariate gating (i.e., gating strategy) is sufficiently robust when dealing with few parameters. However, this strategy is very limited when analyzing multidimensional data and new dimensionality reduction algorithms can be used for an unbiased data analysis approach. These algorithms can be divided into two main categories: (1) *linear dimensionality reduction* that conserve the information at the single cell level preserving the global structure of the data but assuming a linear relationship among parameters such as Principal Component Analysis (PCA) (Newell et al., 2012) and (2) *non-linear dimensionality reduction* in which local distances are prevalent over the global structure of the data such as t-Distributed Stochastic Neighbor Embedding (t-SNE), diffusion maps, and Uniform Manifold Approximation and Projection for Dimension Reduction (UMAP). Each algorithm has different requirements for the size, normalization and/or transformation of the data and should be carefully studied (Amir et al., 2013; Becht et al., 2019). Fig. 3 illustrates a traditional manual gating strategy to define tumor-infiltrating myeloid cells within total $CD45^+$ live cells from a mouse tumor, where only 11 different populations can be defined based on the subjective gating strategy. Below is a brief description of some of the dimensionality reduction, clustering and trajectory methods that can be utilized and how they compare among them and to the traditional gating strategy when applied to the same dataset. See also Table 1 for a list of markers that can be used for this type of analysis and Tables 2–4 for a summary of the methods described below.

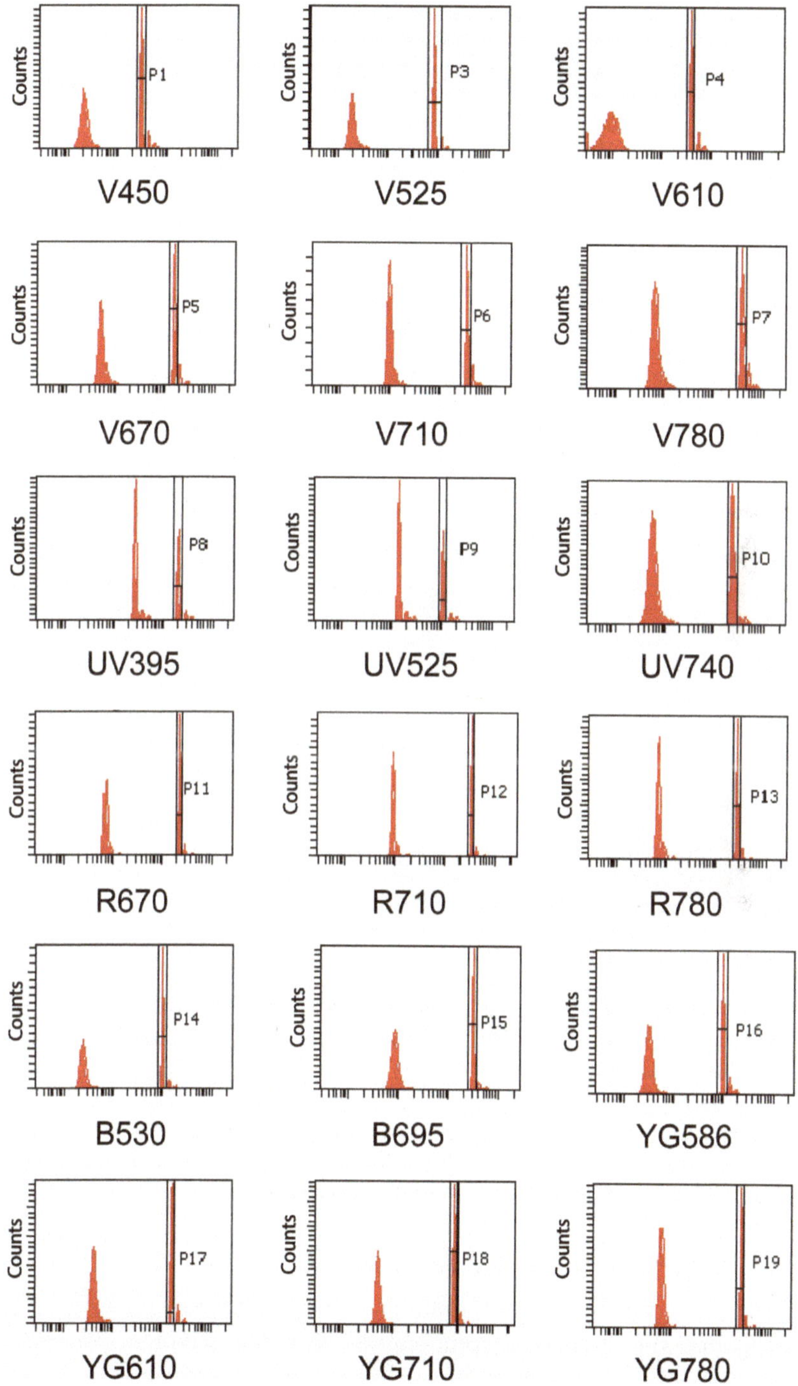

Fig. 2 Target value template for a 20-parameter BD LSR Fortessa cytometer. Histograms depict an example of target value for each parameter.

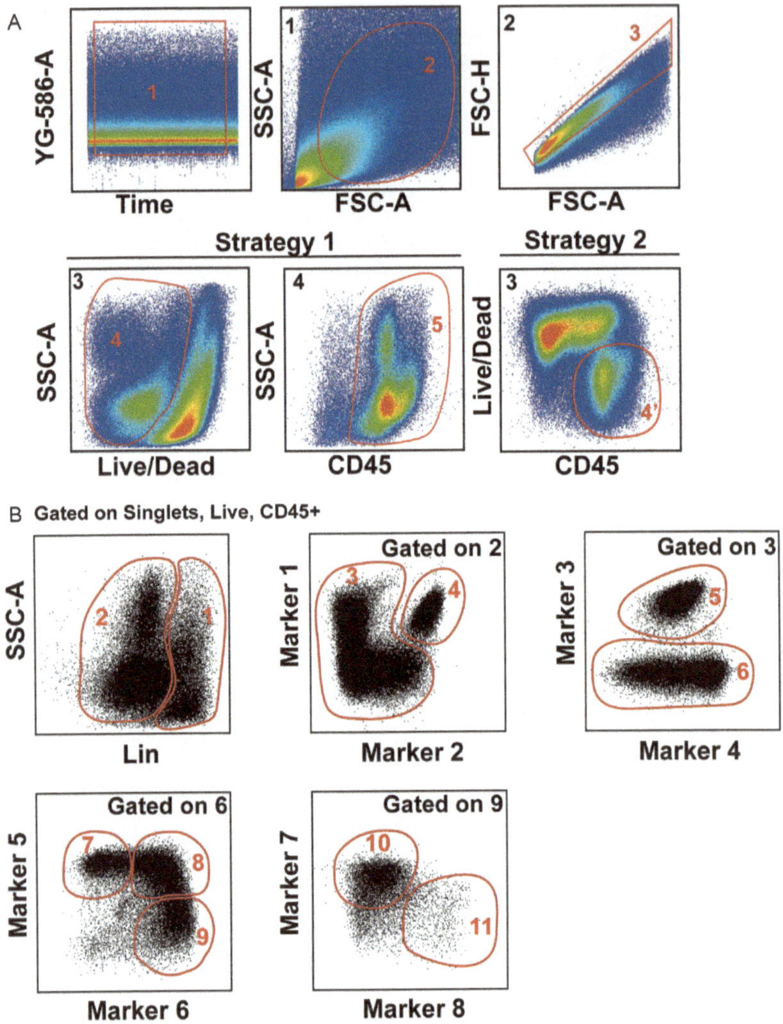

Fig. 3 Definition of populations by traditional manual gating strategy applied to the analysis of 20-parameters flow cytometry data of a subcutaneous mouse tumor. (A) Initial gating strategy starting with Gate 1 to select the events acquired with stable flow stream, Gate 2 to exclude cell debris, and Gate 3 to exclude doublets. Strategy 1 and Strategy 2 depict options to select live CD45 positive cells. (B) Example of manual gating strategy to define specific populations starting from Gate 4′ or 5.

7.1 Linear dimensionality reduction

PCA is based on the distance and relatedness of different populations and aims to find the directions (components) with the maximum variance in the dataset. One advantage of this method is that the distance in the PCA

Table 1 Flow cytometry panel (example).

Marker	Clone	Manufacturer
CD45.2	clone 104	BD Biosciences
Ly6G	1A8	BD Biosciences
CD64	X54-5/7.1	BD Biosciences
F4/80	BM8	BD Biosciences
CD24	M1/69	BD Biosciences
Ly6C	AL-21	BD Biosciences
CX3CR1	SA011F11	BioLegend
CD103	2E7	BioLegend
MHC II	M5/114.15.2	eBioscience
CD68	FA-11	BioLegend
CD11c	N418	eBioscience
TCRb	H57-597	eBioscience
TCRgd	eBioGL3	eBioscience
CD3	145-2C11	eBioscience
NK1.1	PK136	eBioscience
Ter119	TER-119	eBioscience
CD19	eBio1D3	eBioscience
CD135	A2F10.1	BD Biosciences
CD206	C068C2	BioLegend
CD11b	M1/70	eBioscience
Siglec F	E50-2440	BD Biosciences
CD209a	MMD3	Invitrogen
LIVE/DEAD Fixable Dead Cell Stain Kit	–	Invitrogen
CD16/32	2.4G2	Bioxcell
CD103	2E7	BioLegend
CD11b	M1/70	eBioscience
CD11c	N418	eBioscience
CD127	A7R34	BioLegend
CD135	A2F10.1	BD Biosciences
CD16/32	2.4G2	Bioxcell
CD19	eBio1D3	eBioscience

Continued

Table 1 Flow cytometry panel (example).—cont'd

Marker	Clone	Manufacturer
CD206	C068C2	BioLegend
CD209a	MMD3	Invitrogen
CD223/Lag3	C9B7W	BD Biosciences
CD24	M1/69	BD Biosciences
CD25	PC61 5.3	eBioscience
CD274/PDL1	MIH5	BD Biosciences
CD279/PD1	RMP1-30	BD Biosciences
CD3	145-2C11	eBioscience
CD4	GK1.5	BD Biosciences
CD44	IM7	BioLegend
CD45.2	clone 104	BD Biosciences
CD62L	MEL-14	BD Biosciences
CD64	X54-5/7.1	BD Biosciences
CD68	FA-11	BioLegend
CD8a	53-6.7	BioLegend
CX3CR1	SA011F11	BioLegend
F4/80	BM8	BD Biosciences
Foxp3	FJK-16s	eBioscience
KLRG1	2F1	eBioscience
LIVE/DEAD Fixable Dead Cell Stain Kit	–	Invitrogen
Ly6C	AL-21	BD Biosciences
Ly6G	1A8	BD Biosciences
MHC II	M5/114.15.2	eBioscience
NK1.1	PK136	eBioscience
Siglec F	E50-2440	BD Biosciences
TCRb	H57-597	eBioscience
TCRgd	eBioGL3	eBioscience
Ter119	TER-119	eBioscience
TIGIT	1G9	BD Biosciences
Tim3	RMT3-23	eBioscience

Table 2 Visualization methods.

Method	Type of method	Environment	Description	Limitations
PCA	Visualization, relatedness or trajectory	Many different options, R (included in Cytofkit), Phyton, etc.	It establishes distance and relatedness between populations in the linear space Preserve global distances Fast	Non-linear relatedness not considered Overcrowding of data points
tSNE	Visualization	R (included in Cytofkit), Phyton, FlowJo, etc.	Non-linear dimensionality reduction based on the k-neighbor algorithm Not overcrowding of data points Preserve local distances in detriment of global structure	Distance among clusters has no meaning Limited number of events per run Not so fast
UMAP	Visualization, some degree of relatedness	R (included in Cytofkit2 but requires Python) Python, and FlowJo Exchange plug-in	Non-linear dimensionality reduction. Preserve local distances, with some global structure Fast	Loss of resolution among populations with little variation Crowding of similar populations

plot infers relationship among cell populations in the linear space. However, it does not provide information for nonlinear relatedness among the parameters. Fig. 4A right panel depicts the first two components of a PCA analysis of the same dataset shown in Fig. 3.

7.2 Non-linear dimensionality reduction

tSNE is based on the k-neighbor algorithm, where the dataset is visualized as a scatter plot and the location of each event (cell) in the plot indicates the similarities with nearby events. While the distance of the events (dots) within a cluster indicates similarity, unlike PCA, the distance between dots from

Table 3 Clustering methods.

Method	Environment	Description	Limitations
PhenoGraph	R (included in Cytofkit), Phyton, MATLAB™, FlowJo Exchange plug-in	Based on nearest-neighbor graph, followed by definition of communities as interconnected events	Number of clusters not customizable Not so fast
ClusterX	R (defined in Cytofkit)	Based on densities in the tSNE projection map Fast	Limited to *tSNE* ability to define clusters Number of clusters not customizable
FlowSOM	R package (included in Cytofkit), FlowJo Exchange plug-in	Based in Self-Organization Maps (SOM) follow by hierarchical clustering Very fast Number of clusters customizable	For exploratory analysis, hard to determine the number of clusters

Table 4 Differentiation-trajectory methods.

Method	Environment	Description	Limitations
Diffusion Map	R (included in Cytofkit), and Python package	Preserve global structure Pseudotemporal relatedness Discovery of rare populations Fast	Definition of multi-branch pathways Requires moderate down-sampling
pCreode	Phyton package	Preserve global structure Pseudotemporal relatedness Definition of multi-branch pathways	Require extreme down-sampling Loss of rare populations Not so fast

different clusters cannot be used to infer cellular relationship. The same dataset displays more complexity in a tSNE plot than when analyzed by PCA (Fig. 4B). A disadvantage of this method is that only a few hundreds of thousands of events can be run at a given time in most platforms (e.g., Phyton, R, FlowJo) and the required down-sampling of the data may result in the loss of rare populations.

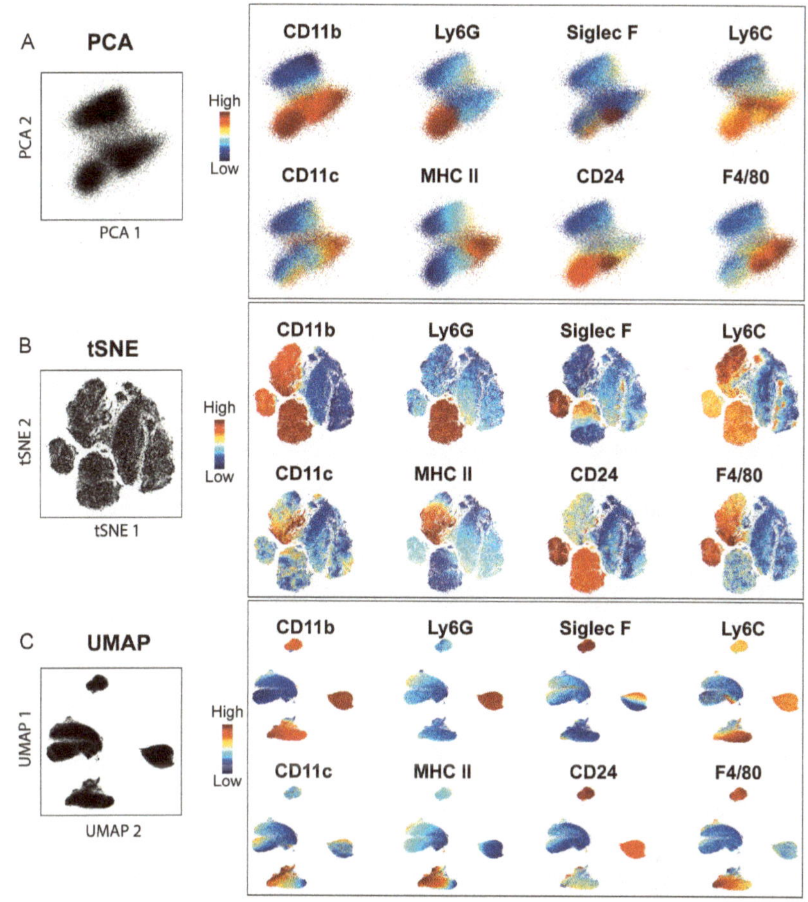

Fig. 4 Comparison of different visualization methods. PCA (A), tSNE using perplexity = 50 and iterations = 1000 (B), and UMAP with nearest neighbor = 30 and min_dist = 0.3 (C) showing the expression level of some of the markers used for the analysis overlaid on each plot.

UMAP is similar to tSNE but allows the use of significantly larger datasets and preserves more of the global structure of the data. As shown in Fig. 4C, UMAP increases the distance among cells that are markedly different (distance between clusters), while grouping more those that bear more similarity (fewer clusters). Because it preserves more of the global structure than a tSNE, it allows to some degree to infer cellular relatedness based on cluster distance.

Importantly, the visual clusters generated by these methods are strongly influenced by several parameters that are specific for each (e.g., perplexity, iterations for tSNE and min_dist, nearest neighbors for UMAP), therefore a clear understanding of them is critical for an accurate analysis (Amir et al., 2013; Becht et al., 2019) (see also Table 2).

In all of these analyses, color can be added as an additional dimension. The example shown in Fig. 4 right panels use this additional dimension to show the expression level (color gradient) of some of the markers used for the analysis, which help identify cell types (see below). Alternatively, color can be used to indicate sample(s) of origin for example when several treatment conditions are combined.

7.3 Clustering methods

The methods described above provide a tool for data visualization following dimensionality reduction, however, they do not provide information regarding cell subpopulations. Additional partitioning methods are used to define clusters within the high-dimensional space in an unbiased manner (Weber & Robinson, 2016). Some of the algorithms used for this purpose are briefly described below (see also Table 3).

PhenoGraph: Partitions high-dimensional data into subpopulations based on finding the k-nearest neighbor for each cell (Levine et al., 2015). It uses the parameters (in this case markers expressed on the cell) to define a point in the high-dimensional space and creates a graph/network that represents the phenotypic similarities between cells (neighbors). These neighbors are then grouped into communities which are stratified into clusters. The results provide a quantification (number of clusters) of the general population structure that is based on different cell types and phenotypes within a same cell type. PhenoGraph clusters are depicted as colored and numbered according to their cluster identity and can be displayed on the two-dimensional tSNE or UMAP data visualization scatter plots (Fig. 5A, upper panels). Cluster annotation (i.e., identification of cell types or subtypes) can then be done by obtaining a heatmap representation of the marker expression level (Fig. 5A, bottom panel) or by coloring the scatter plot by marker expression level rather than cluster number (Fig. 4, right panels). Among the advantages of this method are that it can resolve rare populations and that it allows for population discovery. One of the limitations on the other hand is the need to pre-define number of neighbors.

ClusterX: This method is based on the assumption that cluster centers are characterized by a higher density than the neighbors, and far away from any other points with a higher local density in a pre-defined visualization map

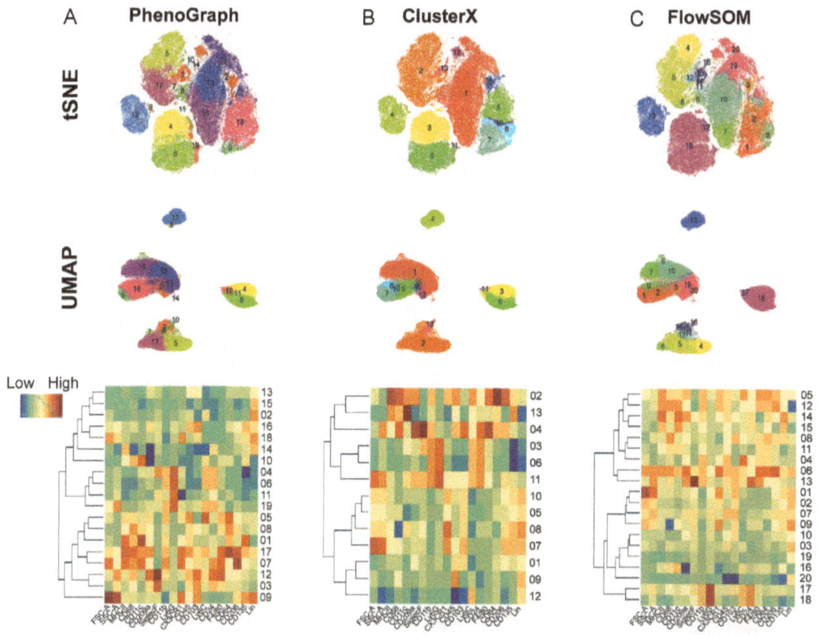

Fig. 5 Clustering methods. Visualization of PhenoGraph (k = 30) (A), Cluster X (B), and FlowSOM (k = 20) (C) in tSNE (n = 100,000 events, perplexity = 50, iterations = 1000; top panels), UMAP (n = 100,000 events, nearest neighbor = 30, min_dist = 0.3; middle panels) and heatmap (normalized by rows; bottom panels).

(Rodriguez & Laio, 2014). The algorithm depends on relative density rather than the absolute values, and clusters are recognized independently of the dimensionality of the space in which the data points are embedded. This density-based clustering was applied to the t-SNE embedded map in the example and might therefore be limited to the power of the tSNE's dimensionality reduction in representing the whole variation of the global data into the 2D representation (Fig. 5B, upper panel). Cluster annotation can be performed as indicated for PhenoGraph (Fig. 5B, bottom panel).

FlowSOM: This is an unsupervised technique for clustering and dimensionality reduction that uses a Self-Organizing Map (SOM) to visualize and interpret the data (Van Gassen et al., 2015). A SOM consists of a grid of nodes, in which a node is a point in the multidimensional input space. The input space is trained so that the nodes closely connected are more similar to each other than to the ones connected through a larger path. During clustering, each point of the dataset is classified with the node that resembles it the best (nearest neighbor). The advantages of this method include the fast runtime, several visualization options, and that it offers one of the best clustering performances allowing users to explore large datasets and the discovery of rare populations. Like with other methods, FlowSOM requires the pre-selection of several parameters and having a clear understanding of them is critical (Van Gassen et al., 2015).

Fig. 5A–C allows for a comparison of the clustering methods mentioned above. The different number of clusters obtained with the different methods may be due to intrinsic limitations of the method or the pre-selected parameters used (e.g., numbers of neighbors and/or clusters pre-defined) (Table 3). All these methods allow for a more granular characterization of the tumor immune microenvironment than the classical bivariate flow cytometry analysis. In addition, one can select specific clusters of interest and run the same or different algorithms for an even more detailed classification.

7.4 Trajectory or differentiation analysis

To infer differentiation paths or cell population relationship, non-linear dimensionality reduction algorithms that preserve the global structure of the dataset are recommended (Table 4). Two options that can be used to analyze cellular relatedness in the tumor microenvironment are:

Diffusion map: Organize the events in non-linear and complex branches of differentiation still allowing the discovery of rare populations (Haghverdi, Buettner, & Theis, 2015). Diffusion map preserves the global structure and pseudo-temporally order of the cells. It is applicable to datasets with high levels of noise (such as flow cytometry data), samples with heterogeneous densities, and with missing or uncertain values. However, this algorithm falls short in depicting multi-branching transitions. Using the same dataset as in previous examples, Fig. 6A depicts an example of a diffusion map run with equal number of events (see Note 28) from selected clusters defined by UMAP and FlowSOM, containing monocytes, macrophages and dendritic cells. There are at least two branches defined by the transitional expression of markers associated to monocytes, macrophages and dendritic cells projected in the diffusionmap 1 and diffusionmap 2 axes (Fig. 6B). One of them represents the differentiation from Ly6C^{hi} monocytes to macrophages (clusters 4-5-6) and the other resemble the differentiation path from dendritic cell precursor to conventional dendritic cells (cDCs) (clusters 14, 15, and 16). Cluster 12 represents DCs derived from monocytes (moDCs). This analysis also allows to dissect what marker or group of markers drive the differentiation path proposed by the algorithm. As an example, Fig. 6C shows

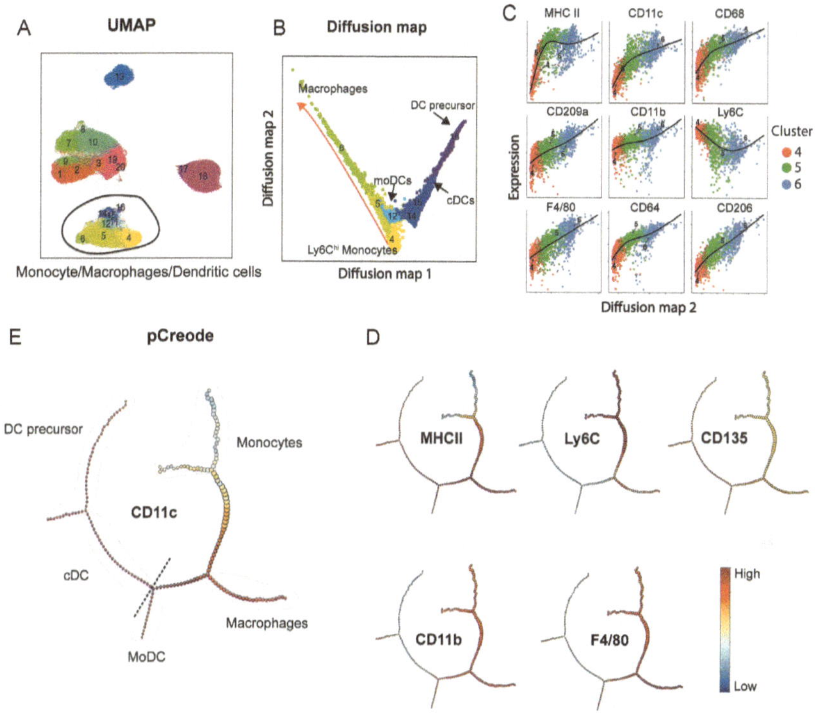

Fig. 6 Trajectory analysis of monocytes, macrophages, and dendritic cells in the TME. (A) UMAP visualization from Fig. 5C showing clusters selected for diffusion map analysis. (B) 2D representation of the diffusion map analysis with an overlay of FlowSOM clusters. (C) Expression level of markers across clusters 4-5-6. (D) pCreode analysis from the same clusters from (B) (plot ranked 0 from n=10 runs). (E) Overlay of markers in pCreode from (D).

increased or decreased levels of expression of different monocyte and macrophage-associated markers across clusters 4-5-6 in the trajectory defined by diffusionmap 2 axis.

pCreode: Is an unsupervised algorithm that compares graphs with different topologies and infers a statistically significant hierarchy of cell dynamics that define a developmental trajectory based on the parameters used in the analysis (Herring et al., 2018). An advantage of pCreode is that it allows for the visualization of multi-branching transitions (Fig. 6D) as well as the overlay of different markers to facilitate the interpretation (Fig. 6E). However, it requires a much stronger down-sampling, which results in the loss of information (rare populations). Another advantage of pCreode over diffusion map is that the node size indicates the abundance of cells within each differentiation transitional state.

Regardless of the robustness or limitations of the method of choice to define differentiation trajectories, the biological meaning of the data needs to be confirmed experimentally.

8. Basic steps for high dimensional flow cytometry data analysis

(1) Export the files in FCS format and open them in any suitable software (e.g., FlowJo, Cytobank).

(2) For each individual sample, plot the parameter in the last position of the laser delay settings (e.g., YG-586-A) against Time. Select the segment of Time in which the data looks constant (as shown in Fig. 3A—gate 1) (see Notes 29 and 30).

(3) From the Time gate, select by forward (FSC) and side scatter (SSC) the gate that defines the populations of interest (Fig. 3A—gate 2).

(4) Exclude doublets via any of the methods available (FSC-H vs FSC-A or SSC-W vs SSC-A and FSC-W vs FCS-A) (Fig. 3A—gate 3) (see Note 31).

(5) Transform all fluorescent markers to biexponential normalization given that this transformation will avoid the artifacts due to logarithmic normalization around zero.

(6) Select the live cells with the L/D marker and $CD45^+$ cells (tumor infiltrating-immune cells, Fig. 3A—Gates 4 and 5 in Strategy 1 or Gate 4′ in Strategy 2).

(7) Perform quality control of the compensation matrix automatically generated during data acquisition by comparing single pairwise markers ($N \times N$ plot in a FlowJo Layout). Fine tune any obviously under or over-compensated markers as needed (see Note 32).

(8) Export the population of interest (e.g., $CD45^+$ Live cells) or continue with the gating strategy using the dump channels to exclude any populations that are not of interest (see Note 33).

(9) Prior to running any algorithms, verify that staining, processing and acquisition have no artifacts or defects. The use of specificity controls is crucial for this step (see Section 5).

(10) Depending on the algorithm of choice, down-sample the properly compensated data to a fix value that will be equal in every sample/group to be analyzed.

(11) Concatenate the data including the same number of events from each sample from every group to be analyzed (see Note 34).

For example, for a total of 100,000 events to be analyzed from two treatment conditions (e.g., untreated vs therapy) with 5 samples per condition: 50,000 events from untreated group = 10,000 events per untreated tumor × 5 samples; 50,000 events from therapy group = 10,000 events per treated tumor × 5 samples (see Notes 35 and 36).

(12) Further steps will depend on the dimensionality reduction, clustering and/or trajectory analysis method of choice.

9. Concluding remarks

One of the biggest challenges in the analysis of high-parametric flow cytometry data is the correct definition of cell populations. While under homeostatic conditions and with few parameters this may seem accurate and indisputable, under pathological situations such as in the TME and with high-parametric data, it can be an arduous job. In this situation, traditional manual gating strategies become time consuming and biased and valuable information can be missed. Thus, an unsupervised approach based on the whole dimension of the data is needed for a comprehensive and objective characterization of the TME.

When applying these algorithms, user-defined settings are important variables that affect the data output. Therefore, a good understanding of each method and what each pre-defined parameter means is critical for the success of the analysis. Additionally, the quality of the compensated data, proper down-sampling, and the markers selected for the dimensionality reduction methods will also influence the outcome. It is likely that several trial runs will be needed to find the optimal conditions for different experimental datasets for the best representations that are biologically relevant. Many of these algorithms are also sensitive to sample-to-sample variations. Therefore, standardized methods should be implemented throughout the whole process: from the tissue harvest and processing through the flow data acquisition. A common practice used in CyTOF analysis that can be applied to flow cytometry, is the barcoding of samples that are then acquired in a single tube minimizing batch effects. Although this is still somehow limited by the less abundant number of parameters available for barcoding in flow cytometry, newer instruments and novel improved fluorochromes will soon allow the detection of up to 50 parameters and reducing compensation difficulties.

An advantage of the unbiased analysis approach for high-dimensional data is the possibility to identify novel cell populations. It is important, however, to keep in mind that such populations may represent a different activation/maturation state rather than a new cell type. Further experimental approaches are needed to formally determine their origin/identity, particularly given our still limited knowledge of the TME. Another useful feature of these analyses is that selecting a cluster from a tSNE and back-gating using the pairwise comparison approach can help select the best 4–6 markers that define a given population to be used for purification (e.g., FACS sorting) and further characterization (e.g., functional and transcriptional analysis).

Because all methods have their advantages and disadvantages, and they may provide different type of information, a combination of different approaches should be used for a more thorough analysis. Here we just exemplified a few of them, however, other algorithms are currently available and new ones are constantly being optimized.

10. Notes

1. Optimal culture conditions should be determined for the specific tumor cell line of interest. Recommendations from the American Type Culture Collection (ATCC, Manassas, VA, USA) can be used as a starting point.
2. Cells should not be used at low (<40%) or high (>80%) confluence as this may lead to the selection of subclones or changes in the metabolic status of the cells, respectively.
3. Porcine Trypsin/EDTA is commonly used to detach adherent cells. TripLE reagents (Gibco-Thermo Fisher Scientific) are animal origin-free recombinant dissociation enzymes that better preserve the integrity of cell surface molecules.
4. The incubation time has to be optimized for each cell line and it should be just enough to have the cells loosely attached.
5. If Trypsin is used, FCS containing medium should be used immediately to inhibit its activity. TripLE reagents can be inhibited by dilution.
6. Non-adherent cells can be collected by gently pipetting and transferring to a sterile tube, centrifuged down and resuspended in fresh medium at the appropriate concentration to be reseeded.
7. The dilution factor will depend on the cell type and size of the tissue culture flask of choice.

8 Careful setup of the tumor model of interest is strongly recommended before any further analysis. This includes determining the cell concentration, volume of injection, timing, etc.

9 Euthanasia should be performed following appropriate institutional guidelines.

10 If extended storage of the tumors before processing is needed (i.e., tissue processing cannot be performed immediately after harvest), it is recommended to collect and keep the tumors in PBS containing higher concentration of FCS (2–10%) to avoid pH fluctuations. Tissues should then be transferred to the digestion medium at the time of processing being careful not to increase the FCS content of the digestion medium as it will block the enzymatic activity. It is important to consider that the longer storage may impact on immune cell populations differently, thus affecting the overall composition of the infiltrate and biasing the results. Therefore, long storage periods should be avoided as much as possible.

11 An example of a digestion cocktail used in our lab is as follows: 200 U/mL of Collagenase IV (Gibco) and 0.1 mg/mL of DNase I (Roche). Tumors are incubated 10 min in a water bath at 37 °C and then transfer to a rotator inside an incubator (37 °C, 5% CO_2) for extra 50 min of digestion.

12 To avoid clogs during flow cytometry, cells should be filtered again passing through a 40–50 μm cell strainer or mesh after staining and prior to running the samples in the flow cytometer.

13 Stopping the enzymatic reaction is an important step to reduce cell death.

14 An example of FACS buffer: PBS/2 mM EDTA/2% FCS.

15 The volume should be determined based on the tumor size/number of cells obtained for the tumor model of interest.

16 Using amount of tissue (e.g., 50 mg of tumor) as a reference rather than cell number helps account for dead cells in the mix and avoid artifacts in the staining.

17 A fluorophore with high spectral overlap into other channels can be used for the dump because that channel will be excluded early on in the analysis.

18 Certain cell types are highly autofluorescent (e.g., myeloid cells). This should be considered when selecting the appropriate fluorochrome for a given marker as a true positive signal may be masked by the autofluorescence if using a weak fluorochrome.

19 Brilliant Violet dyes interact with each other quenching their signal. Using buffers that block the interaction of these dyes is strongly recommended.

20 Information about the spread characteristics of each fluorochrome can be obtained from the supplier's specifications.

21 Other metric to assess reagent brightness is the separation index (Ashhurst et al., 2017).

22 It is recommended to include both positive and negative populations within the same tube. This can be done either with cells, in which case only a fraction of them will express the marker used for the single-color control, or with synthetic compensation beads by mixing positive and negative beads (some companies offer a premade cocktail containing reactive and non-reactive compensation beads).

23 The linear range of each detector can be obtained from the CS&T baseline.

24 If the cytometer baseline changes, a new application setting needs to be determined.

25 The rSD_{EN} is the background noise level of each parameter due to the electronic system.

26 It is recommended to run a full compensation and one fully stained tumor sample to verify there are no excessive spillovers and all parameters are within the linear range. If PMT voltages need to be modified, a new application setting with the updated values should be saved and a new compensation acquired.

27 If needed, PMT voltages can be increased as long as their values stay within the linear range before saving the Application Settings. However, it is not recommended to decrease the values given that they were set to the minimum possible.

28 It is important to keep in mind that by using the same number of events from each cluster, information related to frequency/abundance of each population is lost.

29 Avoid including the beginning and end of the run when the sheath flow is less stable. Also, exclude any defective region, for instance, because of a clog during acquisition. This should be applied when doing manual gating as well.

30 Do not run samples at high speed when they contain too much debris and use a higher FSC threshold during acquisition to record mainly the events of interest.

31 After this step, if the tumor cells are labeled (e.g., GFP, YFP), that channel can be used against FSC or SSC to exclude tumor cells from the analysis.

32 Even when the manual compensation is not recommended, obvious over and under compensations could lead to data misinterpretation and need to be corrected before proceeding with further analysis.

33 It is recommended to run the first analysis with total $CD45^+$ live cells to obtain a big picture of the staining and subsets segregation before further analysis.

34 When comparing more than one treatment condition, the analysis should be run in concatenated groups to avoid batch effects.

35 The number of events per sample will depend on the algorithm of choice.

36 Always keep a computational barcoding of the samples (e.g., in FlowJo adding an extra keyword, as "SampleID") which will be useful to determine any batch effect or sample abnormality, and group identification.

Acknowledgment

This work was supported by the Intramural Research Program of the U.S. National Institutes of Health National Cancer Institute.

References

Amir, E. D., Davis, K. L., Tadmor, M. D., Simonds, E. F., Levine, J. H., Bendall, S. C., et al. (2013). viSNE enables visualization of high dimensional single-cell data and reveals phenotypic heterogeneity of leukemia. *Nature Biotechnology*, *31*(6), 545–552. https://doi.org/10.1038/nbt.2594.

Aran, D., Camarda, R., Odegaard, J., Paik, H., Oskotsky, B., Krings, G., et al. (2017). Comprehensive analysis of normal adjacent to tumor transcriptomes. *Nature Communications*, *8*(1). 1077. https://doi.org/10.1038/s41467-017-01027-z.

Ashhurst, T. M., Smith, A. L., & King, N. J. C. (2017). High-dimensional fluorescence cytometry. *Current Protocols in Immunology*, *119*, 5.8.1–5.8.38. https://doi.org/10.1002/cpim.37.

Becht, E., Mcinnes, L., Healy, J., Dutertre, C., Kwok, I. W. H., Ng, L. G., et al. (2019). Dimensionality reduction for visualizing single-cell data using UMAP. *Nature Biotechnology*, *37*(1), 38–44. https://doi.org/10.1038/nbt.4314.

Bendall, S. C., Nolan, G. P., Roederer, M., & Chattopadhyay, P. K. (2012). A deep profiler's guide to cytometry. *Trends in Immunology*, *33*(7), 323–332. https://doi.org/10.1016/j.it.2012.02.010.

Bindea, G., Mlecnik, B., Tosolini, M., Kirilovsky, A., Waldner, M., Obenauf, A. C., et al. (2013). Spatiotemporal dynamics of intratumoral immune cells reveal the immune landscape in human cancer. *Immunity*, *39*(4), 782–795. https://doi.org/10.1016/J.IMMUNI.2013.10.003.

Binnewies, M., Roberts, E. W., Kersten, K., Chan, V., Fearon, D. F., Merad, M., et al. (2018). Understanding the tumor immune microenvironment (TIME) for effective therapy. *Nature Medicine*, *24*(5), 541–550. https://doi.org/10.1038/s41591-018-0014-x.

Chen, H., Lau, M. C., Wong, M. T., Newell, E. W., Poidinger, M., & Chen, J. (2016). Cytofkit: A bioconductor package for an integrated mass cytometry data analysis pipeline. *PLoS Computational Biology*, *12*(9), 1–17. https://doi.org/10.1371/journal.pcbi.1005112.

Haghverdi, L., Buettner, F., & Theis, F. J. (2015). Diffusion maps for high-dimensional single-cell analysis of differentiation data. *Bioinformatics*, *31*(18), 2989–2998. https://doi.org/10.1093/bioinformatics/btv325.

Hanahan, D., & Coussens, L. M. (2012). Accessories to the crime: Functions of cells recruited to the tumor microenvironment. *Cancer Cell*, *21*(3), 309–322. https://doi.org/10.1016/J.CCR.2012.02.022.

Herbst, R. S., Soria, J.-C., Kowanetz, M., Fine, G. D., Hamid, O., Gordon, M. S., et al. (2014). Predictive correlates of response to the anti-PD-L1 antibody MPDL3280A in cancer patients. *Nature*, *515*(7528), 563–567. https://doi.org/10.1038/nature14011.

Herring, C. A., Banerjee, A., Mckinley, E. T., Gerdes, M. J., Coffey, R. J., Lau, K. S., et al. (2018). Unsupervised trajectory analysis of single-cell RNA-seq and imaging data reveals alternative tuft cell origins in the gut. *Cell Systems*, *6*(1), 37–51.e9. https://doi.org/10.1016/j.cels.2017.10.012.

Kim, J. M., & Chen, D. S. (2016). Immune escape to PD-L1/PD-1 blockade: Seven steps to success (or failure). *Annals of Oncology*, *27*(8), 1492–1504. https://doi.org/10.1093/annonc/mdw217.

Levine, J. H., Simonds, E. F., Bendall, S. C., Downing, J. R., Pe, D., Nolan, G. P., et al. (2015). Data-driven phenotypic dissection of AML reveals progenitor-like cells that correlate with prognosis. *Cell*, *162*(1), 184–197. https://doi.org/10.1016/j.cell.2015.05.047.

Mahnke, Y. D., & Roederer, M. (2007). Optimizing a multicolor immunophenotyping assay. *Clinics in Laboratory Medicine*, *27*(3), 469–485. https://doi.org/10.1016/J.CLL.2007.05.002.

Meinelt, E., Reunanen, M., Edinger, M., Jaimes, M., Stall, A., Sasaki, D., et al. (2012). *Technical bulletin: Standardizing application setup across multiple flow cytometers using BD FACSDiva™ version 6 software*. BD Biosciences.

Newell, E. W., Sigal, N., Bendall, S. C., Nolan, G. P., & Davis, M. M. (2012). Cytometry by time-of-flight shows combinatorial cytokine expression and virus-specific cell niches within a continuum of $CD8^+$ T cell phenotypes. *Immunity*, *36*(1), 142–152. https://doi.org/10.1016/J.IMMUNI.2012.01.002.

Newman, A. M., Liu, C. L., Green, M. R., Gentles, A. J., Feng, W., Xu, Y., et al. (2015). Robust enumeration of cell subsets from tissue expression profiles. *Nature Methods*, *12*(5), 453–457. https://doi.org/10.1038/nmeth.3337.

Pagès, F., Mlecnik, B., Marliot, F., Bindea, G., Ou, F.-S., Bifulco, C., et al.Galon, J., (2018). International validation of the consensus Immunoscore for the classification of colon cancer: A prognostic and accuracy study. *Lancet (London, England)*, *391*(10135), 2128–2139. https://doi.org/10.1016/S0140-6736(18)30789-X.

Pitt, J. M., Vétizou, M., Daillère, R., Roberti, M. P., Yamazaki, T., Routy, B., et al. (2016). Resistance mechanisms to immune-checkpoint blockade in cancer: Tumor-intrinsic and -extrinsic factors. *Immunity*, *44*(6), 1255–1269. https://doi.org/10.1016/J.IMMUNI.2016.06.001.

Rodriguez, A., & Laio, A. (2014). Machine learning. Clustering by fast search and find of density peaks. *Science (New York, NY)*, *344*(6191), 1492–1496. https://doi.org/10.1126/science.1242072.

Sharma, P., Hu-Lieskovan, S., Wargo, J. A., & Ribas, A. (2017). Primary, adaptive, and acquired resistance to cancer immunotherapy. *Cell, 168*(4), 707–723. https://doi.org/10.1016/j.cell.2017.01.017.

Spranger, S. (2016). Mechanisms of tumor escape in the context of the T-cell-inflamed and the non-T-cell-inflamed tumor microenvironment. *International Immunology, 28*(8), 383–391. https://doi.org/10.1093/intimm/dxw014.

Van Gassen, S., Callebaut, B., Van Helden, M. J., Lambrecht, B. N., Demeester, P., Dhaene, T., et al. (2015). FlowSOM: Using self-organizing maps for visualization and interpretation of cytometry data. *Cytometry A, 87*(7), 636–645. https://doi.org/10.1002/cyto.a.22625.

Weber, L. M., & Robinson, M. D. (2016). Comparison of clustering methods for high-dimensional single-cell flow and mass cytometry data. *Cytometry, 89*, 1084–1096. https://doi.org/10.1002/cyto.a.23030.

CHAPTER EIGHTEEN

Phenotyping of tumor infiltrating immune cells using mass-cytometry (CyTOF)

Pauline Maby[a,*], Aurélien Corneau[b], Jérôme Galon[a]
[a]INSERM, Laboratory of Integrative Cancer Immunology, Équipe Labellisée Ligue Contre le Cancer, Sorbonne Université, Sorbonne Paris Cité, Université Paris Descartes, Université Paris Diderot, Centre de Recherche des Cordeliers, Paris, France
[b]Sorbonne-Université, UMS037 PASS, Plateforme de Cytometrie CyPS, Faculté des Sciences Sorbonne-Université, Paris, France
*Correspondence author: e-mail address: mabypau@gmail.com

Contents

Methods in Enzymology, Volume 632
ISSN 0076-6879
https://doi.org/10.1016/bs.mie.2019.07.025

Abstract

The anti-tumor immune response plays a crucial role in cancer patient outcome as well as in response to the growing family of immunotherapeutic treatments. Improving patient prognostic and therapeutic management requires a better knowledge of the tumor microenvironment, for which a deep characterization of tumor-infiltrating immune populations is instrumental. Mass cytometry represents an excellent tool in tumor Immunology, as it allows the simultaneous analysis of >40 markers on single cells. In this chapter, we review challenging technical aspects of the mass cytometry phenotyping of tumor infiltrating immune cells, focusing on fresh human solid tumor samples. We first explain how to design mass cytometry experiments, then provide detailed protocols to isolate mononuclear immune cells from solid tissues and to stain them for an acquisition on a mass cytometer. We also discuss how to optimize the preparation of single immune cell samples, and how to ensure the reproducibility of data generated from distinct fresh human samples. Finally, we provide troubleshooting suggestions for difficult sample acquisitions on a mass cytometer.

1. Introduction

In the last decade, checkpoint inhibitors and other immunotherapies have achieved remarkable results in advanced cancer patients. However, clinical responses to these therapies are very heterogeneous across different types of tumors and across patients. It is ever-growingly accepted that the effectiveness of immunotherapy strategies depends on a baseline host immune response (Galon & Bruni, 2019). Therefore, in order to improve cancer patient care management, it is becoming critical to extend our understanding of the anti-tumor immunity. This involves deep analyses of the immune tumor microenvironment. Studies of this highly complex microenvironment on often limited and precious tumor sample material will require single cell technologies able to quantify numerous features with a minimum of cells.

Mass cytometry is based on the same principle as classical flow cytometry, but it uses metal isotopes, instead of fluorochromes, and a mass spectrometric readout. This substantially increases the number of parameters that can be detected on individual cells (>40). Accordingly, it enables the simultaneous analysis of multiple cell sub-populations with limited sample material. Thus, mass cytometry is an attractive technology to dissect the complexity of the tumor immune contexture and has already been used in several human tumor types, (Chevrier et al., 2017; Lavin et al., 2017). Nevertheless, deep phenotyping of tumor infiltrating immune cells using mass cytometry is still confronted with several challenges. First, fresh human solid tumor samples are limited in number and size, and cell isolation from tissues is a delicate procedure that can alter cell viability, activity and phenotype. Moreover, mass cytometry is a recent and complex technology based on atomic mass spectroscopic signal detection. Understanding data acquisition is essential to properly design experiments. Also, since >40 markers are acquired per cell, mass cytometry generates large multiparametric datasets that require unsupervised automatic analysis which are highly sensitive to experimental variability, thus accentuating the need for inter-assay reproducibility common to every biological study. Finally, cell preparations derived from tissues contain significant amounts of debris, which may affect staining and acquisition procedures. This chapter provides solutions to help investigators overcoming these technical challenges.

2. Mass cytometry: Principle, advantages and limitations of the technology

Mass cytometry is the fusion of two technologies: flow cytometry and mass spectrometry. The instrument at the heart of this technology is the mass cytometer or CyTOF® (cytometry by time-of-flight). Similarly to flow cytometry, mass cytometry enables multiparametric single-cell analyses, but cells are labeled with probes coupled to monoisotopic heavy stable metals, instead of fluorochromes.

Labeled cells are introduced into the mass cytometer, aerosolized into droplets, allowing for cellular individualization, and ionized in an argon-plasma. Then, each ion cloud, that corresponds to an individual cell, passes through a quadrupole for the elimination of low mass ions (<80 Da) and finally reaches a time-of-flight analyzer which separates the ions based on the mass-to-charge ratio. For each channel, the intensity of the recorded signal is proportional to the density of the ion cloud, thereby to the amount of molecules that were labeled with metal-conjugated antibodies.

Mass cytometry has two major advantages over conventional flow cytometry. First, the number of parameters analyzed in a single experiment can be more than doubled. Indeed, the third generation CyTOF®, Helios (Fluidigm), can discriminate with great precision 135 atomic masses from 75 to 209 Da. Currently, Fluidigm commercializes antibodies conjugated with 37 highly purified isotopes (mostly of the lanthanide metal family) and >70 isotopes can be used for different applications in mass cytometry (Table 1). This range is expected to expand, increasing the performance

Table 1 List of metal isotopes currently used for mass cytometry and their principal applications (Panel A). The mass channels in which EQ™ *Four Element Calibration Beads* are detected are in grey in the first column and the mass channels used by different isotopes are underlined. References are given in brackets, and reported in panel B.

Panel A

Mass channel	Isotope	Metal	Metal family	Examples of suppliers	Isotope sensitivity	Conjugated antibodies commercialized	Protocol available to conjugate antibodies	Principal applications	
89	89Y*	Yttrium	Transition metal	Trace Sciences International	Lower	Yes (Fluidigm)	Yes (1)	Protein detection (1)	
96	96Ru	Ruthernium		Electron Microscopy Sciences; Trace Sciences International		No	No	Covalent binding barcoding (2)	
98	98Ru								
99	99Ru								
100	100Ru								
101	101Ru								
<u>102</u>	102Ru								
<u>104</u>	104Ru								
103	103Rh	Rhodium		Fluidigm (in *Cell-ID™ Intercalator-Rh*)			No	Viability	
<u>102</u>	102Pd	Palladium		Fluidigm (in *Cell-ID 20-Plex Pd Barcoding Kit*)			Yes (1; 3)	Protein detection (Mainly for extracellular barcoding, 1; 3)	Covalent binding barcoding (4; 5)
<u>104</u>	104Pd								
105	105Pd								
<u>106</u>	106Pd								
108	108Pd								
<u>110</u>	110Pd								

106	106Cd	Cadmium	Post-transition metal	Fluidigm (in *MaxPar® MCP9 antibody-labelling kit)*			Yes (6)	Protein detection and protein detection with Qdot™ 655 Streptavidin Conjugate for 112Cd (7)
110	110Cd							
111	111Cd							
112	112Cd							
113	113Cd							
114	114Cd							
116	116Cd							
113	113In	Indium		Sigma-Aldrich; Trace Sciences International			Yes (1; 3)	Protein detection (Mainly for extracellular barcoding, 1; 3)
115	115In							
122	122Te	Tellurium	Metalloid	Trace Sciences International			No	Hypoxia measurement (8)
125	125Te							
127	127I	Iodine	Reactive non-metal	Sigma-Aldrich				Cell cycle measurements (9)
139	139La*	Lanthanum	Lanthanides	Trace Sciences International			Yes (1)	Protein detection (1)
140	140Ce	Cerium						
142	142Ce			Fluidigm (in *EQ™ Four Element Calibration Beads)*			No	*EQ™ Four Element Calibration Beads*

Continued

Table 1 List of metal isotopes currently used for mass cytometry and their principal applications (Panel A). The mass channels in which *EQ™ Four Element Calibration Beads* are detected are in grey in the first column and the mass channels used by different isotopes are underlined. References are given in brackets, and reported in panel B.—cont'd

Panel A

Mass channel	Isotope	Metal	Metal family	Examples of suppliers	Isotope sensitivity	Conjugated antibodies commercialized	Protocol available to conjugate antibodies	Principal applications
141	141Pr	Praseodymium		Fluidigm (in *MaxPar® X8 antibody-labelling kit)*		Yes (Fluidigm)	Yes (1; 6)	Protein detection (1)
<u>142</u>	142Nd	Neodymium						
143	143Nd							
144	144Nd							
145	145Nd							
146	146Nd							
147	147Sm	Samarium						
148	148Nd	Neodymium						
149	149Sm	Samarium			Intermediate			
150	150Nd	Neodymium						
151	151Eu	Europium						
152	152Sm	Samarium						
153	153Eu	Europium						
154	154Sm	Samarium						

155	155Gd	Gadolinium						
156	156Gd							
157	157Gd**			Trace Sciences International				
158	158Gd			Fluidigm (in *MaxPar® X8 antibody-labelling kit)*				
159	159Tb	Terbium			Higher			
160	160Gd	Gadolinium						
161	161Dy	Dysprosium						
162	162Dy							
163	163Dy							
164	164Dy							
165	165Ho	Holmium						
166	166Er	Erbium						
167	167Er							
168	168Er							
169	169Tm	Thulium						
170	170Er	Erbium						
171	171Yb	Yterrbium						
172	172Yb							
173	173Yb				Intermediate			
174	174Yb							
175	175Lu	Lutetium						
176	176Yb	Yterrbium						

Continued

Table 1 List of metal isotopes currently used for mass cytometry and their principal applications (Panel A). The mass channels in which *EQ™ Four Element Calibration Beads* are detected are in grey in the first column and the mass channels used by different isotopes are underlined. References are given in brackets, and reported in panel B.—cont'd

Panel A

Mass channel	Isotope	Metal	Metal family	Examples of suppliers	Isotope sensitivity	Conjugated antibodies commercialized	Protocol available to conjugate antibodies	Principal applications	
186	186Os	Osmium	Transition metal	Sigma-Aldrich; Trace Sciences International		No	No	Covalent binding barcoding (2)	
187	187Os								
188	188Os								
189	189Os								
190	190Os								
192	192Os								
191	191Ir	Iridium		Fluidigm (in *Cell-ID™ Intercalator-Ir*)				Nucleic acid detection (10)	
193	193Ir								
194	194Pt	Platinum		Fluidigm (in *Cell-ID™ Cisplatin-194Pt*)	Lower		Yes (11)	Viability	Protein detection (Mainly for extracellular barcoding, 11)
195	195Pt			Fluidigm (in *Cell-ID™ Cisplatin*)			No		
196	196Pt								
198	198Pt			Fluidigm (in *Cell-ID™ Cisplatin-198Pt*)			Yes (11)		Protein detection (Mainly for extracellular barcoding, 11)
209	209Bi	Bismuth	Post-transition metal	Trace Sciences International		Yes (Fluidigm)	Yes (12)	Protein detection and Cell cycle measurements (12)	

Panel B

(1)	Han, G., Spitzer, M. H., Bendall, S. C., Fantl, W. J., & Nolan, G. P. (2018). Metal-isotope-tagged monoclonal antibodies for high-dimensional mass cytometry. *Nat Protoc, 13*(10), 2121-2148. doi:10.1038/s41596-018-0016-7
(2)	Catena, R., Ozcan, A., Zivanovic, N., & Bodenmiller, B. (2016). Enhanced multiplexing in mass cytometry using osmium and ruthenium tetroxide species. *Cytometry A, 89*(5), 491-497.
(3)	Mei, H. E., Leipold, M. D., Schulz, A. R., Chester, C., & Maecker, H. T. (2015). Barcoding of live human peripheral blood mononuclear cells for multiplexed mass cytometry. *J Immunol, 194*(4), 2022-2031. doi:10.4049/jimmunol.1402661
(4)	https://www.fluidigm.com/binaries/content/documents/fluidigm/resources/cell-id-20-plex-pd-barcoding-kit-ug-prd023/cell-id-20-plex-pd-barcoding-kit-ug-prd023/fluidigm%3Afile
(5)	Zunder, E. R., Finck, R., Behbehani, G. K., Amir el, A. D., Krishnaswamy, S., Gonzalez, V. D., Nolan, G. P. (2015). Palladium-based mass tag cell barcoding with a doublet-filtering scheme and single-cell deconvolution algorithm. *Nat Protoc, 10*(2), 316-333. doi:10.1038/nprot.2015.020
(6)	https://www.fluidigm.com/binaries/content/documents/fluidigm/resources/maxpar-antibody-labeling-user-guide-201106-201116/maxpar-antibody-labeling-user-guide-201106-201116/fluidigm%3Afile
(7)	Newell, E. W., Sigal, N., Bendall, S. C., Nolan, G. P., & Davis, M. M. (2012). Cytometry by time-of-flight shows combinatorial cytokine expression and virus-specific cell niches within a continuum of CD8+ T cell phenotypes. *Immunity, 36*(1), 142-152.
(8)	Edgar, L. J., Vellanki, R. N., McKee, T. D., Hedley, D., Wouters, B. G., & Nitz, M. (2016). Isotopologous Organotellurium Probes Reveal Dynamic Hypoxia In Vivo with Cellular Resolution. *Angew Chem Int Ed Engl., 55*, 13159.
(9)	Behbehani, G. K., Bendall, S. C., Clutter, M. R., Fantl, W. J., & Nolan, G. P. (2012). Single-cell mass cytometry adapted to measurements of the cell cycle. *Cytometry A, 81*(7), 552-566. doi:10.1002/ cyto.a.22075
(10)	Ornatsky, O. I., Lou, X., Nitz, M., Sheldrick, W. S., Baranov, V. I., Bandura, D. R., & Tanner, S. D. (2008). Study of Cell Antigens and Intracellular DNA by Identification of Element-Containing Labels and Metallointercalators Using Inductively Coupled Plasma Mass Spectrometry. *Analytical Chemistry, 80*(7), 2539-2547. doi:10.1021/ac702128m
(11)	Mei, H. E., Leipold, M. D., & Maecker, H. T. (2016). Platinum-conjugated antibodies for application in mass cytometry. *Cytometry A., 89*(3), 292-300.
(12)	Han, G., Chen, S.Y., Gonzalez, V.D., Zunder, E.R., Fantl, W.J., & Nolan, G.P. (2017) Atomic mass tag of bismuth-209 for increasing the immunoassay multiplexing capacity of mass cytometry. *Cytometry A*, 91(12):1150–63.10.1002/cyto.a.23283

*89Y and 139La have weaker performances as compared to other isotopes commercialized by Fluidigm. Home conjugated antibodies generate higher background noise and are less stable. Moreover they are detected in lower sensitivity channels. It is recommended to use these isotopes for high abundance markers.

**157Gd is not commercialized by Fluidigm as its isotopic purity is only of 92%. It should be used with specific caution regarding spillover.

of this technology. Second, lanthanide isotope channels are hardly subjected to contaminant signals. Indeed, since lanthanide metals are rare in biological samples, natural contaminants within cells are sparse. Furthermore, overlap between channels is very limited, thus highly reducing the need for compensation, in contrast to flow cytometry. Overall, thanks to high specificity of heavy metal signals, mass cytometry enables the detection and analysis of numerous markers on individual cells.

While a multitude of cellular processes can be studied by mass cytometry, this technology has some limitations that must be considered when designing an experiment.

(i) Because the cells are atomized, living cells cannot be recovered and size and granularity information cannot be easily recorded (Stern, Rahman, & Birtwistle, 2017). Nevertheless, the high number of parameters recorded on single cells allows a deep characterization.

(ii) Given the dynamics of ion flight, the flow rate of mass cytometry (between 250 and 500 events/s) is lower than that of fluorescence-based instruments (10,000–50,000 events/s).

(iii) Some mediators of cellular behaviour, such as metabolites, are difficult or even impossible to measure by mass cytometry. For instance, fluorescent reporters for mitochondrial permeabilization (*e.g.* mitoTrackers, Thermo Fisher Scientific) or cell division, (*e.g.* CFSE; carboxyfluorescein succinimidyl ester) used in flow cytometry do not have metal equivalents.

3. Mass cytometry panel design

The quality of a mass cytometry experiment first relies on the quality of the panel composition. Designing a panel consists in choosing probes (usually antibodies) that target the molecules of interest and pairing each of them with an appropriate isotope tag, in order to obtain the desired signal intensity. The challenge is to optimize the sensitivity of each detection channel, that is, to increase to a maximum the target-specific signal while decreasing to a minimum the undesired background. To that purpose, it is first necessary to estimate the level of expression of each target antigen, on all studied cell types. Moreover, since the sensitivity of a CyTOF® varies across the mass range, it is required to assess the sensitivity of each channel for its corresponding isotope. This can be done by acquiring the *ICP-MS Multi-Element Solution 1* (SPEXCertiPrep) that contains almost all the isotopes currently used for protein detection with mass cytometry. An isotopic sensitivity curve can then easily be generated (Tricot et al., 2015). Usually, the most sensitive isotopes are in the atomic mass range between 159 and 170 (Olsen, Leipold, Pedersen, & Maecker, 2019).

Then, not only probes that target the least expressed markers should be matched with the most sensitive isotopes, but also it is essential to take potential undesired background into consideration. In mass cytometry, very low background noise comes from metal contamination and this can be further limited by technical precautions that are discussed thereafter. Background to be considered while designing a panel is mainly due to three sources of signal overlap from metal-conjugated probes into unspecific channels (see Note 1).

(i). ***Impurities*** are present in isotopes. Indeed, naturally, heavy metal elements exist as mixture of isotopes that requires enrichment to be used as mass tags. Accordingly, commercialized lanthanide isotopes are not perfectly pure. Nonetheless, their purity is usually greater than 97%. Contaminant signals derived from lanthanide impurities are provided by Fluidigm (see Note 2). Trace Sciences International, that commercializes stable isotopes, also provides the percentages of natural abundance and enrichment available for each of these isotopes.

(ii). ***Oxidation***, as a gain of an oxygen atom (^{16}O), adds 16 mass units to the mass of an isotope. As a consequence, oxides contaminate the M + 16 channels (where M is a given mass channel). Oxidation of some of the lanthanide ions in the plasma cannot be avoided, but can be predicted. Oxide formation is not identical for all isotopes. La, Ce, Pr and Nd isotopes are the main source of M + 16 signals, while Eu and Sm are subjected to very little oxidation. Approximation of contaminant signals derived from oxides are provided by Fluidigm (see Note 2), but their exact values depend on instrument performance. Even if complete elimination of oxide formation is not possible, tuning the instrument before each run (see Note 3) enables oxides to remain under 3% of their corresponding non-oxide signal, thus greatly limiting their impact on the output data. Of note, because metals outside the analytical channels can generate oxides that contaminate these channels, we recommend to record signals of the entire mass range, in order to identify these potential contaminations.

(iii). ***The abundance sensitivity*** is the ratio of the contaminant signal in the M $\pm$ 1 mass channel to the specific signal detected in an M mass channel. Because ions from the same isotope have slight differences in initial position and velocity, the M signal mass peak can spread and give undesired signal in the adjacent mass channels (M $\pm$ 1). Since the width of a signal peak is linked to its height, the risk of spillover in the mass

channels M + 1 and M − 1 increases with the strength of the signal in the M mass channel (Olsen et al., 2019). This phenomenon is more important for high-mass isotopes, as they have a broader ion distribution. It should be noted that each mass cytometer, under specific operating conditions, has a fixed abundance sensitivity value. With the improvement in ion optics, this phenomenon decreased from ~2% for the first CyTOF® model to ~0.3% for CyTOF®2™ and Helios™.
To help with panel designing, Fluidigm developed an interactive application (*Maxpar® Panel Designer*; see Note 4). According to the antibodies and mass tags that are selected, it calculates the background expected from signal overlap.

Despite careful design of the antibody panel, interferences across channels may remain. To estimate and correct them, a compensation workflow and an algorithm were recently developed (Chevrier et al., 2018). If some doubts persist about the origin of a signal being specific or due to the spillover from another probe, "metalminus-one" (M − 1) control experiments can be performed: one by one, probes are removed from the panel in order to reveal nonspecific signal.

4. Antibody-metal conjugation

Depending on the study, panels can consist in different types of probes associated with isotope tags. In this chapter, we focus on the most common probes, *i.e.*, antibodies, but mass cytometry is compatible with other types of probe, such as oligonucleotidic probes that target RNA sequences (Frei et al., 2016) and 5-iodo-2-deoxyuridine (IdU) that stains cells in the S phase (Behbehani, Bendall, Clutter, Fantl, & Nolan, 2012).

Fluidigm commercializes antibodies conjugated with yttrium, bismuth and heavy metals of the lanthanide family (Table 1) and *MaxPar® X8 antibody-labelling kit* for custom antibody conjugation with lanthanide isotopes. This kit is provided with an optimized protocol (See ref 10 of Table 1 and Note 4) based on a procedure enabling to first chelate metals with polymers and then link the polymers to partially reduced purified antibodies (Han, Spitzer, Bendall, Fantl, & Nolan, 2018). Of note, purified antibodies must not be of the IgM isotype (only IgG or polyclonal) and must be free of carrier proteins (*i.e.* BSA or gelatin) as these proteins can react with the polymers. To conjugate antibodies with yttrium or bismuth metal, *MaxPar® X8 antibody-labelling kit* can also be used, with adapted protocols (see references 1 and 13 of table 1).

In September 2019, Fluidigm released a new polymer and 7 new metal isotopes: *MaxPar® MCP9* (for Metal-Chelating Polymer 9) *antibody-labelling kit*. This kit allows for custom antibody conjugation with 7 cadmium isotopes, increasing the range of antigens that can be detected on single cells with mass cytometry. Overall, the labeling procedure is similar the one used with the *MaxPar® X8 antibody-labelling kit*, with longer incubation times required at certain steps (See Note 4 and ref 10 of Table 1).

Antibodies can also be conjugated to palladium according to a procedure provided by Mei, Leipold, Schulz, Chester, & Maecker (2015), based on the use of isothiocyanobenzyl-EDTA which chelates metal ions with its EDTA moiety and binds antibodies with its amine-reactive compound. Publications that provide procedures to conjugate antibodies to diverse metal elements are referenced in Table 1.

Newly conjugated antibodies must undergo rigorous validation and titration. Indeed, to reach appropriate signal intensities, it is important to assess the performance of each probe which relies both on its avidity for the antigen and on its metal load. First, it should be verified that compensation beads stained with the antibody give a positive signal in the corresponding channel. Then, the volume of antibody required to stain a desired number of cells to saturation, the titer, should be determined. As for classical flow cytometry, the goal is to have the better separation between background and positive staining (van Vreden et al, 2019). As the quality of the cell preparation will influence probe binding, we recommend to titrate antibodies using cells from similar tissues and extracted in the same conditions as for the planed experiments. Starting from 1 g of antibody for 1 to 3 million cells, it can then be refined in order for the signal from negative cells to remain in the first decade while the signal from positive cells remains in the second or the third decade of the logarithmic scale. Indeed, signals above the third decade are associated with risks of strong spillover in adjacent channels (*cf.* abundance sensitivity). Purchased conjugated-antibodies should also be titrated to ensure the quality of signal in specific experimental conditions.

5. Single cell sample preparation for phenotyping with mass cytometry

To phenotype immune cells infiltrating human solid tumors using mass cytometry, sample preparation procedures are similar as when using flow cytometry. Tumor samples are disaggregated into single-cell suspensions and then immune cells are isolated and stained with probes targeting the molecules of interest, most commonly proteins that are detected with antibodies.

In the next section of this chapter we provide a protocol that we developed to isolate and stain colorectal tumor-infiltrating mononuclear immune cells. This protocol can serve as a starting point but we recommend to carefully adapt each step of the procedure to the tumor sample types, the immune cell sub-populations and the markers of your own interest. To that purpose, we will thereafter discuss (i) technical parameters of the immune cell isolation procedure that can be optimized in order to recover the highest yield of living cells, while limiting their alteration, (ii) aspects of the staining procedure that are specific to mass cytometry experiments, (iii) and solutions to limit the impact of variability across samples and experiments on mass cytometry data.

6. Protocol for the isolation and staining of mononuclear immune cells from tumor tissues

This section provides practical guidance for the experimental procedure to follow prior to a CyTOF® acquisition. An exhaustive list of reagents and solutions can be found in Table 2, while all the steps to obtain a single cell suspension from a tissue sample are listed and described in Table 3. Below, the detailed staining protocol, and all the steps prior to sample acquisition.

6.1 Cell surface staining and extracellular barcoding

1. Resuspend the cells in ***Blocking Buffer*** (3×10^6 cells/100 μL) by pipetting up and down.
2. Incubate for 10 min at 4 °C.
3. Add different CD45 antibodies to the different samples (barcoding), gently vortex samples.
4. Incubate for 15 min at RT under gentle agitation.
5. Wash twice with *MaxPar® Cell Staining Buffer* (centrifugation 400–500 × *g*, 5 min, 4 °C).
6. Resuspend cells in ***Blocking Buffer*** by thoroughly pipetting up and down and pool all the samples in a single tube.
7. Centrifuge at 400–500 × *g* for 5 min at 4 °C. Resuspend cells in ***Blocking Buffer*** (3×10^6 cells/100 μL) by pipetting up and down.
8. Add the ***Surface Antibody Mix***, gently vortex samples.

   (Note: Certain antigens require specific staining conditions and should be stained separately. For example, chemokine receptors are usually optimally stained at RT° or 37 °C and should be stained first, in non-cold blocking buffer.)

Table 2 List of material, reagents and media. Of note, this list does not include the most common cell biology material and media are quoted only once.

Function	Required material
To isolate cells	– Petri dishes – Sterilized forceps and scalpels or scissors – gentleMACS™ Octo Dissociator and gentleMACS™ C Tubes (Miltenyi) – 500, 300, 70 and 40 μm-cell strainers – RPMI 1640 – Ficoll or Lymphocytes separation medium – Gibco™ DPBS, no calcium, no magnesium – **Staining Buffer** = *Gibco™ DPBS, no calcium, no magnesium*, 5% FCS, 0.5 mM EDTA
To stimulate cells	– eBioscience™ Cell Stimulation Cocktail (plus protein transport inhibitors) (500 ×) (ThermoFisher Scientific) – **Complete Medium** = RPMI 1640, 10% FCS (Foetal Calf Serum), 1% NEAA (Non-Essential Amino Acids) 1 ×, Glutamine, Penicillin and Streptomycin
To stain cells (solutions can be prepared during cell stimulation)	– Polypropylene Round-bottom Tubes, 5 mL capacity – MaxPar® Cell Staining Buffer, MaxPar® Fix and Perm Buffer, MaxPar® Cell Water, Cell-ID™ Cisplatin, Cell-ID™ Intercalator-Ir and EQ™ Four Element Calibration Beads (Fluidgm) – Milli Q water – eBioscience™ Foxp3/Transcription Factor Staining Buffer Set (ThermoFisher Scientific) that contains: *Fixation/Permeabilization Concentrate, Fixation/Permeabilization Diluent and Permeabilization Buffer 10 ×* – **Blocking Buffer** = *Gibco™ DPBS, no calcium, no magnesium*, 2% human serum, 5%, mouse serum, 0.5 mM EDTA – **Surface Antibody Mix**: antibodies targeting cell surface antigens should be diluted in **Blocking Buffer** to 50 or 100 μL in a non-autoclaved microcentrifuge tube – **Fixation/Nuclear-Permeabilization Solution** = 375 μL *Fixation/Permeabilization Diluent* + 125 μL *Fixation/Permeabilization Concentrate* – **Nuclear Permeabilization Wash Buffer** = *Permeabilization Buffer 10 ×* diluted 10 times in *Milli Q water* – **Intracellular Antibody Mix**: antibodies targeting intracellular or nuclear epitopes should be diluted in **Nuclear Permeabilization Wash Buffer** to 50 or 100 μL in a non-autoclaved microcentrifuge tube (CD4 may be internalized during stimulation, therefore anti-CD4 antibody should be added to this mix) – **Intercalation Solution** = 1 mL of *MaxPar® Fix and Perm Buffer* + 1 μL of *Cell-ID™ Intercalator-Ir*

Table 3 List of procedures and correspondent protocols to be performed on tissue samples before single cell staining in view of a CyTOF® acquisition.

Procedure	Protocol
1. Tissue disaggregation	1. Transfer the patient tissue samples (non-tumor and tumor) to sterile Petri dishes and cut them into small pieces of 2–4 mm with the help of sterile forceps and scalpels or scissors 2. Transfer the tissues pieces into *Gentlemacs C tubes* filled with 10 mL of *RPMI 1640*. Run the *gentleMACS™ Octo Dissociator* with a program adapted to the softness/hardness of the tissue samples 3. Filter the cell suspensions using a 500 μm, a 300 μm and then a 70 μm-cell strainer over 50 mL polypropylene tubes. Gently help the cell suspension passing through the cell strainers by stirring in it with the large end of a 1000 μL or 200 μL tip and, to recover a maximum of cells, rinse the cell strainers with **Staining Buffer** at RT (Room temperature) 4. Centrifuge at 400–500 × *g*, 10 min, RT 5. Discard supernatants
2. Mononuclear cell isolation	1. Resuspend cell pellets in 30 mL of **Staining Buffer** at RT and very gently layer cell suspensions over 15 mL of *Ficoll* (see Note 5) 2. Centrifuge at 800 × *g*, 20 min, RT, without braking 3. Recover extremely carefully mononuclear cells from interface and transfer them to 50 mL polypropylene tubes containing 10 mL of cold *RPMI 1640*, filtering them through a 40 μm-cell strainer 4. Wash three times with cold *RPMI 1640* (centrifugation 400–500 × *g*, 5 min, 4 °C) 5. Resuspend in 2–4 mL of **Complete Medium** depending of the size of the pellet (see Note 6)
3. Cell count	1. Transfer a 50 μL aliquot of the cell suspension within a well of a 96-well round bottom plate that contains 150 μL of RT ACK, mix well by pipetting up and down 2. Incubate for 5 min at RT in order to lyse the red blood cells that could interfere in the counting 3. Count with the system available in your laboratory, considering the dilution of the cell suspension 4. Centrifuge cell suspension at 400–500 × *g* for 5 min at 4 °C

4. Cell stimulation	1. Resuspend in **Complete Medium** to 1×10^6 cells/mL 2. Place 2 mL of the cell suspensions per well of a 24-well plate (at least 2 wells per tissue) 3. Add 4 µL of *eBioscience*™ *Cell Stimulation Cocktail (plus protein transport inhibitors) (500×)* per well 4. Incubate for 4 h at 37 °C, 5% CO_2. If a standard sample is used, an aliquot can be thawed during this incubation time. Cells should be rested in **Complete Medium** for at least 2 h at 37 °C, 5% CO_2. (In case of high mortality of cells, perform the "mononuclear cell isolation" procedure to enrich in living cells) The standard sample will be then treated as the other samples 5. Resuspend cells by vigorously pipetting up and down and transfer cells in a 15 or 50 mL polypropylene tube. Wash the wells several times with *Gibco*™ *DPBS, no calcium, no magnesium*, in order to recover a maximum of cells 6. Resuspend in 2–4 mL of *Gibco*™ *DPBS, no calcium, no magnesium* (see Note 6)
5. Cell count	Count the cells (as described in step 3) and keep the appropriate number of cells. This number should be defined in advanced according to the number of parameters analyzed on individual cells, the evaluated frequency of the rarest cell population of interest and knowing that typically around 10% to 20% of the cells present at the beginning of the experiment are finally recorded as data events during the CyTOF run. Note that the loss of cells is mainly due to the multiple wash steps and can be limited by careful handling of the cell pellets.
6. Viability staining	1. Wash twice with *Gibco*™ *DPBS, no calcium, no magnesium* (centrifugation 400–500 × *g*, 5 min, 4 °C) 2. Resuspend cells to 10×10^6 cells/mL in "Viability Solution" (=1 µL *Cell-ID*™ *Cisplatin*/mL of *Gibco*™ *DPBS, no calcium, no magnesium*) by pipetting up and down and transfer in a 5 mL polypropylene tube (see Note 7) 3. Incubate for 5 min at RT 4. Wash twice with **Blocking Buffer** (500 µL per 1×10^6 cells; centrifugation 400–500 × *g*, 5 min, 4 °C) to quench the staining

9. Incubate for 20 min at RT under gentle agitation.
10. Wash twice with *MaxPar® Cell Staining Buffer* (centrifugation 400–500 × *g*, 5 min, 4 °C).

6.2 Intracellular/intranuclear staining

1. Vortex sample to thoroughly disrupt the pellet.
2. Add 500 µL of ***Fixation/Nuclear Permeabilization Solution*** and mix well by pipetting up and down.
3. Incubate for 30 min at RT (or 4 °C). Agitate every 5–10 min.
4. Wash twice with at least 2 mL of ***Nuclear Permeabilization Wash Buffer*** (centrifugation 600 × *g*, 5 min, RT, or 4 °C).
5. Resuspend pellet in residual volume with gentle vortexing.
6. Add the ***Intracellular antibody mix*** directly to residual volume and mix well by pipetting up and down.

   Note that polymers used for antibody conjugation can adhere to eosinophils leading to unspecific staining. To study this cell type it is recommended to incubated fixed/permeabilized cells in a solution containing heparin prior to adding intracellular antibodies, as described by Rahman et al. (Rahman, Tordesillas, & Berin, 2016).
7. Incubate for 30 min at RT (or 4 °C). Agitate every 5–10 min.
8. Wash three times with 2 mL of *MaxPar® Cell Staining Buffer* (centrifugation 600 × *g*, 5 min, 4 °C).

6.3 Nucleic acid staining and freezing

1. Thoroughly vortex sample to disrupt the pellet.
2. Add the ***Intercalation Solution***, mix well by pipetting up and down and transfer the cell suspension to a non-autoclaved microcentrifuge tube.
3. Incubate overnight at 4 °C. (Note: Cells can be left at 4 °C in the intercalation solution up to 48 h.)
4. Place cells at −80 °C in a freezing container.

6.4 Samples thawing and preparation before acquisition

1. Take frozen samples from the −80 °C and directly transfer them in ice.
2. Thaw the samples by adding *MaxPar® Cell Staining Buffer* and pipetting up and down.
3. Wash twice with *MaxPar® Cell Staining Buffer* (centrifugation 800 × *g*, 5 min, 4 °C).

4. Wash with *MaxPar*® *Cell Water* (centrifugation 800 × *g*, 5 min, 4 °C). Place cell pellet at 4 °C.
5. Dilute *EQ*™ *Four Element Calibration Beads* 10 times in *MaxPar*® *Cell Water.*
6. Immediately prior to sample acquisition, resuspend the cell suspension in this solution, at the appropriate concentration (0.5×10^6 cells/mL or less if it contains many debris and dead cells) by thoroughly pipetting up and down and filter it through a 40 μm-cell strainer.
7. Vortex and run the sample on CyTOF®.

7. Optimization of the extraction of tumor-infiltrating immune cells

Solid tumors represent a broad spectrum of tissues with various types of neoplasms and degrees of differentiation. Their extremely variable composition makes the procedure to dissociate them difficult to standardize across tumor tissues. Therefore, each step of the tumor sample processing should be optimized to find the appropriate balance between increasing the yield of cells, maintaining their viability and functionality, and conserving the integrity of their surface epitopes. Below we discuss the most current approaches to obtain single immune cell suspensions, ready to be stained for CyTOF® acquisition.

(i). ***Mechanic dissociation*** of tumor specimens can be done manually, by cutting and scratching the tissues with surgical tools or with the help of automatic mechanical dissociators which decrease experimental variability and which can limit processing duration, thus limiting cell mortality. In our hand, semi-automated tissue dissociation with the *gentleMACS*™ *C Tubes* combined with the *gentleMACS*™ *Octo Dissociator* increased the yield of cells by more than two, compared to manual dissociation.

(ii). ***Enzymatic digestion*** can be added to the mechanical dissociation procedure, in order to help tissue disaggregation and increase the yield of functionally viable single-cells. Leelatian et al. have compared multiple combinations of enzymes and dissociation kinetics on freshly resected patient-derived tissues. They reached the best yield of living cells with the combination of mechanical disaggregation with enzymatic dissociation by Collagenase and DNase for 1 h (Leelatian et al., 2017). To set the optimal combination of enzymes and dissociation conditions, depending on the type of tissue to be dissociated, you can refer to a list of lab procedures for cell isolation from many different tissues, available online (http://www.worthington-biochem.com/tissuedissociation/).

The integrity of surface epitopes can be altered by enzymatic treatment and should systematically be verified. Autengruber et al. evaluated, on mice splenocytes, the impact of dissociation enzymes on 48 surface molecules that are frequently used for flow cytometric identification of immune cells. Dispase treatment considerably affected the detectability of the majority of these immunologically relevant surface molecules, while collagenase treatment had only minor, but not null, effects (Autengruber, Gereke, Hansen, Hennig, & Bruder, 2012). Miltenyi commercializes a *Tumor Dissociation Kit for human tissue* that consists in three enzymes. The sensitivity of 200 surface epitopes to these enzymes has been tested. Most epitopes were preserved during enzymatic treatment, but some were altered (A summary of the results is available online https://www.miltenyibiotec.com/upload/assets/IM0014390.PDF). In our hands, treatment with these enzymes prevented recognition of CD8 by the anti-CD8 antibody, clone RPA-T8. On the other hand, Miltenyi tests have revealed that the CD8 epitope recognized by another anti-CD8 antibody, the clone BW135, was not affected by these enzymes. Accordingly, if surface antigen detectability is lost upon enzymatic treatment, different antibody clones should be tested in order to recognize other epitopes.

To obtain single cells as close as possible to their native state, we recommend to restrict the procedure to the mechanical tissue dissociation, thereby avoiding alteration of cells by enzymes. In our hand, adding the enzymes of the *Tumor Dissociation Kit for human tissue* or adding *Collagenase* and *DNase*, to the *gentleMACS*™ (with *C Tubes*) mechanical dissociation, did not increase the yield of living immune cells extracted from colorectal tumor tissues. Similarly, Garaud et al. obtained a better viability and cell recovery using an automatic mechanical dissociator compared to using enzymes to digest breast tumor tissues (Garaud et al., 2014).

(iii). ***Filtration*** through a 70 μm-cell strainer, following tissue disaggregation, is commonly used to discard the biggest debris and increase the homogeneity of cell suspensions. Depending on the size and composition of the initial tissue sample, there is a risk that the cell strainer clogs before the whole cell suspension went through and traps cells of interest. To avoid rapid obstruction of the 70 μm-cell strainer, successive cell strainers, with decreasing mesh sizes, can be used (see the steps 1.3 and 2.3 of the protocol provided in Table 3). Besides, to further separate cells, some experimenters pass samples through a syringe and a 16-gage needle (Lavin et al., 2017).

(iv). ***Immune cell isolation*** from whole cell suspensions is usually done by density gradient centrifugation. It is a simple way to keep immune cells while discarding most of the tumor and red blood cells that will fall into the pellet. Depending on the cells of interest, different types of density gradient can be used, such as *Ficoll* for mononuclear cells (see Note 5) or *PolymorphPrep* for granulocytes. If tumor cells are not to be discarded, it is possible to only lyse red blood cells with a buffer that has minimal effects on leukocytes (*e.g., RBC Lysis Buffer (10×)*, BioLegend).

8. Specificities of the immuno-staining procedures for mass cytometry applications

The cell staining protocol provided above is adapted for the detection, on single cells, of extracellular antigens, intranuclear transcription factors (FOXP3) and intracellular cytokines.

Depending on the molecules of interest, other staining procedures are provided by Fluidigm (see Note 4). Overall, most of the cell sample processing is similar as for flow cytometry experiments, with only few modifications required for mass cytometry.

First, dead cells should be excluded from the analysis as they might unspecifically bind antibodies. For mass cytometry, dead cells are classically stained with *Cell-ID™ Cisplatin* at the beginning of the staining procedure. Cisplatin enters the dead cells through their compromised membranes and form covalent bonds with intracellular nucleophile proteins. It is ideally detected in the 195Pt channel of the CyTOF® system, but also gives signal in the 194 and 198 mass channels. If the cisplatin channels are used for other markers, dead cells can be labeled with rhodium-103 (*Cell-ID™ Intercalator-Rh*, Fluidigm). However, because of its non-covalent interaction with cells, intensity of 103Rh signal can be lost during downstream fixation/permeabilization and wash steps, when intracellular antigens are measured.

As previously mentioned, mass cytometry does not enable cell recognition through their morphology, since cells are vaporized. Singlet nucleated cells are identified according to their nucleic acid content with *Cell-ID™ Intercalator-Ir*, a reagent that binds cellular nucleic acids and contains iridium, detected in the 191 and 193 mass channels. In order to enter the nucleus, *Cell-ID™ Intercalator-Ir* is diluted in a nuclear permeabilization buffer (usually *MaxPar® Fix and Perm Buffer*). This so called "**intercalation solution**" is added at the end of the staining procedure.

Additionally, for mass cytometry, it is possible to freeze stained samples for further acquisition. In our hands, it did not alter cell suspensions. This can greatly simplify management of the instrument use and it does not require additional step to the procedure. After a one-hour or an overnight incubation in **intercalation solution**, simply transfer the tubes in a freezing container to place them at −80 °C, where they can remain up to 2–4 weeks. Samples will then be thawed as described in the protocol provided above (Section 6.4).

Finally, metal contaminants present in the environment can generate undesired background noise. In particular, barium (130–138 Da), an abundant element of the earth's crust, may be present in high quantity in dust fall and in laboratory containers, hence it may contaminate samples through exposition to laboratory environment and to some detergents and reagents. For example, *Gibco™ DPBS* contains high barium concentrations and should be avoided in the last steps of the staining procedure. Instead, it is recommended to use *Maxpar® Cell Staining Buffer*, which has minimum contamination with heavy elements, or to use home-made lab buffers prepared with *Milli Q water* in containers rinsed with distilled water.

Iodine (127 Da), and to a lesser extent tin (116–120 Da), mercury (198–202 Da) and lead (206–208 Da) can also be found in some buffers and commercial reagents. For example, *Ficoll* contains large amounts of iodine. Thus, it should be used with particular care (see Note 5) or substituted with *Percoll* (Cederlan) when possible. Moreover, we recommend to avoid the use of autoclave and laboratory dishwashers which are commonly contaminated with heavy metals (Behbehani, 2019).

Overall, to avoid contamination of samples during immunostaining procedure, each batch of buffer and reagent should be tested for the potential presence of metal contaminants.

Furthermore, before running stained tumor patient samples, it is important to verify the background noise of an aliquot of unstained cells as they can be contaminated with *Cisplatin*, a chemotherapeutic agent, or with *Gadolinium*, a contrasting agent used for MRI (Magnetic Resonance Imaging).

9. Management of the experimental variability across samples

With high numbers of parameters detected on numerous single cells, mass cytometry generates large multiparametric datasets. Analyzing these highly complex datasets manually would be extremely time consuming and would probably lack efficiency. To reveal the multitude of information contained in mass cytometry data, they are analyzed with automate (or semi-automate) unsupervised methods based on *ad hoc* algorithms. These powerful tools analyze all the parameters simultaneously and all the samples at once, as a single dataset. Thus, it is absolutely necessary that the relation between the level of expression of a marker and its corresponding signal intensity remains perfectly stable across samples. For that reason, mass cytometry data analysis is extremely sensitive to undesired variations of signal due to experimental conditions.

An important source of inter-assay variability, called batch effect, relies on discrepancies across experimental conditions for patient samples that are processed on different days. Further variability can be added during sample acquisition, as instrument performances may decline during the course of a CyTOF® run and may also vary over longer periods of time.

Another source of variability is due to discrepancies between compositions of the different tissue samples. Indeed, to analyze tumor-specific immune profiles, it is very common to phenotype immune cells isolated from diverse tissues of a single patient. Because each tissue has a unique composition, it behaves differently upon processing. For example, compared to blood samples, solid tissue samples lead to thicker cell suspensions, containing more debris and dead cells that might trap reagents and thus reduce the amount of antibodies available to label cells.

To compute data from a whole patient cohort and reveal its true biological diversity, it is crucial to control experimental variability at different levels.

9.1 Management of general experimental variability with careful planning

As for every biological study, experimental variability can be minimized with simple but important measures, such as having all the samples processed by the same experimenter, in the same conditions and with the same lot for each reagent. In the case of human fresh tissue samples, a specific care will have to be taken while planning sample transportation from the patient to the experimenter. Indeed, it is critical to preserve samples in identical media and containers and to keep the time elapsed between sample harvesting and processing as consistent as possible across patients. Moreover, it was recently shown that antibody cocktails could be cryopreserved and stored at temperatures below −80°C for at least 9 months, without affecting the staining patterns (Schulz et al., 2019). Using an aliquot of a unique mix of antibodies should reduce experimental variability and improve data consistency.

9.2 Management of inter-sample experimental variability with barcoding

Experimental variability can be further limited by staining several "barcoded" samples at once. Barcoding consists in permanently labeling distinct samples with a unique combination of isotopes or a unique isotope. This procedure allows to pool individual samples for subsequent simultaneous processing, thus reducing reagent consumption, saving time and importantly, avoiding tube-to-tube technical variations during the cell staining procedure (such as antibody cocktail volume discrepancies) as well as during sample acquisition. Moreover, barcoding and pooling samples eliminates sample-to-sample carry-over, *i.e.*, contamination from a tube to another during acquisition. Thus, it removes the need to clean the cytometer between samples, thereby reducing CyTOF® runtime and limiting loss of cells within the instrument.

Bodenmiller et al. developed the first mass cytometry barcoding method using seven unique lanthanide isotopes to generate 128 possible combinations (Bodenmiller et al., 2012). Nonetheless, ideal isotopes for barcoding are outside of, and have minimal overlap with, analytical channels. Accordingly, Fluidigm commercializes a *Cell-ID 20-Plex Pd Barcoding Kit* that uses palladium isotopes combined in 20 different barcodes. Palladium atomic masses (comprised between 102 and 110) are far from the lanthanide mass range that is typically reserved for the analysis panel (Table 1).

When coupled with sample cryopreservation, the barcoding method enables the multiplexing of numerous samples, thus eliminating inter-assay variability over independent experiments. Indeed, immune cells isolated from each patient sample can be cryopreserved and, when a sufficient number of samples is reached, they can all be thawed at the same time, individually barcoded, and pooled together to be simultaneously stained in a single experiment. However, cryopreservation can significantly reduce the level of expression of multiple markers, including important markers of the tumor immune microenvironment. For example, it was shown that cryopreservation induced a profound decrease of the expression of CD62L (Costantini et al., 2003), of MDSC (myeloid-derived suppressor cell) markers (Kadic, Moniz, Huo, Chi, & Kariv, 2017) and of both PD-1 and PD-L1 (Campbell et al., 2009). Consequently, in order to most accurately phenotype tumor infiltrating immune cells, we recommend to stain them directly after their isolation (or stimulation, when required).

Moreover, classical mass cytometry barcoding methods, as the ones cited above, rely on intracellular covalent labeling of cells that involves previous cell fixation and permeabilization. Therefore these barcoding methods are not compatible with the staining of surface markers that are sensitive to fixation and/or permeabilization (Dzangue-Tchoupou, Corneau, Blanc, Benveniste, & Allenbach, 2018).

Extracellular antigen staining barcoding is an attractive alternative which is compatible with the detection of surface markers in their native conformation, on unfixed cells. The cell-surface molecules labeled for barcoding have to be expressed by all the cells of interest, and at a stable and sufficient level, under various experimental conditions. CD45, an ubiquitous marker for hematopoietic cells, is a suitable barcode target to study immune cells (Lai, Ong, Li, & Albani, 2015). Unique combinations of differently tagged CD45 antibodies can be used to barcode numerous samples (Hartmann, Simonds, & Bendall, 2018; Mei et al., 2015). However, mixing CD45 conjugates might result in decreased signal intensities in each barcoding channel. Since these channels are usually in low sensitivity ranges, this might lead to the loss of detectability of $CD45^+$ immune cells during the analysis. Barcoding with unique isotopes limits the number of samples that can be multiplexed, but is sufficient to eliminate the experimental variability due to discrepancies between compositions of the different tissue samples from a single patient that are processed the same day. Indeed, each cell sample can be individually stained with a unique monoisotopic anti-CD45 antibody and samples from the patient can be pooled to be processed together (Lavin et al., 2017).

9.3 Management of the batch effect with a standard sample

Because tissue samples from different patients are harvested on different days, prospective studies on human tumors are by nature subjected to batch effect. Indeed, variation of the experimental conditions between individual experiments cannot be completely avoided.

Nevertheless, batch effect can be controlled during data analysis if a unique standard control was stained and run along with patient samples during every individual experiment (Kleinsteuber et al., 2016). To study immune signatures, a reliable reference sample is frozen PBMCs that were extracted from a large volume of blood from a single healthy donor (or blood pooled from several donors) and that were stimulated in order to express all the markers of interest. At the beginning of each experiment, an aliquot of this standard control should be defrosted to be individually barcoded and pooled with the barcoded patient samples, so that all these samples can then be simultaneously stained and run on a mass cytometer.

9.4 Management of the experimental variations due to instrument performance

Variations of instrument performance, from a run to another, can greatly be limited by respecting the procedure to daily tune the mass cytometer (see Note 3). Besides, the drift of instrument performance that occurs during long runs of thick cell suspension, as it is the case for tumor samples, can be limited by periodic recalibration of the mass cytometer.

Additionally, to enable correction of the remaining signal variability, calibration beads have to be acquired along with the samples. Fluidigm provides *EQ™ Four Element Calibration Beads* that contain lanthanide isotopes that cover the whole mass range and a method (see Note 4) to correct for instrument performances fluctuations both over multiple days of data acquisition and over a single run.

10. Troubleshooting for sample acquisitions on a CyTOF® instrument

Obstruction of the fluidic is the most common problem that occurs during acquisition of samples. To prevent debris and cells (especially adherent cells) from aggregating or attaching to the piping system, cells should be resuspended in *MaxPar® Cell Water*, *Milli Q water* or preferentially *Maxpar® Cell Acquisition Solution* (a specialized operating buffer, Fluidigm), complemented with 2mM EDTA. Moreover, cell suspensions must systematically be filtered on a 40m-cell strainer prior to acquisition. In addition, samples containing more debris and dead cells should be more diluted and run at a slower rate. By preventing fluidic obstruction, these procedures limit loss of cells trapped in the clogs and decrease background noise.

When the flow is blocked by debris within the nebulizer spray, inject a 10% Contrad® 70/10% methanol-solution inside the nebulizer using a *Pearl Nebulizer Cleaner* (Meinhard®).

Another possible issue is the sample carry-over that can happen when the instrument is insufficiently washed between different samples. To discard all potential residues from a sample, we recommend to clean the piping system after each sample with a solution that contains 0.02% hydrofluoric acid or 3% nitric acid. The instrument has then to be rinsed with MilliQ water.

11. Processing of mass cytometry data

Data acquired with a CyTOF® will be first encoded in IMD (integrated mass data) files and then exported in FCS (flow cytometry standard) format files to be treated with suitable softwares. The mathematical correction algorithm in the CyTOF® software (See Note 4) will allow for a first normalization of the data according to the calibration beads (*EQ™ Four Element Calibration Beads*). Then, a "pre-gating" workflow that usually relies on manual gating will consist in the exclusion of the calibration beads (thanks to their unique isotope signature, Table 1) and of the undesired metal contamination, as well as in the selection of live cells (negative for the viability marker, positive for *Cell-ID™ Intercalator-Ir*), non-doublets (cell doublets have twice as much of *Cell-ID™ Intercalator-Ir* as single cells). Furthermore, FCS files may necessitate computational treatment involving de-barcoding and normalization with standard sample.

Describing methods for mass cytometry data analysis is beyond the scope of this chapter and has been reviewed elsewhere (Diggins, Greenplate, Leelatian, Wogsland, & Irish, 2017; Weber & Robinson, 2016).

12. Concluding remarks

This chapter offers guidance to properly plan and perform experimental procedures of a mass cytometry-phenotyping of immune cell infiltrating human solid tumors, by explaining the inherent technical challenges and helping investigators to overcome them with solutions adapted to the need of their experiments. It is clear that, despite technically challenging, this method holds enormous potential to unravel the complexity of the tumor-associated immune landscape. Only a highly detailed characterization of the tumor immune contexture, such as that achievable by CyTOF®, could shed new light on the anti- or pro-tumoral role of multiple immune components, ultimately leading to important improvements in cancer patient care management.

13. Notes

1. Fig. 2 of the Fluidigm "Maxpar Panel Designer, user guide" (https://www.fluidigm.com/binaries/content/documents/fluidigm/search/hippo%3Aresultset/maxpar-panel-designer/fluidigm%3Afile) illustrates the three types of signal overlap in mass cytometry.
2. Lanthanide isotope purity and oxide matrix is provided as the Fig. 2 of the "Guidelines for Mass Cytometry Panel Design" (https://med.virginia.edu/flow-cytometry-facility/wp-content/uploads/sites/170/2015/10/CyTOF-Panel-Design-Guidelines.pdf). The typical values of isotopic enrichment and spillover impurity for a broader spectrum of isotopes is given in Fig. 2 from "Metal-isotope-tagged monoclonal antibodies for high-dimensional mass cytometry" (Han et al., 2018).
3. The mass cytometer has to be set before each run according to a procedure provided by Leipold and Maecker (2012). This minimizes the generation of oxides and allows for the stabilization of the instrument performances overtime.
4. Fluidigm provides various support material, such as:
   - *MaxPar Panel Designer* application.
   - *Labeling Antibodies with Metals Using Maxpar Kits* protocols.
   - Several cell staining protocols.
   - User manuals for the different generations of CyTOF.
   - *Normalization of Mass Cytometry Data using EQ™ Four Element Calibration Beads* user guide.

   Note that an account should be created to gain access to some of these material (http://www.dvssciences.com/mydvs/).
5. To limit contamination of cells by the iodine contained in *Ficoll*, one should avoid to mix cells with the layer of density gradient medium by using SepMate™ or Leucosep™, or otherwise, by layering the cell suspension on the *Ficoll* particularly slowly and gently, by recovering the cells very carefully and by adding an extra step of washing.
6. At the steps 2.5 and 4.6 of the protocol shown in Table 3, we advise to use the same medium that will be employed for the following step of the procedure.
7. As *Cell-ID™ Cisplatin* can bind to serum proteins and BSA that complement common cell staining buffers, it is important to use it diluted in pure PBS and to wash cells in pure PBS prior to this staining.

References

Autengruber, A., Gereke, M., Hansen, G., Hennig, C., & Bruder, D. (2012). Impact of enzymatic tissue disintegration on the level of surface molecule expression and immune cell function. *European Journal of Microbiology & Immunology*, *2*(2), 112–120. https://doi.org/10.1556/EuJMI.2.2012.2.3.

Behbehani, G. K., Bendall, S. C., Clutter, M. R., Fantl, W. J., & Nolan, G. P. (2012). Single-cell mass cytometry adapted to measurements of the cell cycle. *Cytometry Part A*, *81*(7), 552–566. https://doi.org/10.1002/cyto.a.22075.

Behbehani, G. K. (2019). Immunophenotyping by Mass Cytometry. *Methods Mol Biol*, *2032*, 31–51. https://doi.org/10.1007/978-1-4939-9650-6_2.

Bodenmiller, B., Zunder, E. R., Finck, R., Chen, T. J., Savig, E. S., Bruggner, R. V., et al. (2012). Multiplexed mass cytometry profiling of cellular states perturbed by small-molecule regulators. *Nature Biotechnology*, *30*(9), 858–867. https://doi.org/10.1038/nbt.2317.

Campbell, D. E., Tustin, N. B., Riedel, E., Tustin, R., 3rd, Taylor, J., Murray, J., et al. (2009). Cryopreservation decreases receptor PD-1 and ligand PD-L1 coinhibitory expression on peripheral blood mononuclear cell-derived T cells and monocytes. *Clinical and Vaccine Immunology*, *16*(11), 1648–1653. https://doi.org/10.1128/CVI.00259-09.

Chevrier, S., Crowell, H. L., Zanotelli, V. R. T., Engler, S., Robinson, M. D., & Bodenmiller, B. (2018). Compensation of signal spillover in suspension and imaging mass cytometry. *Cell Systems*, *6*(5), 612–620. e615. https://doi.org/10.1016/j.cels.2018.02.010.

Chevrier, S., Levine, J. H., Zanotelli, V. R. T., Silina, K., Schulz, D., Bacac, M., et al. (2017). An immune atlas of clear cell renal cell carcinoma. *Cell*, *169*(4), 736–749. e718. https://doi.org/10.1016/j.cell.2017.04.016.

Costantini, A., Mancini, S., Giuliodoro, S., Butini, L., Regnery, C. M., Silvestri, G., et al. (2003). Effects of cryopreservation on lymphocyte immunophenotype and function. *Journal of Immunological Methods*, *278*(1–2), 145–155.

Diggins, K. E., Greenplate, A. R., Leelatian, N., Wogsland, C. E., & Irish, J. M. (2017). Characterizing cell subsets using marker enrichment modeling. *Nature Methods*, *14*(3), 275–278. https://doi.org/10.1038/nmeth.4149.

Dzangue-Tchoupou, G., Corneau, A., Blanc, C., Benveniste, O., & Allenbach, Y. (2018). Analysis of cell surface and intranuclear markers on non-stimulated human PBMC using mass cytometry. *PLoS One*, *13*(3), e0194593. (doi:ARTN) https://doi.org/10.1371/journal.pone.0194593.

Frei, A. P., Bava, F. A., Zunder, E. R., Hsieh, E. W., Chen, S. Y., Nolan, G. P., et al. (2016). Highly multiplexed simultaneous detection of RNAs and proteins in single cells. *Nature Methods*, *13*(3), 269–275. https://doi.org/10.1038/nmeth.3742.

Galon, J., & Bruni, D. (2019). Approaches to treat immune hot, altered and cold tumours with combination immunotherapies. *Nature Reviews. Drug Discovery*, *18*, 197–218. https://doi.org/10.1038/s41573-018-0007-y.

Garaud, S., Gu-Trantien, C., Lodewyckx, J. N., Boisson, A., De Silva, P., Buisseret, L., et al. (2014). A simple and rapid protocol to non-enzymatically dissociate fresh human tissues for the analysis of infiltrating lymphocytes. *Journal of Visualized Experiments*, *6*(94), e52392. https://doi.org/10.3791/52392.

Han, G., Spitzer, M. H., Bendall, S. C., Fantl, W. J., & Nolan, G. P. (2018). Metal-isotope-tagged monoclonal antibodies for high-dimensional mass cytometry. *Nature Protocols*, *13*(10), 2121–2148. https://doi.org/10.1038/s41596-018-0016-7.

Hartmann, F. J., Simonds, E. F., & Bendall, S. C. (2018). A universal live cell barcoding-platform for multiplexed human single cell analysis. *Scientific Reports*, *8*(1), 10770. https://doi.org/10.1038/s41598-018-28791-2.

Kadic, E., Moniz, R. J., Huo, Y., Chi, A., & Kariv, I. (2017). Effect of cryopreservation on delineation of immune cell subpopulations in tumor specimens as determinated by multiparametric single cell mass cytometry analysis. *BMC Immunology*, *18*(1), 6. https://doi.org/10.1186/s12865-017-0192-1.

Kleinsteuber, K., Corleis, B., Rashidi, N., Nchinda, N., Lisanti, A., Cho, J. L., et al. (2016). Standardization and quality control for high-dimensional mass cytometry studies of human samples. *Cytometry Part A*, *89*(10), 903–913. https://doi.org/10.1002/cyto.a.22935.

Lai, L., Ong, R., Li, J., & Albani, S. (2015). A CD45-based barcoding approach to multiplex mass-cytometry (CyTOF). *Cytometry Part A*, *87*(4), 369–374. https://doi.org/10.1002/cyto.a.22640.

Lavin, Y., Kobayashi, S., Leader, A., Amir, E. D., Elefant, N., Bigenwald, C., et al. (2017). Innate immune landscape in early lung adenocarcinoma by paired single-cell analyses. *Cell*, *169*(4), 750–765, e717. https://doi.org/10.1016/j.cell.2017.04.014.

Leelatian, N., Doxie, D. B., Greenplate, A. R., Sinnaeve, J., Ihrie, R. A., & Irish, J. M. (2017). Preparing viable single cells from human tissue and tumors for cytomic analysis. *Current Protocols in Molecular Biology*, *118*, 25C-21-21–25C-21-23. https://doi.org/10.1002/cpmb.37.

Leipold, M. D., & Maecker, H. T. (2012). Mass cytometry: Protocol for daily tuning and running cell samples on a CyTOF mass cytometer. *Journal of Visualized Experiments*, *2*(69), e4398. https://doi.org/10.3791/4398.

Mei, H. E., Leipold, M. D., Schulz, A. R., Chester, C., & Maecker, H. T. (2015). Barcoding of live human peripheral blood mononuclear cells for multiplexed mass cytometry. *Journal of Immunology*, *194*(4), 2022–2031. https://doi.org/10.4049/jimmunol.1402661.

Olsen, L. R., Leipold, M. D., Pedersen, C. B., & Maecker, H. T. (2019). The anatomy of single cell mass cytometry data. *Cytometry Part A*, *95*(2), 156–172. https://doi.org/10.1002/cyto.a.23621.

Rahman, A. H., Tordesillas, L., & Berin, M. C. (2016). Heparin reduces nonspecific eosinophil staining artifacts in mass cytometry experiments. *Cytometry A*, *89*(6), 601–607. https://doi.org/10.1002/cyto.a.22826.

Schulz, A. R., Baumgart, S., Schulze, J., Urbicht, M., Grützkau, A., & Mei, H. E. (2019). Stabilizing Antibody Cocktails for Mass Cytometry. *Cytometry A*, *95*(8), 910–916. https://doi.org/10.1002/cyto.a.23781.

Stern, A. D., Rahman, A. H., & Birtwistle, M. R. (2017). Cell size assays for mass cytometry. *Cytometry A*, *91*(1), 14–24. https://doi.org/10.1002/cyto.a.23000.

Tricot, S., Meyrand, M., Sammicheli, C., Elhmouzi-Younes, J., Corneau, A., Bertholet, S., et al. (2015). Evaluating the efficiency of isotope transmission for improved panel design and a comparison of the detection sensitivities of mass cytometer instruments. *Cytometry. Part A: The Journal of the International Society for Analytical Cytology*, *87*(4), 357–368. https://doi.org/10.1002/cyto.a.22648.

Weber, L. M., & Robinson, M. D. (2016). Comparison of clustering methods for high-dimensional single-cell flow and mass cytometry data. *Cytometry. Part A: The Journal of the International Society for Analytical Cytology*, *89a*(12), 1084–1096. https://doi.org/10.1002/cyto.a.23030.

van Vreden, C., Niewold, P., McGuire, H. M., & Fazekas de St Groth, B. (2019). Titration of Mass Cytometry Reagents. *Methods Mol Biol*, *1989*, 83–92. https://doi.org/10.1007/978-1-4939-9454-0_6.

CHAPTER NINETEEN

Isolation and characterization of immune cells from the tumor microenvironment of genetically engineered pediatric high-grade glioma models using the sleeping beauty transposon system

Maria Belen Garcia-Fabiani[a,b], Andrea Comba[a,b], Padma Kadiyala[a,b], Santiago Haase[a,b], Felipe Javier Núñez[a,b], David Altshuler[a,b], Pedro Ricardo Lowenstein[a,b], Maria Graciela Castro[a,b,]*

[a]Department of Neurosurgery, University of Michigan Medical School, MSRB II, Ann Arbor, MI, United States

[b]Department of Cell and Developmental Biology, University of Michigan Medical School, MSRB II, Ann Arbor, MI, United States

*Corresponding author: e-mail address: mariacas@med.umich.edu

Contents

Abstract

Gliomas are the most common malignant brain tumors in the pediatric population. Even though great efforts have been made to understand their distinctive molecular characteristics, there has not been any improvements in the median survival in decades. In children, high-grade glial tumors have a median survival of 9–15 months. It has

Methods in Enzymology, Volume 632
ISSN 0076-6879
https://doi.org/10.1016/bs.mie.2019.05.023

recently been demonstrated that pediatric high-grade gliomas (pHGG) are biologically and molecularly different from the adult counterparts, which could explain why conventional treatments universally fail. The development of an in vivo pHGG model harboring the specific genetic alterations encountered in pediatric gliomas is imperative in order to study the molecular basis that drives the progression and aggressiveness of these tumors. It would also enable harnessing these results for the development of novel therapeutic approaches. Our lab has implemented a method to induce brain tumors using transposon-mediated integration of plasmid DNA into cells of the subventricular zone of neonatal mouse brain. One of the main advantages of this method is that tumors are induced by altering the genome of the host cells, allowing us to recapitulate the salient features of the human disease. In this chapter we describe a method to isolate two cell populations from tumors generated in situ in mice, i.e., one population enriched in tumor cells and another population enriched in CD45+ cells. We also present methodologies as to how tumor infiltrating immune cells can be phenotypically characterized using flow cytometry.

1. Introduction

1.1 Pediatric high-grade gliomas

High-grade gliomas in children represent 15–20% of all brain tumors and the leading cause of cancer-related mortality in this age group (Braunstein, Raleigh, Bindra, Mueller, & Haas-Kogan, 2017; Faury et al., 2007). These tumors include anaplastic astrocytoma (World Health Organization (WHO) grade III), glioblastomas (WHO grade IV) and diffuse intrinsic pontine glioma (MacDonald, Aguilera, & Kramm, 2011). Seventy to ninety percent of patients die within 2 years of diagnosis (Paugh et al., 2010) and prognosis has remained largely unchanged despite novel treatment approaches (Braunstein et al., 2017; MacDonald et al., 2011). The spectrum of glioma in pediatric patients is distinctly different than that of adults, suggesting unique pathogenesis and disease progression (Paugh et al., 2010). In children, low grade juvenile pilocytic astrocytomas are the most common brain tumor type (Paugh et al., 2010). Low grade gliomas in the pediatric population rarely progress to a higher grade tumor (Braunstein et al., 2017; Broniscer et al., 2007), whereas the opposite is true in adults where glioblastoma is more common and low grade diffuse gliomas frequently undergo anaplastic progression to a high-grade tumor over time (Furnari et al., 2007; Paugh et al., 2010).

The standard of care for pediatric high-grade glioma (pHGG) includes maximal safe surgical resection followed by focal radiation therapy in children usually older than 3 years of age due to cognitive and developmental sequelae from radiation in childhood (MacDonald et al., 2011). The most important prognostic factors are extent of resection and histologic grade (Gajjar et al., 2015). Adjuvant chemotherapeutics used for adult glioma, such as temozolomide, have performed with disappointing results in the pediatric population (Cohen, Heideman, et al., 2011; Cohen, Pollack, et al., 2011; Lashford et al., 2002). Moreover, results from adult clinical trials cannot be extrapolated to children due to the unique differences in pathogenesis and molecular biology between adult and pHGG (Jones et al., 2017). Important molecular signatures in pHGG include the novel oncogenic driver mutations in histones H3.1 (K27) and H3.3 (K27 and G34) as well as in the activin A receptor, type 1 (ACVR1) (Gajjar et al., 2015; Mackay et al., 2017).

1.2 Pediatric high-grade gliomas harboring H3.3-G34R/V mutation

Recently, high-throughput genomic and epigenomic studies have permitted the classification of pediatric glioma subtypes based on molecular features which, in combination with other characteristics such as clinical data, revealed distinct biological subgroups (Sturm et al., 2014). The analysis of high-throughput data emerging from pHGG samples reveal a number of mutations in genes associated with the regulation of epigenetic processes (Schwartzentruber et al., 2012; Sturm et al., 2012; Wu et al., 2012). In eukaryotes, DNA is packaged around a core octamer of histone proteins which, besides conferring structural integrity to the chromatin, are subjected to posttranslational modifications (PTM) on their N-terminal tail as a mean to regulate the chromatin compaction and gene expression (Jenuwein & Allis, 2001; Strahl & Allis, 2000). Accordingly, mutations in the histone genes themselves or the genes encoding the proteins performing the PTMs have been identified as oncogenic drivers in different types of cancer (Funato & Tabar, 2018).

Two independent studies in 2012 identified mutations in the histone genes H3.1 (*HIST1H3B*) and H3.3 (*H3F3A*) in a large proportion of pHGG (Schwartzentruber et al., 2012; Wu et al., 2012). Mutations in the coding regions of the histone tails at lysine 27 to methionine (K27M) in H3.1 and H3.3 were found in diffuse midline glioma, in children with a median age of diagnosis of 9 years, while mutations in H3.3 at glycine 34 to arginine or valine (G34R/V) were found in about 15% of cerebral hemispheric HGG of children and young adults (Mackay et al., 2017).

It has been demonstrated that K27M mutations exhibit a dominant inhibitory effect on the polycomb repressive complex 2 (PRC2), by blocking the active site of enhancer of zeste homolog 2 (EZH2) (Nagaraja et al., 2017; Piunti et al., 2017). These findings were important for the conduct of drug testing experiments and specific clinical trials (Grasso et al., 2015; Jansen, Van Vuurden, Vandertop, & Kaspers, 2012). However, the exact mechanisms by which H3.3-G34R/V drive malignancy of pHGGs remain largely unknown. A more recent meta-analysis showed that most of the pHGG in the H3.3-G34R/V subgroup harbor inactivating mutations in *TP53* and *ATRX* genes (Schwartzentruber et al., 2012; Sturm et al., 2012). Notably, ATRX performs a role as H3.3 depositor (Voon & Wong, 2016), suggesting a positive interaction between these two mutations in cancer (Haase et al., 2018).

H3.3-G34R/V has been shown to lead to a distinctive binding of the H3K36 trimethylation mark throughout the genome (Bjerke et al., 2013). For example, a differential binding of the activating H3K36me3 mark to upregulate genes associated with cortical precursors and the oncogene MYCN has been observed in H3.3-G34V glioma cells (Bjerke et al., 2013). Moreover, this mutation has been associated to a local down-regulation of H3K36me3 at specific gene loci (Lewis et al., 2013). It has been suggested that a mechanism by which SETD2, the methyltransferase responsible for generating the H3K36me3 mark, is locally inhibited by G34R/V mutations, leads to reduced levels of the K36me3 mark in *cis* (Lewis et al., 2013). In conclusion, despite recent advances in the understanding of the H3.3-G34R/V pHGG biology, it is still imperative to develop accurate pre-clinical models in which to implement mechanistic studies that will guide the translation of research findings to the clinical arena.

2. Sleeping beauty transposase-mediated mouse glioma model

Pediatric high-grade glioma (pHGG) is a highly heterogeneous disease entity. Basic and translational research models that accurately recapitulate the genetic aberrations and tumor microenvironment in pHGG are necessary for effective pre-clinical studies. The Sleeping Beauty (SB) transposase method was developed to induce intrinsic brain tumors in mice with non-viral, transposon-mediated integration of DNA into cells of the sub-ventricular zone of neonatal 1-day old mouse pups (Calinescu et al., 2015; Koschmann et al., 2016; Nunez et al., 2019; Wiesner et al., 2009) (see Note 1). Since the SB transposase is encoded in the same plasmid as the Luciferase gene (pT2/SB100x-Luc), and the latter gets inserted into the host genome, the plasmid uptake and disease progression can be monitored using bioluminescence (Calinescu et al., 2015; Koschmann et al., 2016; Nunez et al., 2019). Moreover, the SB plasmids encode not only for the specific genetic alterations, but also for different fluorescent proteins, so that the tumors can be recognized macroscopically using a stereo-zoom microscope equipped with a fluorescent light (Calinescu et al., 2015; Koschmann et al., 2016; Nunez et al., 2019). This model represents a valuable tool to induce tumors from cells original to the animal and to assess the role of candidate genes in the induction and progression of tumors. Also, this model can be used to test novel therapies, without the need of invasive surgical procedures or the risk of cell rejection by the host immune system.

2.1 Sleeping beauty transposase pHGG model harboring H3.3-G34R mutation

1. Prepare the injection solution by mixing the plasmid DNA with the *in vivo* transfection reagent (Polyethylenimine, PEI) (in vivo-Jet PEI, Polyplus, New York, NY, USA) to create the PEI/DNA complexes for transfection. In our case, we used a 1:2 ratio for the injection of the pT2/SB100x-Luc plasmid and the following plasmids encoding:

H3.3-WT group (control group)	H3.3-G34R (case group)
NRAS(V12) oncogene	NRAS(V12) oncogene
shRNA against *Tp53*	shRNA against *Tp53*
shRNA against *Atrx*	shRNA against *Atrx*
	H3.3-G34R

To generate the PEI/DNA complexes for in vivo transfection, prepare:

a. DNA solution: add 2.85 μg of pT2/SB100x-Luc, 5.75 μg of each plasmid, 10 μL of 10% glucose solution and sterile water to a final volume of 20 μL.
b. In vivo-Jet PEI solution: for the H3.3-WT group, add 2.8 μL of in vivo-Jet PEI, 7.2 μL of sterile water, and 10 μL of 10% glucose. For the H3.3-G34R group, add 3.6 μL of in vivo-Jet PEI, 6.4 μL of sterile water, and 10 μL of 10% glucose (see Note 2).
c. Add the PEI solution to the DNA solution, mix by pipetting, vortex and incubate at room temperature for 20 min. The solution is now ready for injection.

2. Place a 10 μL syringe with a 30G hypodermic needle (12.5° bevel) into the micropump of the stereotaxic frame.
3. With a slurry of dry ice and alcohol, cool the neonatal stereotaxic stage to 2–8 °C.
4. To induce anesthesia, place the 1-day old mouse pup wrapped in gauze on wet ice for 2 min (see Note 3).
5. Fill the syringe with the solution with the PEI/DNA complexes for the in vivo transfection.
6. Immobilize the pup in the stereotaxic frame, ensuring that the dorsal side of the skull is horizontal, parallel to the surface of the frame and wipe the head with 70% ethanol.
7. Lower the needle and adjust stereotaxic coordinates until the needle touches the midline intersection of the parietal and occipital bones (lambda).
8. Lift the needle and adjust stereotaxic coordinates to 0.8 mm lateral and 1.5 mm rostral to the lambda.
9. Lower the needle in dorsal-ventral direction, until it touches and slightly dimples the skin and lower the needle another 1.5 mm (see Note 4).
10. Using the automatic injector, inject 0.75 μL of the transfection solution at a rate of 0.5 μL/min.
11. Keep the pup on the frame for another minute and, in the meantime, anesthetize another pup, for the next injection.
12. Lift the needle gently and slowly, and place the pup under a heating lamp. Monitor breathing and activity ensuring first breathing.
13. After 3–5 min, when the pup is warm, looks rosy in color, is breathing regularly and active, return it to its mother (see Notes 5 and 6). Continue with the procedure until all pups have been injected.

3. Mouse perfusion and tumor dissection

1. Two to six weeks after SB injections animals will start developing macroscopic tumors. Monitor tumor progression in mice every week by bioluminescence imaging and look for signs of intracranial tumor burden (see Note 7).
2. When a mouse displays signs of tumor burden, anesthetize it with an intra-peritoneal injection of 120 mg/kg of Ketamine and 0.5 mg/kg of Dexmedetomidine. Examine the animal by toe pinch and tail pinch. When no responses are observed in both tests, consider that the animal is in a deep anesthetic state.

3. Immobilize the mouse on a dissection board, open the abdominal cavity by an incision, dissect the diaphragm and open the thoracic cavity completely by cutting the ribs to visualize the heart.
4. Insert a blunt 20-gauge cannula into the left ventricle of the heart. Connect the cannula with plastic tubing passing through a peristaltic pump to a container with oxygenated and heparinized Tyrode's solution (0.8% NaCl, 0.0264% $CaCl_2$ $2H_2O$, 0.005% NaH_2PO_4, 0.1% glucose, 0.1% $NaHCO_3$, 0.02% KCl). Tyrode's solution may be kept at 4°C for up to 1 week.
5. Make a small incision into the right atrium of the heart to permit the drain of blood and Tyrode's. Perfuse the mouse for 10 min with cold Tyrode's solution.
6. Decapitate the mouse and carefully dissect the brain from the skull. Remove musculature and skeletal matter from around the cranium. Remove the cranium by first flicking off the occipital, then temporal, then frontal cranium.
7. Remove the brain from skull and place it in a petri dish on ice.
8. Place the brain under a stereo-zoom microscope equipped with a fluorescent light illumination system. Dissect the tumor from the brain using scalpels.
9. Place the tumor pieces immediately into a 50 mL conical tube with 10 mL of cold Neurosphere (NS) media (DMEM/F12 supplemented with anti-anti, B27 supplement (1 ×), N2 supplement (1 ×) (all from Gibco, Thermo Scientific, Waltham, MA, USA), Normocin (100 μg/mL) (InvivoGen, San Diego, CA, USA), EGF and FGF (20 ng/mL) (PeproTech, Rocky Hill, NJ, USA)).

4. Tumor homogenization and magnetic labeling of CD45+ cells

1. For tumor homogenization, transfer the dissected tumor sections onto a 70 μm nylon strainer. Use a pestle to disaggregate the tumor and obtain a single-cell suspension. Add 10–15 mL of cold NS media through strainer to wash it and to resuspend the cells. Repeat the washing 3 times.
2. Centrifuge the cells at 300 × *g* for 4 min at room temperature.
3. For red blood cell lysis, remove the supernatant and add 1 mL of cold Red Blood Cell Lysis Buffer 1 × (BioLegend, San Diego, CA, USA) to the cells. Incubate 1 min on ice. Immediately add 9 mL of Dulbecco's

Phosphate Buffered Saline 1× (DPBS) (Gibco, Thermo Scientific, Waltham, MA, USA) to stop the red blood cell lysis reaction (see Note 8).

4. Centrifuge the cells at 300 × *g* for 5 min at 4 °C.

For these next steps work quickly, keep the cells cold at all times, and use pre-cooled solutions. This will prevent capping of antibodies on the cell surface and non-specific cell labeling.

5. Completely remove media and resuspend the cells in 90 μL of cold BSA-EDTA buffer (0.5% Bovine serum albumin (BSA), 2 mM Ethylenediaminetetraacetic acid (EDTA) in phosphate buffered saline pH 7.2) (Miltenyi Biotec, Bergisch Gladbach, Germany). Keep buffer cold (4–8 °C).
6. Add 10 μL of CD45 MicroBeads (Miltenyi Biotec, Bergisch Gladbach, Germany) per 10^7 total cells (see Notes 9 and 10).
7. Mix the cell suspension well and incubate for 15 min at 4 °C (not on ice).
8. Wash cells by adding 1 mL of cold BSA-EDTA buffer and centrifuge at 300 × *g* for 5 min at 4 °C.
9. Pipette off supernatant and resuspend cell pellet in 500 μL of cold BSA-EDTA buffer. Proceed to magnetic column separation.

5. Magnetic separation of tumor and CD45+ enriched cell fractions

1. Place MACS column in the magnetic field of a suitable MACS Separator (Miltenyi Biotec, Bergisch Gladbach, Germany) (see Note 11).
2. Precondition the column by rinsing with appropriate amount of BSA-EDTA buffer. Let all the solution go through the column (see Note 12).
3. Put a 30 μM nylon mesh on the top of the column and apply cell suspension onto it (see Note 13).
4. Collect unlabeled cells which flow through the column (500 μL for large columns). Wash the column once with BSA-EDTA buffer (3 mL for large columns) and collect the cells that eluate in the same tube. This is the first tumor cell-enriched eluate (CD45− cells).
5. Apply the first tumor cell-enriched eluate into a second preconditioned column with the filter on the top and collect the second tumor cell-enriched eluate. Wash once with BSA-EDTA buffer and collect the eluate in the same tube. If using large columns, the total volume should be approximately 6.5 mL. This is the tumor cell-enriched fraction.

6. Save 500 μL of the tumor cell-enriched fraction for flow cytometry analysis (keep them on ice or at 4 °C), centrifuge the rest for 5 min at 300 × *g* at 4 °C and discard the supernatant. The pellet can now be used for downstream applications; such as flow cytometry or RNA purification.
7. To collect the CD45+ cells that are magnetically adhered to the first column, first wash this column twice with BSA-EDTA buffer (see Note 14).
8. Remove the column from the magnetic field and place it on a suitable collection tube, such as a 15 mL tube.
9. Add BSA-EDTA buffer on top of the column and immediately flush out this fraction with the magnetically labeled CD45+ cells by firmly applying the plunger supplied with the column. In the case of large columns, 5 mL are used for the elution step.
10. Save 500 μL of the CD45+ cell fraction for flow cytometry analysis (keep them on ice or at 4 °C), centrifuge the rest for 5 min at 300 × *g* at 4 °C and discard the supernatant. The pellet can now be used for downstream applications; such as flow cytometry or RNA purification.

6. Flow cytometry analysis of cell fractions: CD45 and Ter119 assessment

1. Pellet all the samples that are going to be tested and resuspend them in 200 μL of 2% FBS in Hanks' Balanced Salt Solution (HBSS, no calcium, no magnesium, no phenol red) (Gibco, Thermo Scientific, Waltham, MA, USA) (flow buffer) (see Note 15).
2. Add 1 μL of Fc blocking (see Note 16).
3. Incubate on ice for 15 min.
4. Spin cells down at 300 × *g*, for 5 min at 4 °C.
5. Resuspend the pellets in 600 μL of flow buffer and divide them into 3 wells of a 96 V-bottom multiwell plate (200 μL per well) (see Note 17):
   a. Unstained control
   b. CD45 only
   c. Ter119 only
6. For b and c:
   a. Add 1 μL of anti-mouse CD45 antibody (BioLegend, San Diego, CA, USA) (see Note 18) to well b.
   b. Add 1 μL of anti-mouse Ter119 antibody (BioLegend, San Diego, CA, USA) (see Note 18) to well c.

7. Incubate on ice for 30 min, protected from light.
8. Centrifuge cells down at 300 × *g*, for 5 min at 4 °C.
9. Wash twice with 100 μL of flow buffer, centrifuging in between at 300 × *g*, for 5 min at 4 °C.
10. Resuspend all wells in 500 μL of flow buffer and pass the samples to a flow cytometry tube.
11. Before running the samples, add 2.5 μL of Propidium Iodide (PI) Staining Solution (Invitrogen, Carlsbad, CA, USA) for cell viability assessment (see Note 19). Record at least 20,000 events of the singlet population.

7. Flow cytometry analysis of the tumor microenvironment

1. Wash the cells that were purified from the brain tissue by magnetic sorting with 600 μL of flow buffer. Check cell viability by performing a cell count using trypan blue and a hemocytometer.
2. After the cell count, resuspend the cells at a concentration of 2.0×10^6 cells in 200 μL of flow buffer.
3. Plate the cells at a density of 1.5×10^6 cells in 150 μL volume in a 96 well V-bottom plate for immunostaining.
4. Combine the remaining 50 μL of cells from all samples and plate 150 μL of the resuspension in a 96 wells V-bottom multiwell plate for staining with the isotype control.
5. Pellet the cells by centrifuging the plate at 300 × *g* for 5 min at 4 °C.
6. Discard the supernatant and add 100 μL of Fc block diluted in flow buffer at a 1:200 ratio to each sample. Then, incubate the plate on ice for 15 min (see Note 16).
7. Centrifuge the plate at 300 × *g* for 5 min at 4 °C. Discard the supernatant and wash the cells 2 times with flow buffer before staining with the cell surface markers.
8. To identify various immune cells within the tumor microenvironment, prepare fluorophore conjugated primary antibody mixes with the combination of key markers listed in Table 1. Antibodies and isotype controls are diluted in flow buffer at a 1:200 ratio. Catalog numbers and company names of the antibodies used can be found in Table 2.
9. Add 100 μL of fluorophore conjugated primary antibody mix to the appropriate well and pipette up and down to mix.
10. Incubate the plate for 30 min on ice protected from light.

Table 1 Mouse surface and intracellular markers for the most common immune cell types.

Immune cell	Subtype	Mouse surface markers	Mouse intracellular markers
Macrophages		$CD45^+$, $F4/80^+$	–
	M1	$CD45^+$, $F4/80^+$, $CD206^-$	$IFN\gamma^+$
	M2	$CD45^+$, $F4/80^+$, $CD206^+$	$Arg1^+$
Dendritic cells		$CD45^+$, $CD11c^+$, MHC II^+	–
	Conventional	$CD45^+$, $CD11c^+$, MHC II^+, $B220^-$	–
	Plasmacytoid	$CD45^+$, $CD11c^+$, MHC II^+, $B220^+$	–
T cells		$CD45^+$, $CD3^+$	–
	Cytotoxic T cells	$CD45^+$, $CD3^+$, $CD8^+$	$IFN\gamma^+$, Granzyme B^+
	Helper T cells	$CD45^+$, $CD3^+$, $CD4^+$	–
NK cells	N/A	$CD45^+$, $CD3^-$, $NK1.1^+$	–
MDSC		$CD45^+$, $Gr1^+$, $CD11b^+$	–
	Granulocytic	$CD45^+$, $Gr1^+$, $CD11b^+$, $Ly6G^+$, $LygC^{lo}$	–
	Monocytic	$CD45^+$, $Gr1^+$, $CD11b^+$, $Ly6G^-$, $LygC^{hi}$	–

11. Centrifuge the plate at 300 × *g* for 5 min at 4 °C. Discard the supernatant and wash the cells 3 times with flow buffer.
12. Fix the cells by adding 100 μL of Cytofix/Cytoperm (BD Biosciences, San Jose, CA, USA) solution to each well and incubate the plate for 20 min on ice in the dark (see Note 20).
13. Wash the cells 2 times with 1 × Perm/Wash (BD Biosciences, San Jose, CA, USA) solution before staining with the intracellular markers (see Note 21).
14. To evaluate the expression of intracellular markers in the immune cells, prepare fluorophore conjugated primary antibody mix with the combination of key markers listed in Table 1. Antibodies and isotype controls are diluted in 1 × Perm/Wash solution at a 1:200 ratio.

Table 2 Catalog numbers and company names of the antibodies used.

Protein	Catalog number	Company
CD45 (Pacific blue conjugated)	103126	Biolegend
F4/80	123128	Biolegend
CD206	141734	Biolegend
CD11c	117319	Biolegend
MHCII	107614	Biolegend
B202	103223	Biolegend
CD3	100328	Biolegend
CD8	100722	Biolegend
CD4	100421	Biolegend
NK1.1	108708	Biolegend
Gr1	108412	Biolegend
CD11b	101225	Biolegend
Ly6G	127606	Biolegend
Ly6C	128018	Biolegend
IFNγ	113603	Biolegend
Arg1	IC5868P	R&D Systems
Granzime B	372205	Biolegend
Ter119	116232	Biolegend
CD45 (APC conjugated)	103112	Biolegend

15. Add 100 μL of fluorophore conjugated primary antibody intracellular mix to the appropriate well.
16. Incubate the plate for 30 min on ice protected from the light.
17. Centrifuge the plate at 300 × *g* for 5 min at 4 °C. Discard the supernatant and wash the cells 3 times with 1 × Perm/Wash solution.
18. Resuspend the cell pellets in 200 μL of 1 × Perm/Wash solution and transfer the cell suspension to flow tubes.
19. For flow cytometry acquisition, obtain at least 30,000 events of the singlet population.

8. Concluding remarks

This method represents a valuable tool to obtain two distinct cell populations from genetically engineered high-grade glioma models: one population enriched in tumor cells and the other enriched in immune cells. High-grade gliomas are generated intrinsically in 1-day old mouse pups through the injection of a combination of plasmids encoding for the desired genetic lesions in combination with the SB transposase (Fig. 1A and B). Two advantages of this method are that the plasmid uptake and tumor growth can be monitored using bioluminescence and the tumors can easily be identified and excised using a stereo-zoom microscope equipped with a fluorescent light (Fig. 1C). When a tumor bearing mouse displays signs of tumor burden (ataxia, impaired mobility, hunched posture, seizures, and scruffy fur), the mouse is perfused with Tyrode's solution. After extracting the brain from of the skull, the tumor is dissected to be processed following the described protocol, which is depicted in Fig. 2.

This protocol allowed us to isolate two cell populations, enriched in tumor or immune cells. By Flow Cytometry, we check the purity of these isolated populations, by assessing the quantity of CD45+ and Ter119+ cells within the live cell population (Fig. 3A and B). In both fractions, the proportion of viable cells is between 80% and 90%. In the tumor cell enriched fraction, the amount of CD45+ cells is 3% on average, whereas in the tumors without any separation the amount of CD45+ cell is between 10% and 15% in the case of the H3.3-WT or H3.3-G34R glioma mouse models (Fig. 4). On the contrary, we observed that the amount of CD45+ cells in the immune cell fraction is eightfold higher in comparison to the amount of CD45+ cells in the tumor homogenate without any enrichment (Fig. 4). The amount of Ter119+/CD45− cells (erythroid cells) is 0.4% on average in both cell fractions after the procedure (Fig. 3A and B).

In conclusion, this method allowed us to obtain a high yield separation of tumor cells from immune cells from a high-grade glioma mouse model, with a substantial amount of viable cells. These enriched cell fractions can be employed in downstream applications such as RNA purification, DNA purification, flow cytometry analysis, cell culture, and functional studies among others. This procedure can be also applied for isolating immune cells from tumor cells in any other tumor model.

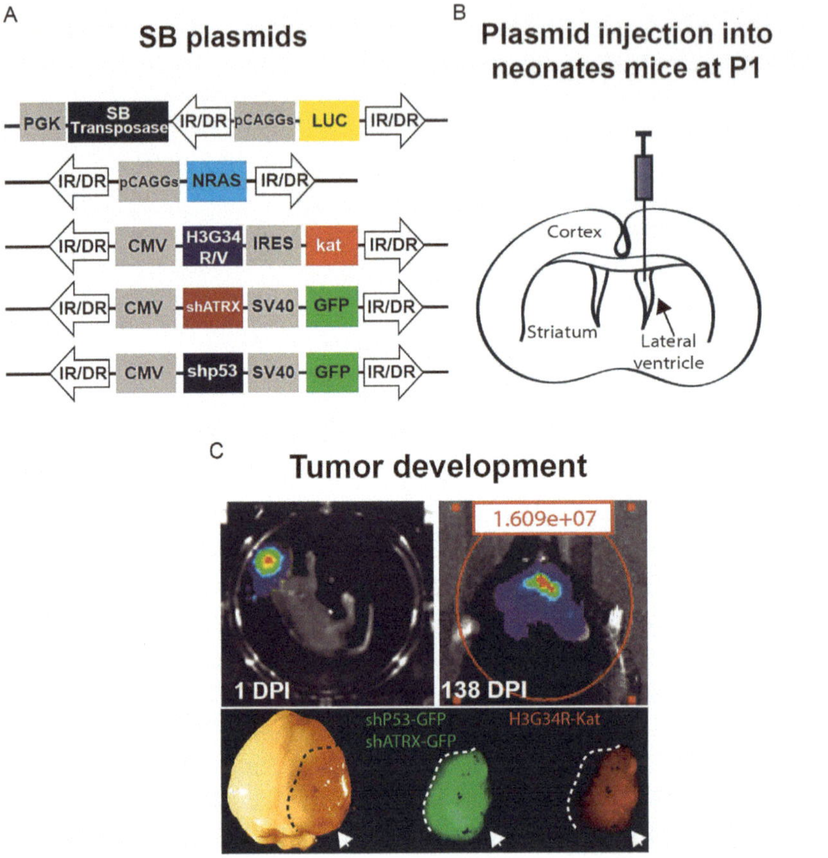

Fig. 1 Generation of a mouse glioma model using the Sleeping Beauty (SB) Transposase System. (A) Diagram representing the plasmid constructs encoding for the SB transposase with the luciferase and the genetic lesions to be inserted into the host cell genome via the SB transposase (SB plasmids). (B) Diagram indicating the brain ventricles and the site for plasmid injection (lateral ventricle), which contains brain stem cells from which the tumor is originated. (C) (Upper panel) Plasmid uptake efficiency 1 day after injection and tumor development monitoring by bioluminescence. (Lower panel) Tumor tissue (white arrows and dashed lines) is identified macroscopically by the expression of fluorophores associated with the respective genetic lesions. H3.3-G34R tumors express both green fluorescent protein (GFP) and Katushka protein (Kat) (a red fluorescent protein), since the shP53 and shATRX plasmids are coupled with GFP coding sequence, whereas the H3.3-G34R coding sequence is coupled with Kat sequence.

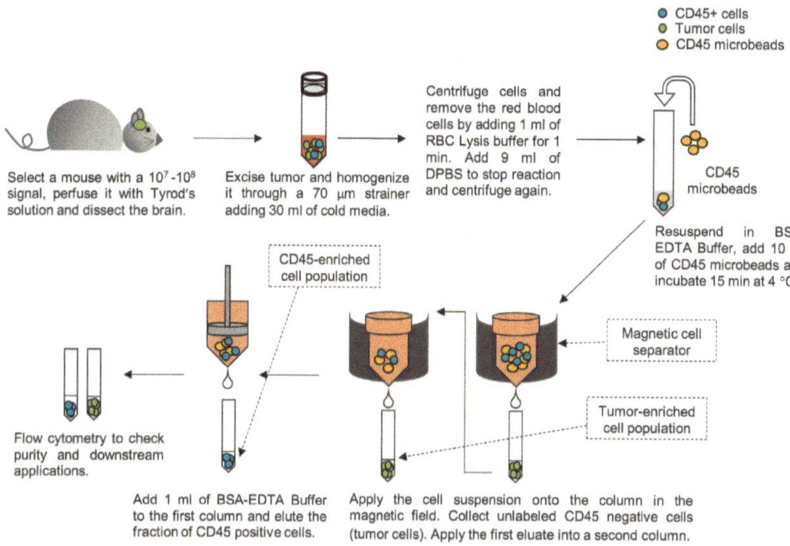

Fig. 2 Schematic summarizing the technique to obtain two cell fractions from a genetically engineered mouse glioma model. One cell fraction will be enriched in tumor cells and the other in CD45+ cells (immune cells).

9. Notes

1. For a more detailed protocol, including the maps of the plasmids used, the cloning strategy and the intra-ventricular injections, see Calinescu et al. (2015).
2. The optimal ratio of nitrogen residues in the polyethyleneimine (PEI) to the phosphate residues in the DNA (N/P ratio) is 7, which ensures the formation of positively charged particle complexes that will bind anionic moieties on cell surfaces and will be endocytosed. The in vivo jet-PEI solution has a concentration of 150 mM expressed as nitrogen residues. Given that 1 μg DNA has 3 nmol of anionic phosphate, the amount of transfection reagent necessary can be calculated with the formula: μL of in vivo jet-PEI = [(μg DNA × 3) × N/P ratio]/150.
3. Pups receive continuous anesthesia when placed on the chilled stereotaxic frame during the plasmid injection. Maintaining the temperature of the frame above freezing, between 2 and 8 °C, will prevent thermal burn injuries for the pups.
4. The needle will pierce the skin and skull, and pass through the cortex to enter the lateral ventricle.

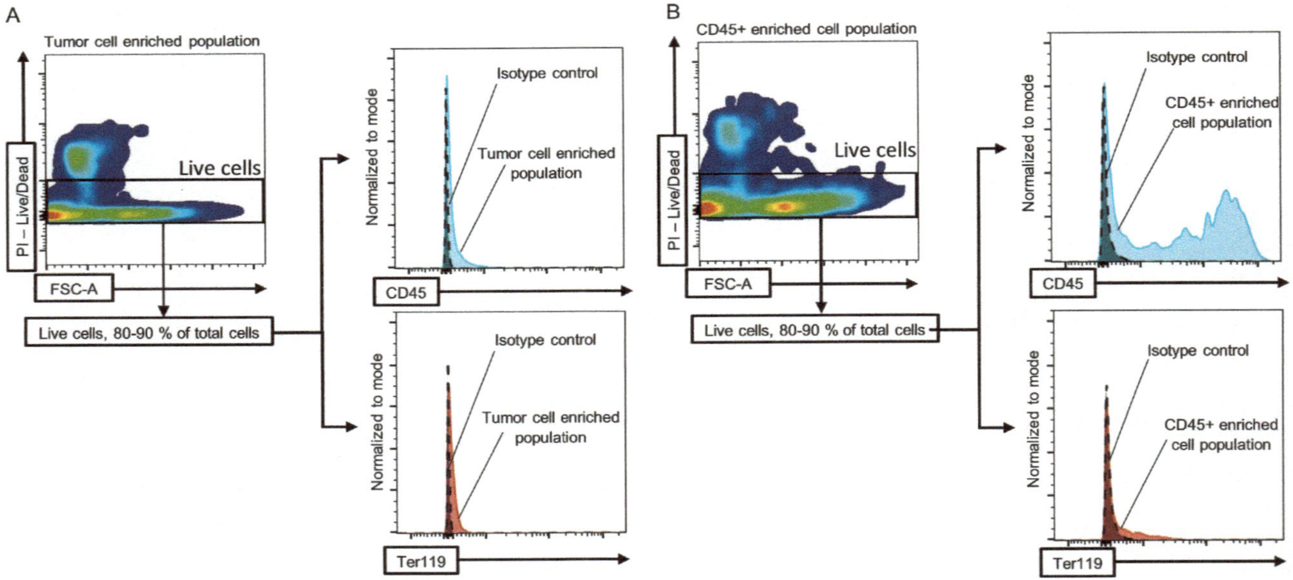

Fig. 3 Flow cytometry plots and histograms showing live, CD45+ and Ter119+ cells. Cell populations enriched in (A) tumor cells or (B) immune cells were isolated as described in the protocol and flow cytometry was performed to analyze the quantity of CD45+ (immune cells) and Ter119+ (erythroid cells) cells in both fractions.

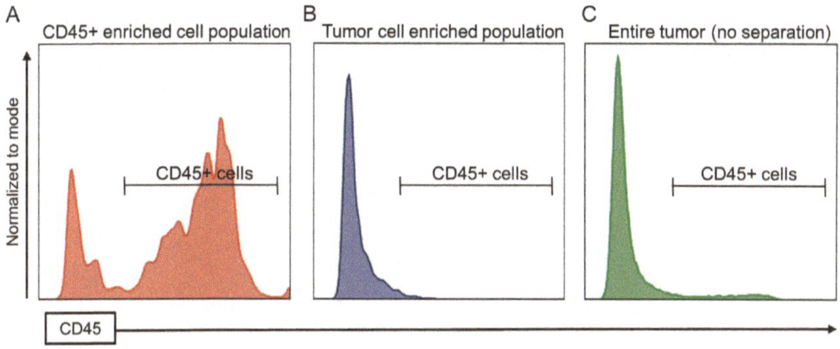

Fig. 4 Surface expression of CD45 marker in the entire tumor homogenate, tumor cell enriched population and CD45+ cell enriched population. The expression of CD45 was studied to analyze the extent of CD45 depletion or enrichment in the tumor cell enriched population or in the CD45+ cell enriched population, respectively.

5. Keep all the pups together under the heating lamp until all the injections are done. If injections take longer than 30 min, return the pups to the cage after the first 30 min.
6. Monitor that the mother is responsive and nurturing to her pups. If necessary, a surrogate mother can be placed in the cage. Also, ensure that the interval between placing the pup on ice and placing it under the warming lamp is <10 min. Pups survival is better when the procedure is carried out quickly. Usually, the time it takes to anesthetize and inject a pup is 4 min.
7. Signs of end point intracranial tumor burden consist of weight loss, ruffled fur, hunched posture, impaired mobility or other abnormalities in movement. At this time point the bioluminescence would be usually 10^7–10^8 photons/s/cm^2/sr.
8. Longer incubations in the Red Blood Cell Lysis Buffer could cause excessive lysis of the cells.
9. Microbeads are conjugated to monoclonal anti-non-human primate CD45 antibodies (Isotype: mouse IgG1, Clone: MB4-6D6), (Miltenyi Biotec Inc., Auburn, CA, USA).
10. The volumes of reagents listed in the protocol are for up to 1×10^7 cells. When working with higher cell numbers, scale up all reagent volumes accordingly.
11. The size of the column used depends on the size of the tumor and the number of cells collected. The SB method usually generates large

tumors and more than 1×10^7 cells can be collected, so large columns were preferentially used.

12. The volume of BSA-EDTA buffer used for preconditioning and washing the column depends on the size of the column used. We used large columns and followed the manufacturer's instructions.
13. It is important to have a suspension of single cells for the optimal performance of the magnetic separation and for preventing the clogging of the column.
14. CD45+ cells should be collected from the first columns only, after washing.
15. Using HBSS buffer yielded better cell viability compared to the use of DPBS buffer.
16. Cells contain numerous Fc receptors that could contribute to non-specific binding and background fluorescence, therefore it is important to block the Fc receptor on the cells.
17. Due to the low amount of CD31+ cells (endothelial cells, less than 3%) in the tumor microenvironment, we did not include this marker in the regular assessment of the cell fractions' purity (data not shown).
18. We used APC conjugated anti-mouse CD45 and Pacific Blue conjugated anti-mouse Ter119.
19. PI is a dye that binds to double stranded DNA by intercalating between base pairs. Viable cell membranes are impermeant, so the dye is excluded from live cells.
20. Fixing the cells minimizes deterioration of the sample.
21. Staining of intra-cellular markers is carried out in 1× Perm/Wash solution to maintain the cells in a permeabilized state.

Acknowledgments

This work was supported by NIH/NINDS Grants, R37-NS094804, R01-NS105556 and 1R21NS107894 to M.G.C.; NIH/NINDS Grants R01-NS076991, and R01-NS096756 to P.R.L.; NIH/NIBIB: R01-EB022563; NIH/NCI U01CA224160; the Department of Neurosurgery, the Rogel Cancer Center, Program in Cancer Hematopoiesis and Immunology (CHI), the ChadTough Foundation and Leah's Happy Hearts to M.G.C. and P.R.L. and RNA Biomedicine Grant F046166 to M.G.C.

M.B.G.F., A.C., P.R.L. and M.G.C. conducted and designed the experiments, analyzed the data, and wrote the manuscript. P.K. performed experiments and contributed to the final manuscript. S.H., F.J.N. and D.A. contributed to the final manuscript and figure design. P.R.L. and M.G.C. directed the research and generated the funding. All the authors read and edited the manuscript.

References

Bjerke, L., Mackay, A., Nandhabalan, M., Burford, A., Jury, A., Popov, S., et al. (2013). Histone H3.3. Mutations drive pediatric glioblastoma through upregulation of MYCN. *Cancer Discovery*, *3*(5), 512–519. https://doi.org/10.1158/2159-8290.cd-12-0426.

Braunstein, S., Raleigh, D., Bindra, R., Mueller, S., & Haas-Kogan, D. (2017). Pediatric high-grade glioma: Current molecular landscape and therapeutic approaches. *Journal of Neuro-Oncology*, *134*(3), 541–549. https://doi.org/10.1007/s11060-017-2393-0.

Broniscer, A., Baker, S. J., West, A. N., Fraser, M. M., Proko, E., Kocak, M., et al. (2007). Clinical and molecular characteristics of malignant transformation of low-grade glioma in children. *Journal of Clinical Oncology*, *25*(6), 682–689. https://doi.org/10.1200/jco.2006.06.8213.

Calinescu, A. A., Núñez, F. J., Koschmann, C., Kolb, B. L., Lowenstein, P. R., & Castro, M. G. (2015). Transposon mediated integration of plasmid DNA into the subventricular zone of neonatal mice to generate novel models of glioblastoma. *Journal of Visualized Experiments*, (96), e52443. https://doi.org/10.3791/52443.

Cohen, K. J., Heideman, R. L., Zhou, T., Holmes, E. J., Lavey, R. S., Bouffet, E., et al. (2011). Temozolomide in the treatment of children with newly diagnosed diffuse intrinsic pontine gliomas: A report from the Children's Oncology Group. *Neuro-Oncology*, *13*(4), 410–416. https://doi.org/10.1093/neuonc/noq205.

Cohen, K. J., Pollack, I. F., Zhou, T., Buxton, A., Holmes, E. J., Burger, P. C., et al. (2011). Temozolomide in the treatment of high-grade gliomas in children: A report from the Children's Oncology Group. *Neuro-Oncology*, *13*(3), 317–323. https://doi.org/10.1093/neuonc/noq191.

Faury, D., Nantel, A., Dunn, S. E., Guiot, M. C., Haque, T., Hauser, P., et al. (2007). Molecular profiling identifies prognostic subgroups of pediatric glioblastoma and shows increased YB-1 expression in tumors. *Journal of Clinical Oncology*, *25*(10), 1196–1208. https://doi.org/10.1200/jco.2006.07.8626.

Funato, K., & Tabar, V. (2018). Histone mutations in cancer. *Annual Review of Cancer Biology*, *2*(1), 337–351. https://doi.org/10.1146/annurev-cancerbio-030617-050143.

Furnari, F. B., Fenton, T., Bachoo, R. M., Mukasa, A., Stommel, J. M., Stegh, A., et al. (2007). Malignant astrocytic glioma: Genetics, biology, and paths to treatment. *Genes & Development*, *21*(21), 2683–2710. https://doi.org/10.1101/gad.1596707.

Gajjar, A., Bowers, D. C., Karajannis, M. A., Leary, S., Witt, H., & Gottardo, N. G. (2015). Pediatric brain tumors: Innovative genomic information is transforming the diagnostic and clinical landscape. *Journal of Clinical Oncology*, *33*(27), 2986–2998. https://doi.org/10.1200/jco.2014.59.9217.

Grasso, C. S., Tang, Y., Truffaux, N., Berlow, N. E., Liu, L., Debily, M. A., et al. (2015). Functionally defined therapeutic targets in diffuse intrinsic pontine glioma. *Nature Medicine*, *21*(6), 555–559. https://doi.org/10.1038/nm.3855.

Haase, S., Garcia-Fabiani, M. B., Carney, S., Altshuler, D., Nunez, F. J., Mendez, F. M., et al. (2018). Mutant ATRX: Uncovering a new therapeutic target for glioma. *Expert Opinion on Therapeutic Targets*, *22*(7), 599–613. https://doi.org/10.1080/14728222.2018.1487953.

Jansen, M., Van Vuurden, D., Vandertop, W., & Kaspers, G. (2012). Diffuse intrinsic pontine gliomas: A systematic update on clinical trials and biology. *Cancer Treatment Reviews*, *38*(1), 27–35.

Jenuwein, T., & Allis, C. D. (2001). Translating the histone code. *Science*, *293*(5532), 1074–1080.

Jones, C., Karajannis, M. A., Jones, D. T. W., Kieran, M. W., Monje, M., Baker, S. J., et al. (2017). Pediatric high-grade glioma: Biologically and clinically in need of new thinking. *Neuro-Oncology*, *19*(2), 153–161. https://doi.org/10.1093/neuonc/now101.

Koschmann, C., Calinescu, A.-A., Nunez, F. J., Mackay, A., Fazal-Salom, J., Thomas, D., et al. (2016). ATRX loss promotes tumor growth and impairs nonhomologous end joining DNA repair in glioma. *Science Translational Medicine*, *8*(328). 328ra328.

Lashford, L. S., Thiesse, P., Jouvet, A., Jaspan, T., Couanet, D., Griffiths, P. D., et al. (2002). Temozolomide in malignant gliomas of childhood: A United Kingdom Children's Cancer Study Group and French Society for Pediatric Oncology Intergroup Study. *Journal of Clinical Oncology*, *20*(24), 4684–4691. https://doi.org/10.1200/jco.2002.08.141.

Lewis, P. W., Muller, M. M., Koletsky, M. S., Cordero, F., Lin, S., Banaszynski, L. A., et al. (2013). Inhibition of PRC2 activity by a gain-of-function H3 mutation found in pediatric glioblastoma. *Science*, *340*(6134), 857–861. https://doi.org/10.1126/science.1232245.

MacDonald, T. J., Aguilera, D., & Kramm, C. M. (2011). Treatment of high-grade glioma in children and adolescents. *Neuro-Oncology*, *13*(10), 1049–1058. https://doi.org/10.1093/neuonc/nor092.

Mackay, A., Burford, A., Carvalho, D., Izquierdo, E., Fazal-Salom, J., Taylor, K. R., et al. (2017). Integrated molecular meta-analysis of 1,000 pediatric high-grade and diffuse intrinsic pontine glioma. *Cancer Cell*, *32*(4), 520–537, e525. https://doi.org/10.1016/j.ccell.2017.08.017.

Nagaraja, S., Vitanza, N. A., Woo, P. J., Taylor, K. R., Liu, F., Zhang, L., et al. (2017). Transcriptional dependencies in diffuse intrinsic pontine glioma. *Cancer Cell*, *31*(5). https://doi.org/10.1016/j.ccell.2017.03.011. 635–652.e636.

Nunez, F. J., Mendez, F. M., Kadiyala, P., Alghamri, M. S., Savelieff, M. G., Garcia-Fabiani, M. B., et al. (2019). IDH1-R132H acts as a tumor suppressor in glioma via epigenetic up-regulation of the DNA damage response. *Science Translational Medicine*, *11*(479), eaaq1427. https://doi.org/10.1126/scitranslmed.aaq1427.

Paugh, B. S., Qu, C., Jones, C., Liu, Z., Adamowicz-Brice, M., Zhang, J., et al. (2010). Integrated molecular genetic profiling of pediatric high-grade gliomas reveals key differences with the adult disease. *Journal of Clinical Oncology*, *28*(18), 3061–3068. https://doi.org/10.1200/jco.2009.26.7252.

Piunti, A., Hashizume, R., Morgan, M. A., Bartom, E. T., Horbinski, C. M., Marshall, S. A., et al. (2017). Therapeutic targeting of polycomb and BET bromodomain proteins in diffuse intrinsic pontine gliomas. *Nature Medicine*, *23*(4), 493–500. https://doi.org/10.1038/nm.4296.

Schwartzentruber, J., Korshunov, A., Liu, X. Y., Jones, D. T., Pfaff, E., Jacob, K., et al. (2012). Driver mutations in histone H3.3 and chromatin remodelling genes in paediatric glioblastoma. *Nature*, *482*(7384), 226–231. https://doi.org/10.1038/nature10833.

Strahl, B. D., & Allis, C. D. (2000). The language of covalent histone modifications. *Nature*, *403*(6765), 41.

Sturm, D., Bender, S., Jones, D. T., Lichter, P., Grill, J., Becher, O., et al. (2014). Paediatric and adult glioblastoma: Multiform (epi)genomic culprits emerge. *Nature Reviews Cancer*, *14*(2), 92–107. https://doi.org/10.1038/nrc3655.

Sturm, D., Witt, H., Hovestadt, V., Khuong-Quang, D. A., Jones, D. T., Konermann, C., et al. (2012). Hotspot mutations in H3F3A and IDH1 define distinct epigenetic and biological subgroups of glioblastoma. *Cancer Cell*, *22*(4), 425–437. https://doi.org/10.1016/j.ccr.2012.08.024.

Voon, H. P., & Wong, L. H. (2016). New players in heterochromatin silencing: Histone variant H3.3 and the ATRX/DAXX chaperone. *Nucleic Acids Research*, *44*(4), 1496–1501.

Wiesner, S. M., Decker, S. A., Larson, J. D., Ericson, K., Forster, C., Gallardo, J. L., et al. (2009). De novo induction of genetically engineered brain tumors in mice using plasmid DNA. *Cancer Research*, *69*(2), 431–439.

Wu, G., Broniscer, A., McEachron, T. A., Lu, C., Paugh, B. S., Becksfort, J., et al. (2012). Somatic histone H3 alterations in pediatric diffuse intrinsic pontine gliomas and non-brainstem glioblastomas. *Nature Genetics*, *44*(3), 251–253. https://doi.org/10.1038/ng.1102.

CHAPTER TWENTY

Analysis of the immune status from peripheral whole blood with a single-tube multicolor flow cytometry assay

Anna-Jasmina Donaubauer, Ina Becker, Paul F. Rühle, Rainer Fietkau, Udo S. Gaipl*,†, Benjamin Frey†

Department of Radiation Oncology, Universitätsklinikum Erlangen, Friedrich-Alexander-Universität Erlangen-Nürnberg, Erlangen, Germany

*Corresponding author: e-mail address: udo.gaipl@uk-erlangen.de

Contents

Abstract

Supplementation of standard cancer therapies (radiotherapy, chemotherapy, surgery) with immunotherapies has revolutionized cancer treatment. In order to include recent improvements of multimodal therapies into clinical routine, knowledge about the immune status, the immune dynamics and the detailed composition and activation of patient's immune system is required. The here presented single-tube multicolor flow cytometry assay allows the discrimination of 20 clinically relevant immune cell subsets and their activation status in peripheral whole blood. It includes 15 different antibodies and can be established on a common 3 laser and 10 color flow-cytometer. Furthermore,

† Contributed equally as senior authors.

Methods in Enzymology, Volume 632
ISSN 0076-6879
https://doi.org/10.1016/bs.mie.2019.03.003

this assay is easy to set-up and to perform as well as fast with only 40 min of sample preparation time. Moreover, only 100 μL of whole blood are sufficient for this precise determination of the individual immune status. It is already applied in translational programs of clinical studies and trials and can further be adapted for future ones.

1. Introduction

The determination of patients' immune status in clinical settings is becoming increasingly important. Especially in cancer therapy, the composition and condition of the patient's immune system is quite relevant (Mittal, Gubin, Schreiber, & Smyth, 2014). In the last decade, cancer therapy has been revolutionized by the supplementation of surgery, chemotherapy (CT) and/or radiotherapy (RT) with immunotherapy (IT) (Galon, Angell, Bedognetti, & Marincola, 2013). Even if IT is a promising addition to the classical cancer treatments, the inclusion of it into preexisting treatment protocols is not yet standardized. This is due to the lack of information on interactions between the different treatment modalities and on how multimodal therapies modulate the phenotype of cancer cells and the composition and activation status of the patient's immune system (Derer, Frey, Fietkau, & Gaipl, 2016; Galon et al., 2013).

By addressing these questions, information about the optimal time point in therapy progression for inclusion of IT should be gained. Moreover, predictive and prognostic immunological markers have to be defined, which forecast the responsiveness, therapeutic success or occurrence of side effects of the multimodal therapies (Balermpas et al., 2014; Ordonez et al., 2014). Therefore, the repetitive screening of the current composition of the immune system has to be performed (Dovedi et al., 2014).

Ideally, biopsies from the affected tissue should be used for the determination of the individual and current immune status. Nonetheless, repetitive biopsies from these tissues are not feasible in most cases. Therefore, liquid biopsies, such as peripheral blood, are the most suitable alternative for the determination of the immune status. Peripheral blood reveals important information on the current immune status, as the immune cells circulate through the blood stream to reach their target tissue (Bjoern, Brimnes, Andersen, Thor Straten, & Svane, 2011; Tokuno et al., 2009). In addition, the burden of repetitive blood withdrawals is much lower for the patients than the burden of other tissue biopsies. The possibility of screening peripheral blood at different time points allows the determination of the immune status over the whole course of a therapy.

Flow cytometry is one of the most suitable technologies for the detailed screening of various immunological parameters (Ruhle, Fietkau, Gaipl, & Frey, 2016). Modern flow cytometers offer the possibility to differentiate between 10 and more colors which allow the differentiation of numerous parameters on single cell level (Chattopadhyay & Roederer, 2012).

The here presented panel allows the determination of a comprehensive immune status, which covers the major immune cell types (T cells, B cells, natural killer (NK) cells, dendritic cells (DCs), monocytes, neutrophil-, eosinophil- and basophil granulocytes), as well as several immune cell subsets and their activation status. Further, this assay requires a total blood volume of only 100 μL which is feasible with every routine blood withdrawal and therefore easily adaptable into clinical routine. Another advantage of the assay is the fact that the immune cells can be directly stained from whole blood. Most of the published immunophenotyping assays on the other hand, require an isolation of peripheral blood mononuclear cells (PBMCs) prior to analysis (Moncunill, Han, Dobano, McElrath, & De Rosa, 2014). This excludes polymorphonuclear cells (eosinophil-, basophil- and neutrophil granulocytes) from the analysis. However, these innate immune cells have already been shown to be a prognostic marker in some tumor entities and are of great importance for the overall immune status (Moreira, Waltraud, Schuler, & Heinzerling, 2017). In addition, the isolation of PBMCs is a potential source of errors when analyzing the expression of activation markers and makes the immunophenotyping more extensive.

The presented single-tube flow cytometry assay for immunophenotyping is simple, fast and can easily be standardized. The hands-on time for sample preparation requires only about 40 min and also the time for data acquisition is very short with about 10 min. This is why this assay is appropriate for the adaption into clinical routine, as well as for the application in clinical trials. It discriminates between eight major leukocyte cell types, together with about 20 immune cell subsets. The major leukocytes are separated by the examination of the expression of specific surface markers, which are molecules that are solely expressed on the surface of one certain cell type. The immune cell subsets can be analyzed by using more antibodies than colors available for the discrimination of the cells. Antibodies coupled to the same fluorochrome are exclusively used for surface markers that are restricted to a certain immune cell type. Nonetheless, a proper discrimination between cell types is possible, by sticking to a strict gating strategy. In order to further extend the informative content, the expression of three common activation markers (CD69, CD25 and HLA-DR) is finally analyzed on all identified cell subsets. The assay can be established on every 3 laser and 10 color flow cytometer. Detailed descriptions on the sample preparation, the staining of cells, as well as on the data analysis will be given in the following sections.

The panel is easily adaptable to individual requirements by the replacement of several antibodies with different markers. For such individual applications only the pan markers should be kept for the identification of the major immune cells and must not be replaced, as they are necessary for the gating strategy. We decided to not include a dead cell marker, since it would occupy a further fluorescence channel. Preceding dead cell marker investigations showed positivity of this signal with the lowered FSC events. This is why all cells with low FSC values are excluded from the analysis in this assay. Additionally, the samples can be stored safely for up to 4h after preparation including staining. However, extending the storage of the samples up to 6h caused a loss of the quality of the analysis. Finally, this assay screens only for the major immune cell types, as well as for common cell subsets being present in the peripheral blood. As the sample volume is restricted, low value subpopulations can be lost. For a deeper analysis of more infrequent immune cell types, the sample volume should be increased.

It should be emphasized that the assay can also be extended by adding a counting tube for the determination of absolute leukocyte counts from whole blood. Such an analysis is especially valuable for longitudinal analyses, in which the immune status of patients is repeatedly analyzed over a longer period of time (Ruhle et al., 2017). Often, only some cell types are in- or decreased and these fluctuations alter the percentages of other cell types in the blood, even though there are no changes in the absolute cell count of those cell subsets. To quantify the changes in the cell numbers, it is necessary to determine an absolute leukocyte count with a counting tube (TruCount tube). The following protocol will also give instructions on the sample preparation, the staining of the cells and the data analysis of the latter, as already described in more detail previously (Donaubauer et al., 2019).

2. Materials

The following materials are needed for the immunophenotyping assay. Please use ultra-pure water for the preparation of all solutions. All solutions have to be stored at 4 °C.

1. Phosphate-buffered saline (PBS): pH 7.4.
2. Solution A (red blood cell lysing solution): 0.12% formic acid solution: Add 1200 μL of >98% formic acid solution to 1 L of water and mix. The solution should be filter sterilized afterwards.
3. Solution B (re-buffering solution): 3.0 g of sodium carbonate, 7.25 g of sodium chloride and 15.55 g of sodium sulfate in 500 mL of water: Dilute the chemicals in 500 mL of water. The solution should be filter sterilized afterwards. For the preparation of solution B, water-free sodium carbonate is required. If the water-free chemical is not available, monohydrate is also sufficient. In this case, use 3.5 g of sodium carbonate monohydrate.

The pH value of solution A and B has to be adjusted after preparation of both solutions. Therefore, solution A and B are combined in the same ratio as later for the lysis of the erythrocytes. Mix 2400 μL of solution A with 1060 μL of solution B in a 15 mL tube. The pH should range between 6.5 and 7.5. If the pH is too acidic, use water for corrections. If the pH is too basic, use formic acid for further adjustments.

1. Solution C (fixation solution): Prepare a solution of 1% of formaldehyde in PBS (pH: 7.4)

The solutions A, B and C can be stored at 4 °C. We recommend storing the solutions no longer than 2 months.

2. Antibody master mix for the general immune status: Mix the required antibodies for the determination of the immune cell types as indicated in Table 1. The volumes indicated in the table are sufficient for the preparation of one blood sample.
3. Optional: Antibody master mix for the TruCount tube: Prepare a mix of all the antibodies in PBS, as indicated in Table 2. The mix is sufficient for the preparation of one blood sample. This mix is only required, if the absolute leukocyte count should be analyzed with the general immune status.

The antibody mixes can be prepared for multiple blood samples in advance. Therefore, the time for the preparation of the mix can be saved up for many stainings and a potential source of errors is eliminated. The mixes can be stored at 4 °C in light proof tubes for up to 1 month.

Further material:

1. ETDA blood collection tube
2. Polypropylene FACS tubes with a volume of 5 mL: The advantage of polypropylene tubes is their special surface, which inhibits the adhesion of cells to the wall of the tube.
3. Optional: TruCount tube (BD Biosciences)

Table 1 Antibody mix.

nm	Specificity	Fluorochrome	Clone	Vendor	Cat. No.	Amount (μL)
488	CD28	BB515	28.2	BD Biosciences	564,492	2.5
	CD16	PE	3G8	BD Biosciences	555,407	0.1
	CD16	PE	B73.1	BD Biosciences	561,313	2.5
	CD25	PED594	M-A251	BioLegend	356,126	5.0
	CD8	PCC5.5	HIT8a	BioLegend	300,924	0.5
	CD14	PCC5.5	M5E2	BioLegend	301,824	0.5
	CD19	PCC5.5	HIB19	BD Biosciences	561,295	0.5
	CD20	PCC5.5	2H7	BD Biosciences	560,736	0.1
	CD127	PEV770	MB15-18C9	Miltenyi Biotec	130-099-719	2.0
638	CD4	APC	RPA-T4	BD Biosciences	555,349	0.2
	CD123	APC	7G3	BD Biosciences	560,087	2.0
	CD56	APCR700	NCAM16.2	BD Biosciences	657,886	2.5
	HLA-DR	APCV770	REA332	Miltenyi Biotec	130-104-825	1.0
405	CD69	BV421	FN50	BD Biosciences	562,884	5.0
	CD3	KO	UCHT1	Beckman Coulter	B00068	4.0
						Total: 28.4 μL

Note: There are two different clones of CD16 in the assay, which either bind to epitopes on neutrophils (3G8) or on NK cells and monocytes (B73.1). The structure of the CD16-molecule differs between these cell subsets. The monocytes and NK cells express the isoform A of the FcRγIII, but the neutrophils express FcRγIIIB. Even though both isoforms are to 95% identical, the antibody clone B73.1 specifically binds to the FcRγIIIA. In contrast, the clone 3G8 may bind both isoforms, but has a higher binding capacity to FcRγIIIB. Consequently, even though the binding affinity of the clone 3G8 might be sufficient enough in many samples to simultaneously bind monocytes, NK cells and neutrophils, the staining is strongly improved when both clones are included in the assay.

Table 2 Antibody mix for the TruCount tube.

nm	Specificity	Fluorochrome	Clone	Vendor	Cat. No.	Amount (μL)
488	CD14	PCC5.5	M5E2	BioLegend	301,824	0.5
	CD19	PCC5.5	HIB19	BD Biosciences	561,295	0.5
	CD20	PCC5.5	2H7	BD Biosciences	560,736	0.1
638	CD56	APCR700	NCAM16.2	BD Biosciences	657,886	2.5
405	CD3	KO	UCHT1	Beckman Coulter	B00068	4.0
						Total: 7.6 μL
						Add: 12.4 μL PBS

4. Optional: TQ Prep workstation (Beckman Coulter): This workstation performs an automatic lysis of erythrocytes, as well as the fixation of the leukocytes. All of these steps can also be performed manually.
5. Ten color and three laser flow cytometer: We use a Gallios flow cytometer (Beckman Coulter) in the standard filter configuration for the assay. The cytometer settings can be found in Table 3.

3. Staining procedure and sample preparation

The sample preparation and the staining of the cells are performed at room temperature.

1. Pour 100 μL of EDTA-treated whole blood into a 5 mL polypropylene FACS tube. Blood samples can be stored up to 24 h at room temperature prior to staining. If this is the case, the increased amount of dead cells (with lowered FSC values) should be excluded from the analysis with appropriate gating. Invert the blood collection tube one to two times before pipetting the blood into the FACS tube to guarantee an even distribution of the blood cells. Blood smears on the tube wall have to be removed. Otherwise, these cells will appear as dead/dying cells in the analysis later.

Table 3 Cytometer settings.

Laser	Power	Detector	Filter	Range	Voltage	Gain	Fluorochrome
488 nm	22 mW	FSC	–	–	130	7.5	–
		SSC	–	–	90	7.5	–
		FL1	525 BP 40	505–545	430	1	BB515
		FL2	575 BP 30	560–590	520	1	PE
		FL3	620 BP 30	605–635	500	1	PED594
		FL4	695 BP 30	680–710	630	1	PCC5.5
		FL5	755 LP	>755	650	1	PEV770
638 nm	25 mW	FL6	660 BP 20	650–670	700	1	APC
		FL7	725 BP 20	715–735	600	1	APCR700
		FL8	755 LP	>755	500	1	APCV770
405 nm	40 mW	FL9	450 BP 50	425–475	470	1	BV421
		FL10	550 BP 40	530–570	400	1	KO

Abbreviations: *BB515*, Brilliant Blue 515; *PE*, Phycoerythrin; *PED594*, PE-Dazzle594; *PCC5.5*, PerCP-Cy5.5; *PEV770*, PE-Vio770; *APC*, Allophycocyanin; *APCR700*, APC-R700; *APCV770*, APC-Vio770; *BV421*, Brilliant Violet 421; *KO*, Krome Orange.

2. Add the antibody master mix (28.4 μL) to the FACS tube and mix by vortexing.
3. Optional: If using the TruCount tube, the beads at the bottom of the tube must be dissolved by adding the TruCount antibody mix directly on it. After dissolving the beads, 50 μL of whole blood are added directly onto the dissolved beads. For the calculation of the absolute cell count, it is necessary that the exact amount of 50 μL of blood is drawn into the TruCount tube. Therefore, we recommend adding the blood by reverse pipetting: Press the pipette knob down to the second stop point, while the blood is drawn into the pipette tip. When releasing the fluid into the tube only press the pipette knob down to the first stop point. Do not vortex. Also, the Lot number and the bead count of the used TruCount tube/package must be noted, as the bead count is relevant for the calculation of the absolute cell number later.
4. Both samples have to be incubated at room temperature in the dark for 20 min.
5. Directly after incubation, the lysis of the erythrocytes and the fixation of the leukocytes are performed for both tubes. The lysis and the fixation

are automated processes, if a TQ Prep Workstation can be used. Otherwise these steps can also be executed manually: Add 600 µL of solution A to the stained cells and vortex for 10 s. Immediately add 265 µL of solution B and vortex again for 10 s. For fixation, add 100 µL of solution C and vortex for 10 s. These steps are also performed for the TruCount tube but to a lesser extend of vortexing, in order to keep the metal retainer in the tube intact.

6. Afterwards, the fixed cells are washed twice. Add 3 mL of PBS to the FACS tube and centrifuge the cells at 300 × *g* for 5 min. Discard the supernatant and dissolve the cells after the second washing step in 200 µL of PBS. After discarding the supernatant, place the tube upside down onto a paper towel. This way, excess fluid is soaked by the towel. Do not invert the tube multiple times; otherwise the cell pellet can be lost. If the samples cannot be measured directly after the preparation, store the samples in the dark at 4 °C until data acquisition. The samples can be stored safely for up to 4 h after preparation.
7. Optional: The TruCount tube requires no washing steps, as cells would be lost during it, which makes it impossible to calculate an absolute leukocyte count. Therefore, the TruCount tube is stored in the dark at 4 °C until the other samples are washed twice.
8. Analyze the stained leukocytes by flow cytometry.

4. Data analysis

The data analysis of this panel can be separated into three main analyzation steps. First of all, the main immune cell types (e.g., T cells or monocytes) are defined. Therefore, the scatter attributes and the expression of the respective panel marker is evaluated on the cells, in order to define their type. All of the selected panel markers are molecules that are well described in the literature and are solely expressed on one specific cell type. After the main immune cell types are defined, these cells are further subdivided into immune cell subtypes. For this purpose, the (co-)expression of further antigens is examined on the earlier defined cell types. In the last step, the expression of the activation markers CD69, CD25 and HLA-DR is analyzed on the immune cell subsets. An overview of all identified cell subsets, as well as their characterization by the specific markers can be found in Table 4. Moreover, the optional analysis of the TruCount tube can be added, if a calculation of absolute cell counts is desired.

Table 4 Definition of cell subsets.

Cell subset	Definition
Leukocytes	FSC vs. SSC
T cells	$CD3^+$
Cytotoxic T cells (T_C)	$CD3^+/CD8^+/CD4^-$
TC8hi	$CD3^+$/CD8hi/$CD4^-$
TC8lo	$CD3^+$/CD8lo/$CD4^-$
T helper cells (T_H)	$CD3^+/CD4^+/CD8^-$
T_{REG}	$CD3^+/CD4^+/CD8^-$/CD25hi/$CD127^-$/lo
$CD28^-$ T cells	$CD3^+/CD28^-$
$CD28^-$ TC	$CD3^+/CD8^+/CD4^-/CD28^-$
$CD28^+$ TC	$CD3^+/CD8^+/CD4^-/CD28^+$
$CD28^-$ TH	$CD3^+/CD4^+/CD8^-/CD28^-$
$CD28^+$ TH	$CD3^+/CD4^+/CD8^-/CD28^+$
DPT	$CD3^+/CD4^+/CD8^+$
DNT	$CD3^+/CD4^-/CD8^-$
B cells	$CD19^+$ or $CD20^+$
NK cells	$CD3^-/CD56^+$
NK1	$CD3^-/CD56^+/CD16^+$
NK2	$CD3^-$/CD56hi/$CD16^-$
NK3	$CD3^-$/CD56lo/$CD16^-$
NKT	$CD3^+/CD56^+$ or $CD3^+/CD16^+$
Neutrophils	SSChi/$CD16^+$
Eosinophils	SSChi/$CD16^-$
Basophils	$CD3^-/CD14^-/(CD16^-)/$ $CD19^-/CD20^-/CD56^-/HLADR^-/CD123^+$
Monocytes	$CD14^+$/SSCint
Mo1	$CD14^+/CD16^-$
Mo2	CD14hi/$CD16^+$
Mo3	CD14lo/$CD16^+$
Mo4	CD14lo/$CD16^-$

Table 4 Definition of cell subsets.—cont'd

Cell subset	Definition
Dendritic cells (DCs)	$CD3^-/CD14^-/(CD16^-)/CD19^-/CD20^-/CD56^-/HLADR^+$
mDC	$CD3^-/CD14^-/(CD16^-)/CD19^-/CD20^-/CD56^-/HLADR^+/CD123^-$/lo
pDC	$CD3^-/CD14^-/(CD16^-)/CD19^-/CD20^-/CD56^-/HLADR^+$/CD123hi

The following table provides an overview of all the cell subsets mentioned in the chapter together with their expression patterns used for characterization.

Table 5 Definition of Boolean gates.

Boolean gate	Definition
Rest1	"All Cells" AND (NOT "T cells")
Rest2	Rest1 AND (NOT (Eosinophils OR Neutrophils))
Rest3	Rest2 AND (NOT Monocytes)
Rest4	Rest3 AND (NOT "NK cells")
Rest5	Rest4 AND (NOT "B cells")
Rest6	Rest5 AND (NOT "Dendritic Cells")
Rest7	Rest6 AND (NOT Basophils)

The table provides an overview of the necessary gates, required for the detection of the immune cell subsets. For the transfer of activation markers on other subsets or for the analysis of co-expression, further gates would be necessary.

In the immunophenotyping panel, several antibodies with different specificities coupled to the same fluorochrome are combined. Therefore, it is essential to stick to a strict gating strategy in order to separate the different cell types stained with those antibodies. With every gating step, the identified cell types are excluded from the upcoming gating step. Therefore, with every gating step the following input gate consists of fewer events. The analysis starts with the definition of the most frequent cells and subsequently the infrequent ones are analyzed, because the best results can be obtained by using this strategy. An overview of all the required Boolean gates for this gating strategy is provided in Table 5.

4.1 Determination of the major cell types

All identified main immune cell types are depicted as an overview in Fig. 1 (big cells). The characteristic markers for each immune cell type are written over the arrows in Fig. 1. For all analysis steps, dot plots are created for the definition of the cell types. A histogram plot is used solely for the examination of the event count. In the plots, the expression of one single marker is usually plotted against the scatter characteristics (SSC), unless indicated otherwise.

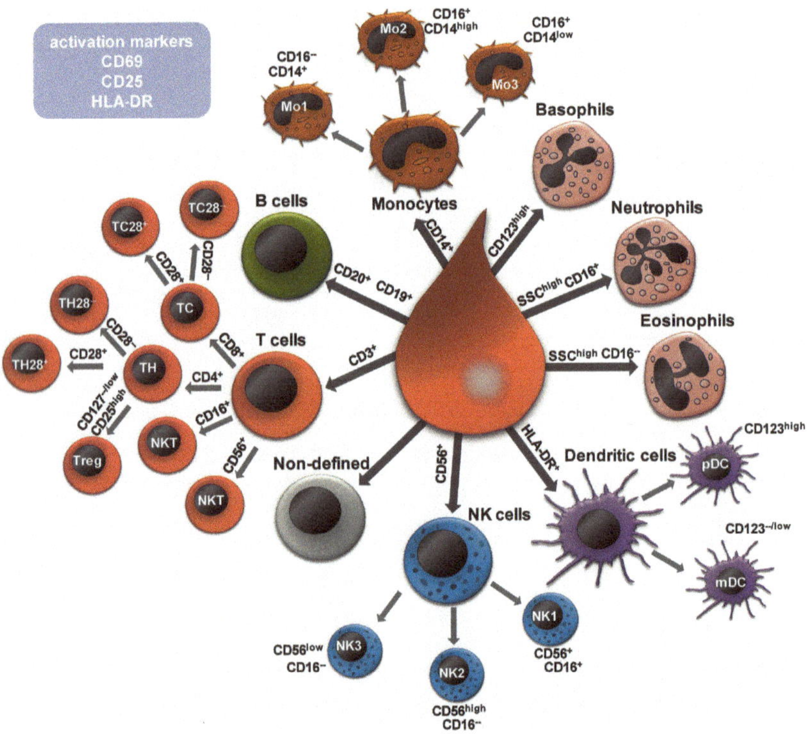

Fig. 1 Schematic overview of the immune cell subtypes identified by the single-tube multicolor flow cytometry assay. The main immune cell types (T cells, B cells, NK cells, dendritic cells, monocytes, eosinophils, basophils and neutrophils) are depicted as big cells. The labeling of the arrows indicates the respective panel marker of the main cell types, used for the determination of the cells from whole blood. Most cell types can be further subdivided into immune cell subtypes, which are indicated as smaller cells. Again, the labeling of the arrows indicates the expression of surface markers, used for subset identification. Eventually, all cell subsets can be analyzed for their expression of the common activation markers CD25, CD69 and/or HLA-DR.

1. First, the event count is examined over time in order to exclude irregularities from the analysis.
2. Next, doublets are excluded. For this purpose, the forward scatter signal (FSC) is examined for the peak as well as the time of flight (TOF) in comparison to the integral attributes. By performing these two exclusion steps, more doublets can be detected.
3. The remaining events are then investigated for their FSC vs. SSC characteristics to define the leukocytes. Set a gate around those cells (*all cells gate*) for further analysis. All cells with lower FSC values should be excluded, as these cells are dying/dead. Dead cell marker investigations showed most positivity of the signal in the lowered FSC events. This is why all cells with low FSC values are excluded from the analysis. In general dead cells are no problem in fresh blood samples. Also, the addition of a dead cell marker would occupy a further fluorescence channel, thus such a marker is left out of the panel.
4. The earlier defined leukocytes are used as the input for the further analysis. Examine the cells of the *all cells gate* for their expression of CD3. All cells that are positive for this marker are defined as T cells (Jarry, Cerf-Bensussan, Brousse, Selz, & Guy-Grand, 1990). Draw a gate around the respective cells (*T cells gate*). As these cells are frequent and their expression of CD3 is quite high, they are easy to determine first.
5. A Boolean gate is set to define the remaining cells (*rest 1 gate*).
6. The events of *rest 1 gate* are the input for the further analysis. Examine the events of the Boolean gate for their expression of CD16. Eosinophilic and neutrophilic granulocytes can be identified by their high SSC characteristics and the differential expression of CD16. A gate is drawn around the SSC high and CD16 positive neutrophils (*neutrophils gate*) and around the SSC high and CD16 negative eosinophils (*eosinophils gate*) (Ravetch, 2001).
7. A new Boolean gate (*rest 2 gate)* is generated for all cells that are neither T cells nor eosinophils or neutrophils. By now, most of the events are already defined.
8. Analyze the cells of *rest 2 gate* for their expression of CD14 and their SSC characteristics. Monocytes are $CD14^+$ (Pan et al., 2017) and show intermediate SSC characteristics. The CD14 antibody carries the same fluorochrome (PCC5.5) as other antibodies (CD19, CD20 and CD8) in the panel. Hence, it is necessary to define only the cells with intermediate SSC values as monocytes, in order to avoid false positive events. Draw a gate around this population (*monocytes gate*).

9. The remaining events are defined as *rest 3*, by creating a new Boolean gate.
10. The cells of rest 3 are now analyzed for the expression of CD56. All the positive events are defined as NK cells (Knudsen, Court-payen, Kjærsgaard, & Christensen, 2009). Normally, NK cells are defined as $CD16^+$ cells, as many other cell types. Thus, the CD56 expression is also analyzed. Unfortunately, the expression of CD56 is very weak, so the possibility to measure false positive events is quite high. Therefore, two different exclusion steps can be performed. As a first step, the $CD123^-$ and/or $CD56^+$ cells are monitored and then all SSC high cells are excluded. In the following step, the NK cells are then defined as $CD56^+$ and/or $CD16^+$ (Hong et al., 2010).
11. A Boolean gate covering the remaining events is created (*rest 4 gate*).
12. Examine the events of *rest 4 gate* for their expression of CD19 and CD20. If the cells are either positive for CD20 and/or CD19, they are defined as B cells (Monserrat et al., 2013). Define a gate covering all B cells (*B cells gate*).
13. Set a Boolean gate around the rest of the events and define it as *rest 5*.
14. Now, the remaining cells of *rest 5* are examined for their expression of HLA-DR, in order to define the dendritic cells. All cells, which are $HLA\text{-}DR^+$ and show intermediate SSC characteristics, are defined as dendritic cells (O'Doherty et al., 1994). The DCs can only be identified after the definition of the previous cell types, as the marker HLA-DR is expressed by many cell types. If cells are falsely gated, this can lead to false positive events in the *DCs gate*. Consequently, only cells with intermediate SSC characteristics are defined as DCs. Set a gate around the population (*DCs gate*).
15. Again, a new Boolean gate, covering the remaining cells, is created (*rest 6*).
16. Finally, the basophils are separated from cells of gate *rest 6*. As for the DCs, the exclusion of all previous cell types is necessary for the determination of the basophils, as these cells do also not express a pan marker. The cells are analyzed for the expression of CD123 and the $CD123^{high}$ cells are defined as basophils. Under certain circumstances, the expression of CD123 is downregulated on basophils (Chirumbolo, Bjorklund, & Vella, 2017). Thus, it is recommended to create a further gate, covering only the $CD123^{low}$ basophils (*BAS_123low gate*). Set a gate around the population (*basophils gate*).
17. The rest of the cells is defined as *rest* 7 and consists of non-defined immune cell subsets, as well as hematopoietic stem and progenitor cells.

4.2 Determination of cell subsets

When the major immune cell types are defined, the analysis of the immune cell subtypes is performed based on the previous analysis steps. Therefore, the co-expression of several markers is analyzed on the major immune cells. An overview of the identified immune cell subsets is given in Fig. 1 (small cells), together with the respective markers for their definition. Exemplarily, an overview of the subgating of the T cells is given in detail in Fig. 2.

1. First, the T cells are further subdivided, as for these cells the most subsets are defined. Thus, the earlier defined $CD3^+$ T cells are chosen as an input gate for the analysis. Examine the expression of CD4 and CD8 on the T cells, in order to define $CD8^+/CD4^-$ T killer cells and $CD4^+/CD8^-$ T helper cells. Quadrants are set into the plot to separate the cell types. Define gates for the double positive cells (*DPT gate*), double negative cells (*DNT gate*), T helper cells (*TH gate*) and the cytotoxic T cells (*TC gate*). Double positive as well as double negative T cells differ from the classical view of circulating T cells. However, we record these cells in few numbers in all blood samples and analyze them. Double positive and negative T cells for example are elevated in the progression of autoimmune diseases, making them relevant in pathological processes (Crispin et al., 2008; Quandt, Rothe, Scholz, Baerwald, & Wagner, 2014). Also, a further distinction can be performed for the T killer cells, as these cells show lowered expression levels of CD8 under certain circumstances (Trautmann et al., 2003). We therefore suggest separating the cells into $CD8^{high}$ cells ($T8^{hi}$) and a smaller $CD8^{low}$ subset ($T8^{lo}$) (see Fig. 2B).
2. To define regulatory T cells, the cells of the *TH gate* are further subdivided. Examine the cells for the expression of CD127 and CD25. Cells showing a $CD127^{-/low}$ and $CD25^{high}$ phenotype are defined as regulatory T cells (T_{regs} *gate*) (see Fig. 2C) (Liu et al., 2006; Seddiki et al., 2006). Sometimes, the regulatory T cell population is not easily distinguishable from the other $CD4^+$ cells. Therefore, the expression level of CD127 and CD25 can be examined on the level of the $CD8^+$ T cells. As regulatory T cells are not $CD8^+$, the respective gate for the T_{regs} should not contain many events. On the level of the $CD8^+$ cells, the gate should be adjusted, so that nearly all T cells are outside of this gate. This gating step can then be transferred to the level of the $CD4^+$ cells.
3. The T cells are also analyzed for their CD28 expression. Therefore, the cells of the *T cells gate* are used as input. The CD28 expression can be examined easily, when it is analyzed together with the expression of CD25.

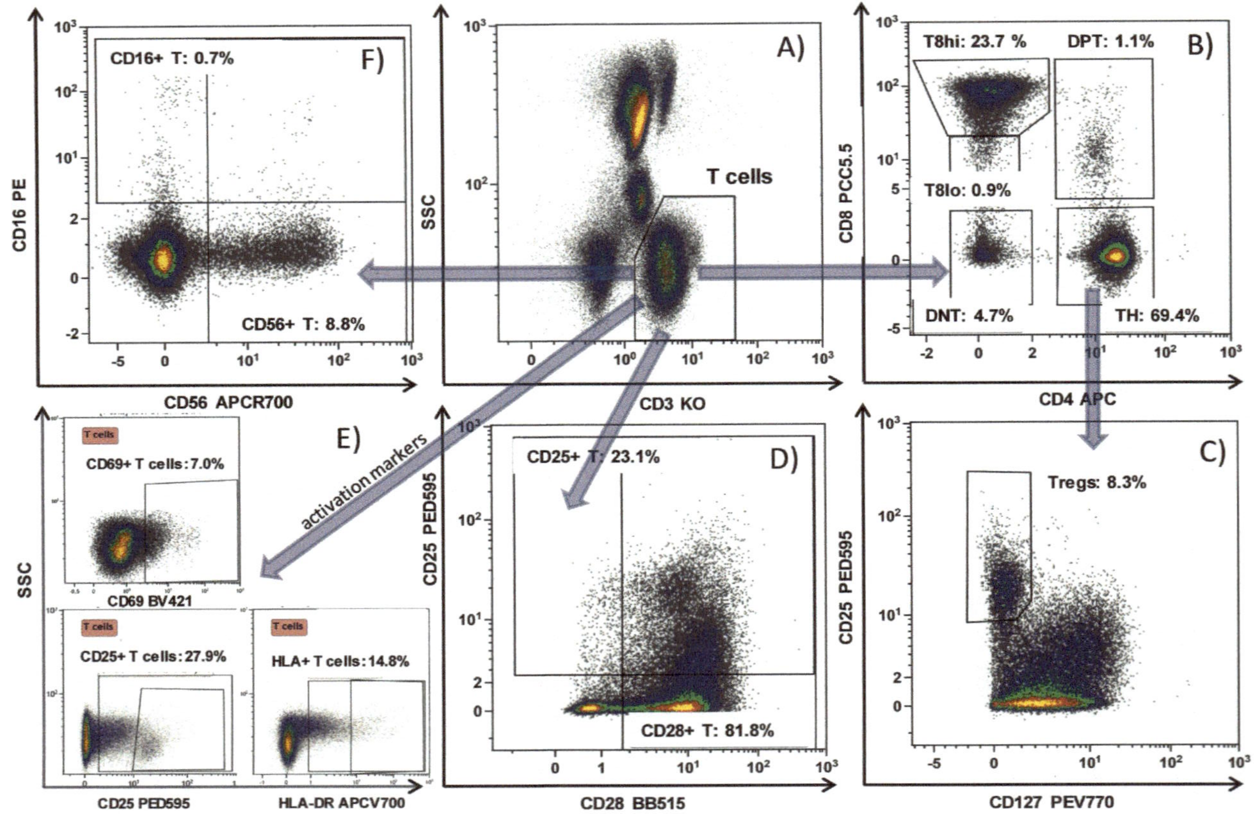

Fig. 2 See legend on opposite page.

By setting quadrants, the population can be separated (see Fig. 2D). Also, the gating can be transferred to the level of T helper or T killer cells, by using Boolean gates.

4. Again, the T cells are picked as input for the analysis. Even though NKT cells do not belong to the T cells, they express T cell markers and are determined from the *T cells gate*. As NKT cells also express NK cell markers, the expression of CD56 is analyzed together with the expression of CD16 (Al Omar, Marshall, Middleton, & Christmas, 2012; Bjorkstrom et al., 2008; Michel et al., 2007). Set a gate around all $CD56^+$ and/or $CD16^+$ cells (*NKT cells gate*) (see Fig. 2F). NKT cells differ in their expression of CD8 and CD4 (Michel et al., 2007). Thus, the NKT cell gating can additionally be performed on the level of $CD4^+$ T cells and $CD8^+$ T cells and not only on the level of $CD3^+$ cells. By this, variations in the expression level can be analyzed.
5. Next, a separation of monocytes into three subtypes (Mo1–Mo3) is performed. The cells of the *monocytes gate* are selected as an input and the expression of CD14 and CD16 is evaluated on the cells. The Mo1 cells ($CD14^+/CD16^-$), the Mo2 cells ($Cd14^{high}/CD16^+$) and the Mo3 cells ($CD14^{low}/CD16^+$) are separated by different gates (*Mo1, Mo2 and Mo3 gate*) (Guilliams et al., 2014). There is a very rare fourth subtype of

Fig. 2 Subgating of the T cell subsets and their activation status. After the determination of the main immune cell types, immune cell subsets can easily be determined, using the earlier defined main cell types as input for the further analysis. For the subgating of the T cell subsets, all $CD3^+$ cells (T cells) (A) are chosen as an input for the analysis. By examining the expression of CD4 and CD8 on the T cells, the population can be subdivided into T helper and T killer cells. Moreover, double negative or double positive cells can be recorded, as well as those T killer cells which express only low levels of CD8 (B). Next, the just defined T helper cells can be used as an input to define those T helper cells, which are regulatory T cells. Therefore, the expression of CD25 and CD127 is analyzed on all $CD4^+$ T cells/T helper cells. Cells showing a $CD127^{-/low}$ and $CD25^{high}$ phenotype are defined as regulatory T cells (C). By choosing the $CD3^+$ cells again as an input, the T cells can be analyzed for their expression of CD28. The positive cells are most easy to determine when the expression of CD28 is analyzed together with CD25 (D). The expression of CD25 can also be analyzed on the level of T helper and T killer cells, if desired. The NKT cells can also be determined from the $CD3^+$ cells, by analyzing the cells for their expression CD16 and/or CD56 (F). Finally, activation status of the T cells is analyzed. Therefore, the expression of CD25, CD69 and HLA-DR is analyzed in the $CD3^+$ cells (E). For a more differentiated activation status of the T cells, the expression of the activation markers can also be determined on the level of each cell subtype. As exemplarily shown here for the T cells, the subgating can also be performed similarly for the other immune cell types.

monocytes, which is characterized by a $CD14^{low}/CD16^{-}$ phenotype (Thomas, 1994). These cells are called Mo4 or preDCs and can additionally be monitored.

6. Our analyses performed during monitoring of patients in several clinical trials revealed a notable co-expression of CD8 with HLA-DR on monocytes, although there are no descriptions of this subtype in the literature to date. Nevertheless, we recommend examining this expression pattern in an explorative setting. The expression of HLA-DR and CD8 is therefore analyzed on the cells of the *monocytes gate*. By setting quadrants into the plot, the populations can be separated.
7. Then, the NK cells are also subdivided. The *NK cell gate* is selected for the analysis and the cells are examined for their expression of CD56 and CD16. Different gates (*NK1, NK2 and NK3 gate*) are drawn to determine the $CD56^{+}/CD16^{+}$ NK1 cells, the $CD56^{high}/CD16^{-}$ NK2 cells and the $CD56^{low}/CD16^{-}$ NK3 cells (Montaldo et al., 2013).
8. Also, a subgating step is performed for the DCs. The cells can be differentiated into plasmacytoid DCs (pDCs) ($CD11c^{-}/CD123^{high}$) and myeloid DCs (mDCs) ($CD11c^{high}/CD123^{-/low}$) by their co-expression of CD123 and CD11c (O'Doherty et al., 1994; Robinson et al., 1999). In this panel, the CD123 has been already included for the identification of basophils. We renounced the investigation of CD11c, since after exclusion of all other cells, almost all $CD123^{high}$ cells were $CD11c^{-}$ and almost all $CD11c^{+}$ cells were $CD123^{-/low}$. Therefore, in order to separate the subtypes, the expression of HLA-DR is plotted against the expression of CD123. The cells of the *DCs gate* are selected for the analysis and the populations are separated by gates around the $CD123^{-/low}$ mDCs (*mDCs gate*) and the $CD123^{high}$ pDCs (*pDCs gate*).

In this panel, no further subgating is performed for the remaining cell types (B cells, eosinophils, neutrophils and basophils). The exemplary subgating of the T cell subsets, as well as the gating of the activation markers on the T cell level is depicted in Fig. 2 together with the desired gates as a graphical demonstration of the complex gating strategy. If desired, immune cell subsets can be left out or replaced by other immune cell types and respective markers.

4.3 Determination of the activation status

As a final step, the activation status of the earlier analyzed immune cell types is examined. For this purpose, the expression of CD69, CD25 and HLA-DR on the major immune cells is analyzed in parallel. Unlike the previous analysis steps, it is not necessary to stick to a strict gating strategy. Many cells (in this case: T cells, B cells, NK cells, basophils and DCs) show similar expression patterns. Therefore, we applied the same gate setting for all cells, which show the same auto- fluorescence characteristics. The gating of the activation markers on the T cells is exemplarily shown in Fig. 2E. In general, the activation status can be examined on any desired, earlier defined cell subtype to gain further data. Therefore, the respective input gate has to be chosen for the analysis. As the gating is often difficult for the expression of activation markers, the gate settings were evaluated by different controls during the assay development. The controls include fluorescence minus one (FMO) controls, single stainings of whole blood, as well as isotype stainings. These stainings were performed with the blood of normal healthy donors (NHDs). We suggest performing such controls in order to evaluate the optimal gate settings for the activation markers. Table 6 provides an overview of the expression patterns of the different activation markers on the earlier defined cell types.

1. First the expression level of CD25 is analyzed on the different cell types. The marker is examined on B cells, T cells, NK cells, DCs, basophils

Table 6 Expression patterns of activation markers.

	CD25		CD69	HLA-DR	
	Positive	**High**	**Positive**	**Positive**	**High**
T cells	+	+	+	+	+
T_H	+	+	+	+	+
T_C	+	+	+	+	+
B cells	+	+	+	+	+
NK cells	+	–[a]	+	+	+
Basophils	+	–[a]	+	–[b]	–[b]
DCs	+	–[a]	+	–[c]	–[c]
Monocytes	–[d]	–[d]	+	+	+
Neutrophils	+	–[a]	+	+	–[a]
Eosinophils	+	–[a]	+	+	–[a]

[a]Positive expression, but no distinction of high expression pattern.
[b]Per definition negative.
[c]Per definition positive.
[d]Not available as isotype indicated unspecific binding properties.

and eosinophils. Use the respective input gate for the analysis and create dot plots comparing the SSC characteristics to the expression of CD25 for each cell type, except monocytes, because isotype controls showed unspecific binding properties. Gate borders are defined based on the controls. Additional high expression gates can be defined for the CD25 expression on B cells and T cells to increase the informative content.

2. Afterwards, HLA-DR expression is analyzed on T cells, B cells, NK cells, DCs and monocytes. Even though, B cells, monocytes and DCs express HLA-DR constantly, it was included in the panel as an activation marker. Under some circumstances, there are fluctuations in the expression level of HLA-DR on those cells in response to certain stimuli (Holmannova et al., 2016). Dot plots are created for each cell type and the respective input gate is selected. The SSC characteristics are compared to the expression of HLA-DR. The gate borders are set based on the controls, to define the HLA-DR$^+$ cells. The gating for the HLA-DR positive monocytes should be performed carefully, as these cells showed an unspecific binding to the antibody. Nonetheless, we observed similar expression patterns with other antibodies. As the specific signal was still much higher than the unspecific isotype signal, the expression of HLA-DR on monocytes was included in the panel.
3. CD69 is the final activation marker to be analyzed. Again, dot plots are created to examine the expression on T cells, B cells, NK cells, DCs, basophils and neutrophils. Use the respective input gate and compare the expression of CD69 to the SSC characteristics of the cells. As before, draw new gates based on the controls.

The gating of the activation marker CD25 on T cells is exemplarily depicted in Fig. 3. In order to show the suitability of the assay in clinical applications, the differential expression of CD25 on T cells of two independent cancer patients undergoing IT with Nivolumab is shown over the course of therapy. In comparison, the expression of CD25 on the T cells of a NHD is presented. Nivolumab is a therapeutic antibody, targeting the checkpoint molecule PD1 on T cells. By this, the immunosuppressing interaction of PD1 with PD-L1 or PD-L2 is prevented. Therefore, T cells are not suppressed by tumor cells and an antitumor immune response can be stimulated in patients (Brahmer et al., 2015). CD25 is a common activation marker that expressed by many leukocytes (e.g., T and B cells, NK cells and basophils). During an active state, CD25 is upregulated on these immune cells. As changes in the activation status of patients undergoing

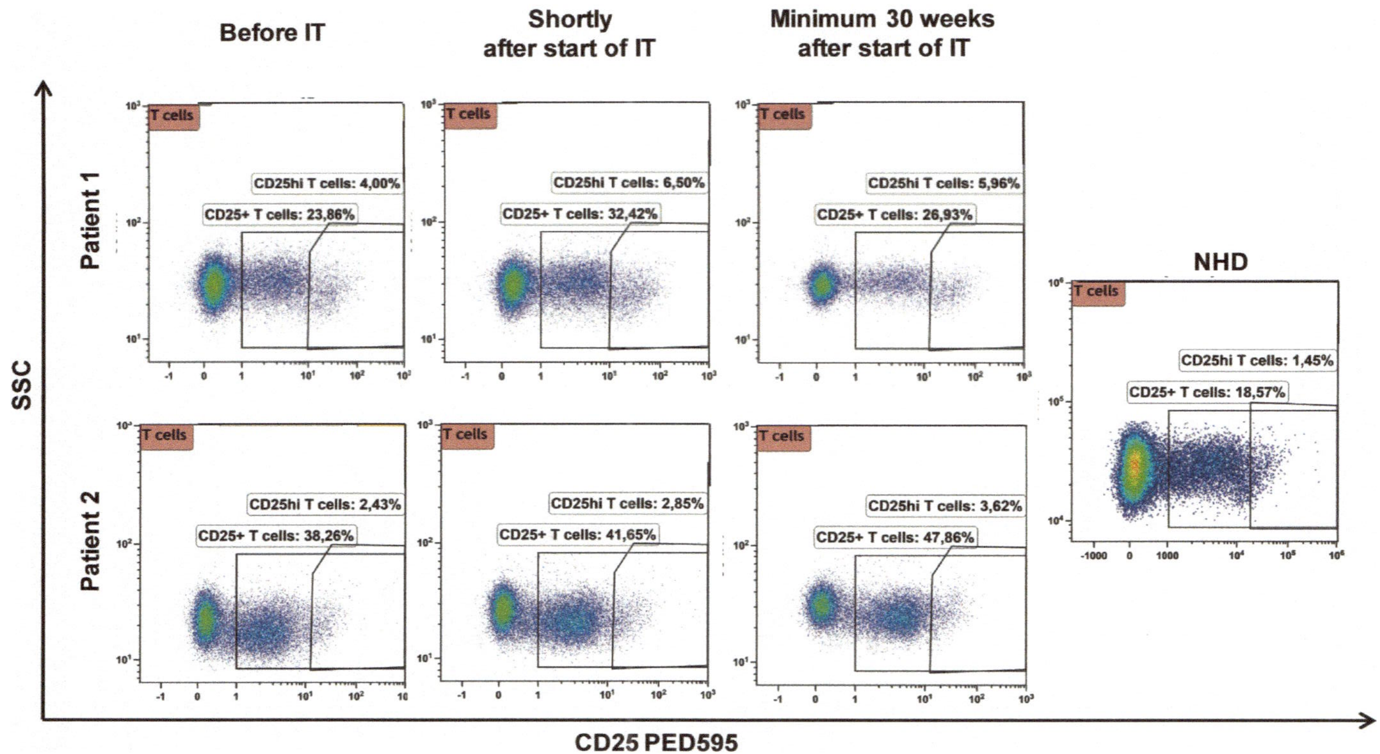

Fig. 3 See figure legend on next page.

IT are expected, we analyzed the expression of CD25 on the T cells of the patients. By this screening of the immune and activation status, therapeutic or prognostic markers for IT can be found in peripheral blood, as already demonstrated for HLA-DR expression on monocytes (Krieg et al., 2018). As seen in Fig. 3, the expression level of CD25 is altered over the course of therapy in the patients. Patient 1 shows an increase in the expression level of CD25 after onset of IT. During therapy, the expression level of CD25 slightly decreases again. On the other hand, patient 2 has a continuous increase in the expression level of CD25 on T cells while receiving IT. In contrast to the patients under IT, the NHD has less active T cells in the peripheral blood. Such minor alterations can easily be measured by the single-tube immunophenotyping assay, pointing out the utility of this assay for clinical applications.

5. Determination of the absolute cell count

A calculation of absolute cell counts from the panel described above is not possible due to the washing steps, which cause a loss of cells. The TruCount tube contains counting beads, which allow the determination of exact cell numbers from whole blood. Similar to the panel above, the analysis of the TruCount tube requires a strict gating strategy.

1. First, the event count over time is analyzed, to exclude irregularities from the analysis.

Fig. 3 Suitability of the assay for the application in clinical settings. As the assay is fast, easy to perform and generates a detailed immune status from whole blood, it is very suitable for the application in clinical settings. This method allows detecting also minor changes of the immune status. Here, exemplarily the immune status two independent, metastasized cancer patients during immunotherapy with Nivolumab was monitored bevor onset and during therapy in order to find prognostic and therapeutic markers for the responsiveness to the immune therapy (IT). Exemplarily, changes in the activation status of T cells (as determined by surface expression of CD25) are depicted. Patient 1 (upper row) shows an increase in the expression level of CD25 after onset of IT. During therapy, the expression level of CD25 slightly decreases again. Patient 2 in contrast (lower row) has a continuous increase in the expression level of CD25 on T cells while receiving IT. In comparison to the patients under IT, the expression level of CD25 on the T cells of a normal healthy donor (NHD) (plot on the right) is shown. The T cells of the NHD are less activated than the cells of the patients undergoing IT. As such minor fluctuations in expression of activation markers can be detected from whole blood with this method, it is particularly suitable for longitudinal analyses of the immune status in different settings.

2. Then, the FSC signal is analyzed for the peak and the TOF characteristics in comparison to the integral attributes. Thereby, the doublets can be excluded from the analysis.
3. The leukocytes are defined by their FSC vs. SSC characteristics. As no washing steps were necessary for the TruCount tube, more dead cells and cell fragments can be seen in the plot. These cells/fragments have low SSC and FSC characteristics and must be excluded from the further analysis. Set a gate around the leukocytes (*all cells gate*). Events with high SSC and low FSC characteristics in this plot belong to the counting bead population. Define the population by setting a gate (*beads gate*).
4. Next, the T cells are defined by their expression of CD3 (Jarry et al., 1990). A gate is drawn around the population (*T cells gate*).
5. Create a Boolean gate, covering all cells except the earlier defined T cells (*rest 1 ($CD3^-$) gate*).
6. The cells of *rest 1 ($CD3^-$) gate* are used as an input to analyze the granulocytes. Therefore the SSC values are plotted against the expression of CD14/CD19/CD20. These markers are detected in the same fluorescence channel. The cell population showing high SSC characteristics is defined as granulocytes (*granulocytes gate*).
7. A new Boolean gate is created, covering the remaining, undefined leukocytes (*rest 2 (R1-PMN) gate*).
8. Use the cells of *rest 2 (R1-PMN) gate* for further analysis. Next, the B cells and the monocytes are defined. Therefore, the expression of CD14/CD19/CD20 is plotted against the SSC values again. The $CD14^+$ monocytes (Pan et al., 2017) can be separated from the B cells by their intermediate SSC characteristics. Draw a gate around the monocytes (*monocytes gate*), as well as around the $CD19^+/CD20^+$ B cells (Monserrat et al., 2013) (*B cells gate*).
9. Again, a Boolean gate is defined, covering the remaining events (*rest 3 (R2-Mo, B) gate*).
10. The remaining events are analyzed for the expression of CD56, to define the NK cells (*NK cell gate*) (Al Omar et al., 2012; Michel et al., 2007). The rest of the events remain undefined.
11. Finally, the counting beads are further characterized. The beads can be characterized by their high autofluorescence, which can be detected in the channels 1, 2, 4 and 8. For each fluorescence channel a dot plot is created, comparing the SSC characteristics to the respective fluorescence signal. Use the *beads* gate as an input for the analysis. The beads population shows high SSC characteristics, as well as high fluorescence

values. In each fluorescence channel a gate is drawn, covering the respective bead population (*beads FL1*, *beads FL2*, *beads FL4* and *beads FL8 gate*). The counting beads can stick together, creating bead doublets. These doublets show a stronger fluorescence signal than the singlets. Therefore, there is the possibility to create different gates for the bead singlets and bead doublets in the four fluorescence channels examined. For the calculation of the bead count the number of the bead doublets must be counted twice to the number of the bead singlets. This guarantees a more precise bead count. When the bead count was determined in all four fluorescence channels, the bead count recorded in the four channels should be compared. If the bead count differs strongly between the channels, the gating strategy should be reevaluated. For a more precise bead count, the mean of the four channels can be calculated and used in the formula stated below.

12. For the calculation of the final cell count for the different populations the following formula must be used. The absolute bead count varies between different lot numbers of the TruCount tubes. Therefore, it is crucial to document the bead count of the used tube.

$$\frac{\text{cells acquired}}{\text{beads acquired}} \text{x} \frac{\text{beads absolute}}{\text{blood volume}}$$

6. Concluding remarks

Based on this single-tube multicolor flow cytometry-based immunophenotyping panel, one can easily measure the general immune status in peripheral blood in an easy, standardized and fast manner. Not only in cancer, but also in other disease settings, the screening of the immune status can be of great value (Frey et al., 2017). The panel is easily adaptable to individual requirements by the replacement of several antibodies with different markers. For such individual applications only the pan markers should be kept for the identification of the major immune cells and must not be replaced, as they are necessary for the gating strategy.

Acknowledgment

This work has been in part funded by the Bundesministerium für Bildung und Forschung (GREWIS and GREWIS-alpha, 02NUK017G and 02NUK050E, respectively).

References

Al Omar, S. Y., Marshall, E., Middleton, D., & Christmas, S. E. (2012). Increased numbers but functional defects of CD56+CD3+ cells in lung cancer. *International Immunology*, *24*(7), 409–415. https://doi.org/10.1093/intimm/dxr122.

Balermpas, P., Rodel, F., Liberz, R., Oppermann, J., Wagenblast, J., Ghanaati, S., et al. (2014). Head and neck cancer relapse after chemoradiotherapy correlates with CD163+ macrophages in primary tumour and CD11b+ myeloid cells in recurrences. *British Journal of Cancer*, *111*(8), 1509–1518. https://doi.org/10.1038/bjc.2014.446.

Bjoern, J., Brimnes, M. K., Andersen, M. H., Thor Straten, P., & Svane, I. M. (2011). Changes in peripheral blood level of regulatory T cells in patients with malignant melanoma during treatment with dendritic cell vaccination and low-dose IL-2. *Scandinavian Journal of Immunology*, *73*(3), 222–233. https://doi.org/10.1111/j.1365-3083.2010.02494.x.

Bjorkstrom, N. K., Gonzalez, V. D., Malmberg, K. J., Falconer, K., Alaeus, A., Nowak, G., et al. (2008). Elevated numbers of fc RIIIA+ (CD16+) effector CD8 T cells with NK cell-like function in chronic hepatitis C virus infection. *The Journal of Immunology*, *181*(6), 4219–4228. https://doi.org/10.4049/jimmunol.181.6.4219.

Brahmer, J., Reckamp, K. L., Baas, P., Crino, L., Eberhardt, W. E., Poddubskaya, E., et al. (2015). Nivolumab versus docetaxel in advanced squamous-cell non-small-cell lung Cancer. *The New England Journal of Medicine*, *373*(2), 123–135. https://doi.org/10.1056/NEJMoa1504627.

Chattopadhyay, P. K., & Roederer, M. (2012). Cytometry: Today's technology and tomorrow's horizons. *Methods*, *57*(3), 251–258. https://doi.org/10.1016/j.ymeth.2012.02.009.

Chirumbolo, S., Bjorklund, G., & Vella, A. (2017). Using a CD45dim/CD123bright/HLA-DRneg phenotyping protocol to gate basophils in FC for airway allergy. CD123 does not decrease. *Advances in Respiratory Medicine*, *85*(4), 193–201. https://doi.org/10.5603/ARM.2017.0032.

Crispin, J. C., Mohammed, O., Bayliss, G., Cohen, R. A., Beek, V., Christine, A., et al. (2008). Expanded double negative T cells in patients with systemic lupus erythematosus produce IL-17 and infiltrate the kidneys. *Journal of Immunology*, *181*(12), 8761–8766.

Derer, A., Frey, B., Fietkau, R., & Gaipl, U. S. (2016). Immune-modulating properties of ionizing radiation: Rationale for the treatment of cancer by combination radiotherapy and immune checkpoint inhibitors. *Cancer Immunology, Immunotherapy*, *65*(7), 779–786. https://doi.org/10.1007/s00262-015-1771-8.

Donaubauer, A. J., Ruhle, P. F., Becker, I., Fietkau, R., Gaipl, U. S., & Frey, B. (2019). One-tube multicolor flow cytometry assay (OTMA) for comprehensive immunophenotyping of peripheral blood. *Methods in Molecular Biology*, *1904*, 189–212. https://doi.org/10.1007/978-1-4939-8958-4_8.

Dovedi, S. J., Adlard, A. L., Lipowska-Bhalla, G., McKenna, C., Jones, S., Cheadle, E. J., et al. (2014). Acquired resistance to fractionated radiotherapy can be overcome by concurrent PD-L1 blockade. *Cancer Research*, *74*(19), 5458–5468. https://doi.org/10.1158/0008-5472.CAN-14-1258.

Frey, B., Ruckert, M., Deloch, L., Ruhle, P. F., Derer, A., Fietkau, R., et al. (2017). Immunomodulation by ionizing radiation-impact for design of radio-immunotherapies and for treatment of inflammatory diseases. *Immunological Reviews*, *280*(1), 231–248. https://doi.org/10.1111/imr.12572.

Galon, J., Angell, H. K., Bedognetti, D., & Marincola, F. M. (2013). The continuum of cancer immunosurveillance: Prognostic, predictive, and mechanistic signatures. *Immunity*, *39*(1), 11–26. https://doi.org/10.1016/j.immuni.2013.07.008.

Guilliams, M., Ginhoux, F., Jakubzick, C., Naik, S. H., Onai, N., Schraml, B. U., et al. (2014). Dendritic cells, monocytes and macrophages: A unified nomenclature based on ontogeny. *Nature Reviews. Immunology*, *14*(8), 571–578. https://doi.org/10.1038/nri3712.

Holmannova, D., Kolackova, M., Kunes, P., Krejsek, J., Mandak, J., & Andrys, C. (2016). Impact of cardiac surgery on the expression of CD40, CD80, CD86 and HLA-DR on B cells and monocytes. *Perfusion*, *31*(5), 391–400. https://doi.org/10.1177/0267659115612905.

Hong, H. S., Eberhard, J. M., Keudel, P., Bollmann, B. A., Ahmad, F., Ballmaier, M., et al. (2010). Phenotypically and functionally distinct subsets contribute to the expansion of CD56-/CD16+ natural killer cells in HIV infection. *AIDS*, *24*(12), 1823–1834. https://doi.org/10.1097/QAD.0b013e32833b556f.

Jarry, A. C.-B., Cerf-Bensussan, N., Brousse, N., Selz, F., & Guy-Grand, D. (1990). Subsets of CD3+ (T cell receptor alp or y/6) and CD3-lymphocytes isolated from normal human gut epithelium display phenotypical features different from their counterparts in peripheral blood. *European Journal of Immunology*, *20*(5), 1097–1103. https://doi.org/10.1002/eji.1830200523.

Knudsen, J. H., Court-payen, M., Kjærsgaard, E., & Christensen, N. J. (2009). Lymphocyte subset composition in peripheral blood from normal subjects may be influenced by both spleen size and plasma norepinephrine. *Scandinavian Journal of Clinical and Laboratory Investigation*, *55*(7), 643–648. https://doi.org/10.3109/00365519509110264.

Krieg, C., Nowicka, M., Guglietta, S., Schindler, S., Hartmann, F. J., Weber, L. M., et al. (2018). High-dimensional single-cell analysis predicts response to anti-PD-1 immunotherapy. *Nature Medicine*, *24*(2), 144–153. https://doi.org/10.1038/nm.4466.

Liu, W., Putnam, A. L., Xu-Yu, Z., Szot, G. L., Lee, M. R., Zhu, S., et al. (2006). CD127 expression inversely correlates with FoxP3 and suppressive function of human CD4+ T reg cells. *The Journal of Experimental Medicine*, *203*(7), 1701–1711. https://doi.org/10.1084/jem.20060772.

Michel, J. J., Turesson, C., Lemster, B., Atkins, S. R., Iclozan, C., Bongartz, T., et al. (2007). CD56-expressing T cells that have features of senescence are expanded in rheumatoid arthritis. *Arthritis and Rheumatism*, *56*(1), 43–57. https://doi.org/10.1002/art.22310.

Mittal, D., Gubin, M. M., Schreiber, R. D., & Smyth, M. J. (2014). New insights into cancer immunoediting and its three component phases—Elimination, equilibrium and escape. *Current Opinion in Immunology*, *27*, 16–25. https://doi.org/10.1016/j.coi.2014.01.004.

Moncunill, G., Han, H., Dobano, C., McElrath, M. J., & De Rosa, S. C. (2014). OMIP-024: Pan-leukocyte immunophenotypic characterization of PBMC subsets in human samples. *Cytometry. Part A*, *85*(12), 995–998. https://doi.org/10.1002/cyto.a.22580.

Monserrat, J., de Pablo, R., Diaz-Martin, D., Rodriguez-Zapata, M., de la Hera, A., Prieto, A., et al. (2013). Early alterations of B cells in patients with septic shock. *Critical Care*, *17*(3), R105. https://doi.org/10.1186/cc12750.

Montaldo, E., Del Zotto, G., Della Chiesa, M., Mingari, M. C., Moretta, A., De Maria, A., et al. (2013). Human NK cell receptors/markers: A tool to analyze NK cell development, subsets and function. *Cytometry. Part A*, *83*(8), 702–713. https://doi.org/10.1002/cyto.a.22302.

Moreira, A., Waltraud, L., Schuler, G., & Heinzerling, L. (2017). Eosinophilic count as a biomarker for prognosis of melanoma patients and its importance in the response to immunotherapy. *Immunotherapy*, *9*(2), 115–121. https://doi.org/10.2217/imt-2016-0138.

O'Doherty, U., Peng, M., Gezelter, W., Swiggard, W. J., Betjes, M., Bhardwaj, N., et al. (1994). Human blood contains two subsets of dendritic cells, one immunologically mature and the other immature. *Immunology*, *82*(3), 487–493.

Ordonez, R., Henriquez-Hernandez, L. A., Federico, M., Valenciano, A., Pinar, B., Lloret, M., et al. (2014). Radio-induced apoptosis of peripheral blood CD8 T lymphocytes is a novel prognostic factor for survival in cervical carcinoma patients. *Strahlentherapie und Onkologie*, *190*(2), 210–216. https://doi.org/10.1007/s00066-013-0488-x.

Pan, T., Zhou, T., Li, L., Liu, Z., Chen, Y., Mao, E., et al. (2017). Monocyte programmed death ligand-1 expression is an early marker for predicting infectious complications in acute pancreatitis. *Critical Care*, *21*(1), 186. https://doi.org/10.1186/s13054-017-1781-3.

Quandt, D., Rothe, K., Scholz, R., Baerwald, C. W., & Wagner, U. (2014). Peripheral CD4CD8 double positive T cells with a distinct helper cytokine profile are increased in rheumatoid arthritis. *PLoS One*, *9*(3), e93293. https://doi.org/10.1371/journal.pone.0093293.

Ravetch, J. V., & Silvia, B. (2001). IgG Fc receptors. *Annual Review of Immunology*, *19*, 275–290. https://doi.org/10.1146/annurev.immunol.19.1.275.

Robinson, S. P., Steven, P., English, N., Davies, D., Knight, S. C., & Reid, C. D. L. (1999). Human peripheral blood contains two distinct lineages of dendritic cells. *European Journal of Immunology*, *29*(9), 2769–2778. doi:10.1002/(SICI)1521-4141(199909)29:09<2769::AID-IMMU2769>3.0.CO;2-2.

Ruhle, P. F., Fietkau, R., Gaipl, U. S., & Frey, B. (2016). Development of a modular assay for detailed immunophenotyping of peripheral human whole blood samples by multicolor flow cytometry. *International Journal of Molecular Sciences*, *17*(8). https://doi.org/10.3390/ijms17081316.

Ruhle, P. F., Goerig, N., Wunderlich, R., Fietkau, R., Gaipl, U. S., Strnad, A., et al. (2017). Modulations in the peripheral immune system of glioblastoma patient is connected to therapy and tumor progression-a case report from the IMMO-GLIO-01 trial. *Frontiers in Neurology*, *8*, 296. https://doi.org/10.3389/fneur.2017.00296.

Seddiki, N., Santner-Nanan, B., Martinson, J., Zaunders, J., Sasson, S., Landay, A., et al. (2006). Expression of interleukin (IL)-2 and IL-7 receptors discriminates between human regulatory and activated T cells. *The Journal of Experimental Medicine*, *203*(7), 1693–1700. https://doi.org/10.1084/jem.20060468.

Thomas, R., & Lipsky, P. E. (1994). Human peripheral blood dendritic cell subsets. Isolation and characterization of precursor and mature antigen-presenting cells. *The Journal of Immunology*, *153*, 4016–4028.

Tokuno, K. H., Shoichi, H., Yoshino, S., Yoshida, S., & Oka, M. (2009). Increased prevalence of regulatory T-cells in the peripheral blood of patients with gastrointestinal cancer. *Anticancer Research*, *29*(5), 1527–1532.

Trautmann, A., Beate, R., Schmid-Grendelmeier, P., Niederer, E., Bröcker, E.-B., Blaser, K., et al. (2003). Human CD8 T cells of the peripheral blood contain a low CD8 expressing cytotoxic/effector subpopulation. *Immunology*, *108*(3), 305–312.

Further reading

Perussia, B., Trinchieri, G., Jackson, A., Warner, N. L., Faust, J., Rumpold, H., et al. (1984). The Fc receptor for IgG on human natural-killer cells—Phenotypic, functional, and comparative studies with monoclonal-antibodies. *Journal of Immunology*, *133*(1), 180–189.

CHAPTER TWENTY-ONE

Ex vivo assessment of *in vivo* DC-targeted antibodies in pre-clinical models

Qingrong Huang[a,b], **Niroshana Anandasabapathy**[a,b,*]
[a]Department of Dermatology, Weill Cornell Medical College, New York, NY, United States
[b]Meyer Cancer Center, Weill Cornell Medical College, New York, NY, United States
*Corresponding author: e-mail address: nia9069@med.cornell.edu

Contents

Abstract

APCs play a key role at initiating adaptive immune responses by presenting antigens to lymphocytes and DCs are professional APCs. It is critical to understand the differential antigen capture and presentation ability of different DC subsets, which is important for DC-targeted immunotherapy. In this section, we give a brief introduction to different antigen presentation pathways and introduce the key concept of cross-presentation, the major antigen presentation pathway used for anti-viral and anti-tumoral immune responses. CD205, a DC restricted receptor, is highly expressed on certain DCs subsets. We find CD205-mediated antigen uptake to be a useful model for studying antigen uptake and defects. These methods provide an introduction to CD205-mediated pre-clinical delivery of antigens to cross-presenting DCs, which can be adapted to the study of targeting to multiple receptors and other C-type lectins. This is a promising strategy to detect the antigen capture capacity and to study the key players orchestrating tolerance and immunity *ex vivo*.

Methods in Enzymology, Volume 632
ISSN 0076-6879
https://doi.org/10.1016/bs.mie.2019.11.007

1. Introduction

Antigen-presenting cells (APCs) are a heterogeneous group of cells that include macrophages, B cells, dendritic cells (DCs). These specialized populations initiate adaptive immune responses by presenting antigens captured from the environment to B and T lymphocytes. Differences in antigen capture ability have been reported among APC types and subsets, and multiple strategies aim to deliver antigens directly to one or more APC subsets in order to expand immune priming. In this section, we will briefly review the antigen presentation pathway, focusing on cross-presentation—the major mechanism by which DCs present tumor and viral antigens. We will then address a strategy to deliver antigens to cross-presenting DCs in a pre-clinical setting, which can be adapted to other DCs or myeloid subsets.

Direct delivery of antigens to specific APC subsets is an effective method—proven both to dampen and to expand a T cell driven immune response and to regulate antibody responses (Cheong et al., 2010; Hawiger et al., 2004; Kastenmuller et al., 2014; Tesfaye et al., 2019). In the absence of immune adjuvants targeted immunization can drive tolerance to self-antigens (Hawiger et al., 2004). When instead antigens are administered with an appropriate licensing stimuli, targeted immunization can increase the efficacy of immunization. The targeting methodology can also be used to evaluate the contribution of antigen presentation to specific APC subsets. As such, we have found this to be a helpful strategy to evaluate subset specialization, adjuvant efficacy and routes of immunization (Anandasabapathy et al., 2014; Chappell et al., 2014). Reported strategies include delivery of antigens to unique receptors *in vivo* by directed ligands, using antibodies conjugated to antigen, nanoparticles, lentiviruses, or recombinant adenovirus engineered to contain antigens (Tacken, Torensma, & Figdor, 2006). Given their broad availability, safety, selectivity, high affinity and avidity, mAbs are a highly popular method for antigen-targeting.

Steinman and Cohn first found a group of phagocytic cells in 1973, which could prime and activate naïve T cells, in mouse spleen and named them "dendritic" due to common morphology with the dendrites on a neuron or branched arborization on a tree (Steinman & Cohn, 1973; Steinman & Witmer, 1978). DCs are the major professional APC for initiating adaptive T-cell immunity (Steinman, 2010, 2012) and as such are a rationale choice for targeting antigens. Many studies have demonstrated that antigens can be effectively directed to DCs and presented on both major histocompatibility complex class I and II (MHC I and MHC II) molecules.

When delivered together with co-stimulatory signals or adjuvant to promote maturation of DCs this method is capable of successfully initiating anti-viral and anti-cancer immune responses (Macri et al., 2016; Ueno et al., 2011). To better understand DC-target immunotherapy and vaccinology, studying the *in vivo* uptake and presentation of antigens by DCs in a pre-clinical is necessary.

2. Antigen presentation and DCs

Based on their capacity to activate naïve T cells and their dependence on specific transcription factors in their ontogeny, DCs are subdivided into different subsets including conventional/classical DCs (cDCs), plasmacytoid DCs (pDCs), Langerhans cells and monocyte-derived DCs. Based on differences in function, phenotype markers and tissue localization, these DCs are also divided into migratory DCs (migDCs), which actively traffic and deliver antigens from tissues to draining lymph nodes (LNs), or LN/spleen resident DCs (cDC and pDC), who spend their entire lifespan in the lymphoid organs. Each DC subset has a distinct role in immunity, infection, tumor presentation, or in the maintenance of self-tolerance because of their different intrinsic abilities to capture, process and present antigens on their MHC I and MHC II molecules (Villadangos & Schnorrer, 2007). MigDCs can actively capture and present antigens located in peripheral tissues and transport antigens to the draining lymph nodes. There, they may present antigens directly or transfer those antigens resident CD8 + DCs for presentation. This kind of cooperation between migDCs and resident DC subsets allows for the exploitation of varied functionality between different DC subsets and helps to optimize the capacity of the DCs network in responding to different infections. MigDCs and cDCs also capture protein and particulate antigens arriving at the draining LN through the lymphatics without active transport by a carrier cell. As such the spatio-temporal regulation of DC behavior is complex and needs to be better dissected. As an additional layer of complexity the half-life and turn-over rate of DCs varies both by subset and by location, again highlighting the dynamic and complex spatio-temporal behavior of these populations (Kamath et al., 2002).

2.1 Models of antigen presentation: An overview

Three main different antigen-presenting pathways exist, including MHC I, MHC II pathway and cross-presentation pathway. Antigens that are synthesized intracellularly (for example, those from viral or tumor) are presented by MHC I molecules to activate cytotoxic CD8+ T cells. Intracellular antigens, are presented by MHC I molecules in the endoplasmic reticulum (ER) after the antigens translocate from the cytosol. These are derived from proteins degraded mainly in the cytosol, including pathogens replicating in the cytoplasm (such as viruses) or endogenous proteins (produced by the cells themselves) (Blum, Wearsch, & Cresswell, 2013; Heath et al., 2004).

In contrast, extracellular antigens are presented by MHC II molecules to activate CD4 + T helper cells. Extracellular antigens, which are presented by MHC II molecules in an endocytic and phagocytic pathways, are derived from proteins degraded in the endosomal compartments, including exogenous materials, which are endocytosed from extracellular environment, and endogenous molecules, such as plasma membrane proteins, components of the endocytic pathway and cytosolic proteins which gain access to the endosomes through autophagy.

Extracellular antigens can also be presented to stimulate CD8+ T cells via the MHC I pathway; however, this often requires presentation by DCs specialized to take up dying cells. This is required for immunity against viruses that do not infect APCs directly, or against tumor antigens that are not endogenously expressed by DCs. Specific subsets of DCs are particularly adept at this activity termed "cross-presentation." These DCs who have the unique ability to deliver exogenous antigens to the MHC I molecules are therefore crucial in anti-virus and anti-tumor immune responses. Cross-presentation is critically important to prime CD8+ T cell-mediated primary immune response (Blum et al., 2013). In cross-presenting DCs, MHC II and MHC I cross-presentation pathways might compete for exogenous antigens. The endocytic mechanism involved in internalization of a given antigen could also affect whether it is preferentially delivered to the MHC II pathway or the MHC I cross-presentation pathway. Therefore understanding of the differential modes of antigen capture (by differential endocytic mechanisms and receptors) and the application of an antigen-targeting method will help us better explore the key roles played and balanced by DC subsets to cross-present antigens.

2.1.1 Cross-presentation pathway

Cross-presentation is the process by which exogenous antigens are processed and presented in the MHC I pathway (Bevan, 1976a, 1976b). Both *in vivo* and *in vitro* studies suggest MHC I-mediated antigen cross-presentation to CD8+ T cells is associated with cell-associated antigens, which are a physiological substrate for cross-presentation (Gutierrez-Martinez et al., 2015; Schulz & Reis e Sousa, 2002; Thacker & Janssen, 2012).

While multiple APCs are reported to be able to cross-present antigens, DCs are the most efficient cells *in vivo* at this process (Barrio et al., 2012; Leiriao, del Fresno, & Ardavin, 2012; Milo et al., 2013). DCs are ubiquitous in the tissue, and uniquely able to not only pick up dead cells, but to migrate to the lymph nodes, where they can cross-present cell-associated antigens to CD8 + T cells (Gutierrez-Martinez et al., 2015). The ability of DCs to cross-present antigens has inspired many studies finding strategies to target DCs, enhance their cross-presentation, and improve anti-tumor and anti-viral immunity for the treatment of cancer or infectious disease.

MHC I-mediated antigen cross-presentation pathway contains five basic steps: antigen acquisition, antigen tagging for destruction and proteolysis, antigen transporting to ER, antigen loading to MHC I molecules, antigen-MHC I complexes displaying on the cell surface (Fehres et al., 2014).

MHC I-mediated cross-presentation requires the processing and trimming of the endocytosed antigenic proteins. This generally takes place in two pathways: *the cytosolic and the vacuolar pathways. In the cytosolic pathway*, internalized antigens need to be transferred to the cytosol from the endosomal compartment. They are then degraded by the proteasome, followed by an antigen processing (TAP)-dependent transporting to the ER or endosomes, where final peptide trimming and MHC I-peptide loading take place (Thacker & Janssen, 2012). Studies suggest inhibition of TAP in endosomes or inhibition of endosomal trafficking to the cell membrane abrogates the cross-presentation of soluble OVA (Burgdorf et al., 2008). While the majority of published studies report the cytosolic pathway as the major cross-presentation pathway, there is also substantial evidence indicating antigen cross-presentation in an proteasome- and TAP-independent pathway, through the vacuolar pathway. *In the vacuolar pathway, antigens are processed and loaded onto MHC I molecules in the endosomal/lysosomal compartment.* This does not require antigens to exit the endosomal compartment. Instead, antigenic proteins are degraded by lysosomal enzymes and peptides are locally generated and trimmed in the endosome. These then further bind onto MHC I molecules directly. The comparatively contribution of the cytosolic and vacuolar pathways to tumor antigen cross-presentation *in vivo* is not entirely clear however (Embgenbroich & Burgdorf, 2018; Sanchez-Paulete et al., 2017). Further experiments are needed to investigate the relative importance of both pathways *in vivo*.

2.1.2 Endocytic pathway

Endocytosis involves internalization of the molecules via cell membrane invagination and pinching off to form the early endosome, which then forms the late endosome, and then the lysosome. The endocytosis of extracellular antigens by APCs is mainly achieved through three pathways: *receptor-mediated endocytosis, phagocytosis, and pinocytosis* (Fehres et al., 2014).

The receptor-mediated endocytosis is a process by which receptors are applied for transferring antigenic proteins from extracellular matrix into the cells. It is known as clathrin-mediated endocytosis, because clathrin is crucial for this cellular process. During this endocytosis pathway, antigens derived from extracellular proteins bind onto specific receptors and then moves into clathrin-coated pit. The clathrin-coated pit gathers and concentrates extracellular antigenic proteins. Different receptors responsible for the receptor-mediated endocytosis of antigens, such as low density lipoprotein, growth factors, antibodies and so on. When these receptor carrying their respective antigens, reach the clathrin-coated pit, the pit folds inward and then, together with related membrane part, detaches itself from the cell membrane and forms a closed vesicle. These vesicles move antigens inside the cell and mark them at the same time. When the receptor-carried antigens are pathogens, opsonization system would be activated, which means those antigens are tagged for destruction. After opsonization tagging, the clathrin coat is detached to allow the vesicle to merge into early endosomes, which helps to separate the antigens from their receptors. After antigens are separated multiple chemical changes occur within the early endosomes to form late endosomes, which then split into two parts, with one part containing the antigens, and the other part with receptors. The antigens-containing endosome part then combines and merges with lysosomes, which contain multiple digestive enzymes to break up antigenic proteins. The receptors in the other endosome are then recycled and sent back to the cell surface (Kaksonen & Roux, 2018).

Phagocytosis is a process by which larger, extracellular, solid particles, such as bacteria are captured into phagosomes. It is sometimes referred to as "cell eating." It involves the application of larger membrane areas than receptor-mediated endocytosis. Phagocytosis is an active and highly regulated process that involves specific cell surface receptors and signaling cascade. This process is performed only by special types of cells such as macrophages, monocytes, and neutrophils. In mammals, it is largely concerned with the elimination of foreign materials and pathogens. Viruses use phagocytosis to enter cells. Phagocytic cells are usually very specific in the particles they engulf.

Pinocytosis is a process of internalizing extracellular fluids along with their solute molecules or antigens into pinosomes. It is also referred to as "cell drinking" compared to phagocytosis. In contrast to phagocytosis usually, pinocytic cells are not specific in the molecules they uptake. Pinocytosis includes several unrelated endocytic mechanisms, in which micropinocytosis is the most common reported one. Except for micropinocytosis, there is also a group of pinocytic pathways that use different proteins to coat the endocytic vesicles derived from the plasma membrane. In addition, some pinocytic pathways involve different membrane lipids or lipid-modifying enzymes (Guermonprez et al., 2002).

The main differences between phagocytosis and pinocytosis are as follows. First, phagocytosis ingests comparatively large solid particles, like bacteria and amoeboid protozoans, whereas pinocytosis ingests liquid into the cell by budding a small vesicle from the cell membrane. Second, pinocytosis is not substrate-specific and the cell takes in all kinds of surrounding fluids with solutes, while phagocytosis involves specificity in substrate uptake. Third, the purpose of pinocytosis is to uptake of materials, including enzymes, hormones, amino acids, sugars, etc. Phagocytosis is for defensive purpose and ingesting dust, foreign particles, harmful bacteria and viruses and is performed by neutrophils, macrophages, and protozoans.

Although there are differences in the efficiency of these pathways among DCs, B cells, and macrophages, macrophages seem to be best at phagocytosis, whereas DCs prefer receptor-mediated endocytosis (Fehres et al., 2014). Furthermore, DC heterogeneity dictates differences in the relative efficiency of processing and presentation after antigen capture which vary by DC subtype and by route. Prior work measured the internalization and presentation capacity of DCs to MHC I and MHC II-restricted T cells by administering OVA to conventional DCs through either receptor-mediated endocytosis, or pinocytosis, or phagocytosis (Kamphorst et al., 2010). They found that $CD8^-$ DCs are more efficient than any other type of APCs tested at presenting antigens to MHCII-restricted T cells, irrespective of the route of antigen capture. In contrast, both subsets of splenic DCs are highly effective in cross-presenting antigens to CD8+ T cells. And also, DCs and activated monocytes could cross-present antigens delivered by DEC205-mediated endocytosis and pinocytosis.

There are many types of receptors which could mediate endocytosis of antigens, such as B cell receptor (specific for B cells), Fc receptors, heat-shock protein receptors, scavenger receptors, and the C-type lection receptors (CLRs). Overall, these receptors mediate internalization of related antigens to endosomes. However, the nature of the related endosomes and theirs fate seem to vary for different receptor types involved and finally, also their ability in inducing antigen cross-presentation (Fehres et al., 2014).

3. Monitoring receptor-mediated antigen uptake *in vivo*

Delivering antigens *in vivo* by coupling them to antibodies specific for unique receptors and pathways of internalization and presentation on DCs is a highly promising approach for conditioning immunogenic or tolerogenic immune responses. However in addition to their use as a means to regulate immune responses, antigen-targeting systems can also be a useful method to investigate the function of DC subsets and the cellular mechanisms involved (Chappell et al., 2014).

DCs have a number of specialized receptors for antigen uptake which include Fcγ receptors and the macrophage mannose receptors, typically shared with other cells, and C-type lectins such as CD205, CD207, and CD209, which can be more DC restricted, depending on species. CD205 (DEC-205) is an endocytic receptor reported to have special ability to uptake and subsequently direct antigens to processing and loading onto both MHC II and MHC I molecules (Bonifaz et al., 2002, 2004). CD205 is highly expressed on certain DC subsets including cross-presenting and tissue migratory DCs in mouse (Anandasabapathy et al., 2014). Anti-CD205 has been used to target antigens specifically to DCs in mice (Mukherjee et al., 2013) and is in current use in human cancer clinical trials. Moreover targeting CD205 and other C-type lectins is a useful model by which to study the spatio-temporal regulation of and subset specific antigen capture by DCs. Prior studies using this strategy have highlighted important differences in the kinetic capture ability of DCs present in the draining LN having migrating to the draining LN from peripheral tissues, and those which are always present in the draining LN without having surveyed the tissues (Anandasabapathy et al., 2014). Such differences are linked to important functional outcomes for immunity and tolerance *in vivo*.

To study antigen capture on CD205 *in vivo, labeled* anti-CD205 antibodies to CD205 receptors on DCs can be used and this method is provided below (Jiang et al., 1995). Anti-CD205 antibodies are rapidly taken up by receptor-mediated endocytosis into coated pits and vesicles, and then transferred to a multi-vesicular endosomal compartment, which resembles the MHC class II-containing vesicles involved in antigen presentation (Jiang et al., 1995). CD205 direct antigens to the late endosomes on human cross-presenting DCs (Cohn et al., 2013). Anti-CD205 antibodies provide an efficient method for targeting DCs to evaluate receptor-based antigen uptake under various conditions and by different cell types. This is a useful measure of cross-presentation capacity and the study of tolerance in the steady state or immunity when administered with agents to enable DC maturation (Bonifaz et al., 2002).

4. Protocol for anti-CD205-mediated antigen uptake *in vivo*

4.1 Materials

- 0.5 mL Insulin Syringes with the Micro-Fine™ Needle (28G needle)
- Six-well plate
- 10 cm Petri dish
- 3 mL syringe with 25G needle
- curved forcep, straight forcep
- 15 mL conical tube, 50 mL conical tube
- short Pasteur pipettes with cotton top
- 70 μm cell strainer

4.2 Reagents

- Isoflurane
- Fluorophore conjugated anti-mouse CD205 (DEC-205) Antibody
- Fluorophore conjugated Isotype Antibody
- 70% EtOH
- R5 medium

Components of R5 medium	Volume
RPMI	500 mL
FBS	30 mL
L-Glutamine	5 mL
Pen Strep (10,000 U/mL)	5 mL
HEPES (1 M)	5 mL
Gentamicin (10 mg/mL)	2.5 mL
2-Mercaptoethanol (1000 ×)	500 μL

- 1 × DPBS
- Hanks Balanced Salt Solution (HBSS) with Calcium chloride and Magnesium chloride (HBSS +/+)
- HBSS without Calcium chloride and Magnesium chloride (HBSS −/−)
- CollagenaseD and DNase I
- 0.5 M EDTA (quenches collagenase and dissociates T cell/DC complexes to enhance the isolation)
- FACS buffer (1 × DPBS with 2% FBS)
- Anti-CD16/CD32 FC block

4.3 Process (Fig. 1)

4.3.1 Footpad injection

- Prepare 2.5 μg antibodies in 100 μL of sterile PBS for each mouse.
- Anesthetize mice with isoflurane and give 2.5 μg antibodies to each mouse through footpad injection in 50 μL per footpad. Administer isotype control antibodies to the control group separately in parallel.

4.3.2 Tissue harvest

- 1, 3, and 5 h after antibodies injection, harvest popliteal lymph nodes (LNs), inguinal LNs, spleen, brachial LNs, cervical LNs, from each individual mouse.
- Put harvested tissues into R5 medium before processing.

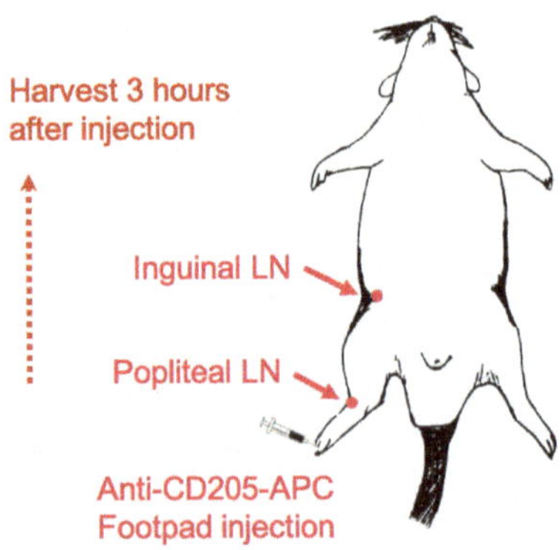

Fig. 1 Anti-CD205 antibody is administered by footpad injection. Popliteal and inguinal LNs are harvested and analyzed 3 h later.

4.3.3 DCs preparation from tissues

4.3.3.1 Spleen and LN

- Sacrifice animals, soak in 70% ethanol or spray them with 70% ethanol.
- Dissect the mice and get LNs and spleen, put them in the individual six-well plates (contain 5 mL R5 media in each well).
- Wash and transfer them in new wells of the six-well plates (containing 4.5 mL of HBSS +/+ in each well), add 500 μL collagenase (the final concentration would be 2.5 mg/mL) to each well right before teasing/ballooning. Balloon spleens with medium containing collagenase D. When spleen turns "white," tease spleen into small pieces.
- Using curved forceps and 3 mL syringe with a 25 g needle, tease apart LNs into small pieces (in medium containing 2.5 mg/mL collagenase D). Tease very well to the point when no floating pieces of stroma are left.
- Incubate LNs and spleen at 37 °C (incubator) for 22–28 min (22 min for spleen and 28 min for LN).
- Use a glass pipet to bulb pipet up and down multiple times, then add 100 μL of 0.5 M EDTA to stop the digestion, pipet up and down again, then swirl dish and put back in incubator for 5 min more.
- Add a full volume of 2% FACS buffer or R5 medium to stop the EDTA after 5 min.
- Transfer and filter cells into 15 mL conical tube through 0.7 μm strainer using short cotton glass Pasteur pipet and then spin down at 1500 rpm for 5 min.
- For LNs, vacuum off supernatant and resuspend the cells in 5 or 10 mL R5 medium.
- For the spleen, vacuum off supernatant and resuspend in 1 mL ACK lysis buffer (each spleen), lysis for 15 s, then add FACS buffer to stop the lysis, spin at 1500 rpm for 5 min. Vacuum supernatant and resuspend in 10 mL R5 medium.
- Count the cells from LN and spleen.

4.3.4 DCs staining to check antigen uptake through flow cytometry analysis

4.3.4.1 Surface marker staining

- Plate the cells into a 96-well plate V bottom plate for staining on ice in PBS or FACS buffer as indicated.
- Staining cocktail should leave the fluorophore channel empty which was used for antibody targeting.

- 250 μL staining system, up to 5×10^6 cells (the most) per 250 μL.
- Wash cells twice with PBS.
- Stain cells with aqua (11000) dilution in PBS for 30 min on ice.
- Wash with DPBS once and then with FACS buffer once.
- Block the cells in FC block (anti-CD16/CD32, 1:200), 250 μL/each well, and Incubate in FC block for 20–25 min.
- Wash with FACS twice.
- Prepare staining cocktail in FACS buffer and incubate stained cells for 45 min on ice.
- Wash cells with FACS buffer twice.
- Fix the cells using BD Fix/Perm for 10 min on ice (100 L/per well).
- Wash cells with BD 1 × Perm/wash buffer, twice.
- For the spleen and LN, resuspend cells in FACS, transfer cells to FACS tube.

4.3.4.2 Intracellular marker staining

- After the surface marker staining mentioned above.
- Wash cells with 1 × BD Perm/wash buffer twice.
- Prepare intracellular staining cocktail in 1 × BD Perm/wash buffer and incubate cells for 30 min on ice.
- Wash twice with 1 × BD Perm/wash buffer.
- Resuspend cells in FACS, and then transfer cells to FACS tube.
- Running samples and analyze the data. Compare the relative antigen capture of anti-CD205 antibody to isotype control by site specific uptake for each.

Acknowledgments

The authors are grateful to the ongoing wisdom of Ralph M. Steinman whose passion for Dendritic Cells was contagious.

References

Anandasabapathy, N., et al. (2014). Classical Flt3L-dependent dendritic cells control immunity to protein vaccine. *The Journal of Experimental Medicine*, *211*(9), 1875–1891.

Barrio, M. M., et al. (2012). Human macrophages and dendritic cells can equally present MART-1 antigen to CD8(+) T cells after phagocytosis of gamma-irradiated melanoma cells. *PLoS One*, 7(7) e40311.

Bevan, M. J. (1976a). Cross-priming for a secondary cytotoxic response to minor H antigens with H-2 congenic cells which do not cross-react in the cytotoxic assay. *The Journal of Experimental Medicine*, *143*(5), 1283–1288.

Bevan, M. J. (1976b). Minor H antigens introduced on H-2 different stimulating cells cross-react at the cytotoxic T cell level during in vivo priming. *Journal of Immunology*, *117*(6), 2233–2238.

Blum, J. S., Wearsch, P. A., & Cresswell, P. (2013). Pathways of antigen processing. *Annual Review of Immunology*, *31*, 443–473.

Bonifaz, L., et al. (2002). Efficient targeting of protein antigen to the dendritic cell receptor DEC-205 in the steady state leads to antigen presentation on major histocompatibility complex class I products and peripheral CD8+ T cell tolerance. *The Journal of Experimental Medicine*, *196*(12), 1627–1638.

Bonifaz, L. C., et al. (2004). In vivo targeting of antigens to maturing dendritic cells via the DEC-205 receptor improves T cell vaccination. *The Journal of Experimental Medicine*, *199*(6), 815–824.

Burgdorf, S., et al. (2008). Spatial and mechanistic separation of cross-presentation and endogenous antigen presentation. *Nature Immunology*, *9*(5), 558–566.

Chappell, C. P., et al. (2014). Controlling immune responses by targeting antigens to dendritic cell subsets and B cells. *International Immunology*, *26*(1), 3–11.

Cheong, C., et al. (2010). Improved cellular and humoral immune responses in vivo following targeting of HIV Gag to dendritic cells within human anti-human DEC205 monoclonal antibody. *Blood*, *116*(19), 3828–3838.

Cohn, L., et al. (2013). Antigen delivery to early endosomes eliminates the superiority of human blood BDCA3+ dendritic cells at cross presentation. *The Journal of Experimental Medicine*, *210*(5), 1049–1063.

Embgenbroich, M., & Burgdorf, S. (2018). Current concepts of antigen cross-presentation. *Frontiers in Immunology*, *9*, 1643.

Fehres, C. M., et al. (2014). Understanding the biology of antigen cross-presentation for the design of vaccines against cancer. *Frontiers in Immunology*, *5*, 149.

Guermonprez, P., et al. (2002). Antigen presentation and T cell stimulation by dendritic cells. *Annual Review of Immunology*, *20*, 621–667.

Gutierrez-Martinez, E., et al. (2015). Cross-presentation of cell-associated antigens by MHC class I in dendritic cell subsets. *Frontiers in Immunology*, *6*, 363.

Hawiger, D., et al. (2004). Immunological unresponsiveness characterized by increased expression of CD5 on peripheral T cells induced by dendritic cells in vivo. *Immunity*, *20*(6), 695–705.

Heath, W. R., et al. (2004). Cross-presentation, dendritic cell subsets, and the generation of immunity to cellular antigens. *Immunological Reviews*, *199*, 9–26.

Jiang, W., et al. (1995). The receptor DEC-205 expressed by dendritic cells and thymic epithelial cells is involved in antigen processing. *Nature*, *375*(6527), 151–155.

Kaksonen, M., & Roux, A. (2018). Mechanisms of clathrin-mediated endocytosis. *Nature Reviews. Molecular Cell Biology*, *19*(5), 313–326.

Kamath, A. T., et al. (2002). Developmental kinetics and lifespan of dendritic cells in mouse lymphoid organs. *Blood*, *100*(5), 1734–1741.

Kamphorst, A. O., et al. (2010). Route of antigen uptake differentially impacts presentation by dendritic cells and activated monocytes. *Journal of Immunology*, *185*(6), 3426–3435.

Kastenmuller, W., et al. (2014). Dendritic cell-targeted vaccines—Hope or hype? *Nature Reviews. Immunology*, *14*(10), 705–711.

Leiriao, P., del Fresno, C., & Ardavin, C. (2012). Monocytes as effector cells: Activated Ly-6C(high) mouse monocytes migrate to the lymph nodes through the lymph and cross-present antigens to CD8+ T cells. *European Journal of Immunology*, *42*(8), 2042–2051.

Macri, C., et al. (2016). Targeting dendritic cells: A promising strategy to improve vaccine effectiveness. *Clinical & Translational Immunology*, *5*(3) e66.

Milo, I., et al. (2013). Dynamic imaging reveals promiscuous crosspresentation of blood-borne antigens to naive CD8+ T cells in the bone marrow. *Blood*, *122*(2), 193–208.

Mukherjee, G., et al. (2013). DEC-205-mediated antigen targeting to steady-state dendritic cells induces deletion of diabetogenic CD8(+) T cells independently of PD-1 and PD-L1. *International Immunology*, *25*(11), 651–660.

Sanchez-Paulete, A. R., et al. (2017). Antigen cross-presentation and T-cell cross-priming in cancer immunology and immunotherapy. *Annals of Oncology*, *28*(Suppl. 12), xii44–xii55.
Schulz, O., & Reis e Sousa, C. (2002). Cross-presentation of cell-associated antigens by CD8alpha+ dendritic cells is attributable to their ability to internalize dead cells. *Immunology*, *107*(2), 183–189.
Steinman, R. M. (2010). Some active areas of DC research and their medical potential. *European Journal of Immunology*, *40*(8), 2085–2088.
Steinman, R. M. (2012). Decisions about dendritic cells: Past, present, and future. *Annual Review of Immunology*, *30*, 1–22.
Steinman, R. M., & Cohn, Z. A. (1973). Identification of a novel cell type in peripheral lymphoid organs of mice. I. Morphology, quantitation, tissue distribution. *The Journal of Experimental Medicine*, *137*(5), 1142–1162.
Steinman, R. M., & Witmer, M. D. (1978). Lymphoid dendritic cells are potent stimulators of the primary mixed leukocyte reaction in mice. *Proceedings of the National Academy of Sciences of the United States of America*, *75*(10), 5132–5136.
Tacken, P. J., Torensma, R., & Figdor, C. G. (2006). Targeting antigens to dendritic cells in vivo. *Immunobiology*, *211*(6–8), 599–608.
Tesfaye, D. Y., et al. (2019). Targeting conventional dendritic cells to fine-tune antibody responses. *Frontiers in Immunology*, *10*, 1529.
Thacker, R. I., & Janssen, E. M. (2012). Cross-presentation of cell-associated antigens by mouse splenic dendritic cell populations. *Frontiers in Immunology*, *3*, 41.
Ueno, H., et al. (2011). Targeting human dendritic cell subsets for improved vaccines. *Seminars in Immunology*, *23*(1), 21–27.
Villadangos, J. A., & Schnorrer, P. (2007). Intrinsic and cooperative antigen-presenting functions of dendritic-cell subsets in vivo. *Nature Reviews. Immunology*, 7(7), 543–555.

CHAPTER TWENTY-TWO

Flow cytometry-based assessment of direct-targeting anti-cancer antibody immune effector functions

Michelle L. Miller, Olivera J. Finn*
Department of Immunology, University of Pittsburgh School of Medicine, Pittsburgh, PA, United States
*Corresponding author: e-mail address: ojfinn@pitt.edu

Contents

Abstract

Monoclonal antibody-based therapies are increasingly being used to treat cancer. Some mediate their therapeutic effects through modifying the function of immune cells globally, while others bind directly to tumor cells and can recruit immune effector cells through their Fc regions. As new direct-binding agents are developed, having the ability to test their Fc-mediated functions in a high-throughput manner is important for selecting antibodies with immune effector properties. Here, using monoclonal anti-CD20 antibody (rituximab) as an example and the $CD20^+$ Raji cell line as tumor target, we describe flow cytometry-based assays for determining an antibody's capacity for mediating antibody-dependent cellular cytotoxicity (ADCC), antibody-dependent cellular phagocytosis (ADCP) and complement-dependent cytotoxicity (CDC). These assays are sensitive, reliable, affordable and avoid the use of radioactivity.

Methods in Enzymology, Volume 632
ISSN 0076-6879
https://doi.org/10.1016/bs.mie.2019.07.026

1. Introduction

1.1 Fc-dependent mechanisms of antibody effector functions

Antibodies can exert effector functions through their Fab domains by binding to their targets on cells and blocking key signaling pathways or, in certain settings, by activating certain pathways, either through mimicking the natural ligand or though crosslinking their target antigens. Based on this, antibodies can be classified as antagonistic or agonistic. The majority of effector functions of antibodies, however, are mediated through their Fc-regions that either bind to Fc receptors on immune effector cells, such as NK cells or macrophages (ADCC and ADCP, respectively), or to effector molecules that can initiate the complement cascade (CDC, Fig. 1). Tumor-targeting antibodies such as rituximab (anti-CD20, Rituxan), alemtuzumab (anti-CD52, CAMPATH) and many others were initially thought to cause direct cytotoxicity through disruption of cell signaling pathways, but more recently studies have determined that they operate primarily through their Fc-mediated immune effector functions (reviewed in Golay & Introna, 2012; Lindorfer, Wiestner, Zent, & Taylor, 2012). One important determinant of Fc-mediated functions is the antibody isotype, as this dictates which Fc receptors (FcRs) the antibody interacts with.

Antibodies undergo class-switching from IgM to IgD, IgG, IgE and IgA Fc isotypes. In addition to these main isotypes, just within IgG there are four subclasses of Fc isotypes, IgG1, IgG2, IgG3, and IgG4 in humans and IgG1, IgG2b, IgG3 and IgG2a or IgG2c in mice. Binding of Ig Fc domains to their corresponding Fc receptors initiates and coordinates a targeted immune attack against many different classes of antigens expressed on a large variety of target cells. Most cancer-targeting therapeutic antibodies are of the IgG1 class, which is the most abundant IgG isotype in the serum that binds to Fc receptors with high affinity.

1.2 Fc optimization

Extensive work has been done to improve antibody effectiveness by optimizing their Fc regions, especially IgG1 Fc, through mutations and glycoengineering (reviewed in Bruhns & Jönsson, 2015; Rose et al., 2013). There are several known mutations that enhance ADCC, ADCP, and CDC (Lee et al., 2017; Moore, Chen, Karki, & Lazar, 2010), (also see table 1 in Wang, Mathieu, & Brezski, 2018), as well as selective binding to FcγRs

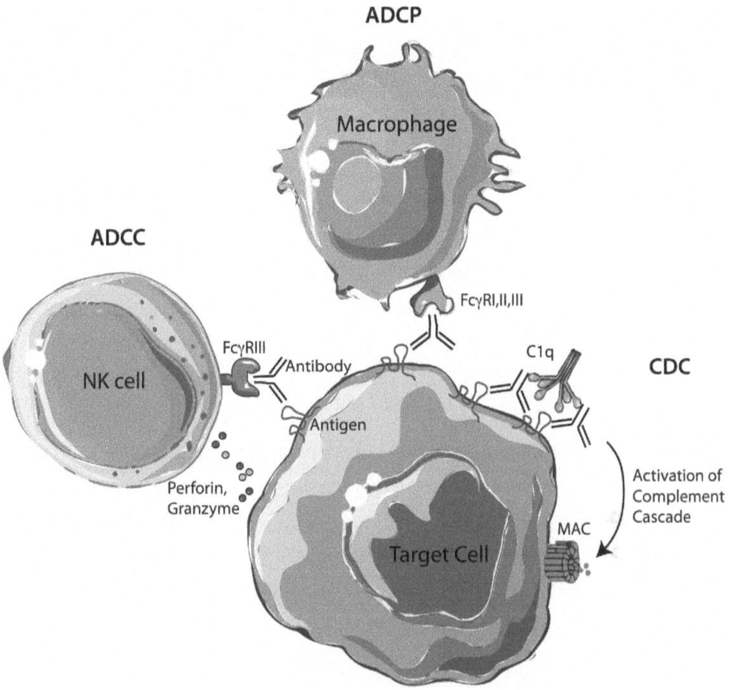

Fig. 1 Antibodies mediate immune effector functions through their Fc domains. Antibodies bound to their antigens on target cells with their Fab domains and that also bind through their Fc domains, activating FcγRs on NK cells (such as FcγRIIIA) can mediate antibody-dependent cellular cytotoxicity (ADCC). Antibodies that opsonize target cells and also bind to activating FcγRs on macrophages can mediate antibody-dependent cellular phagocytosis (ADCP), in which the target cell is then engulfed by the macrophage. Antibodies that bind C1q in a set arrangement can facilitate activation of the complement cascade, leading to the formation of the membrane attack complex (C5-C9, the MAC), and complement-dependent cytotoxicity (CDC).

expressed on dendritic cells, which can enhance soluble antigen uptake and promote a vaccine-like effect (Bruhns & Jönsson, 2015).

One modification that improves binding to FcγRIIIA and enhances ADCC is lack of fucosylation at the Fc N297 glycosylation site (Shields et al., 2002). A recent study used directed evolution to identify mutations on aglycosylated Fc domains that still retain ADCC function (Jo, Kwon, Lee, Lee, & Jung, 2018). There are also engineered mutations (E345K and E430G and E345R/E430G/S440Y) that disrupt a salt bridge between CH2 and CH3 in the Fc domain (de Jong et al., 2016). Mutations at K326W, K326W/E333S, S267E, S239D/S267E/H268F/S324T and S239D/H268F/S324T/I332E increase C1q binding affinity (Idusogie et al., 2001; Moore et al., 2010;

Tammen et al., 2017). Certain modifications that enable selective binding of C1q, such as aglycosylated (T299L) K320E/Q386R and aglycosylated (T299L) L245K/G246M/G247R/L351Q, can promote complement-dependent cell-mediated cytotoxicity (CDCC) and complement-dependent cell-mediated phagocytosis (CDCP) (Lee et al., 2017).

Modifications that abolish selective antibody effector functions are also known, such as the N297A mutation that eliminates an N-glycosylation site critical for FcR binding or production of antibodies in cells that do not glycosylate, such as *E. coli* (Simmons et al., 2002; Walker, Lund, Thompson, & Jefferis, 1989). A couple of mutations that have been identified that diminish binding to FcγRIIA and FcγRIIB, such as R292A and K414A (Shields et al., 2001) could improve the therapeutic antibody effect through selective reduction of binding to the inhibitory FcγRIIB while retaining binding to the activating FcγRIIIA. Although yet untested, insight into why these mutations may be effective come from data showing that a blocking antibody that binds FcγRIIB synergized with anti-CD20 antibodies *in vivo* (Roghanian et al., 2015), and that a SHIP-1 inhibitor diminished FcγRIIB signaling to increase ADCC/ADCP (Burgess et al., 2017). Another example is the addition of an A330L mutation to a S239D/I332E Fc variant that abolished CDC while retaining enhanced ADCC (Lazar et al., 2006). Known polymorphisms in Fc receptors, including FcγRIIA 131H and FcγRIIIA 158V with their higher affinity for IgG1 Fc correlate with better therapeutic outcomes (Cartron et al., 2002; Ferris, Jaffee, & Ferrone, 2010; Musolino et al., 2008; Yin, Albers, Smith, Riddell, & Richards, 2018), recently reviewed in Mellor, Brown, Irving, Zalcberg, and Dobrovic (2013).

1.3 Regulation of antibody effector functions

In addition to Fc domains and Fc receptors, other molecules can regulate ADCC, ADCP and CDC. Some cancer cells have intrinsic resistance to ADCC by their differential gene expression. This includes c-Abl (Murray et al., 2014), WEE1 kinase (Friedman et al., 2018), caveolin-1 (Sekhar et al., 2013) and oncogenic RAS (Kasper et al., 2013; Nakadate et al., 2014). Others have alterations in apoptotic factors including overexpression of survivin, Bcl-X_L and YY1 that correlate with resistance to antibody-mediated effects (Dalle et al., 2009; de Haart et al., 2016). CD74 overexpression, increased histone- and interferon-related gene expression, and downregulation of HSPB1 and target antigen expression have also been found in ADCC-resistant cells (Aldeghaither et al., 2019). Surface expression of cell adhesion molecules is decreased in resistant cells and hyaluronan, an extracellular matrix protein, is overexpressed, suggesting the importance of a close synapse between effector and target cells (Aldeghaither et al., 2019; Singha et al., 2015). Similarly, disruption of E-cadherins increases susceptibility to ADCC through its disruption of tumor cell-tumor cell contacts (Green, Karlsson, Ravetch, & Kerbel, 2002). Other cancer cells express NK inhibitory ligands, such as HLA-E and HLA-G that could contribute to ADCC resistance (Diepstra et al., 2008; Levy et al., 2009; Lin et al., 2007). Finally, expression and cell-surface distribution of molecules that interact with the target (such as MUC1 and MUC4 with HER2 or EGFR) has been shown to cause resistance to ADCC (Aldeghaither et al., 2019; Mercogliano et al., 2017; Namba et al., 2019).

Some tumor cells upregulate CD47, the "don't eat me signal," which can lower ADCP by binding to SIRPα on macrophages (Chao et al., 2010). Adding blocking anti-CD47 antibodies increases phagocytosis *in vitro* and *in vivo* (Chao et al., 2010). Reduced ADCC/ADCP is also seen in macrophages with increased signaling through the inhibitory FcγRIIB (Burgess et al., 2017). Elevated amounts of immune complexes, as can occur during chronic viral infections and in autoimmunity, have been shown to inhibit Fc-mediated functions (Ahuja et al., 2011; Wieland et al., 2015; Yamada et al., 2015).

Increased production by tumor cells of complement regulatory factors including CD46, CD55 (decay-accelerating factor, DAF), CD59 and Factor H may limit tumor cells' susceptibility to CDC (reviewed in Gancz & Fishelson, 2009). These factors may also be regulated by the tumor microenvironment (Kesselring et al., 2014). Downregulating these factors with siRNAs or blocking their function with antagonistic antibodies has been shown to restore CDC (Geis et al., 2010; Zell et al., 2007). Other mechanisms of CDC resistance involve increased anti-apoptotic factors (Dalle et al., 2009; Hussain et al., 2007), changes in sialylation in target cells (Bordron et al., 2018) and increased levels of heat shock proteins (Fishelson, Hochman, Greene, & Eisenberg, 2001). Low pH in the tumor microenvironment may also contribute to resistance to CDC (Dantas et al., 2016). The levels of surface antigen expression and the epitope's proximity to the cell surface also affect CDC, with lower antigen levels and more distal epitopes being less favorable (Cleary, Chan, James, Glennie, & Cragg, 2017; Golay et al., 2001; Loeff et al., 2017; Moreno et al., 2019; Ragupathi et al., 2005).

On the effector cell side, STING agonists have been shown to increase Fc gamma receptor expression on NK cells (Dahal et al., 2017). Similarly,

TLR agonists, NKGD ligands and cytokines GM-CSF, G-CSF, IFN-α, IL-2, IL-12, IL-15 and IL-21 that activate effector cells can increase FcR expression (Cornet et al., 2016; Dahal, Gadd, Edwards, Cragg, & Beers, 2018; Shah et al., 2013). The addition of blocking antibodies for inhibitory receptors or agonistic antibodies to activating receptors on effector cells can augment effector functions (reviewed in Kohrt et al., 2012). Finally, effector cells may be exhausted by high circulating levels of therapeutic antibody, thus requiring dose and schedule optimization (Kennedy et al., 2004; Lindorfer et al., 2012).

2. Antibody-dependent cellular cytotoxicity (ADCC)

ADCC is mediated by the Fc-regions of antibodies bound to their cognate antigens on the target cells binding to activating Fc receptors on effector cells, such as NK cells and macrophages. Thus activated, effector cells release cytotoxic granules containing perforins and granzymes as well as cytokines such as IFNγ, that can kill the target cells. ADCC is primarily mediated by IgG1 and IgG3 antibodies that bind to FcγRIIIA and IIIB. IgG3 has been shown to have higher affinity for activating FcγRs than IgG1 but a shorter half-life *in vivo*. Most FDA-approved therapeutic antibodies are of the IgG1 class.

2.1 Experimental protocol for ADCC

2.1.1 Reagents

- The medium used in all procedures is "complete RPMI": RPMI-1640 with L-glutamine (Corning, 10-040CV) supplemented with 1% non-essential amino acids, 1% sodium pyruvate, 1% penicillin-streptomycin and 10% fetal bovine serum.
- Lymphocyte Separation Medium (LSM) gradient (Ficoll) (MP Biomedicals)
- Red Blood Cell Lysis buffer (Sigma)
- Dimethyl Sulfoxide (if using cryopreserved PBMCs)
- 24- or 48-well flat-bottom tissue culture plates
- 96-well round-bottom tissue-culture plates
- Items for magnetically enriching NK cells, including magnets, magnetic columns (LS columns, Miltenyi), antibodies, and magnetic particles
- 1× Phosphate Buffered Saline (PBS)
- Buffer for enriching cells, "MACS buffer," 0.5% Bovine Serum Albumin (BSA) in 1× PBS with 2 mM EDTA
- Buffer for staining cells, "FACS buffer" (1% BSA in 1× PBS)

- rhIL-2 (Miltenyi)
- CellTrace Yellow (ThermoFisher) or another amine-labeling cell dye, such as carboxyfluorescein succinimidyl ester (CFSE)
- Ghost Dye Red 780 (Ghost 780, Tonbo Biosciences) or another cell viability dye, such as 7-aminoactinomycin D (7-AAD) or propidium iodide (PI)

2.1.2 Assay procedure and analysis

2.1.2.1 Isolation and culture of NK cells from PBMCs

1. To isolate PBMCs from buffy coats, add 15 mL of Lymphocyte Separation Medium (LSM) gradient (Ficoll) to the bottom of a 50 mL tube and carefully and slowly overlay 35 mL of buffy coat cells on top either dropwise or slowly letting liquid run down the side of the tube so as to not disturb the interface.
2. Centrifuge at 400 × *g* for 30 min at room temperature with no acceleration and no brake in order to preserve the interface between the two liquid densities.
3. Discard some of the volume on top of the interface leaving about 10 mL and gently remove cells from the interface with a pipette into a new tube, trying to avoid taking the Ficoll. Red blood cells pellet to the bottom of the tube.
4. Wash the cells with 40 mL 1× PBS by centrifuging at 400 × *g* for 15 min with full acceleration and brake.
5. Resuspend in 2 mL of hypotonic buffer, such as ACK lysis buffer or Red Blood Cell Lysis buffer (Sigma) for 5 min at room temperature to get rid of the few contaminating red blood cells that may have come along during isolation of cells at the interface.
6. Wash the cells two more times with 40 mL 1× PBS at 300 × *g* for 10 min.
7. Aspirate supernatant and resuspend pellet in 5 mL 1× PBS.
8. Count the cells in a hemacytometer at 40× magnification.
9. Proceed to NK cell isolation. The data shown in Fig. 2 used Miltenyi's protocol for magnetic isolation of NK cells through negative enrichment on Miltenyi LS magnetic columns. Other enrichment protocols can also be used (Cosentino & Cathcart, 1987; Dittel, 2010; Shore, Melewicz, & Gordon, 1977), although the viability and effector function of NK cells post-enrichment was found to be better with magnetic enrichment (Hietanen, Pitkänen, Kapanen, & Kellokumpu-Lehtinen, 2016).

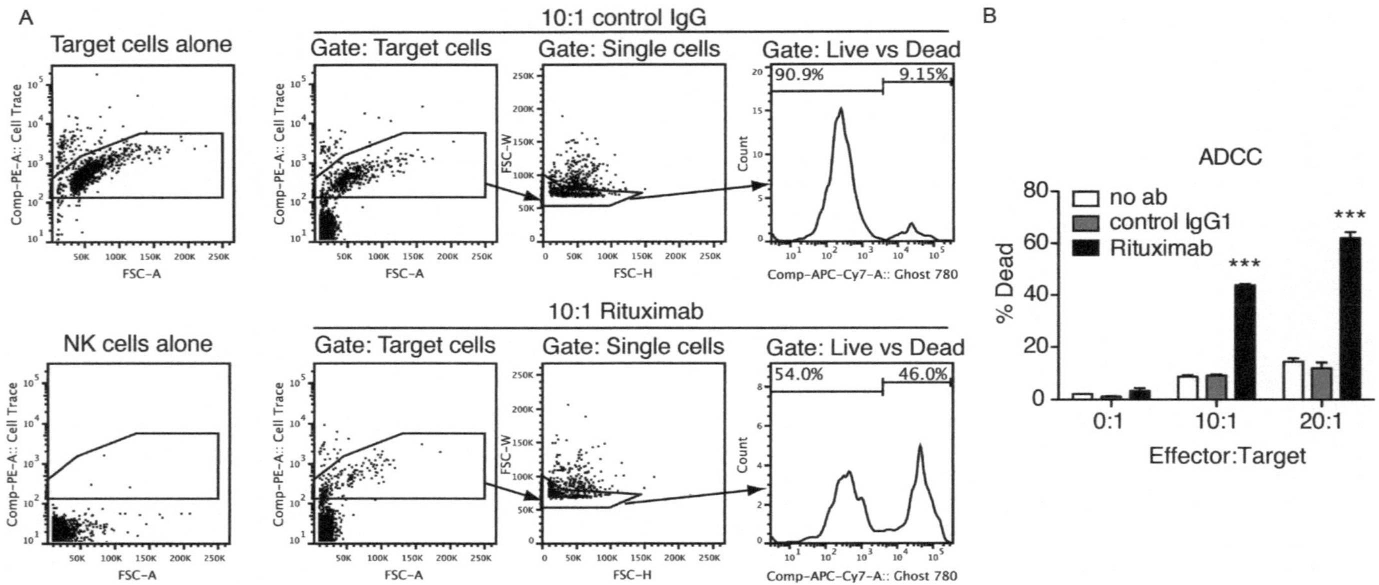

Fig. 2 Assessment of ADCC by flow cytometry. (A) Representative gating strategy, showing CellTrace Yellow-labeled target cells alone and human NK cells on the left and two examples of gating co-incubated samples, 10:1 Effector:Target (E:T) cell ratio when incubated with control IgG1 mAb (Herceptin, top right) or rituximab (bottom right three panels). (B) Bar graph depicting the percentages of dead cells (% Ghost 780^+) in each condition. Two-way ANOVA with Bonferroni correction was used to determine statistical significance, $^{***}P < 0.001$.

10. Resuspend the desired number of cells in MACS buffer (0.5% BSA in 1× PBS + 2 mM EDTA), 400 μL per 100 million cells. Add 100 μL of NK-cell biotin antibody cocktail and incubate at 4 °C for 5 min.
11. Add an additional 300 μL of MACs buffer and 200 μL of NK cell microbead cocktail per 100 million cells and incubate for 10 min at 4 °C.
12. During this incubation prime the LS column with 3 mL of MACS buffer, which can be collected in a waste container or in a 15 mL tube labeled "bound fraction."
13. Switch collection tubes. Place a new tube labeled "unbound fraction" underneath the column.
14. Add the 1 mL of sample to the column, collecting the flowthrough.
15. Add an additional 3 mL of MACS buffer to wash the column.
16. If the leftover cells are needed, switch back to the "bound cells" collection tube. Remove the column from the magnet, placing it at the top of the 15 mL tube. Add 5 mL of MACS buffer and quickly flush the liquid through the column with the accompanying plunger.
17. Count the cells in the "unbound" fraction and multiply the cells/mL by the total volume, in this case 4 mL for 100 million starting PBMCs.
18. Before centrifuging the cells, take a 50 μL aliquot to check enrichment purity by flow cytometry, using fluorescently conjugated antibodies, anti-human CD56 and anti-human CD16 (FcγRIIIA). Compare these percentages to those in pre-enrichment or "unbound" fraction samples.
19. Centrifuge the cells at 300 × *g* for 5 min, discard supernatant and resuspend pellet in complete RPMI supplemented with 100 IU/mL of rhIL-2 at 5 million cells/mL. Plate cells as close to 2 million/cm^2 as possible. Typically a 48-well or a 24-well plate offers the best surface-area to volume flexibility. Keep overnight in a 5% CO_2 incubator.

If fresh PBMCs are not available for NK cell isolation, frozen PBMCs may be used. We have successfully used PBMCs cryopreserved in 10% DMSO and 90% FBS at 50 million cells/mL for NK cell isolation. After overnight culture in 100 IU/mL rhIL-2, these thawed NK cells perform well in the assay.

2.1.2.2 Preparing effector cells and labeling target cells

1. Count the NK cells after overnight incubation, as this number will determine how many wells can be set up in the assay. The number of NK cells is usually the limiting factor, as they are used in ratios of 20:1 or 10:1 to target cells. If cell numbers allow, duplicate or triplicate wells are set up for each condition. After determining an optimal cell

ratio and antibody concentration, multiple serial dilutions may not always be necessary, but for initial experiments, especially with new antibodies, they can be quite useful.

2. Count the target cells. In our experience, the assay can be reliably performed with as few as 5000 target cells/well. Take at least twice the number of cells needed for the assay because some will be lost during washes. For the experiment in Fig. 2 we used the human lymphoma cell line Raji that grows in suspension. If using adherent target cells, trypsinize and wash the cells before counting.
3. If labeling 10 million target cells or fewer, wash cells once in 10 mL of 1× PBS and resuspend in 500 μL of 1× PBS. The CellTrace dyes bind free amines. Therefore, the cells need to be in a solution without free proteins such as FBS or BSA. Make up a second tube with 500 μL of 1× PBS and add 0.2 μL of CellTrace Yellow, constituting a "2×," 2 μM solution. Add this to the cells with rapid mixing or vortexing to promote even labeling in a final 1 μM concentration solution of CellTrace. A 10 μM concentration is commonly used for labeling for cell division tracking and can also be used here if no increased toxicity from the higher concentration of dye is noted.
4. Incubate the cells for 4 min in the dark in a 37 °C water bath. Gently mix the cells by flicking or pulse vortex them to continue promoting even labeling.
5. Incubate the cells for another 4 min. IMPORTANTLY, incubation past 8–9 min total will result in significant cell death and should be avoided.
6. At the end of the labeling incubation, immediately quench the reaction by adding 10 mL of pre-warmed complete RPMI containing at least 10% FBS and incubate the cells for 5 min in the 37 °C water bath. Then centrifuge the cells at 300 × *g* for 5 min.
7. Discard the supernatant and resuspend the cells in 0.5 mL of complete RPMI and count. Resuspend the cells such that the desired number can be added to each well of a 96-well plate in 50 μL.

2.1.2.3 Setting up the co-incubation assay

1. Make up antibody dilutions as "4×" concentrations, for adding 50 μL/well. For the example experiment, a 40 μg/mL or 10 μg/mL final concentration was used. This concentration has been shown to be effective for many antibodies, but a titration of lower antibody concentrations can be useful to find optimal effective dilutions, allowing for potential conservation of antibodies if lower concentrations retain

high efficacy. Furthermore, a titration curve may reveal additional information about each antibody and its immune effector capabilities (Tóth, Szöllősi, & Vereb, 2017).

2. Aliquot labeled target cells into a 96-well round bottom plate. Round bottom plates allow for higher cell densities in the center, and more efficient interaction between effectors and targets than flat bottom plates.
3. Aliquot antibodies into the 96-well plates. Incubate the plate at room temperature for 15 min to allow antibodies to bind to the target cells before adding NK cells.
4. Add NK cells in 100 μL/well. If interested in determining if the antibodies being tested have any direct cytotoxicity, leave one condition without NK cells.
5. Incubate the plate for 4 h at 37 °C in a 5% CO_2 incubator.

2.1.2.4 Measuring cell viability

1. Centrifuge the plate for a 30 s pulse spin in a table-top centrifuge. If pulse spinning is unavailable as an option, spin at 300 × *g* for 5 min.
2. Invert the plate rapidly to remove supernatant.
3. Wash the cells with 200 μL/well of FACS buffer to remove residual unbound antibodies. Centrifuge the plate again as in step 1 and afterwards invert the plate to remove supernatant.
4. Resuspend the pelleted cells in their residual volume by racking the bottom of the plate and add 60 μL per well of 1:1000 Ghost Dye Red 780 in 1× PBS, or similar cell viability dye. If using this dye, it must be diluted in PBS as it is amine-reactive and needs to be in a buffer without free proteins in order to bind efficiently to cells. The Ghost dye will lightly label membrane proteins on live cells, but it will have access to all cytoplasmic proteins in dead cells, resulting in much greater labeling of dead cells.
5. For Ghost dye staining, incubate the plate at 4 °C for 15 min. After this incubation, add 140 μL of FACS buffer/well to bring up the volume to 200 μL/well, and pulse centrifuge.
6. OPTIONAL: To determine the amount of remaining primary antibody bound to the surface of cells at the end of the assay, a fluorescently-conjugated secondary antibody recognizing the human IgG Fc domain can be used. This is another advantage that the flow cytometry-based assay affords over radiographic or other release-based protocols such as calcein dye or lactate dehydrogenase (LDH).

7. Resuspend the cells in the desired volume and proceed to analysis on a flow cytometer. Remember to include controls for setting up the voltages and determining all channels are working properly. At a minimum this includes unlabeled target cells, unlabeled effector cells, CellTrace labeled target cells and Ghost dye 780 labeled target cells. If also determining the NK cell purity post-enrichment and overnight culture, include single color controls for each of the markers used for that analysis. It is important to keep all samples on ice and to run individual replicates first before running duplicates or triplicates of the same condition to prevent any unwanted variation in the data during sample acquisition.
8. While we cannot provide a complete protocol for flow cytometry analysis, we can provide some helpful tips for users familiar with this analysis. Voltages for FSC-A and SSC-A need to be set such that both effector cells and target cells are visible in the same panel. Smaller, dead target cells should still be above the FSC-A threshold and visible on the graph. If the target cells are much bigger than NK cells, as can be the case for some epithelial tumor cell lines, it is better to have the target cells centered since the final analysis will be on the target cell viability. Although most first gates are made on a plot of FSC-A by SSC-A, much better resolution of target cells can be achieved if gating first on a FSC-A by CellTrace plot (see Fig. 2A). Note, we typically do not include in this first gate of target cells those that appear small by FSC-A and very CellTrace$^+$ as these cells were dead before CellTrace labeling rather than in the co-incubation assay. The next gate is on single cells, to remove doublets from the analysis, and the final gate is on the Ghost 780$^+$ cells. Therefore the "%Dead" values in Fig. 2B represent the percentages of Ghost 780$^+$ cells of single target cells that were killed during the co-incubation.
9. The background killing by NK cells in the no antibody (no ab) control wells can be subtracted from the ones with antibody when reporting the extent of ADCC. In Fig. 2 they are shown separately to emphasize the similarity to the control IgG1 group.

3. Antibody-dependent cellular phagocytosis (ADCP)

ADCP occurs though the Fc-domain of an antibody binding to and activating FcRs on effector cells such as macrophages and dendritic cells that then engulf target cells. The importance of this antibody-mediated function in cancer control has only recently been appreciated (reviewed in Gül & van

Egmond, 2015). For example, an *in vivo* study of alemtuzumab (anti-CD52, CAMPATH) in human CD52-Tg mice showed that the major mechanisms of action were through ADCC and ADCP, not through CDC, as was previously thought from *in vitro* studies (Hu et al., 2009). Intravital imaging revealed ADCP by liver Kupffer cells could mediate a substantial amount of depletion of CD20-expressing B cells in the presence of rituximab (Grandjean et al., 2016), while increased slan+ monocytes were shown to correlate with better rituximab-mediated ADCP against diffuse large B-cell lymphoma (DLBCL) (Vermi et al., 2018). Phagocytosis has also been shown to be the dominant mechanism for anti-CD20 depletion *in vivo* (Burgess et al., 2017; Uchida et al., 2004) and an important therapeutic mechanism for trastuzumab, daratumumab (anti-CD38) and elotuzumab (anti-SLAMF7) (Kurdi et al., 2018; Overdijk et al., 2015; Shi et al., 2015; Yin et al., 2018). A key clue pointing to the role of phagocytic cells in these monoclonal antibody therapies is the requirement for and modulation by certain Fc receptors expressed primarily on these cell types (Clynes, Towers, Presta, & Ravetch, 2000; Minard-Colin et al., 2008).

3.1 Experimental protocol for ADCP

3.1.1 Reagents

Complete RPMI.

1× PBS.

FACS buffer, 1% BSA in 1× PBS.

CellTrace Yellow, or another amine-reactive cell labeling dye.

CellTrace Violet, or another amine-reactive cell labeling dye in a second color.

Ghost 780, or another cell viability dye, such as 7AAD or PI.

Target cells.

Effector cells, here THP-1 cells, a human macrophage/monocyte cell line. As stocks of THP-1 cells can sometimes vary between laboratories, primary monocyte-derived macrophages can also be differentiated from human PBMCs with M-CSF and used as effectors (Herter et al., 2014).

3.1.2 Assay procedure and analysis

3.1.2.1 Labeling effector cells and target cells

1. Count THP-1 cells. If cell number allows, duplicate or triplicate wells are set up for each condition.
2. Count the target cells. Take at least twice the number of cells needed for the assay, because some cells will be lost throughout washes. In ADCP

assays target cell number is usually the limiting factor as they are used at up to 5 times the number of effector THP-1 cells. For the example in Fig. 3, we used Raji cells that grow in suspension. If using adherent cells, trypsinize and wash the cells before counting. In our experience, the assay can be reliably performed with as few as 10,000 THP-1 cells and 10,000 target cells at a 1:1 ratio.

3. Follow steps 3–7 of the ADCC cell labeling protocol. Use two separate colors, one for targets and one for effectors. In Fig. 3, CellTrace Yellow (on the PE channel) and CellTrace Violet (on the Pacific Blue channel) were used.
4. In addition to resuspending the target cells in a concentration where they can be added at 50 μL/well, resuspend the effector THP-1 cells at a concentration such that they can be added to the plate in 100 μL/well.

3.1.2.2 Setting up the co-incubation assay

1. Follow steps 1–3 of the ADCC co-incubation assay setup protocol for labeling and plating target cells with antibodies.
2. Add the THP-1 cells in 100 μL/well.
3. Incubate the plate for 1 h at 37 °C in a 5% CO_2 incubator.

3.1.2.3 Measuring phagocytosis

1. Centrifuge the plate (a 30 s pulse spin in a table-top centrifuge is sufficient).
2. Invert the plate rapidly only once to remove supernatant.
3. Wash the cells with 200 μL/well of FACS buffer to remove residual unbound antibodies. Centrifuge the plate again as in step 1. Afterward, invert the plate to remove supernatant.
4. Resuspend the cells in the desired volume, place all of the samples on ice and proceed to analysis on a flow cytometer. Remember to bring along controls for setting up the voltages and determining all channels are working properly, at a minimum: unlabeled target cells, unlabeled effector cells, and CellTrace-labeled target cells and effector cells by themselves.
5. Set the voltages for FSC-A and SSC-A ideally such that both effectors and targets are visible in the same panel. If one cell type is much larger than the other, as can be the case for some epithelial cell lines, it is better here to have the THP-1 cells centered and the target cells perhaps out of range, since the final analysis will be on the percentages of THP-1 cells that took up target cells.

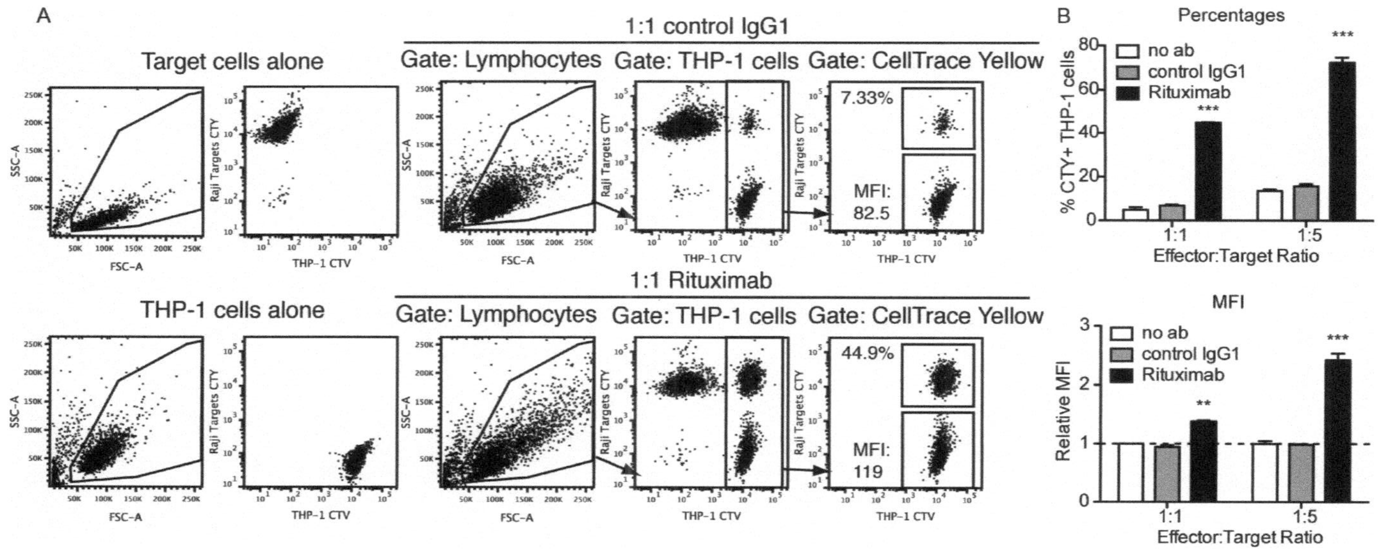

Fig. 3 Measurement of ADCP by flow cytometry. (A) Representative gating strategy, showing CellTrace Yellow (CTY)-labeled target Raji cells alone and CellTrace Violet (CTV)-labeled THP-1 macrophages on the left and two examples of gating co-incubated samples, 1:1 Effector:Target (E:T) cell ratio when incubated with control IgG1 mAb (Herceptin, top right) or rituximab (bottom right three panels). (B) Bar graphs depicting the percentages of CellTrace Violet^{+} THP-1 cells (top) in each condition, and also showing the relative mean fluorescent intensities of CellTrace Violet^{low} THP-1 cells for each condition. The MFIs from each group were normalized to the no ab control, which was set to 1. The data were analyzed by Two-way ANOVA with Bonferroni correction to determine statistical significance, $^{**}P < 0.01$; $^{***}P < 0.001$.

6. Here the first gates are made on a plot of FSC-A by SSC-A. The next plots look at CellTrace Violet (THP-1 cells) by CellTrace Yellow (target cells), in order to gate only on all of the THP-1 cells. The final gate takes the percentage of CellTrace Yellow$^+$ THP-1 cells, which represent either entire target cells that were phagocytosed or cell conjugates. Because flow cytometry alone cannot distinguish this (microscopy and flow-based microscopy can), in single THP-1 cells the MFI of CellTrace Yellow is also reported as a readout of uptake of target cell fragments. These values are reported in Fig. 3B. A very recent paper used a pH-sensitive dye to label target cells, so that it only fluoresced when the target cell was internalized in the phagocyte (Kamen et al., 2019). Others have used antibodies against additional cell surface proteins on target cells to identify phagocytosed targets as those lacking this additional staining, versus cell conjugates which would stain positively with the added target cell marker (Herter et al., 2014).

4. Complement-dependent cytotoxicity

CDC occurs when antibodies array themselves around a target antigen on the cell surface in an orientation that allows for C1q binding, clustering and initiation of the complement cascade (Taylor & Lindorfer, 2016). There are many components of the complement cascade that get recruited to ultimately form a membrane attack complex, which forms a pore in the cell membrane, killing the cell. To prevent spontaneous complement-mediated cytotoxicity on normal host cells, there are several complement regulatory factors that can inhibit the pathway.

4.1 Experimental protocol for CDC

4.1.1 Reagents

Complete RPMI with and without FBS (serum-free), see Section 2.
Non-heat inactivated human AB serum.
1× PBS.
FACS buffer: 1% BSA in 1× PBS.
Ghost 780 dye, or another cell viability dye.
Target cells of interest.
Antibodies of interest.

4.1.2 Assay procedure and analysis

4.1.2.1 Preparing the target cells

1. Count the numbers of target cells, as here this number will determine how many wells can be set up in the assay.
2. Centrifuge the cells at 300 × *g* and resuspend the cells in *serum-free* media, at a concentration of 50 μL cells/ well. We have been able to use as few as 5000 cells/well.

4.1.2.2 Setting up the co-incubation assay

3. Make up antibody dilutions as "4×" concentrations in *serum-free* media, for adding 50 μL/well. For the example experiment, a 40 μg/mL or 10 μg/mL final concentration was used. As with the other assays, a titration curve may be useful.
4. Aliquot labeled target cells into a 96-well round bottom plate.
5. Aliquot antibodies into the 96-well plates. Incubate the plate at room temperature for 15 min to allow the antibodies to pre-bind to the target cells before adding the serum.
6. Add 30% human AB serum in 100 μL/well, thus the final concentration will be 15% human serum. Be careful to keep the serum at 4 °C or on ice up until its addition to cells. Complement factors are heat-labile and can lose activity. Even with constant 4 °C storage, we have noticed loss in activity over several weeks. It is best to carefully thaw the serum and aliquot it into smaller vials that can be refrozen once and thawed just prior to the experiment. To further show it is complement-mediated killing, a separate aliquot of serum can be heat-inactivated at 56 °C for 30 min, and after it has cooled, added to the cells as an extra control.
7. Incubate the plate for 30 min at 37 °C in a 5% CO_2 incubator.

4.1.2.3 Measuring cell viability

1. Follow steps 1–5 of the ADCC cell viability measurement protocol.
2. Resuspend the cells in the desired volume and proceed to analysis on a flow cytometer. If desired, one could also measure surface bound C1q and C4b by flow cytometry (Broyer, Goetsch, & Broussas, 2013). Remember to bring along controls for setting up the voltages and determining all channels are working properly, at a minimum: unlabeled target cells and Ghost dye 780 labeled target cells. It is important to keep all samples on ice and to run individual replicates first before running

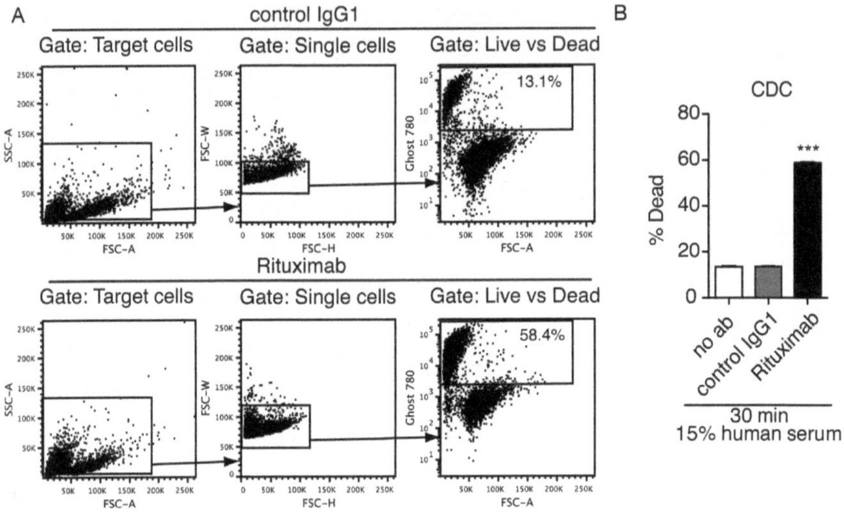

Fig. 4 Determination of CDC by flow cytometry. (A) Representative gating strategy, showing two examples of gating samples incubated with 15% human AB serum for 30 min and control IgG1 mAb (Herceptin, top) or rituximab (bottom three panels). (B) Bar graph showing the percentages of dead cells (% Ghost 780^{+}) in each condition. One-way ANOVA with Bonferroni correction was used to determine statistical significance, $***P < 0.001$.

duplicates or triplicates of the same condition to prevent any unwanted variation in the data from coming through during sample acquisition.

3. For the analysis shown in Fig. 4, the first gates are made on a plot of FSC-A by SSC-A, including both the normal lymphocyte gate and smaller cells that are likely dead. The second gate uses FSC-H and FSC-W to select single cells, and the final gate determines the percentages of cells that are dead (Ghost 780^{+}). Therefore the "%Dead" values in Fig. 4B represent the percentages of Ghost 780^{+} cells of single target cells. Due to the potential that dead cells could be fractionated into multiple smaller cell fragments that may or may not retain the ability to stain positively with the viability dye, an alternative analysis can focus on live cells to avoid any potential inflation or deflation of dead cell percentages. In this way, the starting numbers of live cells plated per well would be taken into account and normalization to no antibody or to heat-inactivated controls could be used. To quantify numbers of live cells accurately by flow cytometry, counting beads can be spiked into each well to account for any differences in flow rate during sample acquisition. As this method requires calculations across wells, particular care must be taken to plate equivalent numbers of cells in each well and acquire samples at a uniform speed.

Some target cell lines may be more susceptible to fragmentation than others. For the example in Fig. 4B, the percentages of CDC (% dead cells versus % reduction of live cells from no ab controls) is 13.4%, 13.5% and 58.8% versus 0%, 6.1%, and 53.5% for no ab, control IgG and rituximab conditions, respectively.

5. Concluding remarks

We described here our specific flow cytometry-based methods for determining three separate antibody immune effector functions: ADCC, ADCP and CDC. Each of these methods utilizes fluorescent molecules to distinguish target cells from effector cells and to measure cell viability as indicator of efficiency of tumor killing. Several other groups have recently published variations on these experimental protocols (Alrubayyi et al., 2018; Carter et al., 2019; González-González et al., 2019; Kamen et al., 2019; Tanaka et al., 2019; van der Haar Àvila, Marmol, Cany, Kiessling, & Pico de Coaña, 2019; Yamashita et al., 2016). These techniques have several advantages over other methods, including the ability to simultaneously measure the amount of target antibody bound to the cells and cell viability. Importantly, these experiments avoid the use of radioactivity. The assays are reliable and sensitive even if the number of target cells are limiting. As more anti-cancer therapeutic antibodies are developed, fast, reproducible assays such as these will help identify the candidates with the greatest likelihood of therapeutic efficacy.

Acknowledgments

We would like to thank members of the Finn lab for helpful discussions. This work was supported by NIH grants T32 CA82084 and F32 CA236457 to M.L.M. and R35 CA210039 to O.J.F.

References

Ahuja, A., Teichmann, L. L., Wang, H., Dunn, R., Kehry, M. R., & Shlomchik, M. J. (2011). An acquired defect in IgG-dependent phagocytosis explains the impairment in antibody-mediated cellular depletion in lupus. *Journal of Immunology (Baltimore, Md.: 1950)*, *187*(7), 3888–3894. https://doi.org/10.4049/jimmunol.1101629.

Aldeghaither, D. S., Zahavi, D. J., Murray, J. C., Fertig, E. J., Graham, G. T., Zhang, Y.-W., et al. (2019). A mechanism of resistance to antibody-targeted immune attack. *Cancer Immunology Research*, 7(2), 230–243. https://doi.org/10.1158/2326-6066.CIR-18-0266.

Alrubayyi, A., Schuetz, A., Lal, K. G., Jongrakthaitae, S., Paolino, K. M., Ake, J. A., et al. (2018). A flow cytometry based assay that simultaneously measures cytotoxicity and monocyte mediated antibody dependent effector activity. *Journal of Immunological Methods*, *462*, 74–82. https://doi.org/10.1016/j.jim.2018.08.012.

Bordron, A., Bagacean, C., Mohr, A., Tempescul, A., Bendaoud, B., Deshayes, S., et al. (2018). Resistance to complement activation, cell membrane hypersialylation and relapses in chronic lymphocytic leukemia patients treated with rituximab and chemotherapy. *Oncotarget*, *9*(60), 31590–31605. https://doi.org/10.18632/oncotarget.25657.

Broyer, L., Goetsch, L., & Broussas, M. (2013). Evaluation of complement-dependent cytotoxicity using ATP measurement and C1q/C4b binding. *Methods in Molecular Biology (Clifton, N.J.)*, *988*, 319–329. https://doi.org/10.1007/978-1-62703-327-5_20.

Bruhns, P., & Jönsson, F. (2015). Mouse and human FcR effector functions. *Immunological Reviews*, *268*(1), 25–51. https://doi.org/10.1111/imr.12350.

Burgess, M., Mapp, S., Mazzieri, R., Cheung, C., Chambers, L., Mattarollo, S. R., et al. (2017). Increased FcγRIIB dominance contributes to the emergence of resistance to therapeutic antibodies in chronic lymphocytic leukaemia patients. *Oncogene*, *36*(17), 2366–2376. https://doi.org/10.1038/onc.2016.387.

Carter, M. J., Dahal, L. N., Cleary, K. L. S., Marshall, M. J. E., French, R. R., Beers, S. A., et al. (2019). Immunological methods to study monoclonal antibody activity in chronic lymphocytic Leukaemia. *Methods in Molecular Biology (Clifton, N.J.)*, *1881*, 173–184. https://doi.org/10.1007/978-1-4939-8876-1_14.

Cartron, G., Dacheux, L., Salles, G., Solal-Celigny, P., Bardos, P., Colombat, P., et al. (2002). Therapeutic activity of humanized anti-CD20 monoclonal antibody and polymorphism in IgG Fc receptor FcgammaRIIIa gene. *Blood*, *99*(3), 754–758.

Chao, M. P., Alizadeh, A. A., Tang, C., Myklebust, J. H., Varghese, B., Gill, S., et al. (2010). Anti-CD47 antibody synergizes with rituximab to promote phagocytosis and eradicate non-Hodgkin lymphoma. *Cell*, *142*(5), 699–713. https://doi.org/10.1016/j.cell.2010.07.044.

Cleary, K. L. S., Chan, H. T. C., James, S., Glennie, M. J., & Cragg, M. S. (2017). Antibody distance from the cell membrane regulates antibody effector mechanisms. *Journal of Immunology (Baltimore, Md.: 1950)*, *198*(10), 3999–4011. https://doi.org/10.4049/jimmunol.1601473.

Clynes, R. A., Towers, T. L., Presta, L. G., & Ravetch, J. V. (2000). Inhibitory Fc receptors modulate in vivo cytotoxicity against tumor targets. *Nature Medicine*, *6*(4), 443–446. https://doi.org/10.1038/74704.

Cornet, S., Mathé, D., Chettab, K., Evesque, A., Matera, E.-L., Trédan, O., et al. (2016). Pegfilgrastim enhances the antitumor effect of therapeutic monoclonal antibodies. *Molecular Cancer Therapeutics*, *15*(6), 1238–1247. https://doi.org/10.1158/1535-7163.MCT-15-0759.

Cosentino, L. M., & Cathcart, M. K. (1987). A multi-step isolation scheme for obtaining CD16+ human natural killer cells. *Journal of Immunological Methods*, *103*(2), 195–204. https://doi.org/10.1016/0022-1759(87)90290-0.

Dahal, L. N., Dou, L., Hussain, K., Liu, R., Earley, A., Cox, K. L., et al. (2017). STING activation reverses lymphoma-mediated resistance to antibody immunotherapy. *Cancer Research*, 77(13), 3619–3631. https://doi.org/10.1158/0008-5472.CAN-16-2784.

Dahal, L. N., Gadd, A., Edwards, A. D., Cragg, M. S., & Beers, S. A. (2018). UC-1V150, a potent TLR7 agonist capable of activating macrophages and potentiating mAb-mediated target cell deletion. *Scandinavian Journal of Immunology*, *87*(6), e12666. https://doi.org/10.1111/sji.12666.

Dalle, S., Dupire, S., Brunet-Manquat, S., Reslan, L., Plesa, A., & Dumontet, C. (2009). In vivo model of follicular lymphoma resistant to rituximab. *Clinical Cancer Research: An Official Journal of the American Association for Cancer Research*, *15*(3), 851–857. https://doi.org/10.1158/1078-0432.CCR-08-1685.

Dantas, E., Erra Díaz, F., Pereyra Gerber, P., Merlotti, A., Varese, A., Ostrowski, M., et al. (2016). Low pH impairs complement-dependent cytotoxicity against IgG-coated target cells. *Oncotarget*, 7(45), 74203–74216. https://doi.org/10.18632/oncotarget.12412.

de Haart, S. J., Holthof, L., Noort, W. A., Minnema, M. C., Emmelot, M. E., Aarts-Riemens, T., et al. (2016). Sepantronium bromide (YM155) improves daratumumab-mediated cellular lysis of multiple myeloma cells by abrogation of bone marrow stromal cell-induced resistance. *Haematologica*, *101*(8), e339–e342. https://doi.org/10.3324/haematol.2015.139667.

de Jong, R. N., Beurskens, F. J., Verploegen, S., Strumane, K., van Kampen, M. D., Voorhorst, M., et al. (2016). A novel platform for the potentiation of therapeutic antibodies based on antigen-dependent formation of IgG hexamers at the cell surface. *PLoS Biology*, *14*(1), e1002344. https://doi.org/10.1371/journal.pbio.1002344.

Diepstra, A., Poppema, S., Boot, M., Visser, L., Nolte, I. M., Niens, M., et al. (2008). HLA-G protein expression as a potential immune escape mechanism in classical Hodgkin's lymphoma. *Tissue Antigens*, *71*(3), 219–226. https://doi.org/10.1111/j.1399-0039.2008.01005.x.

Dittel, B. N. (2010). Depletion of specific cell populations by complement depletion. *Journal of Visualized Experiments*, (36), e1487. https://doi.org/10.3791/1487.

Ferris, R. L., Jaffee, E. M., & Ferrone, S. (2010). Tumor antigen-targeted, monoclonal antibody-based immunotherapy: Clinical response, cellular immunity, and immunoescape. *Journal of Clinical Oncology: Official Journal of the American Society of Clinical Oncology*, *28*(28), 4390–4399. https://doi.org/10.1200/JCO.2009.27.6360.

Fishelson, Z., Hochman, I., Greene, L. E., & Eisenberg, E. (2001). Contribution of heat shock proteins to cell protection from complement-mediated lysis. *International Immunology*, *13*(8), 983–991.

Friedman, J., Morisada, M., Sun, L., Moore, E. C., Padget, M., Hodge, J. W., et al. (2018). Inhibition of WEE1 kinase and cell cycle checkpoint activation sensitizes head and neck cancers to natural killer cell therapies. *Journal for Immunotherapy of Cancer*, *6*(1), 59. https://doi.org/10.1186/s40425-018-0374-2.

Gancz, D., & Fishelson, Z. (2009). Cancer resistance to complement-dependent cytotoxicity (CDC): Problem-oriented research and development. *Molecular Immunology*, *46*(14), 2794–2800. https://doi.org/10.1016/j.molimm.2009.05.009.

Geis, N., Zell, S., Rutz, R., Li, W., Giese, T., Mamidi, S., et al. (2010). Inhibition of membrane complement inhibitor expression (CD46, CD55, CD59) by siRNA sensitizes tumor cells to complement attack in vitro. *Current Cancer Drug Targets*, *10*(8), 922–931.

Golay, J., & Introna, M. (2012). Mechanism of action of therapeutic monoclonal antibodies: Promises and pitfalls of in vitro and in vivo assays. *Archives of Biochemistry and Biophysics*, *526*(2), 146–153. https://doi.org/10.1016/j.abb.2012.02.011.

Golay, J., Lazzari, M., Facchinetti, V., Bernasconi, S., Borleri, G., Barbui, T., et al. (2001). CD20 levels determine the in vitro susceptibility to rituximab and complement of B-cell chronic lymphocytic leukemia: Further regulation by CD55 and CD59. *Blood*, *98*(12), 3383–3389.

González-González, E., Camacho-Sandoval, R., Jiménez-Uribe, A., Montes-Luna, A., Cortés-Paniagua, I., Sánchez-Morales, J., et al. (2019). Validation of an ADCC assay using human primary natural killer cells to evaluate biotherapeutic products bearing an Fc region. *Journal of Immunological Methods*, *464*, 87–94. https://doi.org/10.1016/j.jim.2018.11.002.

Grandjean, C. L., Montalvao, F., Celli, S., Michonneau, D., Breart, B., Garcia, Z., et al. (2016). Intravital imaging reveals improved Kupffer cell-mediated phagocytosis as a mode of action of glycoengineered anti-CD20 antibodies. *Scientific Reports*, *6*, 34382. https://doi.org/10.1038/srep34382.

Green, S. K., Karlsson, M. C. I., Ravetch, J. V., & Kerbel, R. S. (2002). Disruption of cell-cell adhesion enhances antibody-dependent cellular cytotoxicity: Implications for antibody-based therapeutics of cancer. *Cancer Research*, *62*(23), 6891–6900.

Gül, N., & van Egmond, M. (2015). Antibody-dependent phagocytosis of tumor cells by macrophages: A potent effector mechanism of monoclonal antibody therapy of cancer. *Cancer Research, 75*(23), 5008–5013. https://doi.org/10.1158/0008-5472.CAN-15-1330.

Herter, S., Birk, M. C., Klein, C., Gerdes, C., Umana, P., & Bacac, M. (2014). Glycoengineering of therapeutic antibodies enhances monocyte/macrophage-mediated phagocytosis and cytotoxicity. *The Journal of Immunology, 192*(5), 2252–2260. https://doi.org/10.4049/jimmunol.1301249.

Hietanen, T., Pitkänen, M., Kapanen, M., & Kellokumpu-Lehtinen, P.-L. (2016). Post-irradiation viability and cytotoxicity of natural killer cells isolated from human peripheral blood using different methods. *International Journal of Radiation Biology, 92*(2), 71–79. https://doi.org/10.3109/09553002.2016.1115136.

Hu, Y., Turner, M. J., Shields, J., Gale, M. S., Hutto, E., Roberts, B. L., et al. (2009). Investigation of the mechanism of action of alemtuzumab in a human CD52 transgenic mouse model. *Immunology, 128*(2), 260–270. https://doi.org/10.1111/j.1365-2567.2009.03115.x.

Hussain, S.-R. A., Cheney, C. M., Johnson, A. J., Lin, T. S., Grever, M. R., Caligiuri, M. A., et al. (2007). Mcl-1 is a relevant therapeutic target in acute and chronic lymphoid malignancies: Down-regulation enhances rituximab-mediated apoptosis and complement-dependent cytotoxicity. *Clinical Cancer Research: An Official Journal of the American Association for Cancer Research, 13*(7), 2144–2150. https://doi.org/10.1158/1078-0432.CCR-06-2294.

Idusogie, E. E., Wong, P. Y., Presta, L. G., Gazzano-Santoro, H., Totpal, K., Ultsch, M., et al. (2001). Engineered antibodies with increased activity to recruit complement. *Journal of Immunology (Baltimore, Md.: 1950), 166*(4), 2571–2575.

Jo, M., Kwon, H. S., Lee, K.-H., Lee, J. C., & Jung, S. T. (2018). Engineered aglycosylated full-length IgG Fc variants exhibiting improved FcγRIIIa binding and tumor cell clearance. *MAbs, 10*(2), 278–289. https://doi.org/10.1080/19420862.2017.1402995.

Kamen, L., Myneni, S., Langsdorf, C., Kho, E., Ordonia, B., Thakurta, T., et al. (2019). A novel method for determining antibody-dependent cellular phagocytosis. *Journal of Immunological Methods, 486*, 55–60. https://doi.org/10.1016/j.jim.2019.03.001.

Kasper, S., Breitenbuecher, F., Reis, H., Brandau, S., Worm, K., Köhler, J., et al. (2013). Oncogenic RAS simultaneously protects against anti-EGFR antibody-dependent cellular cytotoxicity and EGFR signaling blockade. *Oncogene, 32*(23), 2873–2881. https://doi.org/10.1038/onc.2012.302.

Kennedy, A. D., Beum, P. V., Solga, M. D., DiLillo, D. J., Lindorfer, M. A., Hess, C. E., et al. (2004). Rituximab infusion promotes rapid complement depletion and acute CD20 loss in chronic lymphocytic leukemia. *Journal of Immunology (Baltimore, Md.: 1950), 172*(5), 3280–3288.

Kesselring, R., Thiel, A., Pries, R., Fichtner-Feigl, S., Brunner, S., Seidel, P., et al. (2014). The complement receptors CD46, CD55 and CD59 are regulated by the tumour microenvironment of head and neck cancer to facilitate escape of complement attack. *European Journal of Cancer (Oxford, England: 1990), 50*(12), 2152–2161. https://doi.org/10.1016/j.ejca.2014.05.005.

Kohrt, H. E., Houot, R., Marabelle, A., Cho, H. J., Osman, K., Goldstein, M., et al. (2012). Combination strategies to enhance antitumor ADCC. *Immunotherapy, 4*(5), 511–527. https://doi.org/10.2217/imt.12.38.

Kurdi, A. T., Glavey, S. V., Bezman, N. A., Jhatakia, A., Guerriero, J. L., Manier, S., et al. (2018). Antibody-dependent cellular phagocytosis by macrophages is a novel mechanism of action of elotuzumab. *Molecular Cancer Therapeutics, 17*(7), 1454–1463. https://doi.org/10.1158/1535-7163.MCT-17-0998.

Lazar, G. A., Dang, W., Karki, S., Vafa, O., Peng, J. S., Hyun, L., et al. (2006). Engineered antibody Fc variants with enhanced effector function. *Proceedings of the National Academy of Sciences of the United States of America, 103*(11), 4005–4010. https://doi.org/10.1073/pnas.0508123103.

Lee, C.-H., Romain, G., Yan, W., Watanabe, M., Charab, W., Todorova, B., et al. (2017). IgG Fc domains that bind C1q but not effector Fcγ receptors delineate the importance of complement-mediated effector functions. *Nature Immunology, 18*(8), 889–898. https://doi.org/10.1038/ni.3770.

Levy, E. M., Sycz, G., Arriaga, J. M., Barrio, M. M., von Euw, E. M., Morales, S. B., et al. (2009). Cetuximab-mediated cellular cytotoxicity is inhibited by HLA-E membrane expression in colon cancer cells. *Innate Immunity, 15*(2), 91–100. https://doi.org/10.1177/1753425908101404.

Lin, A., Yan, W.-H., Xu, H.-H., Gan, M.-F., Cai, J.-F., Zhu, M., et al. (2007). HLA-G expression in human ovarian carcinoma counteracts NK cell function. *Annals of Oncology: Official Journal of the European Society for Medical Oncology, 18*(11), 1804–1809. https://doi.org/10.1093/annonc/mdm356.

Lindorfer, M. A., Wiestner, A., Zent, C. S., & Taylor, R. P. (2012). Monoclonal antibody (mAb)-based cancer therapy: Is it time to reevaluate dosing strategies? *Oncoimmunology, 1*(6), 959–961. https://doi.org/10.4161/onci.20368.

Loeff, F. C., van Egmond, H. M. E., Nijmeijer, B. A., Falkenburg, J. H. F., Halkes, C. J., & Jedema, I. (2017). Complement-dependent cytotoxicity induced by therapeutic antibodies in B-cell acute lymphoblastic leukemia is dictated by target antigen expression levels and augmented by loss of membrane-bound complement inhibitors. *Leukemia & Lymphoma, 58*(9), 2185–2195. https://doi.org/10.1080/10428194. 2017.1281411.

Mellor, J. D., Brown, M. P., Irving, H. R., Zalcberg, J. R., & Dobrovic, A. (2013). A critical review of the role of Fc gamma receptor polymorphisms in the response to monoclonal antibodies in cancer. *Journal of Hematology & Oncology, 6*, 1. https://doi.org/10.1186/1756-8722-6-1.

Mercogliano, M. F., De Martino, M., Venturutti, L., Rivas, M. A., Proietti, C. J., Inurrigarro, G., et al. (2017). TNFα-induced mucin 4 expression elicits trastuzumab resistance in HER2-positive breast cancer. *Clinical Cancer Research: An Official Journal of the American Association for Cancer Research, 23*(3), 636–648. https://doi.org/10.1158/1078-0432.CCR-16-0970.

Minard-Colin, V., Xiu, Y., Poe, J. C., Horikawa, M., Magro, C. M., Hamaguchi, Y., et al. (2008). Lymphoma depletion during CD20 immunotherapy in mice is mediated by macrophage FcγRI, FcγRIII, and FcγRIV. *Blood, 112*(4), 1205–1213. https://doi.org/10.1182/blood-2008-01-135160.

Moore, G. L., Chen, H., Karki, S., & Lazar, G. A. (2010). Engineered Fc variant antibodies with enhanced ability to recruit complement and mediate effector functions. *MAbs, 2*(2), 181–189.

Moreno, L., Pérez, C., Zabaleta, A., Manrique, I., Alignani, D., Ajona, D., et al. (2019). The mechanism of action of the anti-CD38 monoclonal antibody isatuximab in multiple myelmoa. *Clinical Cancer Research: An Official Journal of the American Association for Cancer Research, 25*. https://doi.org/10.1158/1078-0432.CCR-18-1597.

Murray, J. C., Aldeghaither, D., Wang, S., Nasto, R. E., Jablonski, S. A., Tang, Y., et al. (2014). c-Abl modulates tumor cell sensitivity to antibody-dependent cellular cytotoxicity. *Cancer Immunology Research, 2*(12), 1186–1198. https://doi.org/10.1158/2326-6066.CIR-14-0083.

Musolino, A., Naldi, N., Bortesi, B., Pezzuolo, D., Capelletti, M., Missale, G., et al. (2008). Immunoglobulin G fragment C receptor polymorphisms and clinical efficacy of trastuzumab-based therapy in patients with HER-2/neu-positive metastatic breast cancer. *Journal of Clinical Oncology, 26*(11), 1789.

Nakadate, Y., Kodera, Y., Kitamura, Y., Shirasawa, S., Tachibana, T., Tamura, T., et al. (2014). KRAS mutation confers resistance to antibody-dependent cellular cytotoxicity of cetuximab against human colorectal cancer cells. *International Journal of Cancer*, *134*(9), 2146–2155. https://doi.org/10.1002/ijc.28550.

Namba, M., Hattori, N., Hamada, H., Yamaguchi, K., Okamoto, Y., Nakashima, T., et al. (2019). Anti-KL-6/MUC1 monoclonal antibody reverses resistance to trastuzumab-mediated antibody-dependent cell-mediated cytotoxicity by capping MUC1. *Cancer Letters*, *442*, 31–39. https://doi.org/10.1016/j.canlet.2018.10.037.

Overdijk, M. B., Verploegen, S., Bögels, M., van Egmond, M., van Bueren, J. J. L., Mutis, T., et al. (2015). Antibody-mediated phagocytosis contributes to the anti-tumor activity of the therapeutic antibody daratumumab in lymphoma and multiple myeloma. *MAbs*, 7(2), 311–320. https://doi.org/10.1080/19420862.2015.1007813.

Ragupathi, G., Liu, N. X., Musselli, C., Powell, S., Lloyd, K., & Livingston, P. O. (2005). Antibodies against tumor cell glycolipids and proteins, but not mucins, mediate complement-dependent cytotoxicity. *Journal of Immunology (Baltimore, Md.: 1950)*, *174*(9), 5706–5712.

Roghanian, A., Teige, I., Mårtensson, L., Cox, K. L., Kovacek, M., Ljungars, A., et al. (2015). Antagonistic human FcγRIIB (CD32B) antibodies have anti-tumor activity and overcome resistance to antibody therapy in vivo. *Cancer Cell*, *27*(4), 473–488. https://doi.org/10.1016/j.ccell.2015.03.005.

Rose, R. J., van Berkel, P. H. C., van den Bremer, E. T. J., Labrijn, A. F., Vink, T., Schuurman, J., et al. (2013). Mutation of Y407 in the CH3 domain dramatically alters glycosylation and structure of human IgG. *MAbs*, *5*(2), 219–228. https://doi.org/10.4161/mabs.23532.

Sekhar, S. C., Kasai, T., Satoh, A., Shigehiro, T., Mizutani, A., Murakami, H., et al. (2013). Identification of caveolin-1 as a potential causative factor in the generation of trastuzumab resistance in breast cancer cells. *Journal of Cancer*, *4*(5), 391–401. https://doi.org/10.7150/jca.6470.

Shah, P., Fatehchand, K., Patel, H., Fang, H., Justiniano, S. E., Mo, X., et al. (2013). Toll-like receptor 2 ligands regulate monocyte Fcγ receptor expression and function. *The Journal of Biological Chemistry*, *288*(17), 12345–12352. https://doi.org/10.1074/jbc.M113.449983.

Shi, Y., Fan, X., Deng, H., Brezski, R. J., Rycyzyn, M., Jordan, R. E., et al. (2015). Trastuzumab triggers phagocytic killing of high HER2 cancer cells in vitro and in vivo by interaction with Fcγ receptors on macrophages. *The Journal of Immunology*, *194*(9), 4379–4386. https://doi.org/10.4049/jimmunol.1402891.

Shields, R. L., Lai, J., Keck, R., O'Connell, L. Y., Hong, K., Meng, Y. G., et al. (2002). Lack of Fucose on human IgG1 N-linked oligosaccharide improves binding to human FcγRIII and antibody-dependent cellular toxicity. *Journal of Biological Chemistry*, *277*(30), 26733–26740. https://doi.org/10.1074/jbc.M202069200.

Shields, R. L., Namenuk, A. K., Hong, K., Meng, Y. G., Rae, J., Briggs, J., et al. (2001). High resolution mapping of the binding site on human IgG1 for Fc gamma RI, Fc gamma RII, Fc gamma RIII, and FcRn and design of IgG1 variants with improved binding to the Fc gamma R. *The Journal of Biological Chemistry*, *276*(9), 6591–6604. https://doi.org/10.1074/jbc.M009483200.

Shore, S. L., Melewicz, F. M., & Gordon, D. S. (1977). The mononuclear cell in human blood which mediates antibody-dependent cellular cytotoxicity to virus-infected target cells. I. Identification of the population of effector cells. *Journal of Immunology (Baltimore, Md.: 1950)*, *118*(2), 558–566.

Simmons, L. C., Reilly, D., Klimowski, L., Raju, T. S., Meng, G., Sims, P., et al. (2002). Expression of full-length immunoglobulins in Escherichia coli: Rapid and efficient production of aglycosylated antibodies. *Journal of Immunological Methods*, *263*(1–2), 133–147.

Singha, N. C., Nekoroski, T., Zhao, C., Symons, R., Jiang, P., Frost, G. I., et al. (2015). Tumor-associated hyaluronan limits efficacy of monoclonal antibody therapy. *Molecular Cancer Therapeutics*, *14*(2), 523–532. https://doi.org/10.1158/1535-7163.MCT-14-0580.

Tammen, A., Derer, S., Schwanbeck, R., Rösner, T., Kretschmer, A., Beurskens, F. J., et al. (2017). Monoclonal antibodies against epidermal growth factor receptor acquire an ability to kill tumor cells through complement activation by mutations that selectively facilitate the hexamerization of IgG on opsonized cells. *Journal of Immunology (Baltimore, Md.: 1950)*, *198*(4), 1585–1594. https://doi.org/10.4049/jimmunol.1601268.

Tanaka, M., Ishige, A., Yaguchi, M., Matsumoto, T., Shirouzu, M., Yokoyama, S., et al. (2019). Development of a simple new flow cytometric antibody-dependent cellular cytotoxicity (ADCC) assay with excellent sensitivity. *Journal of Immunological Methods*, *464*, 74–86. https://doi.org/10.1016/j.jim.2018.10.014.

Taylor, R. P., & Lindorfer, M. A. (2016). Cytotoxic mechanisms of immunotherapy: Harnessing complement in the action of anti-tumor monoclonal antibodies. *Seminars in Immunology*, *28*(3), 309–316. https://doi.org/10.1016/j.smim.2016.03.003.

Tóth, G., Szöllősi, J., & Vereb, G. (2017). Quantitating ADCC against adherent cells: Impedance-based detection is superior to release, membrane permeability, or caspase activation assays in resolving antibody dose response. *Cytometry. Part. A: The Journal of the International Society for Analytical Cytology*, *91*(10), 1021–1029. https://doi.org/10.1002/cyto.a.23247.

Uchida, J., Hamaguchi, Y., Oliver, J. A., Ravetch, J. V., Poe, J. C., Haas, K. M., et al. (2004). The innate mononuclear phagocyte network depletes B lymphocytes through Fc receptor-dependent mechanisms during anti-CD20 antibody immunotherapy. *The Journal of Experimental Medicine*, *199*(12), 1659–1669. https://doi.org/10.1084/jem.20040119.

van der Haar Àvila, I., Marmol, P., Cany, J., Kiessling, R., & Pico de Coaña, Y. (2019). Evaluating antibody-dependent cell-mediated cytotoxicity by flow cytometry. *Methods in Molecular Biology (Clifton, N.J.)*, *1913*, 181–194. https://doi.org/10.1007/978-1-4939-8979-9_13.

Vermi, W., Micheletti, A., Finotti, G., Tecchio, C., Calzetti, F., Costa, S., et al. (2018). slan$^+$ monocytes and macrophages mediate CD20-dependent B-cell lymphoma elimination via ADCC and ADCP. *Cancer Research*, *78*(13), 3544–3559. https://doi.org/10.1158/0008-5472.CAN-17-2344.

Walker, M. R., Lund, J., Thompson, K. M., & Jefferis, R. (1989). Aglycosylation of human IgG1 and IgG3 monoclonal antibodies can eliminate recognition by human cells expressing Fc gamma RI and/or Fc gamma RII receptors. *The Biochemical Journal*, *259*(2), 347–353.

Wang, X., Mathieu, M., & Brezski, R. J. (2018). IgG Fc engineering to modulate antibody effector functions. *Protein & Cell*, *9*(1), 63–73. https://doi.org/10.1007/s13238-017-0473-8.

Wieland, A., Shashidharamurthy, R., Kamphorst, A. O., Han, J.-H., Aubert, R. D., Choudhury, B. P., et al. (2015). Antibody effector functions mediated by Fcγ-receptors are compromised during persistent viral infection. *Immunity*, *42*(2), 367–378. https://doi.org/10.1016/j.immuni.2015.01.009.

Yamada, D. H., Elsaesser, H., Lux, A., Timmerman, J. M., Morrison, S. L., de la Torre, J. C., et al. (2015). Suppression of Fcγ-receptor-mediated antibody effector function during persistent viral infection. *Immunity*, *42*(2), 379–390. https://doi.org/10.1016/j.immuni.2015.01.005.

Yamashita, M., Kitano, S., Aikawa, H., Kuchiba, A., Hayashi, M., Yamamoto, N., et al. (2016). A novel method for evaluating antibody-dependent cell-mediated cytotoxicity by flowcytometry using cryopreserved human peripheral blood mononuclear cells. *Scientific Reports*, *6*, 19772. https://doi.org/10.1038/srep19772.

Yin, J., Albers, A. J., Smith, T. S., Riddell, G. T., & Richards, J. O. (2018). Differential regulation of human monocytes and NK cells by antibody-opsonized tumors. *Cancer Immunology, Immunotherapy*, *67*(8), 1239–1250. https://doi.org/10.1007/s00262-018-2179-z.

Zell, S., Geis, N., Rutz, R., Schultz, S., Giese, T., & Kirschfink, M. (2007). Down-regulation of CD55 and CD46 expression by anti-sense phosphorothioate oligonucleotides (S-ODNs) sensitizes tumour cells to complement attack. *Clinical & Experimental Immunology*, *150*(3), 576–584. https://doi.org/10.1111/j.1365-2249.2007.03507.x.

CHAPTER TWENTY-THREE

Enzyme-linked immunospot assay to monitor antigen-specific cellular immune responses in mouse tumor models

Donatella Negri[a,*], Paola Sestili[b], Martina Borghi[a], Maria Ciccolella[c], Laura Bracci[c,*]

[a]Department of Infectious Diseases, Istituto Superiore di Sanità, Rome, Italy
[b]National Centre for the Control and the Evaluation of Medicines, Istituto Superiore di Sanità, Rome, Italy
[c]Department of Oncology and Molecular Medicine, Istituto Superiore di Sanità, Rome, Italy
*Corresponding authors: e-mail address: donatella.negri@iss.it; laura.bracci@iss.it

Contents

Abstract

Critical to the advancement of tumor immunotherapy is the reliable identification of responders and the quantification of the tumor-specific immune response elicited by treatments. In this regard, Enzyme-Linked Immunospot assay (ELISpot) is an ideal monitoring technique due to its high sensitivity, ease of execution and cost-effectiveness. Originally developed for the enumeration of B cells secreting antigen-specific antibodies, ELISpot assay has been adapted to detect and quantify cytokine-secreting immune cells present at low frequency in a variety of biological samples, including blood, in response to antigen-specific stimuli. The above-mentioned features emphasize the role of ELISpot as valuable assay for translational research and clinical

Methods in Enzymology, Volume 632
ISSN 0076-6879
https://doi.org/10.1016/bs.mie.2019.05.058

applications. In the present chapter, we will focus on the use of ELISpot assay for monitoring the tumor-specific effector responses induced by different treatments in preclinical models and will provide some protocols and technical hints for its application.

1. Introduction

It is now well recognized that the immune system plays a pivotal role in tumor development and, if properly stimulated, can represent an additional weapon against cancer. Monitoring of immune responses over time is therefore an essential step both in translational cancer research and in clinical settings. In preclinical settings, Enzyme-Linked Immunospot (ELISpot) assay is frequently used to assess the immunogenicity of experimental treatments such as immunogenic chemotherapy (Song et al., 2018), adoptive immunotherapy (Bracci et al., 2007), vaccination against virus-associated oncogenic proteins (Grasso et al., 2013) and to validate candidate neoantigens whose activity has been predicted in silico (Johanns et al., 2016). Similarly, ELISpot assay has been introduced in several clinical trials designed to preliminary assess or follow up the immune response of patients undergoing experimental treatments. At first, monitoring the immune response was restricted to cancer patients undergoing immuno-stimulating treatments, such as anticancer vaccination (Dangoor et al., 2010; Kjeldsen et al., 2018; Malyguine, Strobl, Dunham, Shurin, & Sayers, 2012). More recently, ELISpot assay was also used to verify antitumor lymphocyte function in patients undergoing radiotherapy (Domouchtsidou et al., 2018) and to assess the prognostic relevance of circulating tumor-specific T cells (Safi et al., 2017). Most importantly, in some cases tumor-specific immune responses assessed by ELISpot correlated with clinical benefit (Safi et al., 2017), thus making this assay a valuable tool for patient monitoring and follow up in different prognostic and therapeutic settings.

2. Principle and application of the ELISpot assay

Originally established for the enumeration of antigen-specific antibody-secreting B cells in 1983 (Czerkinsky, Nilsson, Nygren, Ouchterlony, & Tarkowski, 1983) the technology has been expanded to the analysis of cytokine secretion by T cells (Malyguine et al., 2012), NK cells (Das et al., 2017; Lion, Smits, Berneman, & Van Tendeloo, 2009) and monocytes

(Kouwenhoven et al., 2001). The ELISpot assay for cytokines is a quantitative and qualitative immunoassay that measures the frequency of cytokine-secreting cells in a given sample based on a modified sandwich enzyme-linked immunosorbent (ELISA) technique. The principle of the assay is illustrated in Fig. 1. The cells are seeded in plates with PVDF (polyvinylidene difluoride) membrane coated with a monoclonal antibody (mAb) against the cytokine of interest. Cytokines secreted from the cells in presence of appropriate stimuli are captured by the membrane-bound mAb in the local environment of the cytokine-secreting cell. A biotinylated secondary Ab specific for the cytokine of interest is added after several washes to completely remove the cells. To make the sandwich reaction visible, an enzymeconjugated to streptavidin and an insoluble enzyme-specific substrate are added. The bound cytokine is visible as colored precipitate forms and appears as spot representing an individual cytokine-secreting cell. The spots can be counted with automated ELISpot reader systems. The spot forming cell (SFC) number estimates the frequency of cells specifically stimulated in the sample.

The assay is cost-effective and is easy to perform compared to other functional assays including intracellular staining (ICS) and cytotoxicity assays,

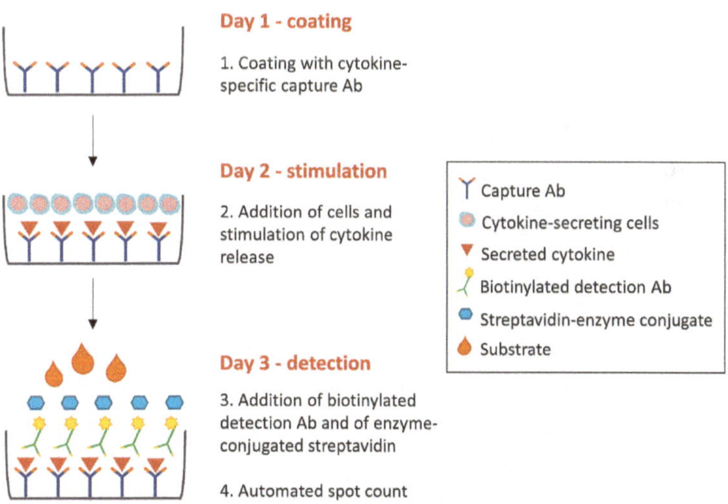

Fig. 1 Principle and protocol of the ELISpot assay. The ELISpot assay consists of three phases, which are carried out in three consecutive days. *Day 1: coating.* The plate membrane is coated with a cytokine-specific capture antibody (Ab). *Day 2: stimulation.* The cells of interest are placed into the wells and properly stimulated to release the cytokine of interest. *Day 3: detection.* The presence of cytokine-secreting cells is revealed by enzymatic reaction.

allowing to screen high number of samples with high sensitivity. Suitable for large-scale trials, the ELISpot assay is widely used in preclinical models and humans for monitoring the antigen-specific immune response to pathogens and vaccines as well as in cancer immunotherapy. In particular, IFNγ, a pleiotropic cytokine produced manly by effector T cells and NK cells represents the most common cytokine measured in the context of tumor immunity studies. Production of IFNγ by helper $CD4^{+}$T cells as well as cytotoxic $CD8^{+}$T cells is a hallmark of the Th1-type response, typically associated with effective host defense against intracellular pathogens and tumor cells. In addition, due to the fact that naïve T cells do not produce IFNγ within the 18–24h time frame of an ELISpot incubation (Dutton, Bradley, & Swain, 1998), it is clear that this assay measures IFNγ release by pre-activated or memory T cells. However, it is important to note that the IFNγ ELISpot assay alone may not be sufficient to identify an effective response to tumor cells, since some non-cytotoxic cells can also secrete IFNγ. To this regard, Granzyme B secretion can be analyzed by ELISpot as well, allowing for a more specific evaluation of the cytotoxic activity of T lymphocytes and NK cells (Shafer-Weaver et al., 2004). Also, the analysis of a single immunological parameter is not sufficient to provide solid and comprehensive information about the complex immunological mechanisms underlying the interactions between effector and tumor cells. To have a more comprehensive picture of the anti-tumor immune responses, it is strongly suggested to measure several immunological parameters by combining different assays, including various modifications of ELISpot assays. There is the possibility to measure more than one cytokine in the same assay using dual ELISpot assays and the recently developed FluoroSpot assay (Körber, Behrends, Hapfelmeier, Protzer, & Bauer, 2016; Schulten & Sette, 2018). FluoroSpot method employs fluorescence instead of chromogenic detection. Using a biotinylated detection Ab for one analyte and a tag-labeled detection Ab for the second and the third one enables analysis of more analytes in the same well (Janetzki, 2015). The IFNγ ELISpot assay is the most commonly used in cancer research, but the presence of other cytokine-producing cells with immunomodulatory activity affecting the efficacy of a given treatment may also be investigated. Among these, there are ELISpot kits to enumerate the frequency of IL-10-secreting cells (e.g., T-regulatory cells) or IL-17-secreting cells (i.e., Th17 cells). IL-5 (Ikutani et al., 2012; Zaynagetdinov et al., 2015) and IL-13 (Bruchard & Ghiringhelli, 2019; Zhou, Qian, Gu, Chen, & Xiang, 2013) among the Th2 cytokines can also be investigated, and many other cytokines can be monitored as well, depending on the known or supposed immune-based mechanisms associated to the selected tumor model and treatment under investigation.

Based on the feasibility and cost-effectiveness, the ELISpot assay should be performed to initially screen the response to treatments aimed at inducing and/or improving the anti-tumor immunity. Other functional tests, including multiparametric ICS by flow cytometry and killing assay can be performed to better characterize the phenotype, polyfunctionality and cytotoxic activity of the antigen-specific secreting cells previously identified by ELISpot.

3. Sample preparation

Sample preparation is a critical step for reducing variability of results. In fact, abundance of erythrocytes, granulocytes or platelets as well as high dead/apoptotic cell rates in test samples may cause a variety of artificial signals and misinterpretations. For this reason, samples must be processed according to protocols optimized depending on the type and source of the responder cells.

3.1 Source of cells

Different sample types can be used for the evaluation of antigen-specific immune responses in tumor bearing mice. Most commonly, the presence of cytokine-producing cells by ELISpot is analyzed in cell suspensions from explanted lymphoid tissues (spleen, lymph nodes or bone marrow).

A brief description of the protocol for the isolation of cells from mouse spleen is given here.

3.1.1 Protocol for spleen processing

- Prepare ACK lysing buffer by dissolving 8.26 g ammonium chloride (NH_4Cl), 1 g potassium bicarbonate ($KHCO_3$) and 0.037 g EDTA disodium dihydrate in 1 L of deionized water.
- Check pH with a pH-meter (pH should be 7.2–7.4).
- Filter sterile by passing on 0.22 µm pore membranes and store at 4 °C until use.

    Note: the ACK lysing solution is stable for 6 months
- After mouse euthanasia, surgically remove the spleen (keep sterile!) and place it in a 5 mL polystyrene tube containing 3 mL of cold PBS. Keep in ice until processing.

- Transfer the spleen onto a cell strainer (70–100 μm pore size) and place it in a sterile petri dish or lean it on a 50 mL tube and gently pressed with the plunger of a sterile 2–5 mL syringe to grind the tissue.
- Add 3 mL of ACK lysing solution to lyse red blood cells (3 min at RT).
- Add 7 mL of RPMI 1640 supplemented with 5–10% heat inactivated FBS or PBS + 5% FBS to neutralize the ACK buffer activity.
- Centrifuge the cells 290 RCF for 7 min at RT.
- Discard supernatant and resuspend the cells in 5 mL of RPMI 1640 medium +10% FBS + 2 mM L-glutamine +1% penicillin/streptomycin +50 mM 2-mercaptoethanol, thereafter referred to as complete medium.
- Count the cells in trypan blue 0.4% solution and aliquot the cell suspensions at 2×10^5 cells/well in the ELISpot plate.

Although more cumbersome than for lymphoid organs, ELISpot assay can also be performed from tumor-infiltrating lymphocytes (TILs) isolated from excised tumor mass according to protocols that use either enzymatic (Liu et al., 2013) or non-enzymatic (Swartz et al., 2018) dissociation procedures, usually followed by leucocyte enrichment or isolation steps and, eventually, by rounds of growth factor in vitro supplementation (e.g., IL-2 for T lymphocytes) or antigenic restimulation (Johanns et al., 2016). The analysis of cytokine-secreting cells from lymphoid organs or tumor, however, implies the euthanasia of animals. This means including a high number of animals per experiment to analyze different time points after the tumor implant and/or treatment.

Recently, we developed a protocol for the use of blood as an alternative source of cells and successfully applied this method to longitudinally monitor the frequency of circulating antigen-specific IFNγ secreting cells induced by a given treatment (e.g., vaccine) (Gallinaro et al., 2018; Grasso et al., 2013).

3.1.2 Protocol for blood processing

- Collect blood samples (150–200 μL/mouse) from tumor-bearing mice according to National regulations in EDTA-coated 1 mL tubes.
- Centrifuge blood samples at 515 RCF for 10 min to separate blood fractions.
- Remove plasma and eventually store it at −80 °C for further analysis.
- Dilute blood in pre-warmed (at 37 °C) ACK lysing buffer (8 mL of ACK/100 μL of blood) and incubate for 5 min at RT to lyse red blood cells.
- Add 7 mL of RPMI 1640 supplemented with 5–10% heat inactivated fetal bovine serum (FBS) or PBS + 5% FBS to neutralize the ACK buffer activity.

- Centrifuge cells at 290 RCF for 7 min at RT.
- Discard supernatant and resuspend cells in 1 mL of complete medium.
- Count cells and aliquot $1–2 \times 10^5$ cells/well in triplicate in ELISpot plate.

Blood samples processed according to this protocol are viable, deprived of erythrocytes and have a cellular composition comparable to splenocytes in terms of T cell and NK cell abundancy (Fig. 2). Given the possible discrepancies between the frequency of cytokine-secreting cells in tumor tissue and in blood, we recommend performing the ELISpot assay from blood as a screening approach before proceeding further with more invasive investigations that require the euthanasia of the animals. In addition, based on the complexity of tumor mass composition, to analyze the anti-tumor T cell response at the tumor site, we would definitely recommend performing an ICS and flow-cytometry-based analysis, which allows an easier identification of specific cell populations.

3.2 Cell stimulation

ELISpot assay is very plastic with respect to the choice of the antigen for cell stimulation. Depending on the nature and intensity of the stimulation, a wide range of soluble factors can be assayed. Polyclonal stimuli such as Phorbol 12-myrstate 13-acetate (PMA) plus Ionomycin (Iono) or Concanavalin A (Con A) are widely used as positive controls as they induce the secretion of a variety of cytokines from leucocytes. When stimulated for 18–24 h with Con A for example, blood cells and splenocytes secrete IFNγ proportionally to the number of plated cells (Fig. 3A and B). When assaying tumor-specific responses, purified immunodominant MHC class I and II-restricted peptides are commonly used to stimulate cytokine release from bulk, unsorted cell preparations. In this case, the assay retrieves information on the type of antigen-specificity and, indirectly, on the responding cell population (e.g., $CD4^+$ T vs $CD8^+$ T cells). Fig. 3C shows an example of kinetics of antigen-specific T cell response evaluation in tumor-bearing mice treated with a T cell vaccine. C57BL/6J mice were injected subcutaneously with EG7-OVA tumor cells, a lymphoma cell line expressing ovalbumin (OVA), and vaccinated with an integrase-defective lentiviral vector delivering OVA (IDLV-OVA) when a large tumor mass was developed. The OVA-specific T cell response was measured by IFNγ ELISpot in blood cells at different time points up to 8 months after tumor injection. The versatility of the ELISpot assay can be exploited to perform the

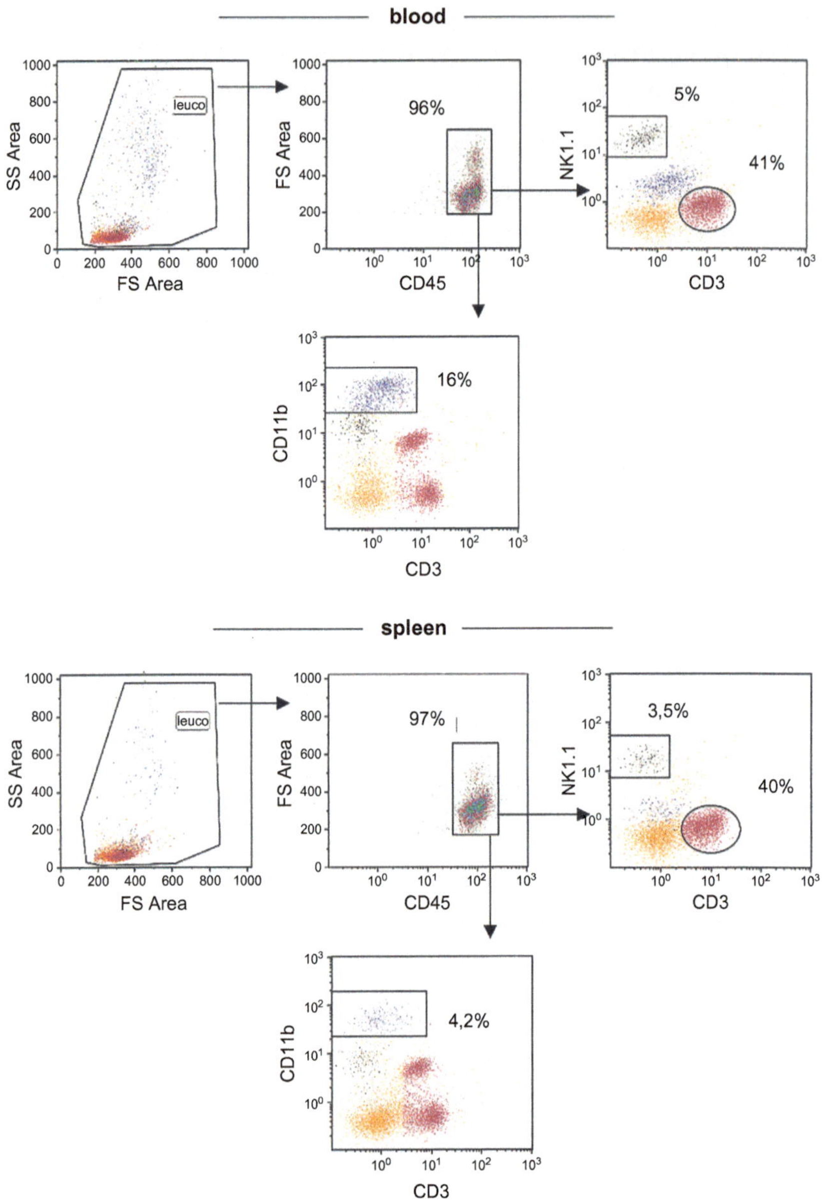

Fig. 2 Comparative analysis of blood leucocyte and splenocyte composition. Blood and spleen were taken from a C57BL/6J mouse and processed in ACK lysing solution according to the described protocols. Cell suspensions were stained with fluorochrome-labeled Abs binding the most relevant immune-cell populations and analyzed on a Gallios™ flow cytometer (Bechman Coulter). Data were plotted by using Kaluza Software. One representative experiment out of two with similar results is shown.

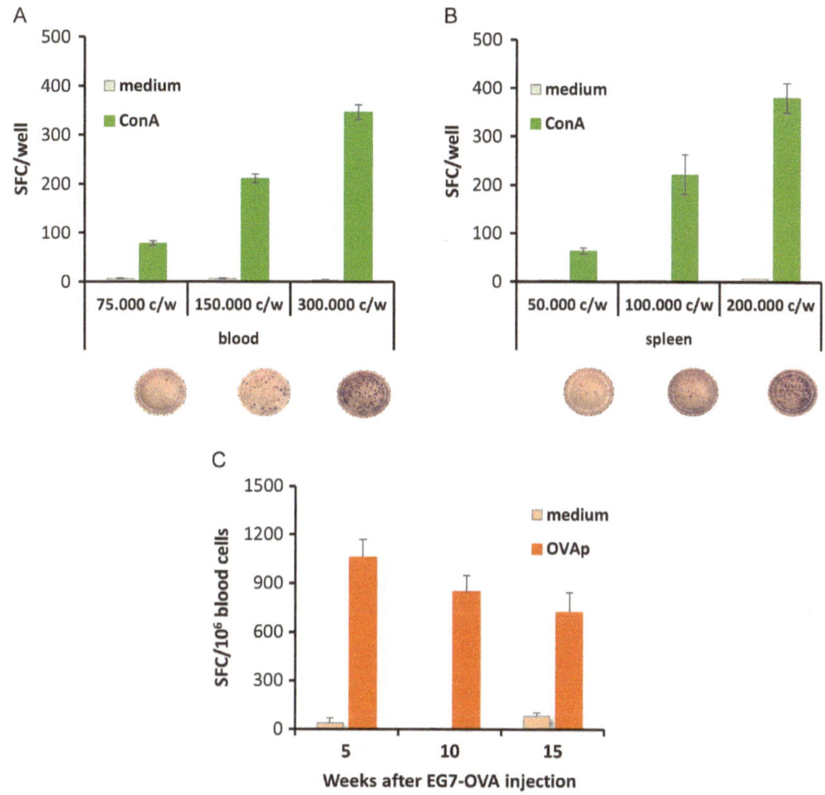

Fig. 3 T cell activity measured by IFNγ ELISpot assay. Leucocytes from blood (A) and spleen (B) from C57BL/6J mice were aliquoted in pre-coated ELISpot plates at different cell densities (see ELISpot protocol for details) and stimulated for 20 h with 5 μg/mL of Concanavalin A (Con A) to induce IFNγ release from T lymphocytes. Data are expressed as mean spot forming cells (SFC)/well ± standard deviation (SD) of triplicates. (C) C57BL/6J mice were injected subcutaneously with EG7-OVA tumor cells, a lymphoma cell line expressing Ovalbumin (OVA) and vaccinated with integrase-defective lentiviral vector (IDLV) expressing OVA when a large tumor mass was developed. The OVA-specific T cell response was measured by IFNγ ELISpot in blood cells at different time points up to 8 months after tumor injection, using the MHC class I-restricted peptide epitope of ovalbumin (OVAp) as a specific stimulation. Results are expressed as average of specific SFC/10^6 cells ± SD of seven mice.

epitope-mapping for the identification of new MHC-restricted epitopes of vaccine, pathogen or tumor associated antigens, using pool of peptides overlapping the protein sequence (Fiore-Gartland et al., 2016; Negri et al., 2006). In some experimental settings where the tumor associated antigen and the immunodominant epitopes are unknown or are unavailable, the stimulus can be represented by autologous DC that have been previously incubated with tumor apoptotic bodies (Schiavoni et al., 2011) or by the tumor cells themselves (Fig. 4). In the latter case, tumor cells should be irradiated to inhibit cell proliferation and histologically unrelated tumor cells should be included as specificity control. Interestingly, Fig. 4A shows that irradiation of B16 tumor cells, a cell line known to express negative/low surface MHC class I, induces MHC class I upregulation, thus rendering these cells more susceptible to T cell recognition in the ELISpot assay.

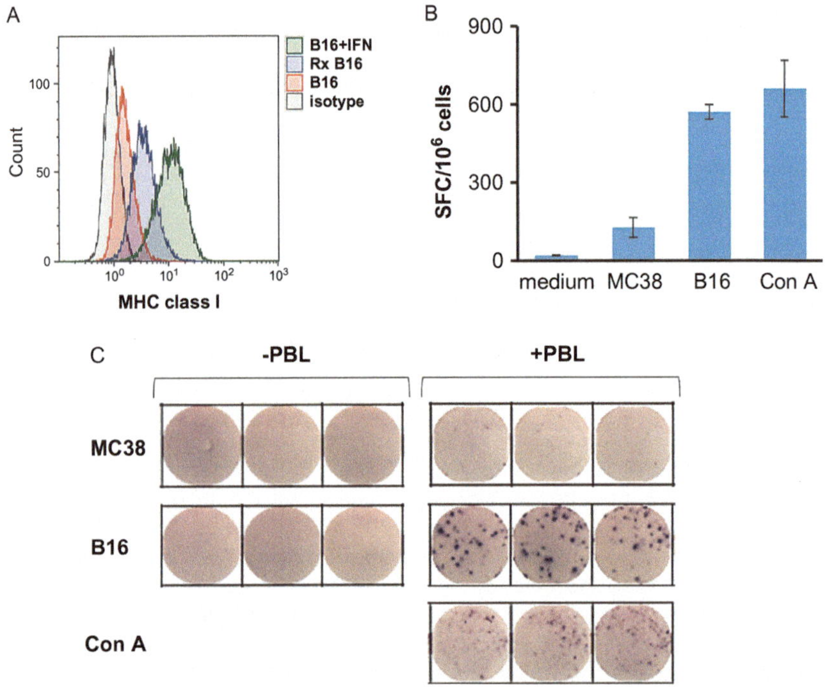

Fig. 4 IFNγ ELISpot of blood cells upon stimulation with a tumor cell line. (A) Surface expression of MHC class I molecules in B16 melanoma cells prior and 24h after irradiation (Rx, 20Gy). Culture of B16 with 1000U of type I IFNs for 24h was used as positive control for MHC class I upregulation. One representative experiment out of three is shown. (B) Blood cells from C57BL/6J mice implanted with B16 melanoma were processed in ACK buffer as described and cocultured for 20h with irradiated B16 melanoma cells or MC38 colon carcinoma cells or with 5μg/mL Con A in precoated ELISpot plates. Some wells were left untreated as negative control. Results are expressed as mean SFC/10^6cells ± SD of triplicates. (C) Representative triplicates from the same assay.

3.3 Fresh vs cryopreserved samples

The use of cryopreserved samples, such as murine splenocytes for translational research or PBMC in clinical trials is preferable, when possible, to allow the synchronous analysis of all samples in the same assay, thus reducing the inter-assay variability. When dealing with cryopreserved samples, cells should be handled very carefully to avoid excessive mortality that could affect the assay outcome. A comparative analysis of protocols for optimal handling of cryopreserved murine splenocytes and human PBMCs for ELISpot has been already published (Gad et al., 2013; Santos et al., 2014). In our experience, the best way to proceed is to resuspend thawed cells at the concentration of 2×10^6 cells/mL in complete medium in 15 mL polypropylene tubes and incubate the tubes for at least 2 h at 37 °C 5% CO_2 with loosely closed caps to allow gas exchange. This step is critical as it allows the death of damaged cells and increases accuracy of viable cell count before plating. Cells are then harvested, counted in trypan blue 0.4% solution and seeded in the ELISpot plate with a cell density ranging between 1 and 3×10^5 cells/well in triplicate. Fig. 5 shows a comparison between ELISpot assays performed using either fresh or cryopreserved splenocytes from the same group of mice in two different experimental settings. A modest increase in spot count is observed in cryopreserved splenocytes as compared to fresh splenocytes regardless of the stimulus applied, thus supporting the use of frozen cells for the analysis of antigen-specific T cell responses.

4. Protocol for IFNγ ELISpot assay

The most used plate format is represented by PVDF membrane 96-well plate. PVDF membrane is superior over mixed cellulose ester format, due to improved binding of coating Ab, but is highly hydrophobic and may necessitate pre-wetting with alcohol prior to addition of the coating Ab, resulting sometimes in poor performance assay if not properly done. The choice of plate size is determined by several reasons, including amount of available cells, the expected frequency of cytokine-secreting cells, the total number of samples and the type of reader, since many ELISpot readers can count only 96-well plates. The use of 384-well plates allows the assay miniaturization which can simultaneously reduce cell requirements while increasing throughput, especially when robotic systems are available (Li Pira et al., 2014).

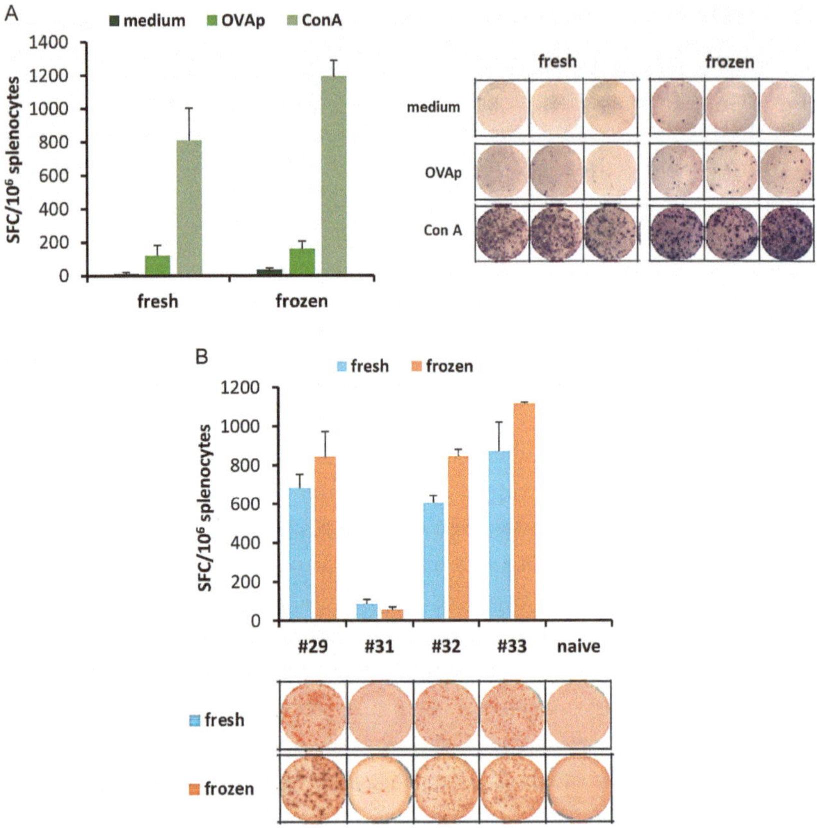

Fig. 5 Comparison between ELISpot assay from fresh and frozen splenocytes. (A) Fresh or frozen splenocytes from the same C57BL/6J mice implanted with EG7-OVA tumor cells were assayed 1 month apart by IFNγ ELISpot. Cells were stimulated for 20h with 10μg/mL of a MHC class I-restricted peptide epitope of ovalbumin (OVAp) or 5μg/mL of Con A or left untreated (medium). Results are expressed as the mean SFC/10^6 cells ± SD of triplicates. Representative wells from the same IFNγ ELISpot assay are shown. (B) Fresh and frozen splenocytes from C57BL/6J immunized with IDLV delivering the Influenza nucleoprotein (NP) or untreated (naïve) were stimulated for 20h with 2μg/mL of MHC class I-restricted NP peptide or left untreated (medium, not shown). Results are expressed as specific mean SFC/10^6 cells ± SD of triplicates, after "medium" count subtraction. Representative wells stimulated with NP peptide from the same ELISpot assay are shown.

Day 0—Coating (sterile conditions required)

(OPTIONAL depending on the plates, as specified in the manufacturer's instructions)

- Prepare PVDF membranes in the 96-well plate by incubating with 30μL of 70% ethanol (in Milli-Q water) for 1min at room temperature.

  Note: When pipetting, avoid touching the membrane in the wells at all times.

- Invert the ELISpot plates to remove the ethanol and wash three times with 150 μL of sterile phosphate buffered saline (PBS); any remaining ethanol can affect cell viability and antibody binding.
- Coat the plate with 100 μL/well of anti-mIFNγ capture antibody (e.g., AN18, 5 μg/mL in sterile PBS).
- Incubate overnight at 4 °C (Seal the plate with Parafilm until use).

Day 1—Cell stimulation (sterile conditions required)

- Invert the plates to remove unbound Ab and wash wells three times with 200 μL/well of PBS.
- Add 200 μL/well of complete medium to saturate free binding sites, and incubate at RT for at least 2 h.
- When ready to seed the cells, invert the plates to remove the medium. Add 100 μL/well of cell suspension in triplicate at the desired concentration ($0.5–2 \times 10^5$ cells/well) in complete medium.

  Note: when using cells previously stimulated in vitro, wash them carefully prior to the seeding to avoid the carryover of cytokines produced by the cells in the culture or of recombinant cytokines added exogenously.
- Consider the following controls (in triplicate): (i) negative control, *i.e.*, unstimulated cells or cells stimulated with an irrelevant antigen; (ii) positive control, *i.e.*, cells stimulated with a mitogen (e.g., Con A or PMA/Iono); (iii) blank, *i.e.*, medium only or in case of using an irradiated tumor cell line as stimulus seed the tumor cells only.
- Add 100 μL/well of stimuli (target cells or purified antigen or mitogen) and incubate the plates at 37 °C, 5% CO_2 for 18–24 h. Avoid moving the plates during incubation.

Day 2—Detection (sterile conditions no longer needed)

- Invert the plates to remove the cells and wash twice with 200 μL/well of deionized water to remove any residual cells.
- Wash three times with 200 μL/well of PBS-T (PBS with 0.05% Tween 20).
- Dilute biotinylated detection anti-IFNγ Ab (e.g., R4-6A2) to 1 μg/mL in PBS-T and add 100 μL/well.
- Incubate at RT for 2 h in the dark.
- Wash each well four times with 200 μL of PBS-T.
- Add 100 μL/well of streptavidin-alkaline phosphatase (or horseradish peroxidase) diluted according to the manufacturer instructions in PBS-T and incubate for 1 h at RT.

  Note: Exceeding 1 h incubation with enzyme conjugate can result in increased background.

- Wash wells four times with 200 μL of PBS-T followed by three washes with PBS.

  Note: the final washes with only PBS are important, because Tween-20 may interfere with spot development.
- Prepare the substrate* according to the manufacturer's instructions. Add 50 μL/well of substrate solution and incubate in the dark for 5–15 min.

  **BCIP/NBT*: 5-bromo-4-chloro-3-indolyl phosphate (BCIP)/nitro blue tetrazolium (NBT) is an insoluble substrate for ALP conjugates, producing blue-purple product. The substrate is commercially available as tablets or liquid solution ready-to-use. Tablets must be dissolved in the indicated volume of deionized water, resulting in the substrate solution to be used.

  TMB: tetramethylbenzidine is a substrate for HRP conjugates, available as a ready-to-use solution. *AEC*: 3-amino-9-ethylcarbazole is a substrate for HRP conjugates, commercially available as a kit, associated with acetate buffer and hydrogen peroxide, to be prepared before use.

  Note: optimization of the time of substrate incubation is critical; over-development will result in increased background.
- Rinse the plates under running tap water.
- Remove the underdrain of the plates and allow complete drying before spot counting.
- Count the spots using an automated ELISpot reader

  Note: The plates can be stored at RT in the dark and analyzed later.

A review of guidelines for automated evaluation of ELISpot assays has been recently published elsewhere (Janetzki et al., 2015). A list of possible problems and solutions during ELISpot assay execution is reported in Table 1.

Table 2 lists existing antibodies pairs (capture and detection) available for murine cytokine ELISpot assay.

5. Data analysis

Each condition should be assayed in triplicate. Blank and negative control wells should be included to ensure the quality of the assay. The spot counts read by an automated ELISpot reader system will be expresses as the number of SFC per each well. Means and standard deviations are calculated out of the counts for each triplicate in each sample and treatment condition. The type of stimuli, the source of cells and the immune activation state have profound effects on the results. Appropriate negative and positive controls

Table 1 ELISpot troubleshooting and tips.

Issue	Possible cause	Solution
High background	Insufficient washing	Wash both sides of the membrane with distilled water before and after color development
	Too many cells secreting cytokine/protein of interest	Reduce the number of cells per well and/or optimize the concentration of stimulant used
	Plate not dried properly	Dry the plate longer before reading, keeping it protected from light
	Over-developed plate	Reduce developing time
No spots/very few spots	Few cells secreting the cytokine/protein of interest	Increase the number of cells. Cell density should range between 0.5 and 2×10^5 cells/well
	Ensure the cells are stimulated correctly	Use a positive control, i.e. a stimulus that you know will induce expression of the cytokine/protein of interest
	Cells not incubated for long enough to respond	Increase the cell incubation time or pre-treat cells with stimulant before the assay
	Sub-optimal primary or secondary Ab concentration	Increase the concentration of the primary and/or secondary antibody
Poorly defined spots	Plate movement during cell incubation	Minimize plate movement during cell incubation. Do not tap the plate after adding cells
	Membrane has dried out at some stage	Ensure membrane does not dry. Always keep membrane moistened
	Cells unevenly distributed	Mix cells gently before pipetting the cells into the wells. Make sure the incubator shelf is not inclined
	Pipette tip touched the membrane	Take care with pipetting steps, particularly with automated washing
False positives	Cell still on the membrane, cell debris	Ensure all the cells are washed away from the membrane before secondary Ab incubation
	Contaminating platelets (when using blood samples)	Optimize sample preparation. Increase the number of washing rounds
	Cell culture contamination (dust and microbial)	Keep reagents as sterile and clean as possible
	Mitogens and other factors in the serum are stimulating the cells	Heat inactivate the serum and filter sterile the culture medium

Table 2 Combination of capture and detection Ab clones for murine cytokines, validated for the ELISpot assay.

Cytokine	Capture Ab	Detection Ab
IL-2	JES6-1A12	JES6-5H4
IL-4	11B11	BVDG-24G2
IL-5	TRFK5	TRFK4
IL-6	20F3	32C11
IL-10	JES5-16E3	JES5-2A5
IL-12	C15.6	C17.8
IL-17	TCH-18H10.1	TC11-8H4
	50101.111	17B7
IFNγ	AN18	R46A2
	XMG1.2	
TNFα	1C8/23C9	11B10
	6B8	MP6-XT22

should be included and positivity criteria needs to be established within the laboratory using the selected reagents, stimuli and source of cells. To define positivity criteria for the quantitative ELISpot assay data two different approaches can be considered, the empirical method and the hypothesis test (statistical approach) (Moodie et al., 2010). The empirical approach is based on ad hoc (Empirical) rules, as described in Dubey paper, showing an example of optimization and validation of the ELISpot assay for the detection of HIV-specific T cell response in humans (Dubey et al., 2007). This method provides a threshold limit (cut-off) and a fold difference between the experimental and negative controls. For instance in our settings, a positive response is based on a threshold minimum of 50 SFC/million cells in the experimental wells and at least a twofold increase of spot number over background. The main weakness of the empirical method is that false positive or false negative results are not properly controlled.

The hypothesis test approach uses the statistical difference between the experimental and negative control wells as the criteria for determining positivity. This approach is based on statistical methods such as *t*-test, Wilcoxon rank sum test and others (Dittrich & Lehmann, 2012; Moodie et al., 2010).

In general, hypothesis test methods are data-driven and objective, but they may detect a statistically significant difference that does not represent a real biological meaning.

In clinical trials the validation of the ELISpot assay is a fundamental step to allow for comparison of data among different clinical sites and give reliable and relevant results with significant clinical impact. To do this, standard operation procedures (SOPs) under good laboratory practice (GLP) guidance need to be followed and shared among collaborators (Cox, Ferrari, & Janetzki, 2006). In preclinical studies, even if not mandatory, it would be a good practice to validate the ELISpot assay within the laboratory by either the empirical or hypothesis test based approach to assure the sensitivity and specificity of the test and the accuracy of the data.

6. Advantages and limitations of the ELISpot assay

ELISpot assays are fast, reliable and cost-effective, as compared to other functional assays (e.g., ICS). They are particularly suitable for screenings due to their high sensitivity and ease of execution. Other advantages are the application to both fresh and cryopreserved samples which makes the assay very attractive for use across multiple centers participating in the same study.

Although it shares some similarities with ELISA assays, ELISpot retrieves the frequency of cells that are secreting the protein of interest, while ELISA only measures the total analyte concentration in the culture medium. In addition, at the end of the assay ELISpot plates can be stored for long time or shipped somewhere else for spot counting, while ELISA assay needs to be read immediately.

Among the disadvantages, the ELISpot assay does not provide information on the phenotype of responding cells. Expression and/or upregulation of co-receptors or lineage markers and cytokine secretion cannot be simultaneously analyzed. However, since the cells are not fixed during the assay, they could be further harvested and analyzed by other techniques. Alternatively, to identify the cytokine-secreting subset in a mixed population (e.g., $CD4^+$ T vs $CD8^+$ T) the subsets can be purified prior to the assay, however this approach may limit the success of antigen presentation due to the lack of proper "helper" function by ancillary cell subsets in culture. Other strategies to indirectly identify responding cell subsets in an unsorted population is by blocking MHC class I or class II presentation with specific antibodies or using, whenever possible, MHC-restricted synthetic peptides for stimulation.

7. Conclusions

The ELISpot assay is a highly sensitive, cost-effective and relatively easy to perform method to enumerate even rare cytokine-producing cells in biological samples upon antigen-specific stimulation. The assay is applicable to translational research studies as well as to clinical settings where monitoring the tumor-specific immune response is gaining much interest due to the clinical revolution introduced by cancer immunotherapy. We encourage the use of ELISpot assays as ideal screening platforms for prognostic purposes and as follow up upon immune-stimulating treatments.

Acknowledgments

The authors acknowledge AIRC (MFAG 13058), Italian Ministry of Health (Ricerca Corrente) and National Institute of Allergy and Infectious Diseases (grant #1P01AI110485-01A1) for financial support. In addition, this project has received funding from the European Union's Horizon 2020 research and innovation program under grant agreement No. 681137.

References

Bracci, L., Moschella, F., Sestili, P., La Sorsa, V., Valentini, M., Canini, I., et al. (2007). Cyclophosphamide enhances the antitumor efficacy of adoptively transferred immune cells through the induction of cytokine expression, B-cell and T-cell homeostatic proliferation, and specific tumor infiltration. *Clinical Cancer Research*, *13*(2), 644–653. https://doi.org/10.1158/1078-0432.CCR-06-1209.

Bruchard, M., & Ghiringhelli, F. (2019). Deciphering the roles of innate lymphoid cells in cancer. *Frontiers in Immunology*, *10*(656). https://doi.org/10.3389/fimmu.2019.00656.

Cox, J., Ferrari, G., & Janetzki, S. (2006). Measurement of cytokine release at the single cell level using the ELISPOT assay. *Methods*, *38*(4), 274–282. https://doi.org/10.1016/j.ymeth.2005.11.006.

Czerkinsky, C. C., Nilsson, L. A., Nygren, H., Ouchterlony, O., & Tarkowski, A. (1983). A solid-phase enzyme-linked immunospot (ELISPOT) assay for enumeration of specific antibody-secreting cells. *Journal of Immunological Methods*, *65*(1–2), 109–121. Retrieved from http://www.ncbi.nlm.nih.gov/pubmed/6361139.

Dangoor, A., Lorigan, P., Keilholz, U., Schadendorf, D., Harris, A., Ottensmeier, C., et al. (2010). Clinical and immunological responses in metastatic melanoma patients vaccinated with a high-dose poly-epitope vaccine. *Cancer Immunology, Immunotherapy : CII*, *59*(6), 863–873. https://doi.org/10.1007/S00262-009-0811-7.

Das, K., Eisel, D., Lenkl, C., Goyal, A., Diederichs, S., Dickes, E., et al. (2017). Generation of murine tumor cell lines deficient in MHC molecule surface expression using the CRISPR/Cas9 system. *PLoS One*, *12*(3), e0174077. https://doi.org/10.1371/journal.pone.0174077.

Dittrich, M., & Lehmann, P. V. (2012). *Statistical analysis of ELISPOT assays* (pp. 173–183). Totowa, NJ: Humana Press. https://doi.org/10.1007/978-1-61779-325-7_14.

Domouchtsidou, A., Barsegian, V., Mueller, S. P., Best, J., Ertle, J., Bedreli, S., et al. (2018). Impaired lymphocyte function in patients with hepatic malignancies after selective internal radiotherapy. *Cancer Immunology, Immunotherapy : CII, 67*(5), 843–853. https://doi.org/10.1007/s00262-018-2141-0.

Dubey, S., Clair, J., Fu, T.-M., Guan, L., Long, R., Mogg, R., et al. (2007). Detection of HIV vaccine-induced cell-mediated immunity in HIV-seronegative clinical trial participants using an optimized and validated enzyme-linked immunospot assay. *Journal of Acquired Immune Deficiency Syndromes, 45*(1), 20–27. https://doi.org/10.1097/QAI.0b013e3180377b5b.

Dutton, R. W., Bradley, L. M., & Swain, S. L. (1998). T cell memory. *Annual Review of Immunology, 16*(1), 201–223. https://doi.org/10.1146/annurev.immunol.16.1.201.

Fiore-Gartland, A., Manso, B. A., Friedrich, D. P., Gabriel, E. E., Finak, G., Moodie, Z., et al. (2016). Pooled-peptide epitope mapping strategies are efficient and highly sensitive: An evaluation of methods for identifying human T cell epitope specificities in large-scale HIV vaccine efficacy trials. *PLoS One, 11*(2), e0147812. https://doi.org/10.1371/journal.pone.0147812.

Gad, E., Rastetter, L., Herendeen, D., Curtis, B., Slota, M., Koehniein, M., et al. (2013). Optimizing the cryopreservation of murine splenocytes for improved antigen-specific T cell function in ELISPOT. *Journal for Immunotherapy of Cancer, 1*(S1), P211. https://doi.org/10.1186/2051-1426-1-S1-P211.

Gallinaro, A., Borghi, M., Bona, R., Grasso, F., Calzoletti, L., Palladino, L., et al. (2018). Integrase defective lentiviral vector as a vaccine platform for delivering influenza antigens. *Frontiers in Immunology, 9*(171). https://doi.org/10.3389/fimmu.2018.00171.

Grasso, F., Negri, D. R. M., Mochi, S., Rossi, A., Cesolini, A., Giovannelli, A., et al. (2013). Successful therapeutic vaccination with integrase defective lentiviral vector expressing nononcogenic human papillomavirus E7 protein. *International Journal of Cancer, 132*(2), 335–344. https://doi.org/10.1002/ijc.27676.

Ikutani, M., Yanagibashi, T., Ogasawara, M., Tsuneyama, K., Yamamoto, S., Hattori, Y., et al. (2012). Identification of innate IL-5-producing cells and their role in lung eosinophil regulation and antitumor immunity. *The Journal of Immunology, 188*(2), 703–713. https://doi.org/10.4049/jimmunol.1101270.

Janetzki, S. (2015). Immune monitoring technology primer: The enzyme-linked immunospot (Elispot) and Fluorospot assay. *Journal for Immunotherapy of Cancer, 3*(1), 30. https://doi.org/10.1186/s40425-015-0074-0.

Janetzki, S., Price, L., Schroeder, H., Britten, C. M., Welters, M. J. P., & Hoos, A. (2015). Guidelines for the automated evaluation of Elispot assays. *Nature Protocols, 10*(7), 1098–1115. https://doi.org/10.1038/nprot.2015.068.

Johanns, T. M., Ward, J. P., Miller, C. A., Wilson, C., Kobayashi, D. K., Bender, D., et al. (2016). Endogenous neoantigen-specific CD8 T cells identified in two glioblastoma models using a cancer immunogenomics approach. *Cancer Immunology Research, 4*(12). https://doi.org/10.1158/2326-6066.CIR-16-0156.

Kjeldsen, J. W., Iversen, T. Z., Engell-Noerregaard, L., Mellemgaard, A., Andersen, M. H., & Svane, I. M. (2018). Durable clinical responses and long-term follow-up of stage III–IV non-small-cell lung cancer (NSCLC) patients treated with IDO peptide vaccine in a phase I study—A brief research report. *Frontiers in Immunology, 9*(2145). https://doi.org/10.3389/fimmu.2018.02145.

Körber, N., Behrends, U., Hapfelmeier, A., Protzer, U., & Bauer, T. (2016). Validation of an IFNγ/IL2 FluoroSpot assay for clinical trial monitoring. *Journal of Translational Medicine, 14*(1), 175. https://doi.org/10.1186/s12967-016-0932-7.

Kouwenhoven, M., Ozenci, V., Teleshova, N., Hussein, Y., Huang, Y.-M., Eusebio, A., et al. (2001). Enzyme-linked immunospot assays provide a sensitive tool for detection of cytokine secretion by monocytes. *Clinical and Vaccine Immunology*, *8*(6), 1248–1257. https://doi.org/10.1128/CDLI.8.6.1248-1257.2001.

Li Pira, G., Ivaldi, F., Starc, N., Landi, F., Locatelli, F., Rutella, S., et al. (2014). Miniaturized and high-throughput assays for analysis of T-cell immunity specific for opportunistic pathogens and HIV. *Clinical and Vaccine Immunology*, *21*(4), 488–495. https://doi.org/10.1128/CVI.00660-13.

Lion, E., Smits, E. L. J. M., Berneman, Z. N., & Van Tendeloo, V. F. I. (2009). Quantification of IFN-γ produced by human purified NK cells following tumor cell stimulation: Comparison of three IFN-γ assays. *Journal of Immunological Methods*, *350*(1–2), 89–96. https://doi.org/10.1016/J.JIM.2009.08.014.

Liu, Y., Chen, K., Wang, C., Gong, W., Yoshimura, T., Wang, J., et al. (2013). Isolation of mouse tumor-infiltrating leukocytes by Percoll gradient centrifugation. *Bio-Protocol*, *3*(17). https://doi.org/10.21769/BioProtoc.892.

Malyguine, A. M., Strobl, S., Dunham, K., Shurin, M. R., & Sayers, T. J. (2012). ELISPOT assay for monitoring cytotoxic T lymphocytes (CTL) activity in cancer vaccine clinical trials. *Cell*, *1*(2), 111–126. https://doi.org/10.3390/cells1020111.

Moodie, Z., Price, L., Gouttefangeas, C., Mander, A., Janetzki, S., Löwer, M., et al. (2010). Response definition criteria for ELISPOT assays revisited. *Cancer Immunology, Immunotherapy*, *59*(10), 1489–1501. https://doi.org/10.1007/s00262-010-0875-4.

Negri, D. R. M., Borghi, M., Baroncelli, S., Macchia, I., Buffa, V., Sernicola, L., et al. (2006). Identification of a cytotoxic T-lymphocyte (CTL) epitope recognized by Gag-specific CTLs in cynomolgus monkeys infected with simian/human immunodeficiency virus. *Journal of General Virology*, *87*(11), 3385–3392. https://doi.org/10.1099/vir.0.81934-0.

Safi, S., Yamauchi, Y., Rathinasamy, A., Stamova, S., Eichhorn, M., Warth, A., et al. (2017). Functional T cells targeting tumor-associated antigens are predictive for recurrence-free survival of patients with radically operated non-small cell lung cancer. *Oncoimmunology*, *6*(11), e1360458. https://doi.org/10.1080/2162402X.2017.1360458.

Santos, R., Buying, A., Sabri, N., Yu, J., Gringeri, A., Bender, J., et al. (2014). Improvement of IFNg ELISPOT performance following overnight resting of frozen PBMC samples confirmed through rigorous statistical analysis. *Cell*, *4*(1), 1–18. https://doi.org/10.3390/cells4010001.

Schiavoni, G., Sistigu, A., Valentini, M., Mattei, F., Sestili, P., Spadaro, F., et al. (2011). Cyclophosphamide synergizes with type I interferons through systemic dendritic cell reactivation and induction of immunogenic tumor apoptosis. *Cancer Research*, *71*(3), 768–778. https://doi.org/10.1158/0008-5472.CAN-10-2788.

Schulten, V., & Sette, A. (2018). *The identification of allergen-derived T cell epitopes*. In *Vol. 1799. Methods in molecular biology* (pp. 153–163). Clifton, NJ. https://doi.org/10.1007/978-1-4939-7896-0_13.

Shafer-Weaver, K. A., Sayers, T., Kuhns, D. B., Strobl, S. L., Burkett, M. W., Baseler, M., et al. (2004). Evaluating the cytotoxicity of innate immune effector cells using the GrB ELISPOT assay. *Journal of Translational Medicine*, *2*(1), 31. https://doi.org/10.1186/1479-5876-2-31.

Song, W., Shen, L., Wang, Y., Liu, Q., Goodwin, T. J., Li, J., et al. (2018). Synergistic and low adverse effect cancer immunotherapy by immunogenic chemotherapy and locally expressed PD-L1 trap. *Nature Communications*, *9*(1), 2237. https://doi.org/10.1038/s41467-018-04605-x.

Swartz, A. M., Reap, E., Norberg, P., Schmittling, R., Janetzki, S., Sanchez-Perez, L., et al. (2018). A simple and enzyme-free method for processing infiltrating lymphocytes from small mouse tumors for ELISpot analysis. *Journal of Immunological Methods*, *459*, 90–93. https://doi.org/10.1016/J.JIM.2018.05.015.

Zaynagetdinov, R., Sherrill, T. P., Gleaves, L. A., McLoed, A. G., Saxon, J. A., Habermann, A. C., et al. (2015). Interleukin-5 facilitates lung metastasis by modulating the immune microenvironment. *Cancer Research*, *75*(8), 1624–1634. https://doi.org/10.1158/0008-5472.CAN-14-2379.

Zhou, R., Qian, S., Gu, X., Chen, Z., & Xiang, J. (2013). Interleukin-13 and its receptors in colorectal cancer (review). *Biomedical Reports*, *1*(5), 687–690. https://doi.org/10.3892/br.2013.132.

CHAPTER TWENTY-FOUR

High-throughput analysis of cell-cell crosstalk in *ad hoc* designed microfluidic chips for oncoimmunology applications

Arianna Mencattini[d,†], Adele De Ninno[b,c,†], Jacopo Mancini[a], Luca Businaro[b], Eugenio Martinelli[d], Giovanna Schiavoni[a,*], Fabrizio Mattei[a,*]

[a]Department of Oncology and Molecular Medicine, Tumor Immunology Unit, Istituto Superiore di Sanità, Rome, Italy
[b]Institute for Photonics and Nanotechnology, Italian National Research Council, Rome, Italy
[c]Department of Civil Engineering and Computer Science, University of Rome Tor Vergata, Rome, Italy
[d]Department of Electronic Engineering, University of Rome Tor Vergata, Rome, Italy
*Corresponding authors: e-mail address: giovanna.schiavoni@iss.it; fabrizio.mattei@iss.it

Contents

† Equal contribution.

Methods in Enzymology, Volume 632
ISSN 0076-6879
https://doi.org/10.1016/bs.mie.2019.06.012

Abstract

Understanding the interactions between immune and cancer cells occurring within the tumor microenvironment is a prerequisite for successful and personalized anti-cancer therapies. Microfluidic devices, coupled to advanced microscopy systems and automated analytical tools, can represent an innovative approach for high-throughput investigations on immune cell-cancer interactions. In order to study such interactions and to evaluate how therapeutic agents can affect this crosstalk, we employed two *ad hoc* fabricated microfluidic platforms reproducing advanced 2D or 3D tumor immune microenvironments. In the first type of chip, we confronted the capacity of tumor cells embedded in Matrigel containing one drug or Matrigel containing a combination of two drugs to attract differentially immune cells, by fluorescence microscopy analyses. In the second chip, we investigated the migratory/interaction response of naïve immune cells to danger signals emanated from tumor cells treated with an immunogenic drug, by time-lapse microscopy and automated tracking analysis. We demonstrate that microfluidic platforms and their associated high-throughput computed analyses can represent versatile and smart systems to: (i) monitor and quantify the recruitment and interactions of the immune cells with cancer in a controlled environment, (ii) evaluate the immunogenic effects of anti-cancer therapeutic agents and (iii) evaluate the immunogenic efficacy of combinatorial regimens with respect to single agents.

1. Introduction

The success of anti-cancer therapies relies on the capacity of the host to develop a tumor-specific immunity. Hence, understanding the interactions occurring between cancerous cells and immune cells becomes essential. The conventional *in vitro* culture methods and *in vivo* models holds limitations in many aspects since neither approach allows to monitor three-dimensional interactions among multiple cell type in real time, while maintaining a tight control of the microenvironment (Boussommier-Calleja, Li, Chen, Wong, & Kamm, 2016). The use of microfluidic devices in biomedical research, including oncoimmunology, has largely increased in the last years (Agliari et al., 2014; Businaro et al., 2013; Lucarini et al., 2017; Sackmann, Fulton, & Beebe, 2014; Vacchelli et al., 2015; Wlodkowic & Cooper, 2010; Wlodkowic & Darzynkiewicz, 2010). Indeed, the growing innovation on microscopy systems, together with advances in materials science adapted for microfabrication (Czarnecki & Bertin, 2018; Selimovic et al., 2011; Wilding, Pfahler, Bau, Zemel, & Kricka, 1994; Wilding, Shoffner, & Kricka, 1994), significantly contributed to the assembly of versatile microfluidic platforms to target specific topics in oncoimmunology.

Here, we show some automated or semi-automatic methods optimized for the use of microfluidic devices in the context of tumor immunology investigations. We demonstrate how these methods allow to extrapolate large numeric datasets for high-throughput analysis of multiple biological features of the crosstalk between immune and cancer cells. We propose two microfluidic chips useful to investigate different aspects of cancer cell vs immune cell crosstalk in 3D or 2D microenvironments. These *ad hoc* fabricated chips are equipped with structural features optimized for the experiments they have been adapted for. In the first chip, we assessed the ability of human peripheral blood mononuclear cells (PBMCs) to migrate preferentially toward SK melanoma cells (SK-MEL) exposed to a single anti-cancer agent or to a combination of drugs. SK-MEL cells were embedded in a Matrigel matrix in order to mimic the presence of extracellular matrix. This setting generates a competitive system to evaluate and quantify the immunogenicity of a combination of drugs, with respect to a single one (Lucarini et al., 2017). SK-MEL and PBMCs were fluorescently labeled with distinct dyes in order to track them. Fluorescence-based analysis at different timepoints was carried out to profile the preferential migratory ability of PBMCs toward either chamber as well as to evaluate ensuing morphological changes occurring in tumor cells.

In the second device, we studied the migratory response of murine immune cells to adherent tumor cells exposed to an immunogenic drug by time-lapse microscopy. The resulting video was analyzed with the Cell-Hunter proprietary software (Mencattini et al., 2019), in order to extract several high-throughput parameters, such as single cell tracking and interaction times between cells. Overall, these parameters may serve as valuable factors for determining immune cell responses to anti-cancer treatments.

2. Fabrication process of microfluidic devices

PDMS is a biocompatible silicone elastomer, exhibits excellent optical qualities, allowing direct observation via microscopy. The manufacture of the PDMS microfluidic assays can be divided into: (i) fabrication of a microstructured epoxy resin mold, (ii) PDMS channel structure realization by standard soft lithographic technique; and (iii) PDMS structure ceiling with microscope slides and functionalization by oxygen plasma treatment.

2.1 Microstructured master and PDMS microfluidic devices

The process is performed by casting a PDMS liquid precursor on silicon wafers with polymeric microstructures complementarily to those wanted on the PDMS soft substrate. The mold used as the negative for PDMS replication is typically referred to as "master." Masters have to be manufactured and functionalized in a clean room facility or foundry equipped with micro and nanoscale instrumentation systems (e-beam lithography, photolithography and thin-film technology, nano-diagnostic tools). Patterns are transferred onto a silicon wafer covered by a radiation sensitive epoxy resin, called resist (SU-8 3000 series, MicroChem Corp, Newton, MA), by two different i-line (365 nm) optical lithography onto the negative resist. The needed chromium masks (Hoya Inc.) were patterned by 100 kV e-beam lithography. Briefly, silicon wafers were spin-coated with a layer of SU-8 3005 (resist thickness 10 μm) and exposed to an i-line UV light source through a photo mask with the microchannel pattern and alignment markers. Next, the second photolithography step, on Su-8 3050 (100 μm thick for Type 2 devices, 150 μm for Type 1 devices) as a second layer, transferred the two gel regions, three chambers and reservoirs for media and cell loading, aligned to the first pattern. Optimization tests were performed to obtain the required exposure dosage for the two step-optical lithography. SU-8 3000 series is capable of producing very high, over 5:1, aspect ratio structures and suited for imaging near vertical sidewalls in very thick films (see Figs. 1A and 2). This obtained silicon wafer can be reused several times as a master to fabricate multiple chip replicas. The master is generated *ad hoc* with defined macro- and microstructures that will feature the chip, according to the desired experiment.

The PDMS process consisted in the points below:

- Attach the 4″ silanized SU-8 master to the bottom of an aluminum-covered glass Petri dish by using kapton tape.
- PDMS monomer is mixed thoroughly with curing agent (Sylgard 184 kit) in the ratio 10:1 to prepare the prepolymer in a plastic cup.

- Degas the prepolymer solution in a vacuum desiccator to remove the entrapped air bubbles for approximately 30 min and then remove any dust on the surface of the silicon wafer by using a nitrogen gun.
- Slowly pour the PDMS prepolymer to obtain the desired thickness. It takes about 20 mL of PDMS to cover the master with a thickness of approximately 5 mm. Leave undisturbed under hood for 10 min at room temperature (RT) before baking. A nitrogen gun with a 0.45 μm filter is used to remove any remaining bubbles.
- Place the master for 1 h 30 min at 110 °C on a hotplate to polymerize PDMS and then let it cool.
- Once cross-linked and cooled down, carefully peel off PDMS from the master with a surgical blade and tweezers. Shape the multiple replica substrates, as desired, and trim away excess PDMS so that the device will fit neatly on your glass slide or culture plastic substrate.
- Next, inlet or outlet fluidic access ports (reservoirs for loading cells and medium) are created using a suite of dermal biopsy punches (Kai Medical, Tedpella). Use 6–8 mm biopsy tools for large wells (Type 2 device) or 4 and 1 mm for medium channels and hydrogel inlets, respectively (Type 1 device).

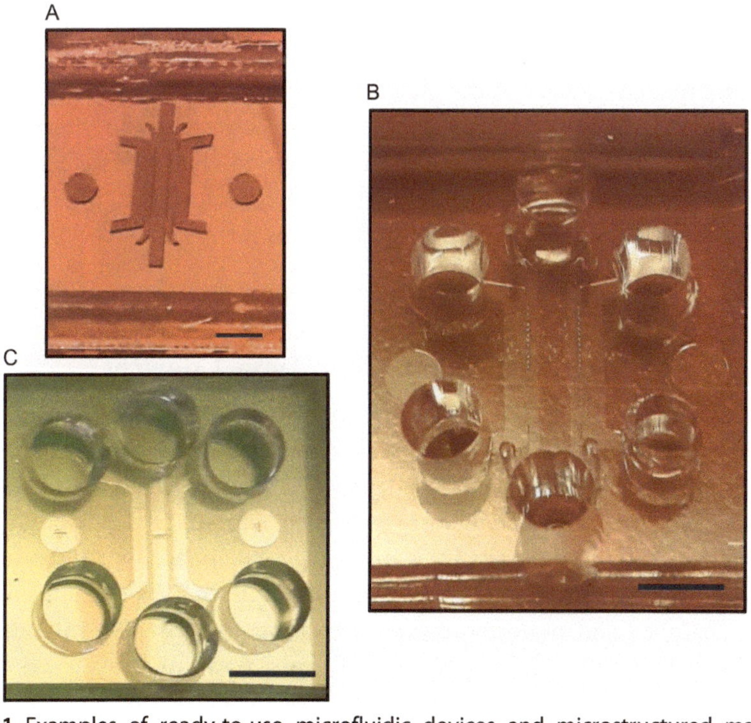

Fig. 1 Examples of ready-to-use microfluidic devices and microstructured master. (A) Example of microstructured master on a silicon wafer to be used for the replica of Type 1 device. (B) Type 1 device depicting the chambers and the loading wells. (C) Type 2 device with the chambers and loading wells clearly visible. Scalebars, 5 mm.

2.2 Plasma bonding of the devices to glass slides or cover slips

Oxygen Plasma activation is the standard method in microfluidics performed for the hydrophilic surface functionalization of PDMS and for irreversible bonding to glass microscope slides, coverslips or optical bottom multiwells by using process at high gas pressures and low RF (radiofrequency) power, in a reactive ion etcher (RIE) or a plasma cleaner. Reactive oxygen radicals attack the methyl groups (Si–CH_3) on the PDMS surface, generating Si–OH silanol groups and glass that undergo condensation reactions to yield Si–O–Si bonds. Plasma-oxidized surfaces brought into conformal contact form a tight permanent seal, improving adhesion of the substrates and pressure tolerance of microchannels. Plasma bonding steps are delineated in the points below:

- Before oxygen treatment, under a chemical hood clean the glass microscope slides (25 mm × 75 mm) or coverslips in Piranha solution (H_2SO_4: H_2O_2 3:1), rinse in distilled H_2O, and dry with a nitrogen gun to remove debris and surface contaminants.
- Place PDMS chips (with micropatterned side facing up) along with the glass substrates inside the RIE chamber (Oxford Plasmalab 80 plus system) at the following RIE settings: 20 W RF power, flux of 60 standard cubic centimeter/minute, 800 mTorr pressure, and 30 s time. Optimize process parameters according to your plasma cleaner machine.
- The treated surfaces of the glass and microfluidic device (channels side faced down) are placed in contact with each other at a clean bench. Visual inspect under an optical microscope to check for defects.
- Post-bake at 80–90 °C for 2 h, to enhance adhesion strength. To restore hydrophobicity (Type 1 devices) dry chips for >4 h at 80–90 °C. Store chips in sterile Petri dishes.

2.3 Plasma functionalization of the Type 1 devices

Selective hydrophilic activation of chamber walls by O_2 plasma treatment was needed to guarantee the correct gel matrix positioning (preserving hydrophobicity of gel regions) and the proper medium filling of connecting-microchannels before biological experiments (Parlato et al., 2017). Mask all access ports by Kapton tape except for two reservoirs in the central compartment. At this point, place assembled chips inside the RIE chamber and perform oxygen activation for few (10–15) seconds. Finally, sterilize the devices under UV light under a laminar hood for 15–30 min and store until use. An example of a ready-to-use Type 1 device is illustrated in Fig. 1B.

2.4 Plasma functionalization and coating of the Type 2 devices

To make PDMS chips hydrophilic and to facilitate medium addition, oxygen plasma-activate the PDMS/glass-assembled substrates in the RIE chamber for 30 s. Immediately pipette 100 μL of distilled H_2O in each reservoir and store until cell loading, in order to maintain hydrophilic surface. Sterilize chips under UV light in a vertical laminar flow biological safety cabinet for 15–30 min. Aspirate the water and add 100 μL of surface coating solution in each reservoir. Various polymer (e.g., Poly-L-lysine) and extracellular matrices (e.g., Fibronectin, Collagen) can be used according to desired cell types. Gently remove the coating solution by vacuum aspiration system and wash three times with sterile distilled H_2O. Wash twice quickly channels with PBS and two–three times with RPMI-1640 medium: Fill the devices with medium, place inside a Petri dish and store in incubator at 37 °C with 5% CO_2. An example of a ready-to-use Type 2 device is illustrated in Fig. 1C.

3. Culture and preparation of tumor and immune cells

3.1 Preparation of cells for Type 1 device

SK-MEL-28 (ATCC; HTB-72) metastatic melanoma cells were cultured in RPMI-1640 (Euroclone) supplemented with high glucose, 10% Fetal Bovine Serum (FBS, EuroClone) and 1% Penicillin/Streptomycin/Amphotericin and 1% L-Glutamine (hereafter complete DMEM). Cells were passaged for a maximum of four times from thawing and routinely tested for morphology, growth curve and absence of mycoplasma contamination. Subconfluent tumor cells were recovered and counted in a hemocytometer (Neubauer counting chamber) by Trypan blue exclusion.

Peripheral blood mononuclear cells (PBMC) were isolated from buffy coat of healthy donors, obtained by Ficoll Hipaque density gradient centrifugation (Lympholyte cell separation media, Cedarlane Labs, Burlington, Canada). The whole blood was diluted in a 1:2 ratio with sterile Dulbecco's phosphate buffered saline (PBS, w/o calcium & magnesium, EuroClone) and then carefully stratified on Ficoll (3:1 ratio). The sample was then centrifuged at 2100 rpm for 30 min at room temperature, without brake function. After centrifugation, the low-density fraction, containing PBMCs, was collected and transferred into a new tube containing 40 mL of sterile PBS. PBMCs were centrifuged for 10 min at 1500 rpm at 20 °C. Supernatant was gently removed; cells were resuspended in 40 mL of PBS and further centrifuged for 10 min at 900 rpm at 20 °C to eliminate eventual platelets. After discarding the supernatant, PBMCs were resuspended in complete DMEM. Living cells were counted in a Neubauer counting chamber by Trypan blue exclusion.

With the purpose of distinguishing PBMCs and SK-MEL-28 in terms of location and morphology within the microfluidic Type 1 device, the two cell populations were labeled with two different fluorescent dyes. Specifically, SK-MEL-28 human tumor cells were stained with PKH67 Green Fluorescent Cell Linker whereas PBMCs were labeled with PKH26 Red Fluorescent Linker (both from Sigma-Aldrich), following manufacturer instructions.

3.2 Preparation of cells for Type 2 device

Mouse MCA-205 fibrosarcoma cells (Merck Millipore; SCC173) were cultured in complete DMEM. Cells were passaged for a maximum of four times from thawing and routinely tested for morphology, growth curve and absence of mycoplasma contamination. Subconfluent, living tumor cells were recovered and counted by Trypan blue exclusion. MCA tumor cells were cultured in complete DMEM containing 25 μM doxorubicin (DOXO) for 4 h at 37 °C and 5% CO_2 to induce tumor apoptosis (Mencattini et al., 2019). Tumor cells were then washed by centrifugation to remove excess of drug and resuspended in complete DMEM.

Splenocytes were purified from naïve 5–7-week-old C57BL/6 female mice (Charles River Laboratories). Spleens were excised, transferred into a Petri dish and mechanically minced with a sterile syringe plunger. The obtained tissue homogenate was collected and filtered through a 70 μm cell strainer (Greiner Bio One) previously placed on a 50 mL tube (Falcon). The unfiltered cell suspension was then washed in RPMI-1640 medium supplemented with 5% FCS, 1% Penicillin/Streptomycin/Amphotericin, and 1% L-Glutamine (hereafter washing medium) by centrifugation at 1500 rpm for 7 min at 4 °C. To eliminate red blood cells, the pellet was resuspended in 3 mL of lysis buffer (155 mM NH_4Cl, 10 mM $KHCO_3$, 0.1 mM EDTA; pH 7.2) and incubated for 3 min at room temperature. The reaction was stopped by adding at least 9 mL (3 × volume) of washing medium. Cells were then centrifuged at 1500 rpm for 5 min at 4 °C. The cell pellet was then resuspended in complete DMEM and enumerated by Trypan blue exclusion.

4. Features, loading mode and analysis of cells in Type 1 device

4.1 Features of the Type 1 device

The Type 1 chip design consists of a central chamber to be used for the floating PBMC intake which is connected at the two sides through a network of narrow arrays of microgrooves (connecting-channels; $10 \times 12 \times 200\,\mu m^3$, H × W × L) to two side tumor-chambers (Figs. 2 and 3A–D). These

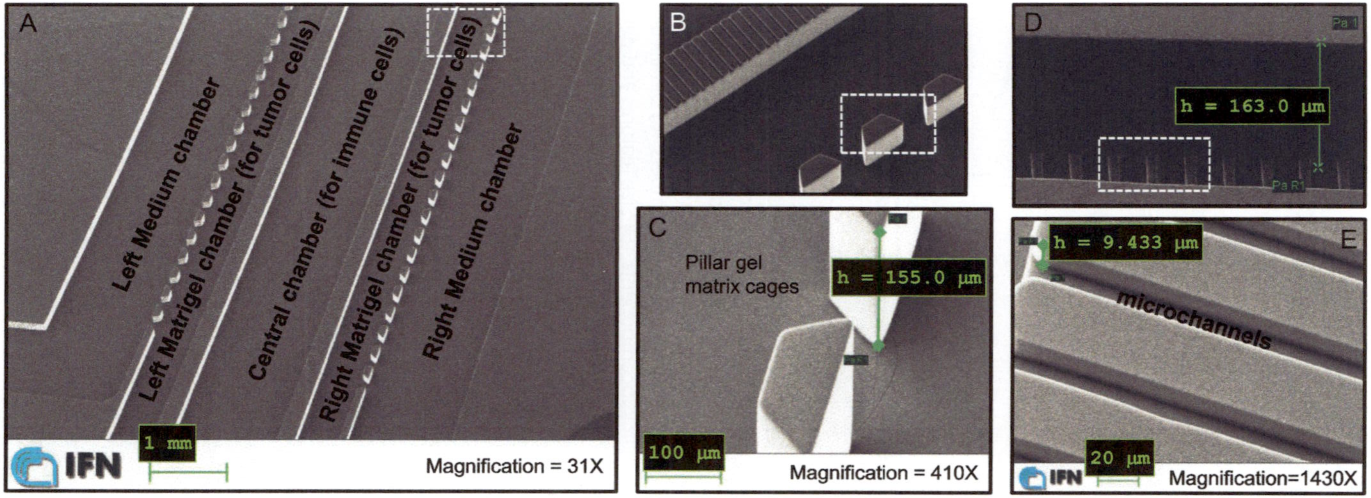

Fig. 2 Details of Type 1 device obtained by scanning electron microscopy (SEM). (A) Overview of a SEM micrograph of a PDMS Type 1 microfluidic device for 3D coculture. (B) Enlarged view of the dashed box in (A) showing the trapezoidal micropillars (about 150–160 μm high). (C) Enlarged view of the dashed box in (B) with details depicting the regularly spaced trapezoidal micropillars suited to hydrogel matrix confinement. (D) Magnified image of negative epoxy master mold with high vertical walls and connecting narrow microgrooves by two step optical lithography (first microchannel layout layer in SU-8 3005 resin; second chambers layer in SU-8 3050, about 160 μm tick). (E) Detailed features of the network of connecting PDMS microchannels (10 μm high) referring to the white dashed box in (D). h, height. Devices and molds were fabricated at IFN-CNR Rome clean room facility.

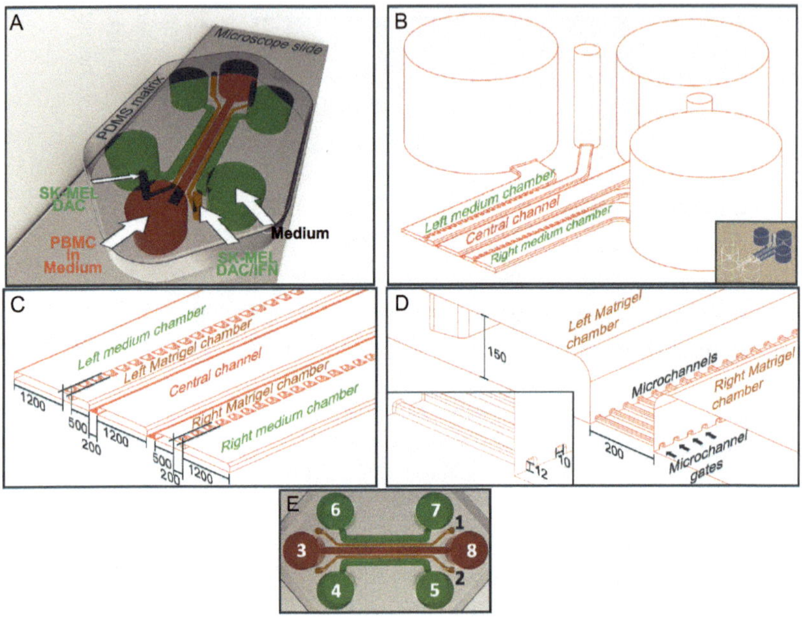

Fig. 3 Type 1 device features and planimetry. Schematic representation of the chip used for the experiments. (A) AutoCAD-rendered image of the chip under the microscope slide, with all the indicated structural components and subcomponents. (B) Detail of the chip with the depicted three principal chambers. (C) Details of the three chambers with size features of their subcomponents. (D) Size features and details of the microchannels and their related subcomponents. (E) Tumor cell and PBMC loading map. 1, Matrigel + SK-MEL + DAC; 2, Matrigel + SK-MEL + DAC/IFN; 3, PBMCs; 4, 5, 6, 7, medium; 8, medium.

microchannel arrays are then connected to the gel regions (one per side, 150–200 μm high), that will contain the two Matrigel mixtures (SK-MEL DAC and SK-MEL DAC/IFN in our experimental setting).

Other detailed features of the Type 1 device are shown in Fig. 3. A set of regularly spaced micropillars (Fig. 3C) connect the Matrigel chamber with two lateral chambers (medium chambers), that may be loaded with medium alone through the appropriate inlet wells. The trapezoidal forms serve as a regular grid to allow the confinement of Matrigel in order to set a gel-liquid boundary with the adjacent chamber by means of a balance between surface tension and capillary forces. This chip can be used to recreate complex tumor microenvironments for 3D co-culture measurements on different cell populations (e.g., stromal components) embedded in gel regions (Lucarini et al., 2017; Nguyen et al., 2018).

4.2 Loading of SK-MEL and PBMC into device

One day before the experiment, a Matrigel (BD Biosciences) stock was removed from the −20 °C freezer and placed on ice into a 4 °C refrigerator, in order to allow liquefaction. Prior to cell loading, the devices were sterilized under UV in a laminar flow hood for 15 min. Next, the devices were placed on ice to avoid Matrigel solidification during the whole procedure of cell loading.

PKH26-labeled SK-MEL-28 cells (2×10^4) were resuspended in 3 μL of Matrigel (2 mg/mL; Corning Life Science) containing 5-aza-2′-deoxycytidine (DAC; 2.5 μM), referred to as SK-MEL DAC. An equal amount of $PKH26^+$ SK-MEL cells were resuspended in 3 μL of Matrigel containing DAC plus and IFN-α2β (IFN; 10^3 U/mL), referred to as SK-MEL DAC/IFN (Lucarini et al., 2017). This operation was performed on ice in order to prevent Matrigel gelification. The two DAC and DAC/IFN Matrigel/tumor cell mixtures were slowly injected into wells 1 and 2, respectively, with a 10-μL micropipette (Fig. 3E).

The device was then placed in an incubator at 37 °C and 5% CO_2 for 30 min to allow gelification of the Matrigel. In the meantime, PKH67-labeled PBMCs (1×10^6 cells) were prepared and resuspended in 10 μL of complete DMEM. After Matrigel solidification, media channels were hydrated with culture medium/coating solution to prevent gel drying in the chips. The chips were kept in incubator until seeding of immune cell suspensions.

The loading process of the device was completed with the following temporal sequence of steps (schematized in Fig. 3E):

1. PBMCs in 10 μL medium into well 3;
2. 100 μL medium into wells 4, 5, 6, and 7;
3. 90 μL medium into well 3;
4. 100 μL medium into well 8.

The integrity of the gel barrier (on each side) avoids the premature flowing of PBMCs in gel or media channels at the starting point of the experiment (0 h) due to volume and pressure initial fluctuations, thus stabilizing the system. Devices were then placed in the incubator at 37 °C and 5% CO_2. Three replica chips per experimental condition were performed.

4.3 Microphotographs acquisition of the cells and data analysis

Using an EVOS-FL fluorescence microscope (Thermofisher Scientific), phase contrast, visible and red/green channels fluorescence microphotographs were acquired at three different timepoints: immediately after cell loading (0 h), after 48 h and after 72 h of incubation (Fig. 4). This allowed evaluating the migratory extent of PBMCs toward the Matrigel chambers containing either SK-MEL DAC or SK-MEL DAC/IFN. A 4× magnification was used in order to acquire the central chamber, the microchannel arrays and the two juxtaposed side channels containing SK-MEL DAC and SK-MEL DAC/IFN.

5. Technical interpretation of the results and analytical procedures used for the Type 1 device

At the 0 h time point, PBMC are uniformly distributed inside the central chamber of the device (Fig. 4A). After 48 h, the PBMCs display a preferential migration toward the melanoma cells containing the DAC/IFN combination. Indeed, a PBMC displacement is clearly visible at 48 h (Fig. 4B) and 72 h timepoints (Fig. 4C) close to microchannel array connected to chamber containing SK-MEL cells (green color) in presence of DAC and IFN. Of note, at this timepoint infiltration of PBMCs inside the SK-MEL DAC/IFN chamber could be observed (Fig. 4C).

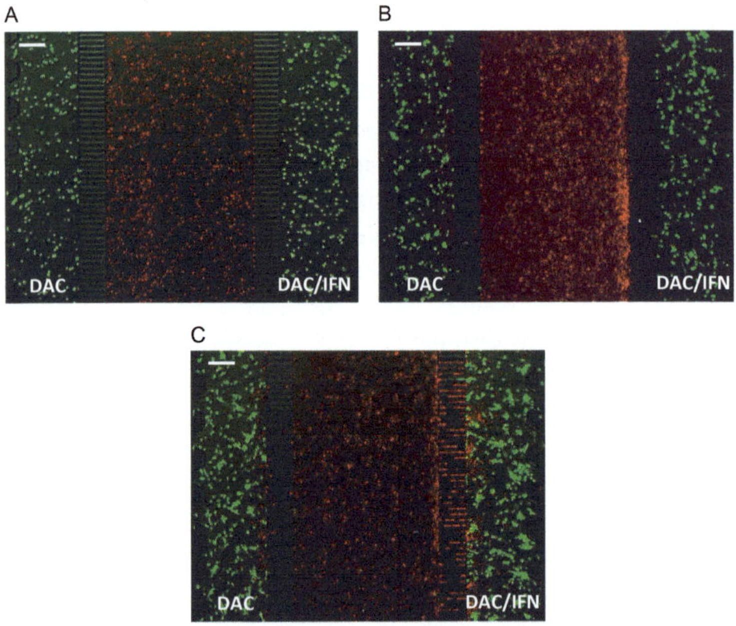

Fig. 4 Acquisition of microphotographs by Fluorescence microscopy. Representative microphotographs for each timepoint. Microphotographs were generated by using an EVOS-FL fluorescence microscope for each experimental condition after cell loading (A), at 48 h (B) and 72 h timepoints (C) by acquiring red (PBMC) and green (SK-MEL) fluorescence channel and visible light. Scalebars, 200 μm.

We then evaluated the PBMC distribution inside the central and side chambers by using the red fluorescence channel. First, the central chamber of three representative images for each time point was subdivided in four regions of interest (ROI) with ImageJ (C_1–C_4 in Fig. 5). These four ROIs were then analyzed as separate entities in order to evaluate the internal homogeneity of PBMC distribution at each time point (Fig. 5A–C). Each ROI was subjected to fluorescence profiling on 8 bit converted, thresholded images. This was made by the "Plot profile" function of ImageJ, separately for each C_1–C_4 ROI, as illustrated by the corresponding fluorescence profiles sets obtained for each time point (Fig. 5A–C). This method was applied for all the acquired microphotographs and the results are shown in Fig. 5. No peak profiles at 0 h is present (Fig. 5D, 0 h), thus suggesting that PBMCs have no migratory preference at this time. At 48 h, a strong peak is present in all right regions of the plots, representing the migration of PBMCs toward SK-MEL DAC/IFN (Fig. 5D, 48 h). After 72 h, these peaks are still present in all the C_1–C_4 ROIs, even though to a minor extent (Fig. 5D, 72 h).

We also evaluated the infiltration of PBMCs inside the two side chambers (SK-MEL DAC vs SK-MEL DAC/IFN, respectively) by quantifying the red fluorescence intensity in these two compartments and by subdividing the chambers in four ideal ROIs, similarly to the procedure used for central chamber calculations. As expected, we observed an evident increase in the PBMC fluorescence ratios at 72 h timepoint, obtained by dividing the fluorescence in the right-side ROIs with the corresponding left-side ones (fluorescence ratios in Fig. 5E). Finally, the dataset of each fluorescence profile was processed for the extrapolation of the linear regression fit by the software Prism GraphPad (Fig. 5F). In these settings, the plotted slope values represent a true estimate of the overall distribution and weight of peaks associated to each profile (Fig. 5G).

To further investigate on peak location displayed in the fluorescence profiles, we performed an automated computation of the area under curve (AUC) allowing the determination of the precise peak position in each PBMC fluorescence profile (Fig. 5H). By setting an appropriate minimum expected peak height for the three timepoints, this function performs a point-to-point AUC evaluation and extrapolate the exact position of each peak displayed in the fluorescence profiles (Fig. 5H and Wang, Ma, George, & Zhou, 2012). Interestingly, the detected peak positions are greater at 48 h and 72 h timepoints, compared to values found for the 0 h timepoint. Conversely, the peak positions after cell loading are

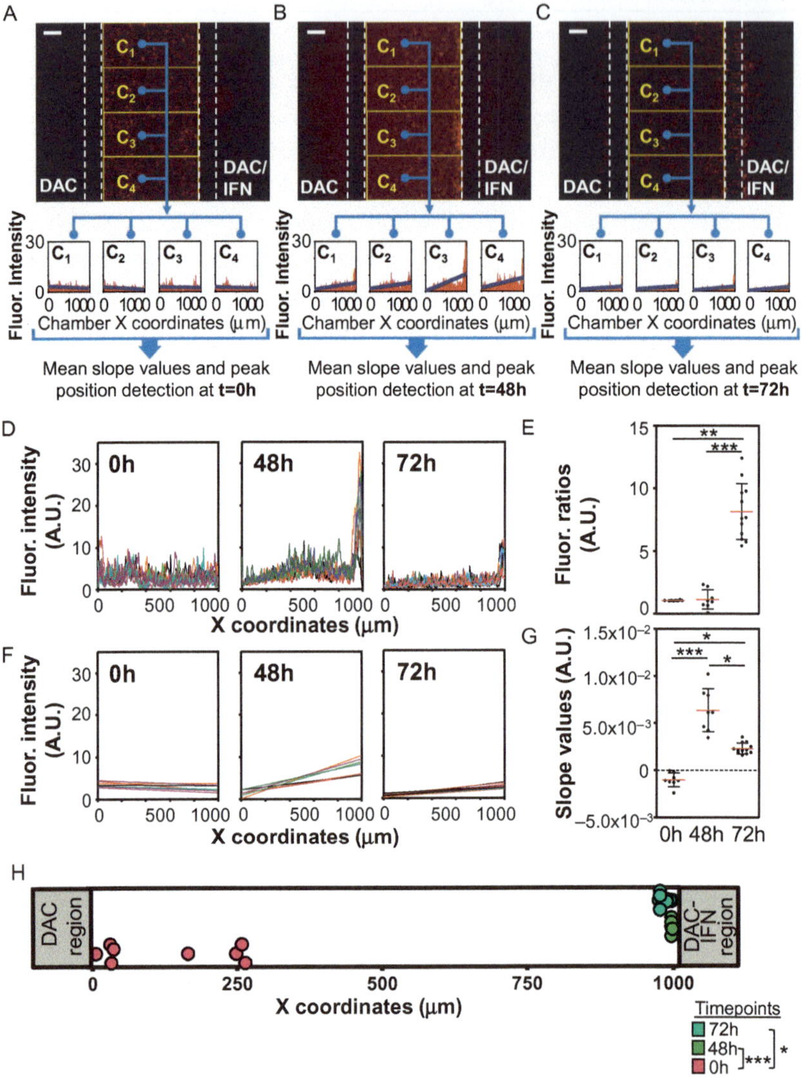

Fig. 5 Quantification of PBMC distribution in the central chamber of the chip after cell loading. Migratory extent of PBMCs was calculated by subdividing the central chamber in four (C1–C4) Region of Interest. Then red fluorescence was calculated and plotted in the histograms for each position of the chamber after cell loading (A), 48 h (B), and 72 h (C) timepoints. Linear regression fit (blue lines) was yielded from each fluorescence profile and then slope of the line was calculated for each fluorescence profile. The slope represents an estimate of the peak distribution inside each fluorescence profile. Automatic peak and area detection were calculated by determination of the area under curve for each fluorescence profile in C1–C4 ROIs. (D–H) Quantification of PBMC fluorescence profiles of the analyzed chips. (D) Fluorescence profiles obtained by the red channel fluorescence, calculated in all the central chambers of the analyzed microfluidic devices at the indicated timepoints. (E) Infiltration of PBMC in the left or right chambers.

(Continued)

markedly located at the center or left side in the same chamber, as depicted in Fig. 5H. This analysis sequence yields a fluorescence profile for each C_1–C_4 ROI, to which unique peak positions and area are also detected that, together, determine an exact quantitative profiling of PBMCs behavior.

In addition, we analyzed the changes in SK-MEL cell morphology over time by selecting the side chambers as ROIs with ImageJ in three representative microphotographs of the three timepoints, as shown in Fig. 6. The visible channel was used as reference for the appropriate sizing of SK-MEL DAC and SK-MEL DAC/IFN chambers. The green fluorescence inside C1 and C2 yellow ROIs was appropriately thresholded with the ImageJ default algorithm (Hartig, 2013). These thresholded images have then been displayed as the corresponding thresholded masks and subjected to morphometric analysis by using the "Particle analysis" function of ImageJ. This allowed extracting morphometric parameters for each SK-MEL cell in the ROI mask. The results of morphometric analysis are depicted in Fig. 6D as plotted datasets, and are subdivided in dimensional (area, perimeter, Feret diameter) and ratio morphometric parameters. The dimensional factors are strictly linked to the cell morphology, whereas the ratio factors are calculated by using the aforementioned dimensional parameters (Boussadia et al., 2018; Suarez-Najera et al., 2018). This semi-automated analytical procedure permits to check possible little/medium morphometric perturbations of melanoma cells with DAC only vs those in presence of DAC plus IFN in the two side chambers (C_1 and C_2, Fig. 6A–C). These data evidenced morphological changes, distinctly comparable in the two C_1–C_2 compartments (Fig. 6). In fact, at 72h SK-MEL DAC/IFN (chamber C2 in Fig. 6C), undergo an evident change in their mean area, with respect to SK-MEL DAC (chamber C1), and this directly reflects changes in the Feret diameter and the Ratio factors (Fig. 6D).

Fig. 5—Cont'd Plots depict the fluorescence ratios obtained by dividing the fluorescence intensities of right ROIs (right chambers) with those from left ROIs (left chmabers). (F) linear regression fits of each of the fluorescence profile shown in panel D. (G) Slope values yielded from the linear regression fits of fluorescence profiles shown in panel D. (H) Automatic peak determination on fluorescence profiles from the indicated timepoints. Points in the graph depict the yielded peak position in X coordinates of the central chamber. $^{*}P<0.05$; $^{**}P<0.001$; $^{***}P<0.0001$. Scalebars, 200 μm.

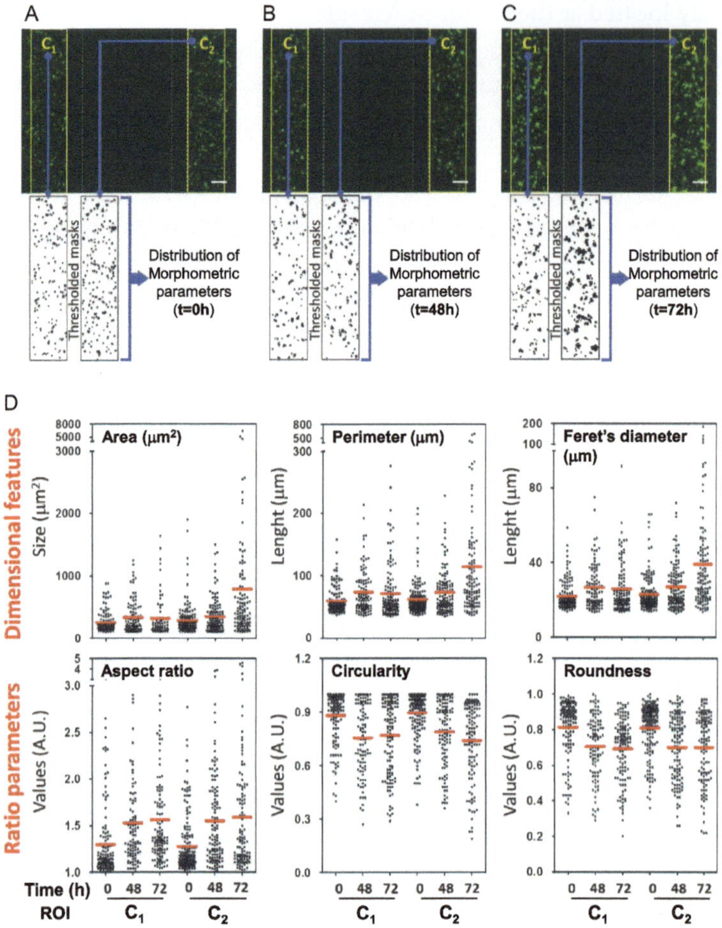

Fig. 6 Evaluation of the SK-MEL tumor cell morphometry. Tumor cells were analyzed by using the green fluorescence channel. The acquired images have subjected to thresholding process and then SK-MEL cells displayed as masks in the C1 and C2 ROIs were analyzed to extract morphometric parameters (area, perimeter, Feret diameter and other morphometric features). Analyses were performed at 0 h (A), 48 h (B), and 72 h (C) timepoints. (D) Morphometric parameters of SK-MEL cells in the two side chambers of the chip at the three timepoints. Dots in graph represent the values of the indicated morphometric parameter of each single particle at the time of cell loading or after 48 and 72 h from the loading. The values were extracted from the C1 (SK-MEL DAC) and C2 (SK-MEL DAC/IFN) ROIs representing the two side chambers of the chip with Matrigel. The red bars depict the mean values for each dot plot group. Scalebars, 200 μm.

6. Features, loading mode and analysis of cells in Type 2 device

6.1 Features of the Type 2 device and loading of mouse MCA-205 tumor cells and splenocytes

This device is characterized by the presence of six wells to be used for the cell loading, as evidenced in the schematization in Fig. 7A. Three chambers are interconnected by two microchannel arrays (Fig. 7B and C). The central chamber ends at the geometric center of the device thus forming two closed chambers (Fig. 7A). These two interrupted chambers serve as indirect pressure and flux stabilization microstructures, blocking immune cells overflowing in tumor compartment at time of loading. This device type is optimized for bidimensional measurements of cell-cell crosstalk and motility of adherent-floating cells (Agliari et al., 2014; Biselli et al., 2017; Businaro et al., 2013; Vacchelli et al., 2015).

MCA-205 tumor cells were adjusted at the concentration of 5×10^4 cells in 100 μL complete DMEM. Splenocytes were adjusted at the concentration of 1×10^6 cells in 100 μL complete DMEM. Prior to cell loading, devices were filled with complete DMEM and incubated at 37 °C 5% CO_2 for 1 h. Then, medium was carefully aspirated and cell loading was performed in the order indicated in Fig. 7A, specifically:

1. DOXO-treated MCA-205 in 100 μL of medium wells 5 and 6 (for a total of 1×10^5 cells)
2. 1×10^6 splenocytes in 100 μL medium into well 1
3. 1×10^6 splenocytes in 100 μL medium into well 2 (for a total of 2×10^6 splenocytes)
4. 100 μL medium into wells 3 and 4

The device was incubated at 37 °C 5% CO_2 for 1 h to stabilize the system prior time-lapse video recording (Fig. 7A). This allows the spontaneous distribution of splenocytes into the intermediate chamber to form a "front" (Fig. 7D), representing the starting point of the experiment.

6.2 Time-lapse microscopy of the device and video analysis

After stabilization, the device was positioned on a July Smart Fluorescence Microscope (Nanoentek) inside an incubator at 37 °C and 5% CO_2. Indeed, the main feature of this microscope is characterized by compact design that allows positioning in a standard CO_2 incubator. By exploiting its built-in software, the microscope is programmed for the time-lapse so that each

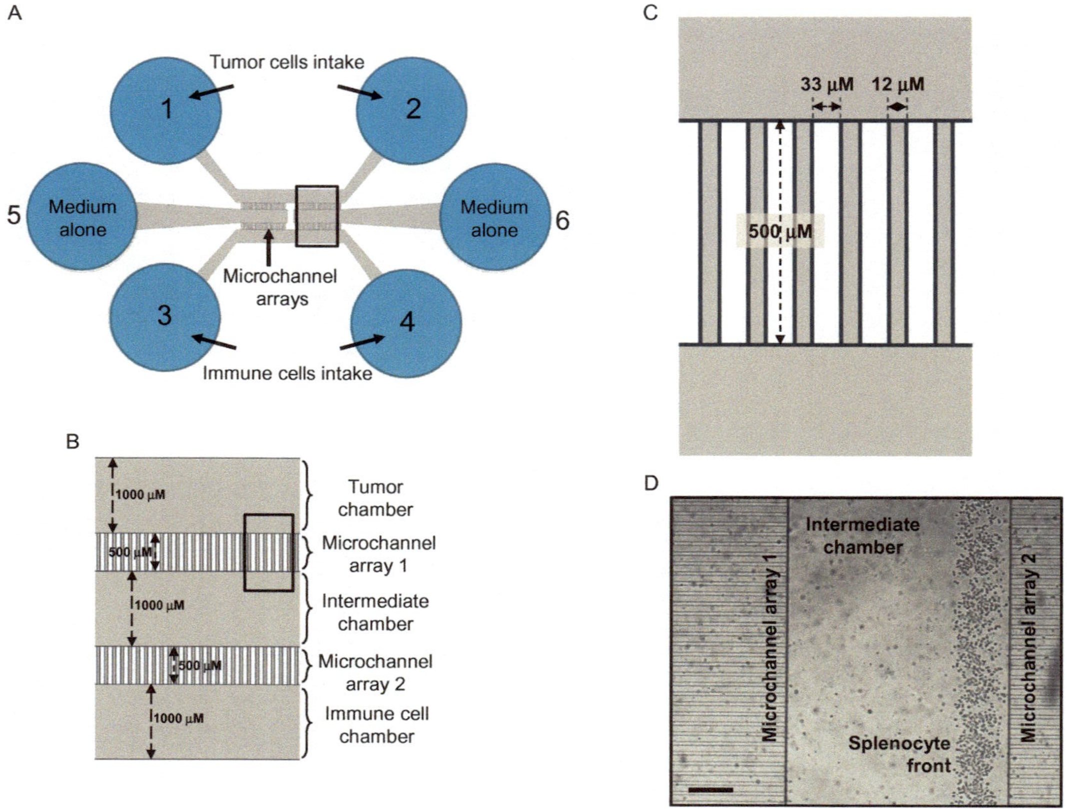

Fig. 7 Type 2 device features and planimetry. (A) Schematic overview of the device used in microfluidic experiments. The numbers in the wells depict the loading order of cells and medium. (B) Detail of the microchannel arrays, showing the sizes of the various substructures of the device. (C) Further details of the microchannel arrays, depicting the size features of each microchannel. (D) Microphotograph acquired before the beginning of the timelapse, depicting the spontaneous formation of a splenocyte front immediately after loading. Scalebar, 200 µm.

image was acquired every 2 min with a spatial resolution equal to 1.33 μm (Biselli et al., 2017). Images were acquired in visible light by centering the microchannel 1 array. At the end of the time-lapse, the frame dataset was converted in a 25 fps uncompressed video file by the function "Import Image Sequence" of the ImageJ software. The generated video file is now ready for cell tracking analysis.

7. Technical interpretation of video analysis data and mathematical procedures used for data extrapolation

7.1 Automatic cell tracking

The approach starts with the acquisition of a video sequence (Supplementary Video 1 in the online version at https://doi.org/10.1016/bs.mie.2019.06.012; Fig. 8A). First, chamber for the analysis was selected (Fig. 8B). Here, tumor chamber contains various events of tumor cells/splenocyte interactions due to migration of splenocytes that crossed the microchannels and approached cancer cells. Then, tumor cells and splenocytes are independently localized under the assumption of circularly shapes, by using the circular Hough transform (CHT) approach (Fig. 8C; Meng, Zhang, Yin, & Ma, 2018). The CHT method is applied using differently tuned range of radii for the two populations. In particular, at a spatial resolution of 1.33 μm, we used a tolerance interval for cancer radii equal to [9.3/17.3] μm and a tolerance range of radii for immune cells equal to [2.7/8.0] μm. We used the software Cell-Hunter, a proprietary software developed for the task of characterizing cell motility in a microfluidic environment (Biselli et al., 2017; Di Giuseppe et al., 2019; Nguyen et al., 2018). Cell tracking was performed using Cell-Hunter by exploiting the coordinates of the cells localized in previous step (see Fig. 8D). Optimal Subpattern Assignment Problem using the Munkres algorithm (Munkres, 1957) was applied to obtain the globally best possible pairing among located objects based on a given assignment cost. The result is a set of trajectories for cancer cells ($n = 67$) and for immune cells ($n = 1322$) during video.

Fig. 8E and F shows some examples of splenocytes and cancer cell trajectories extracted by Cell-Hunter. Fig. 8G illustrates trajectories of a set of immune cells approaching a cancer cell (red dot). The cell tracking analysis can be considered a potent tool for extrapolation of large datasets in which cells are distinguished by their own trajectories and associated parameters, such as trajectory length and speed.

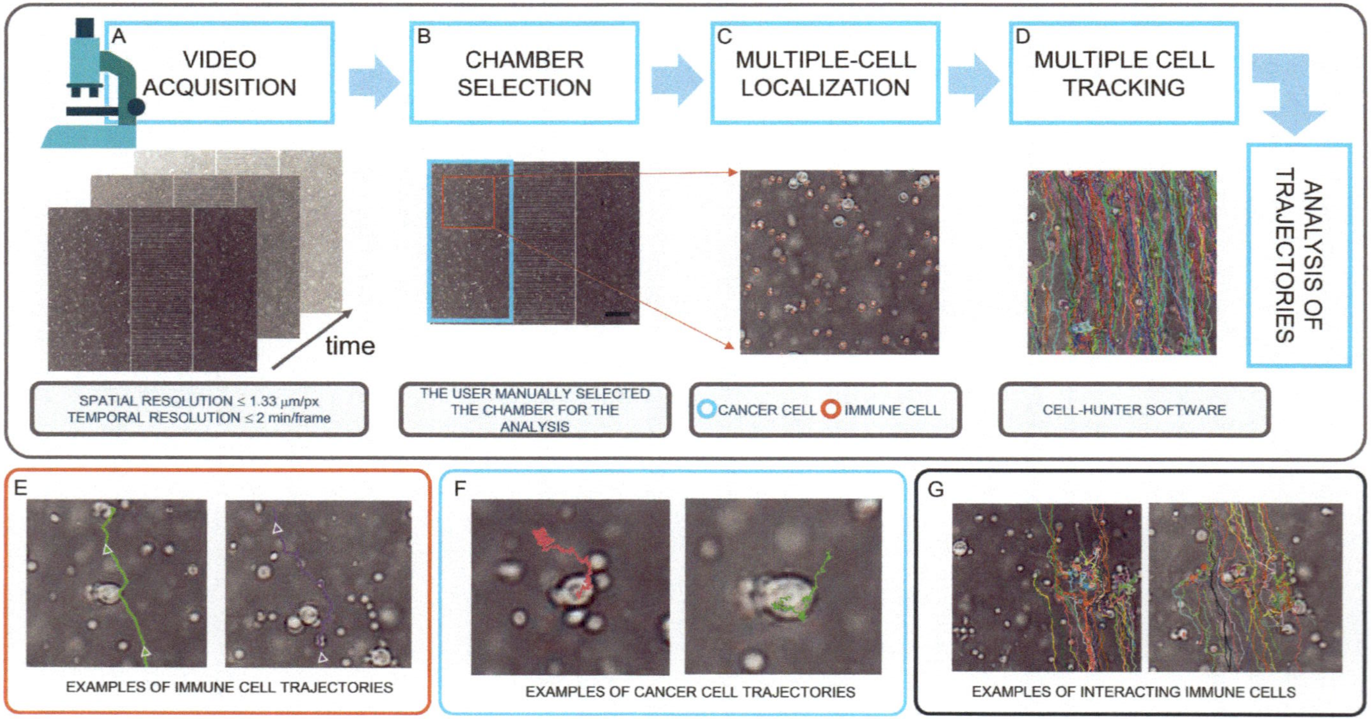

Fig. 8 Cell tracking. (A–D) A flowchart of the whole method used for automatic tracking of cancer (MCA-205) cells pretreated with DOXO and immune cells (splenocytes) in a microfluidic environment. (A) Time-lapse microscopy video acquisition. (B) Manual chamber selection. (C) Automatic cell localization from multiple populations. (D) Automatic cell tracking. Lower panel illustrates some examples of extracted trajectories. (E) Two examples of immune cells trajectories. (F) Two examples of cancer cell trajectories. (G) Two examples of immune cells interacting with cancer cells, with indicated immune cells trajectories approaching. Red dot, location of the cancer cell. Scalebar, 200 μm.

7.2 Automatic extrapolation of cell-cell interaction times

In Fig. 9A, we computed the *interaction time* that corresponds to the time a splenocyte stays within a circle of a given radius centered on the cancer cell. The radius is set as twice the sum of the average radius of immune and cancer cells estimated by the CHT algorithm.

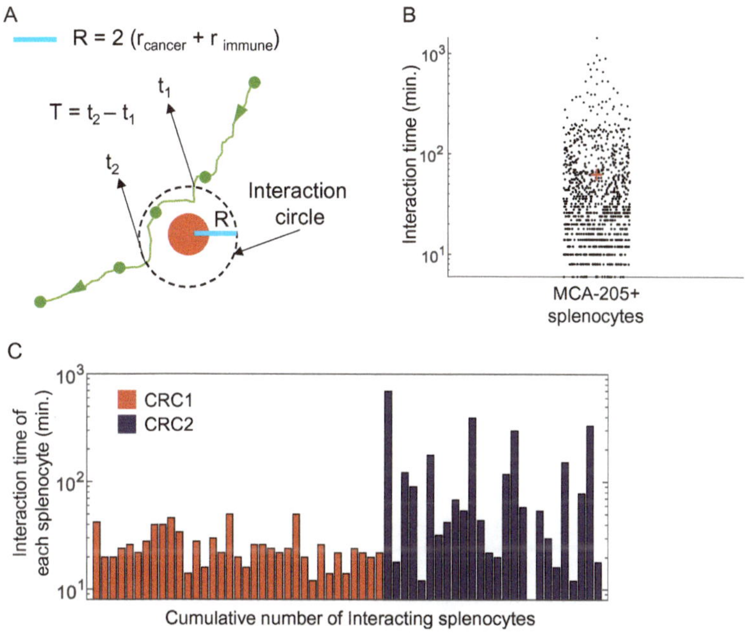

Fig. 9 Computation of Interaction times. (A) Scheme of the computation of time of interaction as the time the immune cell (green cell) remains in the dashed circle, referred as the interaction circle (radius equal to twice the sum of radius of the cancer cell (red dot) and of the immune cell). (B) Boxplot of the distribution of the interaction time computed over the entire video. (C) The interaction time computed for each immune cell for (red) cancer cell CRC1 and (blue) cancer cell CRC2.

Let us denote with $(x_{sp}^i(t), y_{sp}^i(t))$ and with $(x_{crc}^j(t), y_{crc}^j(t))$ the ith and the jth immune and cancer cell coordinates at time t. Let us also indicate with $r_{sp}^i(t)$ and $r_{crc}^j(t)$ the radii of the ith immune cell and of the jth cancer cell located at time t. Let us consider a set of N_{sp} immune cells and a set of N_{crc} cancer cells tracked in the video. Let us denote with $d^{ij}(t)$ the Euclidean distance of immune and cancer cell at time t, given by

$$d^{ij}(t) = \sqrt{\left(x_{sp}^i(t) - x_{crc}^j(t)\right)^2 + \left(y_{sp}^i(t) - y_{crc}^j(t)\right)^2} \tag{1}$$

Let us indicate with D_{int} the radius of the *interaction circle* (see Fig. 8A) equal to twice the sum of the average radius of the immune cells, $\bar{r}_{sp}$, and the average radius of the tumor cells, $\bar{r}_{crc}$, localized. In formula we have

$$D_{int} = 2 \cdot \left(\bar{r}_{sp} + \bar{r}_{crc}\right) \quad (2)$$

with $\bar{r}_{sp} = \frac{1}{N_{sp}} \sum_t \sum_{i=1}^{N_{sp}} \frac{r_{sp}^i(t)}{T_{sp}^i}$ and $\bar{r}_{crc} = \frac{1}{N_{crc}} \sum_t \sum_{j=1}^{N_{crc}} \frac{r_{crc}^i(t)}{T_{crc}^j}$ with T_{sp}^i and T_{crc}^i the temporal duration of the *i*th and of the *j*th trajectory of the immune and cancer cell, respectively.

Then, the interaction time t_{int}^{ij} of *i*th immune cell with the *j*th cancer cell is defined as follows:

$$t_{int}^{ij} = \sum_{T_{sp}^i} 1_E(t) \quad (3)$$

where $1_{E_{ij}}(t)$ is the *indicator function* (i.e., a function defined on a set E that assumes a value equal to one for all the element belonging to the set E and a value equal to zero for elements not belonging to the set E) of the compact set E_{ij}, with $E_{ij} = \{t \mid d^{ij}(t) \leq D_{int}\}$.Note that the interaction time may be due to reciprocal approaching of immune and cancer cells. By applying Eq. (3) to all the localized MCA-205 cancer cells and splenocytes, we collected a dataset ($n = 1042$) of interaction times between these two cell populations (Fig. 9B and C). This represents another example depicting how the application of high-throughput automatic analysis provides extrapolation of large numeric datasets associated to the mutual interaction between cancer cells and immune cells.

These results illustrate that the automatic computation of cell-cell interaction time can represent a valid functional parameter, together with trajectories datasets provided by cell tracking analysis, for the evaluation of cell properties in a label-free manner when morphology-based distinction of cells is possible.

8. Conclusive remarks

Our described approach demonstrate the usefulness of cell-on-chip platforms as a reliable and innovative tool to study different biological features of cancer and immune cells cross-talk, providing novel dataset types strictly associated to intrinsic properties of cells, such as migration, morphology changes and interactions. The major advantages of these platforms is the high degree of customization of the chips and the availability of automatic algorithms capable to perform high-throughput computation of the results. The extrapolated data may represent useful parameters for testing the immunogenicity of drugs, the immunocompetence of the host to therapeutic anticancer treatments, and the efficacy of combinatorial regimens, with respect to single drugs. In perspective, this approach may help clinicians in the choice of the most appropriate treatments and for a personalized medicine.

Acknowledgments

This study was supported by the Italian Ministry of Health (RF-2011-02347120 to F.M.) and the Italian Association for Cancer Research (AIRC; IG 21366 to G.S.).

References

Agliari, E., Biselli, E., De Ninno, A., Schiavoni, G., Gabriele, L., Gerardino, A., et al. (2014). Cancer-driven dynamics of immune cells in a microfluidic environment. *Scientific Reports*, *4*, 6639. https://doi.org/10.1038/srep06639.

Biselli, E., Agliari, E., Barra, A., Bertani, F. R., Gerardino, A., De Ninno, A., et al. (2017). Organs on chip approach: A tool to evaluate cancer-immune cells interactions. *Scientific Reports*, 7(1), 12737. https://doi.org/10.1038/s41598-017-13070-3.

Boussadia, Z., Lamberti, J., Mattei, F., Pizzi, E., Puglisi, R., Zanetti, C., et al. (2018). Acidic microenvironment plays a key role in human melanoma progression through a sustained exosome mediated transfer of clinically relevant metastatic molecules. *Journal of Experimental & Clinical Cancer Research*, *37*(1), 245. https://doi.org/10.1186/s13046-018-0915-z.

Boussommier-Calleja, A., Li, R., Chen, M. B., Wong, S. C., & Kamm, R. D. (2016). Microfluidics: A new tool for modeling cancer-immune interactions. *Trends in Cancer*, *2*(1), 6–19. https://doi.org/10.1016/j.trecan.2015.12.003.

Businaro, L., De Ninno, A., Schiavoni, G., Lucarini, V., Ciasca, G., Gerardino, A., et al. (2013). Cross talk between cancer and immune cells: Exploring complex dynamics in a microfluidic environment. *Lab on a Chip*, *13*(2), 229–239. https://doi.org/10.1039/c2lc40887b.

Czarnecki, S., & Bertin, A. (2018). Hybrid silicon-based organic/inorganic block copolymers with sol-gel active moieties: Synthetic advances, self-assembly and applications in biomedicine and materials science [review]. *Chemistry*, *24*(14), 3354–3373. https://doi.org/10.1002/chem.201705286.

Di Giuseppe, D., Corsi, F., Mencattini, D. A., Comes, M. C., Casti, P., Di Natale, C., et al. (2019). Learning cancer-related drug efficacy exploiting consensus in coordinated motility within cell clusters. *IEEE Transactions on Biomedical Engineering* PMID: 30735982. https://doi.org/10.1109/TBME.2019.2897825.

Hartig, S. M. (2013). Basic image analysis and manipulation in ImageJ. *Current Protocols in Molecular Biology*, *102*, 14.15.1–14.15.12. Chapter 14, Unit 14.15. https://doi.org/10.1002/0471142727.mb1415s102.

Lucarini, V., Buccione, C., Ziccheddu, G., Peschiaroli, F., Sestili, P., Puglisi, R., et al. (2017). Combining Type I interferons and 5-Aza-2′-deoxycitidine to improve antitumor response against melanoma. *The Journal of Investigative Dermatology*, *137*(1), 159–169. https://doi.org/10.1016/j.jid.2016.08.024.

Mencattini, A., Mattei, F., Schiavoni, G., Gerardino, A., Businaro, L., Di Natale, C., et al. (2019). From petri dishes to organ on chip platform: The increasing importance of machine learning and image analysis. *Frontiers in Pharmacology*, *10*, 100. https://doi.org/10.3389/fphar.2019.00100.

Meng, Y., Zhang, Z., Yin, H., & Ma, T. (2018). Automatic detection of particle size distribution by image analysis based on local adaptive canny edge detection and modified circular Hough transform. *Micron, 106*, 34–41. https://doi.org/10.1016/j.micron.2017.12.002.

Munkres, J. (1957). Algorithms for the assignment and transportation problems. *Journal of the Society for Industrial and Applied Mathematics*, *5*(1), 7.

Nguyen, M., De Ninno, A., Mencattini, A., Mermet-Meillon, F., Fornabaio, G., Evans, S. S., et al. (2018). Dissecting effects of anti-cancer drugs and cancer-associated fibroblasts by on-chip reconstitution of immunocompetent tumor microenvironments. *Cell Reports*, *25*(13). https://doi.org/10.1016/j.celrep.2018.12.015. 3884–3893.e3883.

Parlato, S., De Ninno, A., Molfetta, R., Toschi, E., Salerno, D., Mencattini, A., et al. (2017). 3D Microfluidic model for evaluating immunotherapy efficacy by tracking dendritic cell behaviour toward tumor cells. Research Support, Non-U.S. Gov't *Scientific Reports*, 7(1), 1093. https://doi.org/10.1038/s41598-017-01013-x.

Sackmann, E. K., Fulton, A. L., & Beebe, D. J. (2014). The present and future role of microfluidics in biomedical research [review]. *Nature*, *507*(7491), 181–189. https://doi.org/10.1038/nature13118.

Selimovic, S., Piraino, F., Bae, H., Rasponi, M., Redaelli, A., & Khademhosseini, A. (2011). Microfabricated polyester conical microwells for cell culture applications. *Lab on a Chip*, *11*(14), 2325–2332. https://doi.org/10.1039/c1lc20213h.

Suarez-Najera, L. E., Chanona-Perez, J. J., Valdivia-Flores, A., Marrero-Rodriguez, D., Salcedo-Vargas, M., Garcia-Ruiz, D. I., et al. (2018). Morphometric study of adipocytes on breast cancer by means of photonic microscopy and image analysis. *Microscopy Research and Technique*, *81*(2), 240–249. https://doi.org/10.1002/jemt.22972.

Vacchelli, E., Ma, Y., Baracco, E. E., Sistigu, A., Enot, D. P., Pietrocola, F., et al. (2015). Chemotherapy-induced antitumor immunity requires formyl peptide receptor 1. *Science*, *350*(6263), 972–978. https://doi.org/10.1126/science.aad0779.

Wang, X., Ma, J., George, S., & Zhou, H. (2012). Estimation of AUC or partial AUC under test-result-dependent sampling. *Statistics in Biopharmaceutical Research*, *4*(4), 313–323. https://doi.org/10.1080/19466315.2012.692514.

Wilding, P., Pfahler, J., Bau, H. H., Zemel, J. N., & Kricka, L. J. (1994). Manipulation and flow of biological fluids in straight channels micromachined in silicon. *Clinical Chemistry*, *40*(1), 43–47.

Wilding, P., Shoffner, M. A., & Kricka, L. J. (1994). PCR in a silicon microstructure. *Clinical Chemistry*, *40*(9), 1815–1818.

Wlodkowic, D., & Cooper, J. M. (2010). Tumors on chips: Oncology meets microfluidics. *Current Opinion in Chemical Biology*, *14*(5), 556–567. https://doi.org/10.1016/j.cbpa.2010.08.016.

Wlodkowic, D., & Darzynkiewicz, Z. (2010). Microfluidics: Emerging prospects for anti-cancer drug screening. *World Journal of Clinical Oncology*, *1*(1), 18–23. https://doi.org/10.5306/wjco.v1.i1.18.

CHAPTER TWENTY-FIVE

Generation of affinity ranged antigen-expressing tumor cell lines

Amaia Martinez-Usatorre, Pedro Romero*

Department of Oncology UNIL-CHUV, Faculty of Biology and Medicine, University of Lausanne, Epalinges, Switzerland

*Corresponding author: e-mail address: pedro.romero@unil.ch

Contents

Abstract

The interaction strength between $CD8^+$ T cells' TCR and cognate peptide-MHC (pMHC) impacts on the $CD8^+$ T cell response against pathogens and tumors (Martinez-Usatorre, Donda, Zehn, & Romero, 2018; Zehn, Lee, & Bevan, 2009). $CD8^+$ T cell responses against tumors are characterized by the presence of low affinity $CD8^+$ T cells specific for nonmutated tumor associated self-antigens (TAA) and potentially high affinity tumor specific $CD8^+$ T cells recognizing mutated self-antigens (Gros et al., 2016; Kvistborg et al., 2012; McMahan & Slansky, 2007). High affinity T cells display enhanced survival, expansion capacity and tumor control (Martinez-Usatorre et al., 2018; Schmid et al., 2010). In fact, recent clinical trials using neoantigen tumor vaccines showed prolonged progression free survival in melanoma patients (Ott et al., 2017; Sahin et al., 2017), while only modest clinical efficacy was obtained with TAA vaccines (Romero et al., 2016). However, the highly individual nature of neoantigens constitutes a major technical and economical hurdle for routine clinical application. Thus, the characterization of TAA-specific $CD8^+$ T cell responses may reveal new strategies to enhance their

Methods in Enzymology, Volume 632
ISSN 0076-6879
https://doi.org/10.1016/bs.mie.2019.12.001

anti-tumor properties. In parallel, the identification of high affinity antigens and $CD8^+$ T cells may be essential to design effective tumor vaccines and adoptive cell transfer therapies. Therefore, in this chapter, we describe how to generate tumor cell lines with stable expression of affinity-ranged antigens and methods to assess T-cell affinity.

1. Introduction

Assessment and understanding of tumor-specific $CD8^+$ T cell responses has been crucial for the development of successful immunotherapies against cancer (June & Sadelain, 2018; Romero et al., 2016; Sharma & Allison, 2015). However, the identification and monitoring of tumor-specific $CD8^+$ T cells is not trivial in view of the high antigenic overlap between cancer cells and healthy tissues, the negative selection of self-antigen specific T cells in the thymus and the presence of bystander T cells in tumors (Simoni et al., 2018). Generation of TCR transgenic mice and tumor cell lines expressing exogenous antigens has been essential for the deep characterization of tumor-specific $CD8^+$ T cell responses. For instance, the OT-1 transgenic mouse expressing ovalbumin $(OVA)_{257-264}$ specific OT-1 cells (Hogquist et al., 1994), has widely been used to study tumor-specific responses against $OVA_{257-264}$ (SIINFEKL) expressing tumor cell lines (Boczkowski, Nair, Snyder, & Gilboa, 1996; Enamorado et al., 2017; Moore, Carbone, & Bevan, 1988; Sanchez-Paulete et al., 2016). Nonetheless, OT-1 cells recognize SIINFEKL (N4) pulsed $H\text{-}2K^b$ molecules with high affinity (Krummey et al., 2016), while the majority of tumor-reactive T-cells found in cancer patients recognize their cognate antigen with low avidity (McMahan & Slansky, 2007). Therefore, the anti-tumor response of OT-1 cells against N4 expressing tumor cell lines may differ from the anti-tumor response that TAA specific low-avidity T cells exert against tumors. In fact, it has been extensively shown that the TCR-pMHC interaction strength impacts on T-cell priming, expansion, differentiation and survival (Dutoit et al., 2001; Martinez-Usatorre, Donda, Zehn, & Romero, 2018; Moreau et al., 2012; Ozga et al., 2016; Schmid et al., 2010; Zehn, Lee, & Bevan, 2009). Nonetheless, it is recently becoming clear that non-synonymous somatic mutations in tumors generate neoantigens that are unique to the tumor (Castle et al., 2012; Gros et al., 2016; Gubin et al., 2014; Kvistborg et al., 2012; Robbins et al., 2013; van Rooij et al., 2013) and neoantigen-based tumor vaccines prolonged progression free survival in melanoma patients (Ott et al., 2017; Sahin et al., 2017). Neoantigen specific T cells are presumably of high affinity due to the absence of central tolerance.

Thus, high and low affinity tumor-specific $CD8^+$ T cells may co-exist in tumors. Therefore, generation of tumor cell lines expressing affinity ranged antigens represents an ideal tool to track and understand high and low affinity anti-tumor responses. In this chapter, we describe how to generate murine tumor cell lines with stable expression of antigens and propose some selection criteria to assess affinity ranged T-cell responses.

2. Retroviral transduction of tumor cell lines: Considerations

Murine tumor cell lines with stable expression of exogenous antigens can be obtained by retroviral transduction as replicating tumor cells are easily infected by retroviruses. However, lentiviral vectors, which can infect non-dividing cells, can also be used[1]. In the protocol described below, we used a second generation retroviral system, that includes the Murine stem cell virus (MSCV) derived MIGR1 retroviral plasmid (Plasmid #27490, Addgene) containing the antigen[2] of interest in the multiple cloning site (MCS) upstream the IRES-GFP sequence, the packaging vector pMD-gag/pol and the amphotropic envelope plasmid pMD-VSV-G. In this system, the transgene expression is driven by the 5'LTR. Alternative promoters, however, may be depending the target tumor cell line or in case an inducible expression of the transgene is desired (Palmiter, 1994; Qin et al., 2010). The MIGR1 contains the GFP reporter gene but alternative reporter genes such as other fluorescent proteins, luciferase or antibody reporters can also be used to track the expression of the transgene (Kang & Chung, 2008; Roffler et al., 2006). The immunogenicity of fluorescent proteins and luciferase (Baklaushev et al., 2017; Davey et al., 2013; Rosenzweig et al., 2001) as well as tumor growth monitoring options should be considered when choosing the reporter gene.

3. Retroviral transduction of tumor cell lines: Protocol

3.1. On day 1 plate in well 1 (w1) of a 6-well plate, 75,000 and in well 2 (w2) 30,000 HEK 293 T/17[3] cells in 3 mL of 10% FCS (Dominique Dutscher, S1810-500), 50 IU/mL/50 μg/mL Penicilin/Streptomycin (Gibco, 15070-063), 0.05 mM 2ß-mercaptoethanol (Gibco 31350-010), 10 mM HEPES (Amimed, 5-31F00-H), in DMEM with Glutamax (Gibco, 31966-021) (cDMEM)[4].

3.2. Transfection of w1: on day 4 prepare the transfection mix by mixing in an 1.5 mL tube 1 μg of the retroviral construct, 0.5 μg of pMD-VSV-G, 1.5 μg of pMD-gag/pol, 75 μL of 0.5 M $CaCl_2$ (Clontech, 631312) and water up to 150 μL. Add drop by drop 150 μL of 2 × HEBS (Clontech, 631312) to the DNA mix, vortex and incubate the mixture for 20′ at room temperature (RT).

3.3. Replace the media of w1 for 3 mL of fresh cDMEM and add on top the DNA mix. Incubate the cells with the DNA mix for 6 h at 37 °C and replace the media with fresh cDMEM.

3.4. Transfection of w2: on day 5 repeat steps 3.2 and 3.3 but this time adding the DNA mix to w2.

3.5. Plate 75,000 tumor cells per construct[5] in a 6-well plate.

3.6. First round of transduction: on day 6 collect the supernatant of w1 (SN1), filter it through a 0.2 μm filter and add protamine sulfate (P4505, Sigma Aldrich) to a final concentration of 4 μg/mL.

3.7. Remove the media of the tumor cell line and add the filtered SN1 with the protamine sulfate. Centrifuge the cells for 5′ at 400 × *g* and incubate the cells at 37 °C.

3.8. Second round of transduction: on day 7 repeat steps 3.5 and 3.6 but with the supernatant of w2 (SN2) instead of SN1.

3.9. On day 8 transfer transduced cells to a new flask with fresh cDMEM media to expand them.

3.10. On day 11 check transduction efficiency by assessing reporter gene expression. In this example, GFP expression is assessed by flow cytometry. Compare GFP expression to parental untransduced cells to exclude the autofluorescence of tumor cell lines (Corsetti et al., 1988) (Fig. 1A).

4. Generation and selection of clones: Considerations

Generation and selection of transduced cell clones is essential for several reasons. First, retroviral transduction results in random insertion of transgenes. This random process may lead to functional or structural alterations of cellular genes (King, Patel, Lobel, Goff, & Nguyen-Huu, 1985). The risk of transgene insertion next to proto-oncogene is low (Baum et al., 2003). However, localization of the insert can be assessed by PCR with outward-facing primers complementary to the insert and consecutive sequencing (Serrao & Engelman, 2016). As this process is vector and transgene specific it will not be described in the protocol.

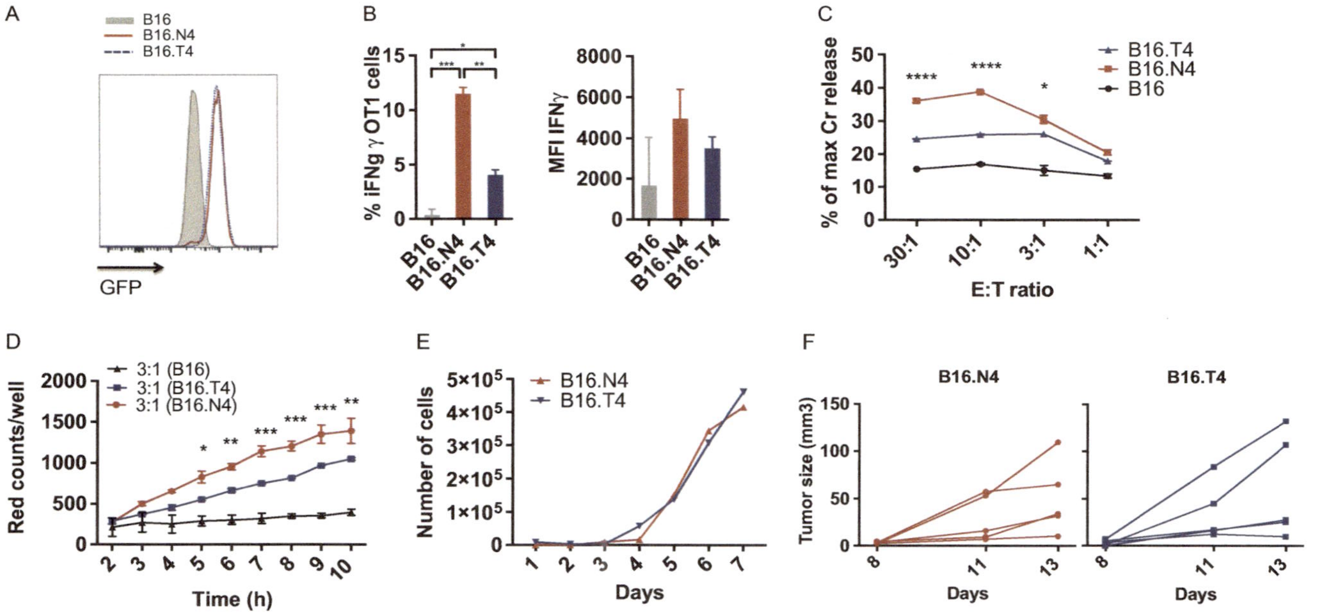

Fig. 1 Validation of B16 tumor cell lines expressing the high and low affinity N4 and SIITFEKL (T4) peptides. (A) Analysis of GFP expression by flow cytometry of selected B16.N4 and B16.T4 cell lines. Parental B16 cells were used as negative control. (B) Percentage of IFNγ^+ OT-1 cells and IFNγ MFI of IFNγ^+ OT-1 cells after *in vitro* stimulation of OT-1 cells with B16.N4 or B16.T4 cell lines at a 4:1 E:T ratio. Bars represent the mean + SD ($n=2$). (C) Analysis of OT-1-mediated B16.N4 and B16.T4 target cell lysis by chromium release assay. Dots represent the mean ± SD of the percentage of maximum Cr^{51} release ($n=2$). (D) Analysis of OT-1-mediated B16.N4 and B16.T4 target cell lysis by live imaging of co-cultures in the presence of Incucyte annexin V Red reagent. A 3:1 effector:target ratio was used. Dots represent the number of dead target cells (red counts) per well ± SD ($n=2$). (E) *In vitro* cell growth curves of B16.N4 and B16.T4 cell lines. Cells were counted manually every day. (F) *In vivo* growth of B16.N4 and B16.T4 cell lines. $2 \cdot 10^5$ cells were subcutaneously engrafted on the right flank of C57BL/6 mice and tumor were measured manually with a caliper ($n=5$). A 1-way ANOVA followed by Tukey's multiple comparison test was performed in panels (B) and (D).

On the other hand, multiple copies of the construct can be inserted in individual target cells, which will then determine transgene expression levels (Kustikova et al., 2003). The amount of antigen plays an important role in $CD8^+$ T cell responses (Utzschneider et al., 2016). Thus, it is very important to select individual transduced clones with similar expression levels of the transgene when comparing affinity ranged cell lines. To reduce the amount of copy numbers per cell it is important to use low multiplicity of infections (MOI). After transduction, exact copy numbers can be measured by quantitative PCR (Kustikova et al., 2003). Nonetheless, as there is a direct correlation between the expression of the reporter gene and transgene (Kustikova et al., 2003), evaluation of GFP expression enables selection of clones with similar transgene expression levels.

5. Generation and selection of clones: Protocol

5.1 Collect transduced tumor cells[6] and wash them with 2% FCS, 2 mM EDTA in PBS (sorting buffer).
5.2 Resuspend cells in sorting buffer and add DAPI to a final concentration of 300 nM just before sorting DAPI– GFP+ cells by flow cytometry. Collect cells in cDMEM.
5.3 Make a limiting dilution of the sorted GFP+ transduced cells and plate single cells on a 96-well plate[7].
5.4 After 1 week in the incubator at 37 °C check for single clones (a single loci should be seen in case of working with adherent cells) and a week later[8] transfer[6] them to a 6-well plate. Discard the wells with multiple clones.
5.5 When ~80% confluency is reached, collect[6] the clones, and wash and resuspend them in sorting buffer to check the GFP expression levels. Select those clones with similar GFI median fluorescent intensity (MFI) for further testing.

6. T-cell avidity assessment: Considerations

There are three parameters that describe the interaction strength between a T cell and antigen presenting cell (APC) or target cell: T cell affinity, avidity, and functional avidity. T cell affinity is the physical binding strength between a single TCR and pMHC molecule, whereas T cell avidity is the interaction strength of multimeric TCR–pMHCs. The functional avidity of a T cell describes its sensitivity to the antigen, which inversely correlates with the antigen dose needed for a T cell response. Although they differ in definition, the three parameters are correlated. T cell avidity normally correlates with T cell affinity (Crawford, Kozono, White, Marrack, & Kappler, 1998) and the functional avidity depends on the T cell affinity. Nonetheless, the functional avidity of a T cell also depends on additional factors such as TCR expression levels, coreceptors, adhesion molecules, and inhibitory and costimulatory molecules. The T cell affinity, defined by the dissociation constant (K_D) between soluble TCR-pMHC molecules, has traditionally been measured by surface plasmon resonance (SPR) (Corr et al., 1994). However, it is a laborious technic requiring the generation of soluble TCRs and pMHC complexes and is not suitable for high-throughput screening of living T cells. More recently, alternative technics have been developed that enable quantification of monomeric TCR-pMHC dissociation kinetics in living cells by flow cytometry (Schmidt et al., 2011) or microscopy (Nauerth et al., 2013). Staining with NTA-His tag-containing multimers (NTAmers) is one of these technics. Like conventional multimers, NTAmers contain pMHC molecules that bind specific TCRs. However, upon imidazole addition they rapidly dissociate into pMHC monomers allowing the measurement of monomeric TCR-pMHC dissociation rates (K_{off} and $t_{1/2}$) (Schmidt et al., 2011). NTAmer dissociation kinetics, unlike SPR measurements, account for the impact of the CD8 coreceptor in the T cell avidity for the pMHC. Nonetheless, there is a significant correlation between SPR and NTAmer measured K_{off} values (Hebeisen et al., 2015). This approach could be used to test the affinity of the query $CD8^+$ T cells to the peptide ligands of interest. 2D TCR affinity measurement technics have also been used to assess TCR affinity (Huang et al., 2010). Nonetheless, functional avidity analysis of the query $CD8^+$ T cells for the antigen is a more accessible technic that provides an idea of the TCR affinity. The antigen concentration needed to reach half of the maximum of a T cell response, termed EC_{50}, is the parameter used to define the functional avidity of a T cell. There are several T cell effector functions that can be assessed: cell proliferation, cytokine production and killing capacity. In all cases, EC_{50} values are calculated by exposing the query T cells to decreasing concentrations of the peptide loaded, when necessary, on target cells. For killing assays of human T cells, HLA-A2 peptides are often loaded on the T2 cell line (Townsend et al., 1989). Usual peptide concentrations range between 10^{-4} and 10^{-13} M (Allard et al., 2017; Gannon et al., 2015; Walker et al., 2008; Zehn et al., 2009). It is important to note that EC_{50} values of a single T cell clone differ among functional assays (Allard et al., 2017). Moreover, comparison of EC_{50} values of a single T cell clone to different peptides is possible, whereas comparison of EC_{50} values of different T cells to a single peptide requires phenotypically similar T cells (*i.e.*, same TCR, co-stimulatory and co-inhibitory molecule expression levels). Additionally, it should be confirmed that the capacity of the peptides to bind MHC-I molecule is equal as the affinity of the peptide to the MHC-I depends on its sequence (Lundegaard, Lund, Buus, & Nielsen, 2010). To address that, MHC-I stabilization can be assessed upon peptide loading on TAP deficient RMA-S (Ljunggren, Ohlen, Hoglund, Franksson, & Karre, 1991) cells for H2-K^b restricted murine peptides (Zehn et al., 2009) and on T2 cells (Cerundolo et al., 1990; Townsend et al., 1989) for HLA-A2 restricted human peptides.

Once the functional avidity of the query $CD8^+$ T cells to the peptides is known, it is important to confirm that the functional avidity of the T cells to the newly generated affinity ranged tumor cell lines correlate with the EC_{50} values calculated by peptide titration. In the following section two type of functional avidity readouts will be described: cytokine production and target cell lysis. Cytokine production of query $CD8^+$ T cells will be assessed by intracellular cytokines staining. For the evaluation of target cell lysis, two technics are described: quantification of ^{51}Cr release upon co-culture with ^{51}Cr-labeled target cell lines or evaluation of apoptosis by live microscopy with IncuCyte Annexin V or IncuCyte Caspase 3/7 reagent on the IncuCyte ZOOM live cell imaging system. The later technic bypasses radioactive material handling and thus it is a safer alternative to the traditional chromium release assay. However, it requires live cell imaging. For the three readouts T cells are cultured with the newly created target tumor cell lines at different effector:target (E:T) cell ratios, comparable to the peptide titration functional assays.

7. T-cell avidity assessment: Protocol

7.1 Cytokine production assessment by intracellular staining

- **7.1.1** For peptide titration, plate 50,000 untransduced tumor cells or alternative peptide presenting cells[9] in a 96 round-bottom well plate in 50 μL T cell media[10] in duplicates. Add on top 50 μL of effector T cells[11] at the desired E:T ratio[12] and 4 × concentrated peptide diluted in 50 μL T cell media[10].
- **7.1.2** For E:T ratio titration, collect transduced tumor cell lines and plate 50,000 cells per well in a 96 round-bottom well plate in

100 μL T cell media[10]. Place on top effector T cells[11] at 30:1, 10:1, 3:1 and 1:1 ratios diluted in 50 μL T cell media[10].

7.1.3 Include some wells with only effector T cells for negative[13] and positive controls in 150 and 100 μL T cell media, respectively.

7.1.4 For positive control wells, add 50 μL of 4× concentrated Phorbol 12-myristate 13-acetate (PMA) (P1585, Sigma-Aldrich) and Ionomycin (I0634, Sigma-Aldrich) to a final concentration of 10 and 500 ng/mL, respectively.

7.1.5 Incubate cells for 30′ at 37 °C in the incubator.

7.1.6 Add 50 μL of BD GolgiPlug and GolgiStop (51-2301KZ and 554724, BD Bioscience) to each well and incubate cells at 37 °C in the incubator for 4 h.

7.1.7 Spin the cells for 5′ at 350 × *g* remove SN and wash them with sorting buffer: add 150 μL of sorting buffer, spin cells for 5′ at 350 × *g* and remove SN.

7.1.8 Resuspend cells in sorting buffer containing the antibodies for surface marker staining of the query $CD8^+$ T cells. Incubate for 20′ at 4 °C.

7.1.9 Wash cells with PBS.

7.1.10 Stain cells with LIVE/DEAD® Aqua fluorescent reactive dye (ThermoFisher, L34965) diluted in PBS for 20′ at 4 °C.

7.1.11 Wash cells with sorting buffer.

7.1.12 Fix cells with Fixation Buffer (420801, Biolegend) for 20′ at 4 °C.

7.1.13 Wash cells with Intracellular Staining Permeabilization Wash Buffer (421002, Biolegend).

7.1.14 Resuspend cells in Intracellular Staining Permeabilization Wash Buffer containing specific antibodies for cytokines such as IFNγ, TNFα and IL-2. Incubate cells for 20′ at 4 °C.

7.1.15 Wash cells with Intracellular Staining Permeabilization Wash Buffer.

7.1.16 Resuspend cells in sorting buffer and analyze by flow cytometry (Fig. 1B). Define EC_{50} as the peptide concentration required for half of the maximum cytokine production or by performing a dose-response curve analysis using Prism software.

7.2 Target cell lysis assessment by ^{51}Cr release assay

7.2.1 Collect target cells[6] and wash them with T cell media[10]. Centrifuge 5′ at 350 × *g* and remove the SN as much as possible for ^{51}Cr labeling.

7.2.2 [14]Resuspend cells in 100 μL of ^{51}Cr and incubate them for 1 h at 37 °C in a water bath. Mix cells 3–4 times during incubation by tilting the 15 mL tube.

7.2.3 Wash cells 3 × by adding 10 mL RPMI and spinning for 4′ at 350 × *g*.

7.2.4 Resuspend cells in 5 mL of T cell media to reach a final concentration of 200,000 cells/mL.

7.2.5 Make a 1:10 dilution of the labeled cells with T cell media to a final concentration to 20,000 cells/mL.

7.2.6 Make 3 × concentrated serial dilutions of the peptide in T cell media for the peptide titration assay.

7.2.7 Prepare the effector T cells[11] for a final E:T ratio of 30:1, 10:1, 3:1 and 1:1 in duplicates in T cell media. Consider that there will be 1000 target cells per well and 50 μL of effector cells will be added to each well.

7.2.8 In a V-bottom 96-well plate 50 μL of labeled target cells (1000 cells) per well. Plate sufficient cells for E:T titration, peptide titration and negative[13] (spontaneous release) and positive controls (maximum release) in duplicates.

7.2.9 Add on top of the target cells 50 μL of the 3 × concentrated peptide for the peptide titration assay wells and incubate 15′ at RT.

7.2.10 Add the effector cells diluted in 50 μL[12] on top of the target cells (together with the peptide in case of the peptide titration assay), except in control wells.

7.2.11 Add 50 μL of T cell media in the wells for E:T titration assay.

7.2.12 Add 100uL of T cell media in the negative control spontaneous release wells.

7.2.13 Add 100uL of 1 M HCl in the positive control wells.

7.2.14 Spin down cells with a short spin.

7.2.15 Incubate cells for 4 h at 37 °C in the incubator.

7.2.16 Spin down cells 5′ at 350 × *g*.

7.2.17 Carefully transfer 50 μL of each well to a LumaPlate™-96 (Packard Bioscience, 6006633).

7.2.18 Let the plate dry overnight.

7.2.19 Next day, cover the plate with a plastic film and read the plate in TopCount NXT Microplate Scintillation and Luminescence Counter (PerkinElmer) or similar instrument able to measure ^{51}Cr radionuclide.

7.2.20 Calculate percentage of specific lysis as 100 × (experimental-negative control)/(positive-negative control)[13] (Fig. 1C). Define EC_{50} as the peptide concentration required for half of the maximum target cell lysis or by performing a dose-response curve analysis using Prism software.

7.3 Target cell lysis assessment by live cell imaging

7.3.1 Collect target cells[6], wash them with T cell media[10] and resuspend them in T cell media to a final concentration of 20,000 cells/mL.

7.3.2 Plate 50 μL of target cells per well in a flat-bottom 96-well plate. Plate sufficient cells for E:T titration, peptide titration and negative[13] (only target cells) controls in duplicates.

7.3.3 Prepare effector T cells[11] for a final E:T ratio[12] of 30:1, 10:1, 3:1 and 1:1 in duplicates in T cell media. Consider that there will be 1000 target cells per well and 50 μL of effector cells will be added to each well.

7.3.4 Add 50 μL of effector cells to the target cells. Include some wells with only effector cells as negative controls.

7.3.5 Make 4 × concentrated serial dilutions of the peptide in T cell media for the peptide titration.

7.3.6 Add 50 μL of the 4 × concentrated peptide on top of the wells with effectors and targets for the peptide titration assay.

7.3.7 Add 50 μL of T cell media to the E:T titration wells and 100 μL to negative controls (only effector cells or only target cells)

7.3.8 Add 50 μL of 4 × concentrated IncuCyte Annexin V fluorescent[15] or IncuCyte Caspase 3/7 reagent[16] for a final dilution of 1:200 and 1:1000, respectively, in T cell media with a final concentration of 1 mM $CaCl_2$ in case of using the IncuCyte Annexin V fluorescent reagent.

7.3.9 Place the plate on the IncuCyte ZOOM or alternative live cell imaging system and acquire images every 2–3 h for 24 h.

7.3.10 Calculate specific apoptosis of target cells[17] by subtracting the amount of fluorescent (green or red) apoptotic cells in the negative controls (target alone) to the experimental wells (Fig. 1D).

8. Growth kinetics assessment of tumor cell lines

As retroviral transduction of tumor cell lines leads to random insertion of transgenes, it is important to verify that the generated tumor cell lines maintain similar kinetics of proliferation and similar *in vivo* engraftment capacity as compared to each other and to the parental cell line. To do that, selected tumor cell clones and parental cell line are cultured *in vitro* and counted every day to compare their growth curve (Fig. 1E). To assess the engraftment capacity of the tumor cell lines, they should be ideally engrafted in immunodeficient mice[18] although wildtype mice are also suitable. Subcutaneous engraftment of tumor cells[19] enables an easy monitoring of the growth by manual tumor volume measurement with a caliper (Fig. 1F). Nonetheless, it may not be suitable for all cell lines. Moreover, depending on the reporter gene inserted in the cell line (*i.e.*, luciferase) alternative injection routes and tumor growth monitoring methods can be applied (*i.e., in vivo* imaging).

9. Concluding remarks

The aforementioned transduction method and selection process was used to generate high and low affinity OT-1 peptide expressing B16 tumor cell lines (Martinez-Usatorre et al., 2018). However, other mouse and even human tumor cell lines could be generated with little modifications in the protocol such as promotor and vector optimization (see Chapter 2). This methodology can be used for basic research addressing the role of T cell affinity in natural and immunotherapy driven $CD8^+$ T cell responses (Martinez-Usatorre et al., 2018) as well as for more translational applications like the assessment of $CD8^+$ T cell responsiveness to TAA and neoantigens before adoptive cell therapy.

10. Notes

1. Lentiviruses and γ-retroviruses use different LTRs and packaging gene (*i.e.*, gag and pol) isoforms. Therefore, lentiviral and retroviral packaging plasmids are not interchangeable. Nonetheless, envelope plasmids such as VSV-G may be compatible in both systems.
2. This method was used to generate two B16 cell lines with high and low affinity ligands for OT-1 cells (Martinez-Usatorre et al., 2018). For that two MIGR1 derived plasmids were generated by the group of Prof. Zehn: one contained the OVA protein and the other one an altered form of OVA bearing the SIITFEKL (T4) peptide in positions 257–264 instead of the native SIINFEKL (N4). It was previously reported that OT-1 cells recognize the T4 peptide with ~10 times lower affinity than the N4 peptide (Krummey et al., 2016).

3. Alternative packaging cell lines may be used according to vector(s) compatibility.
4. Cells should be maintained in the incubator at 37 °C and 5% CO_2 whenever possible.
5. Plated cell numbers were calculated for the B16 cell line but may be adjusted to the tumor cell line of interest according to its growth kinetics.
6. If working with adherent cells like B16 cells, remove the cDMEM from the flask, wash cells with PBS and add 1 × trypsin (15400054, ThermoFisher) for 5′ at 37 °C. Collect the cells in cDMEM and spin them for 5′ at 350 × *g*.
7. For B16 cells, cells were diluted to a final concentration of 1.5 cell/mL in cDMEM and 200 μL of the cell mix was added per well in a 96-well plate to a final 0.3 cells/well concentration.
8. Time lines may be different for each tumor cell line depending on its growth kinetics.
9. Use an untransduced tumor cell line for peptide titration and calculation of EC_{50} values and the newly generated peptide expressing tumor cell lines for E:T titration. Note that certain tumor cell lines express low levels of MHC I molecules and therefore alternative cell lines may be used for the assessment of functional avidity to the peptides.
10. Mouse T-cell media: 10% fetal calf serum (FCS, Dominique Dutscher, S1810-500), 50 IU/mL/50 μg/mL Penicilin/Streptomycin (Gibco, 15070-063), 0.05 mM 2ß-mercaptoethanol (Gibco, 31350-010), 10 mM HEPES (Amimed, 5-31F00-H), 2 mM L-Glutamine (Amimed, 5-10K100-H), 1 mM Sodium Pyruvate (Gibco, 11360-039) and 1 × Non-essential amino acids (Thermo Fisher, 11140-035) in RPMI 1640 with Glutamax (Gibco, 61870-010).

    Human T-cell media: 8% human serum, 100 U/mL Penicillin (Gibco), 100 μg/mL Streptomycin (Gibco), 0.1 mg/mL Kanamycin (Gibco), 2 mM L-glutamine (Gibco), 1% in volume non-essential amino acids (Gibco), 1 mM Na Pyruvate (Gibco), 0.05 mM 2ß-mercaptoethanol (Sigma) in RPMI medium (Gibco).
11. Effector T cells can be obtained by stimulation of the query $CD8^+$ T cells with 1:1 mouse or human αCD3/αCD28 Dynabeads (Thermo Fisher, 11452D and 111163D, respectively) in T cell media[7] with 50 U/mL rhIL-2 (Glaxo-IMB). Functional assays are recommended to be performed when cells are resting, at least 6 days post-activation.
12. For peptide titration assay a 10:1 or 3:1 E:T ratio is recommended.

13. To assess specificity of cell lysis/cytokine production in transduced tumor cell lines, include untransduced tumor cells as additional negative control.
14. From this step on work in an area designated for ^{51}Cr handling, protected with Lead shielding. Wear always a full-length lab coat, gloves and radioactivity dosimeter. Treat all waste as radioactive and place it in a designated receptacle. A Geiger counter must be present in the room during the experiment.
15. IncuCyte Annexin V fluorescent Green or Red reagents (Essen Bioscience, 4641 and 4642) are provided lyophilized and should be solubilized in 100 μL of complete media or PBS before use. Lyophilized form is stable at −20 °C for at least 2 years, whereas soluble form at 4 °C for at least 1 week. Annexin V fluorescent reagents allow detection of apoptotic cells by binging to the phophatidylserine that becomes exposed in apoptotic cells.
16. IncuCyte Caspase 3/7 reagent (Essen Bioscience, 4440) is supplied at 5 mM in DMSO and lasts at least 6 months when stored at 4 °C. This reagent is composed by the activated caspase-3/7 recognition motif (DEVD) and a DNA intercalating dye. It is inert and non-fluorescent in its complete form. Upon addition to the cells it crosses the cell membrane and only in apoptotic cells, activated caspase-3/7 will release the DNA dye that will bind nuclear DNA becoming fluorescent.
17. Target cells can normally be distinguished from effector cells by their size. There are some exceptions like T cell leukemia cell lines. Apply a size mask comparing wells with mixed target and effector cells and target and effector cells alone.
18. Engraftment of tumor cell lines in immunodeficient mice allows growth kinetics assessment without considering the immunogenicity derived from the newly inserted transgene.
19. The initial cell number to inject may be adjusted to each cell line. Usual cell numbers range between 10^5 and $5 \cdot 10^5$. Use the dose normally used with the parental cell line.

References

Allard, M., Couturaud, B., Carretero-Iglesia, L., Duong, M. N., Schmidt, J., Monnot, G. C., et al. (2017). TCR-ligand dissociation rate is a robust and stable biomarker of CD8+ T cell potency. *JCI Insight*, *2*(14).

Baklaushev, V. P., Kilpelainen, A., Petkov, S., Abakumov, M. A., Grinenko, N. F., Yusubalieva, G. M., et al. (2017). Luciferase expression allows bioluminescence imaging but imposes limitations on the orthotopic mouse (4T1) model of breast Cancer. *Scientific Reports*, 7(1), 7715.

Baum, C., Dullmann, J., Li, Z., Fehse, B., Meyer, J., Williams, D. A., et al. (2003). Side effects of retroviral gene transfer into hematopoietic stem cells. *Blood, 101*(6), 2099–2114.

Boczkowski, D., Nair, S. K., Snyder, D., & Gilboa, E. (1996). Dendritic cells pulsed with RNA are potent antigen-presenting cells in vitro and in vivo. *The Journal of Experimental Medicine, 184*(2), 465–472.

Castle, J. C., Kreiter, S., Diekmann, J., Lower, M., van de Roemer, N., de Graaf, J., et al. (2012). Exploiting the mutanome for tumor vaccination. *Cancer Research, 72*(5), 1081–1091.

Cerundolo, V., Alexander, J., Anderson, K., Lamb, C., Cresswell, P., McMichael, A., et al. (1990). Presentation of viral antigen controlled by a gene in the major histocompatibility complex. *Nature, 345*(6274), 449–452.

Corr, M., Slanetz, A. E., Boyd, L. F., Jelonek, M. T., Khilko, S., al-Ramadi, B. K., et al. (1994). T cell receptor-MHC class I peptide interactions: Affinity, kinetics, and specificity. *Science, 265*(5174), 946–949.

Corsetti, J. P., Sotirchos, S. V., Cox, C., Cowles, J. W., Leary, J. F., & Blumburg, N. (1988). Correction of cellular autofluorescence in flow cytometry by mathematical modeling of cellular fluorescence. *Cytometry, 9*(6), 539–547.

Crawford, F., Kozono, H., White, J., Marrack, P., & Kappler, J. (1998). Detection of antigen-specific T cells with multivalent soluble class II MHC covalent peptide complexes. *Immunity, 8*(6), 675–682.

Davey, G. M., Mueller, S. N., van Vliet, C., Gigowski, M., Zaid, A., Davies, B., et al. (2013). Identification of a MHC I-restricted epitope of DsRed in C57BL/6 mice. *Molecular Immunology, 53*(4), 450–452.

Dutoit, V., Rubio-Godoy, V., Dietrich, P. Y., Quiqueres, A. L., Schnuriger, V., Rimoldi, D., et al. (2001). Heterogeneous T-cell response to MAGE-A10(254-262): High avidity-specific cytolytic T lymphocytes show superior antitumor activity. *Cancer Research, 61*(15), 5850–5856.

Enamorado, M., Iborra, S., Priego, E., Cueto, F. J., Quintana, J. A., Martinez-Cano, S., et al. (2017). Enhanced anti-tumour immunity requires the interplay between resident and circulating memory CD8(+) T cells. *Nature Communications, 8*, 16073.

Gannon, P. O., Wieckowski, S., Baumgaertner, P., Hebeisen, M., Allard, M., Speiser, D. E., et al. (2015). Quantitative TCR:pMHC dissociation rate assessment by NTAmers reveals Antimelanoma T cell repertoires enriched for high functional competence. *Journal of Immunology, 195*(1), 356–366.

Gros, A., Parkhurst, M. R., Tran, E., Pasetto, A., Robbins, P. F., Ilyas, S., et al. (2016). Prospective identification of neoantigen-specific lymphocytes in the peripheral blood of melanoma patients. *Nature Medicine, 22*(4), 433–438.

Gubin, M. M., Zhang, X., Schuster, H., Caron, E., Ward, J. P., Noguchi, T., et al. (2014). Checkpoint blockade cancer immunotherapy targets tumour-specific mutant antigens. *Nature, 515*(7528), 577–581.

Hebeisen, M., Schmidt, J., Guillaume, P., Baumgaertner, P., Speiser, D. E., Luescher, I., et al. (2015). Identification of rare high-avidity, tumor-reactive CD8+ T cells by monomeric TCR-ligand off-rates measurements on living cells. *Cancer Research, 75*(10), 1983–1991.

Hogquist, K. A., Jameson, S. C., Heath, W. R., Howard, J. L., Bevan, M. J., & Carbone, F. R. (1994). T cell receptor antagonist peptides induce positive selection. *Cell, 76*(1), 17–27.

Huang, J., Zarnitsyna, V. I., Liu, B., Edwards, L. J., Jiang, N., Evavold, B. D., et al. (2010). The kinetics of two-dimensional TCR and pMHC interactions determine T-cell responsiveness. *Nature, 464*(7290), 932–936.

June, C. H., & Sadelain, M. (2018). Chimeric antigen receptor therapy. *The New England Journal of Medicine, 379*(1), 64–73.

Kang, J. H., & Chung, J. K. (2008). Molecular-genetic imaging based on reporter gene expression. *Journal of Nuclear Medicine*, *49*(Suppl 2), 164S–179S.

King, W., Patel, M. D., Lobel, L. I., Goff, S. P., & Nguyen-Huu, M. C. (1985). Insertion mutagenesis of embryonal carcinoma cells by retroviruses. *Science*, *228*(4699), 554–558.

Krummey, S. M., Martinez, R. J., Andargachew, R., Liu, D., Wagener, M., Kohlmeier, J. E., et al. (2016). Low-affinity memory CD8+ T cells mediate robust heterologous immunity. *Journal of Immunology*, *196*(6), 2838–2846.

Kustikova, O. S., Wahlers, A., Kuhlcke, K., Stahle, B., Zander, A. R., Baum, C., et al. (2003). Dose finding with retroviral vectors: Correlation of retroviral vector copy numbers in single cells with gene transfer efficiency in a cell population. *Blood*, *102*(12), 3934–3937.

Kvistborg, P., Shu, C. J., Heemskerk, B., Fankhauser, M., Thrue, C. A., Toebes, M., et al. (2012). TIL therapy broadens the tumor-reactive CD8(+) T cell compartment in melanoma patients. *Oncoimmunology*, *1*(4), 409–418.

Ljunggren, H. G., Ohlen, C., Hoglund, P., Franksson, L., & Karre, K. (1991). The RMA-S lymphoma mutant; consequences of a peptide loading defect on immunological recognition and graft rejection. *International Journal of Cancer Supplement*, *6*, 38–44.

Lundegaard, C., Lund, O., Buus, S., & Nielsen, M. (2010). Major histocompatibility complex class I binding predictions as a tool in epitope discovery. *Immunology*, *130*(3), 309–318.

Martinez-Usatorre, A., Donda, A., Zehn, D., & Romero, P. (2018). PD-1 blockade unleashes effector potential of both high- and low-affinity tumor-infiltrating T cells. *Journal of Immunology*.

McMahan, R. H., & Slansky, J. E. (2007). Mobilizing the low-avidity T cell repertoire to kill tumors. *Seminars in Cancer Biology*, *17*(4), 317–329.

Moore, M. W., Carbone, F. R., & Bevan, M. J. (1988). Introduction of soluble protein into the class I pathway of antigen processing and presentation. *Cell*, *54*(6), 777–785.

Moreau, H. D., Lemaitre, F., Terriac, E., Azar, G., Piel, M., Lennon-Dumenil, A. M., et al. (2012). Dynamic in situ cytometry uncovers T cell receptor signaling during immunological synapses and kinapses in vivo. *Immunity*, *37*(2), 351–363.

Nauerth, M., Weissbrich, B., Knall, R., Franz, T., Dossinger, G., Bet, J., et al. (2013). TCR-ligand koff rate correlates with the protective capacity of antigen-specific CD8 + T cells for adoptive transfer. *Science Translational Medicine*, *5*(192), 192ra87.

Ott, P. A., Hu, Z., Keskin, D. B., Shukla, S. A., Sun, J., Bozym, D. J., et al. (2017). An immunogenic personal neoantigen vaccine for patients with melanoma. *Nature*, *547*(7662), 217–221.

Ozga, A. J., Moalli, F., Abe, J., Swoger, J., Sharpe, J., Zehn, D., et al. (2016). pMHC affinity controls duration of CD8+ T cell-DC interactions and imprints timing of effector differentiation versus expansion. *The Journal of Experimental Medicine*, *213*(12), 2811–2829.

Palmiter, R. D. (1994). Regulation of metallothionein genes by heavy metals appears to be mediated by a zinc-sensitive inhibitor that interacts with a constitutively active transcription factor, MTF-1. *Proceedings of the National Academy of Sciences of the United States of America*, *91*(4), 1219–1223.

Qin, J. Y., Zhang, L., Clift, K. L., Hulur, I., Xiang, A. P., Ren, B. Z., et al. (2010). Systematic comparison of constitutive promoters and the doxycycline-inducible promoter. *PLoS One*, *5*(5), e10611.

Robbins, P. F., Lu, Y. C., El-Gamil, M., Li, Y. F., Gross, C., Gartner, J., et al. (2013). Mining exomic sequencing data to identify mutated antigens recognized by adoptively transferred tumor-reactive T cells. *Nature Medicine*, *19*(6), 747–752.

Roffler, S. R., Wang, H. E., Yu, H. M., Chang, W. D., Cheng, C. M., Lu, Y. L., et al. (2006). A membrane antibody receptor for noninvasive imaging of gene expression. *Gene Therapy*, *13*(5), 412–420.

Romero, P., Banchereau, J., Bhardwaj, N., Cockett, M., Disis, M. L., Dranoff, G., et al. (2016). The human vaccines project: A roadmap for cancer vaccine development. *Science Translational Medicine*, *8*(334), 334ps9.

Rosenzweig, M., Connole, M., Glickman, R., Yue, S. P., Noren, B., DeMaria, M., et al. (2001). Induction of cytotoxic T lymphocyte and antibody responses to enhanced green fluorescent protein following transplantation of transduced CD34(+) hematopoietic cells. *Blood*, *97*(7), 1951–1959.

Sahin, U., Derhovanessian, E., Miller, M., Kloke, B. P., Simon, P., Lower, M., et al. (2017). Personalized RNA mutanome vaccines mobilize poly-specific therapeutic immunity against cancer. *Nature*, *547*(7662), 222–226.

Sanchez-Paulete, A. R., Cueto, F. J., Martinez-Lopez, M., Labiano, S., Morales-Kastresana, A., Rodriguez-Ruiz, M. E., et al. (2016). Cancer immunotherapy with immunomodulatory anti-CD137 and anti-PD-1 monoclonal antibodies requires BATF3-dependent dendritic cells. *Cancer Discovery*, *6*(1), 71–79.

Schmid, D. A., Irving, M. B., Posevitz, V., Hebeisen, M., Posevitz-Fejfar, A., Sarria, J. C., et al. (2010). Evidence for a TCR affinity threshold delimiting maximal CD8 T cell function. *Journal of Immunology*, *184*(9), 4936–4946.

Schmidt, J., Guillaume, P., Irving, M., Baumgaertner, P., Speiser, D., & Luescher, I. F. (2011). Reversible major histocompatibility complex I-peptide multimers containing Ni(2+)-nitrilotriacetic acid peptides and histidine tags improve analysis and sorting of CD8(+) T cells. *The Journal of Biological Chemistry*, *286*(48), 41723–41735.

Serrao, E., & Engelman, A. N. (2016). Sites of retroviral DNA integration: From basic research to clinical applications. *Critical Reviews in Biochemistry and Molecular Biology*, *51*(1), 26–42.

Sharma, P., & Allison, J. P. (2015). The future of immune checkpoint therapy. *Science*, *348*(6230), 56–61.

Simoni, Y., Becht, E., Fehlings, M., Loh, C. Y., Koo, S. L., Teng, K. W. W., et al. (2018). Bystander CD8(+) T cells are abundant and phenotypically distinct in human tumour infiltrates. *Nature*, *557*(7706), 575–579.

Townsend, A., Ohlen, C., Bastin, J., Ljunggren, H. G., Foster, L., & Karre, K. (1989). Association of class I major histocompatibility heavy and light chains induced by viral peptides. *Nature*, *340*(6233), 443–448.

Utzschneider, D. T., Alfei, F., Roelli, P., Barras, D., Chennupati, V., Darbre, S., et al. (2016). High antigen levels induce an exhausted phenotype in a chronic infection without impairing T cell expansion and survival. *The Journal of Experimental Medicine*, *213*(9), 1819–1834.

van Rooij, N., van Buuren, M. M., Philips, D., Velds, A., Toebes, M., Heemskerk, B., et al. (2013). Tumor exome analysis reveals neoantigen-specific T-cell reactivity in an ipilimumab-responsive melanoma. *Journal of Clinical Oncology*, *31*(32), e439–e442.

Walker, E. B., Haley, D., Petrausch, U., Floyd, K., Miller, W., Sanjuan, N., et al. (2008). Phenotype and functional characterization of long-term gp100-specific memory CD8 + T cells in disease-free melanoma patients before and after boosting immunization. *Clinical Cancer Research*, *14*(16), 5270–5283.

Zehn, D., Lee, S. Y., & Bevan, M. J. (2009). Complete but curtailed T-cell response to very low-affinity antigen. *Nature*, *458*(7235), 211–214.

CHAPTER TWENTY-SIX

Recombinant fusion proteins for targeting dendritic cell subsets in therapeutic cancer vaccine

Stéphanie Corgnac[a,b], Natalia K. Botelho[a], Alena Donda[a,*], Pedro Romero[a]

[a]Department of Fundamental Oncology, Faculty of Biology and Medicine, University of Lausanne, Epalinges, Switzerland

[b]INSERM UMR 1186, Integrative Tumor Immunology and Genetic Oncology, Gustave Roussy, EPHE, PSL, Fac. de Médecine—Univ. Paris-Sud, Université Paris-Saclay, Villejuif, France

[*]Corresponding author: e-mail address: alena.donda@unil.ch

Contents

Abstract

Dendritic cells (DCs) are professional antigen-presenting cells, which are optimal for the priming of a T cell response against pathogens and tumors. Therefore, many efforts are made to develop therapeutic cancer vaccines which preferentially target the antigen to DC subsets. To this aim, we developed two types of recombinant fusion proteins, which favor antigen delivery to pro-inflammatory DCs as well as the crosstalk between specialized subpopulations of DCs. The first approach combines peptide/CpG vaccination with the recruitment of iNKT cells to the tumor site *via* CD1d-antitumor scFv fusion proteins. The second approach is targeting the tumor antigen to cross-presenting Xcr1$^+$ DCs *via* a fusion protein made of Xcl1 fused to a synthetic long peptide followed by an IgG1 Fc fragment. Both strategies allow a potent tumor-specific CD8 T cell response associated with tumor regression or tumor growth delay depending on the model. In the case of iNKT cell activation, the strategy relies on a strong IL-12 release by splenic DCs, while in the second case, the T cell response is strictly dependent on the presence of Xcr1$^+$ cross-presenting DCs.

Methods in Enzymology, Volume 632
ISSN 0076-6879
https://doi.org/10.1016/bs.mie.2019.11.013

1. Introduction

Previous pre-clinical studies in our group have reported the efficacy of peptide-based vaccine formulated with TLR ligands such as Poly I:C (TLR3) or CpG (TLR9) (Perret et al., 2013). However, clinical responses using peptide antigens formulated with various adjuvants have remained disappointing, largely because those selected peptide antigens were mostly self-antigens over-expressed in tumors, which are rather subjected to peripheral tolerance (Klebanoff, Acquavella, Yu, & Restifo, 2011). More recently, mutated tumor neoantigens able to escape tolerance have promoted the development of personalized tumor peptide vaccines (Bezu et al., 2018), which is now an active field of research. Nevertheless, non-mutated tumor antigens remain attractive vaccination targets provided that T cell priming is enhanced by involving pro-inflammatory immune cell populations such as, respectively, CD1d-restricted iNKT cells and cross-presenting DCs, as proposed in this section. The two proposed strategies involving recombinant fusion proteins are still at the pre-clinical stage but represent an active area of research.

1.1 iNKT/CD1d-antitumor immunotherapy combined with peptide/CpG vaccine

Invariant natural killer T cells (iNKT) are non-polymorphic innate-like lymphocytes activated by glycosphingolipids presented by the monomorphic MHC I-like molecule CD1d. CD1d is mainly expressed by B cells, myeloid populations including all DC subsets, Kupffer cells from mouse liver, as well as lymphoid and myeloid-derived malignancies, and some tumors of epithelial origin (Bendelac, Rivera, Park, & Roark, 1997; Bendelac, Savage, & Teyton, 2007). Mouse iNKT cells express the invariant TCRα chain Vα14Jα18 that pairs with a limited number of TCRβ chains (Bendelac et al., 1997; Exley, Garcia, Balk, & Porcelli, 1997). Through their fast IFNγ release, activated iNKT cells promote the maturation of pro-inflammatory DCs, resulting in their potent IL12 secretion, which further activate iNKT and NK cells, and promote the priming of CD4 and CD8 T cells (Cerundolo, Silk, Masri, & Salio, 2009). In view of the adjuvant activity of iNKT cells on the adaptive immune response, several analogs of the agonist iNKT/CD1d ligand α-galactosyl-ceramide (aGalCer) have been synthetized and some exhibited potent adjuvant effects in tumor and malaria vaccination (Vasan & Tsuji, 2010). Importantly, in view of the monomorphic nature of human CD1d, the tumor targeting of CD1d ligand could be universally applied to all individuals. To this aim, we demonstrated the potent antitumor effects of aGalCer-loaded CD1d-anti-HER2 or CD1d-anti-CEA fusion proteins against, respectively, HER2- or CEA-expressing tumors (Corgnac et al., 2013; Stirnemann et al., 2008). In this context, tumor growth inhibition was mainly resulting from the direct cytotoxicity of iNKT and NK cells, although significant T cell expansion was detected. The cellular and molecular scenario is illustrated in Fig. 1.

Practically, the cDNA sequence of the murine β2-microglobuline was fused to the 5′end of the soluble part of CD1d followed by the anti-HER2 scFv, and a 6xHIS tag for purification (Note 1). The β2-microglobuline part is crucial for the folding and stabilization of the CD1d molecule. After the His-tag affinity purification, the CD1d-anti-HER2 fusion protein is loaded with the aGalCer lipid and the ligand excess is eliminated by size-exclusion chromatography. Of note, the anti-HER-2 4D5 scFv retained its ability to bind the HER-2 protein despite being exposed on the C-terminus of the fusion protein and followed by the His-tag (Stirnemann et al., 2008).

In order to evaluate possible additive effects with tumor peptide vaccine, we tested the antitumor efficacy of peptide/CpG-based vaccine combined with CD1d-antitumor protein treatment (Corgnac et al., 2014; Zhang & Donda, 2017).

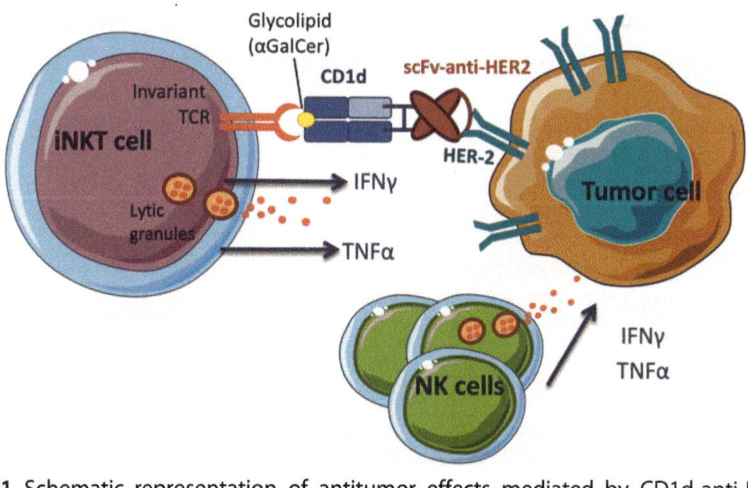

Fig. 1 Schematic representation of antitumor effects mediated by CD1d-anti-HER2 fusion proteins against HER2-expressing tumors.

1.2 Xcl1-synthetic antigen fusion proteins to target cross-presenting DCs

Several DC subsets orchestrate the priming of an adaptive T cell response, among which cross-presenting CD8α^{+} DCs are the most efficient to raise a potent CD8 T cell response in view of their capacity to shunt internalized antigens from the proteasomal compartment to the cytosol where they can be loaded on MHC I molecules (Kurts, Robinson, & Knolle, 2010). Therefore, several strategies have been developed in order to target antigens to specific surface markers expressed by cross-presenting DCs. The most specific receptors for the targeting of mouse and human cross-presenting DCs are CLEC9A and Xcr1, while DEC205 is also expressed on CD8α negative DC subpopulations and on human monocytes (Shrimpton et al., 2009). As an alternative to CLEC9A antigen targeting (Ghinnagow et al., 2017), we opted for the targeting of Xcr1, and chose to use its ligand Xcl1 (Dorner et al., 2002) instead of an antibody as done by others (Hartung et al., 2015). With regard to the peptide antigen, we chose synthetic long peptides (SLP), which contain both MHC I and MHC II epitopes and require internalization and proteasomal processing by professional APCs, unlike the exact length MHC I epitope, that can bind to MHC I molecules expressed by any cell type which may result in tolerance induction (Melief & van der Burg, 2008). The cellular and molecular scenario is depicted in Fig. 2.

Practically, the cDNA sequence of murine Xcl1 was fused to the OVA SLP containing the OT-1 and OT-2 epitopes for, respectively, MHC I and MHC II, followed by the human IgG1 Fc fragment for both stability and purification purposes (Botelho et al., 2019). The whole fusion was cloned in the expression vector pMP-BP for production CHO cells (Note 4).

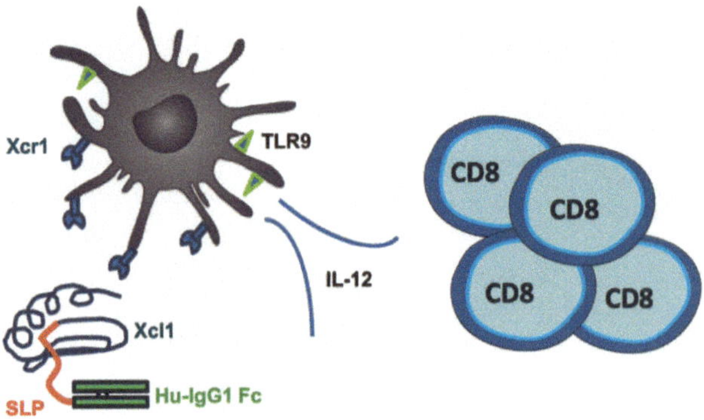

Fig. 2 Schematic representation of Xcr^{+}DC stimulated by the soluble Xcl1-SLP-Fc molecule and TLR9, resulting in CD8 T cell priming.

2. Protocols

2.1 iNKT/CD1d-antitumor immunotherapy combined with peptide/CpG vaccine

The fusion protein of β2M-CD1d-anti-HER2 was cloned in the PEAK8 expression vector (Edge Biosystems), and produced by transient transfection of HEK293-EBNA cells (Thermofisher, Switzerland). The protein production was done at the Protein Production and Structure Core Facility at the EPFL, Switzerland, in mid-scale range (10–30 mg of fusion protein/l of supernatant). Briefly, the human cell line HEK293-EBNA was adapted to serum-free suspension growth in Excell™-293 medium (SAFC Biosciences, Lenexa, KS) in 1-L glass bottles placed on an orbital shaker (Kühner AG, Switzerland). For large scale transfection in suspension cultures, cells were seeded in serum-free RPMI medium (with 25 mM, Hepes, Lonza, Verviers, Belgium) at a density of 2×10^6 cells/mL, and transfected using linear 25-kDa polyethyleneimine, as published (Baldi et al., 2005). Four hours post-transfection, the culture was diluted by adding one volume of Pro-293 s medium (Lonza). After 6 days, the culture was centrifuged, and the supernatant was saved for protein purification.

The expression plasmid is available upon request and if needed, the β2M-CD1d-anti-HER2 can easily be subcloned in other expression plasmids. Cloning, transfection and protein purification is described in Stirnemann et al. (2008), and additional details are provided below.

2.1.1 *β2M-CD1d-anti-HER2 protein purification: General considerations*

A schematic representation is given in Fig. 3.

2.1.2 *Materials and reagents*

Amicon filter 0.2 μm

Filter Sartobind IDA 75 (Sartorius n°85,030–517-46)

Equilibration buffer: Na acetate 0.1 M pH 4.5, 0.5 M NaCl

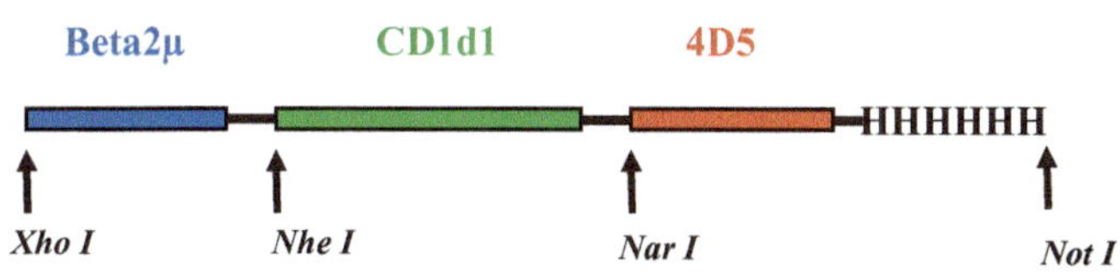

Fig. 3 Scheme of the CD1d-anti-HER2 fusion protein of 694 amino acids (MW 76 kDa); isoelectric point 8.15; abs 0.1% = 1.9 at 280 nm; nucleotide sequence is given in Note 1.

Nickel solution: Nickel sulfate 0.1 M in equilibration buffer
Stripping buffer: Sulfuric acid 1 M
Wash buffer: Tris 10 mM pH 8.0, 250 mM NaCl
Elution buffer: PBS with 250 mM imidazole
Storage buffer: PBS with 20% ethanol
PBS
DMSO
Amicon Ultra-15 centrifugal filter unit ultracel-10 (UFC9010)
AcroPrep 96 filter plates based on Mustang Q anionic exchange from Pall
aGalCer glycolipid (KRN 7000 analog from Alexis ALX-306-027 (www.Axxora.com)
Fast protein liquid chromatography (FPLC)
Column Sephacryl S100 or S200
Peptide of OVA ($OVA_{257–264}$, H-2Kb restricted; amino acid sequence SIINFEKL)
Class B CpG/ODN1826 (Microsynth AG, Balgach, Switzerland).
Vα2 and Vβ5 antibodies (OT-1)
CD3 antibody
aGalCer/CD1d tetramer (made in-house)

2.1.2.1 Mouse strains

C57BL/6J (B6) are obtained from Envigo Laboratories (Gannat, France), and maintained in our animal house. TCR-transgenic OT-I mice are bred in our animal house (Sakuishi et al., 2010). The Vα14Jα18 (T414) transgenic mice were kindly provided by A. Bendelac (University of Chicago, Chicago, USA) (Griewank et al., 2007)

2.1.3 β2M-CD1d-anti-HER2 protein purification: Protocol

The supernatant from the HEK293-EBNA cells is centrifuged and filtered on Amicon 0.2 μm filter bottle to eliminate cells and cell debris. NaCl (250 mM final) and imidazole (15 mM final) is added to the supernatant. We made some modifications to the protocol from Sartorius after discussing with our collaborators from the Protein Production facility from EPFL (https://www.epfl.ch/research/facilities/ptpsp/).

- Filtration on Sartobind IDA 75

Flow rate

Recommended flow rate is 10–30 × membrane volume/min, which corresponds to 10–30 × 2.5 mL. With our pump we reach 22 mL/min which is slightly <10 ×.

Initiation

The cartridge is delivered dried with glycerol.

First wash and equilibrate with the equilibration buffer.

Loading with nickel

Pass 10 membrane volumes, *i.e.* 25 mL, of nickel solution.

Wash away the excess with 25 mL equilibration buffer. Pass through all the supernatant and then leave it on recirculation for 40 min. If up to 200 mL or overnight if 1 L or more.

Wash with 50 mL wash buffer.

Elute with 20 mL elution buffer under recirculation for at least 10 min.

Additional batches of supernatant can be passed without stripping and reloading the nickel. If loading is finished, the cartridge can be washed with equilibration buffer and then Equilibration buffer supplemented with 20% EtOH and kept at 4 °C.

Stripping

After several use, pass 25 mL stripping buffer. Sulfuric acid will remove all the Nickel and Imidazole and allow for a regeneration of the cartridge.

Pass equilibration buffer until pH is back to 4.5 (at least 25 mL). Load again 25 mL nickel solution and proceed as above.

- Yield

Protein concentration of eluates is evaluated by measuring the OD at 280 using the absorption factor of 1.9 for 1 mg/mL defined for the β2M-CD1d-anti-HER2 protein.

Eluates are concentrated down to 1–2 mg/mL on Amicon Ultra-15 centrifugal filter unit with 10 kDa cut off, and dialyzed O/N against 3 L of PBS. Typically, 10–30 mg CD1d-antitumor fusion proteins are obtained from 1 L supernatant.

The purity and absence of degradation products is analyzed on a 1% SDS-PAGE gel.

Aliquots of unloaded fusion proteins can be kept several days at 4C or several months at −80 °C.

- Endotoxin removal

For *in vivo* use, endotoxin removal is performed using anionic exchange on AcroPrep 96 filter plates based on Mustang Q anionic exchange from Pall.

Endotoxin levels are measured by the LAL bioassay (outsourced at AdipoGen Life Sciences, Switzerland), and is <0.5 EU/µg protein.

- Loading with aGalCer or analogs

aGalCer reconstitution

1 mg of aGalcer is solubilized in 4.66 mL PBS 0.5% Tween20 to have a solution of 250 µM. The suspension is warmed up at 65 °C and sonicated, and immediately aliquoted. Aliquots are kept at −20 °C.

The amount aGalCer to be loaded on fusion proteins is calculated according to the following formula:

$$\mathrm{mg(aGalCer)} = \frac{\mathrm{mg(recomb.CD1d)} \times \mathrm{Excess\ aGalCer} \times \mathrm{MW(aGalCer)}}{\mathrm{MW(recomb.CD1d)}}$$

where Excess aGalCer = 3; MW (aGalCer) = 858; MW (sCD1d) = 48,000; MW (sCD1d-anti-HER2) = 76,000; recomb. CD1d = sCD1d or sCD1d-anti-HER2.

Incubate O/N at room temperature (RT).

Next day: eliminate the excess of free aGalCer over FPLC on Sephacryl S100 or S200 depending the sample volume. This step will also allow controlling the monomorphic form of the protein and absence of aggregates and degradation products.

- Peptide/CpG vaccination

The H2-K_b OVA epitope $\mathrm{SIINFEKL}_{257-264}$ is synthetized by the Protein and Peptide synthesis Core facility of the University of Lausanne.

1 mg is solubilized in 20 µL DMSO followed by sequential dilution in PBS until a final concentration of 1 mg/mL. Aliquots are kept at −20 °C.

Class B CpG/ODN1826 is solubilized at 20 mg/mL in DMEM medium and kept in aliquots at −20 °C.

2.1.4 Combined vaccination with β2M-CD1d-anti-HER2 and OVA/CpG vaccine: Considerations

In order to evaluate the potency of this approach, precursors for both iNKT cells and OVA-specific OT-1T cells are co-transferred 1 day before vaccination. iNKT cells are mostly present in liver and spleen (Crowe et al., 2005; Kronenberg, 2005), and more rarely in secondary lymphoid organs. Initially, when OVA-CpG vaccine was administrated s.c. and αGC/CD1d-anti-HER2 fusion by i.v. injection, no synergistic effects were observed regarding either the CD8 T cell priming in the blood or the antitumor effects (unpublished data). These results likely suggest that synergistic effects require interaction between activated iNKT cells and the DCs at the vaccination site. Therefore, we decided to choose the muscle as the site of injection. Most human vaccines demonstrate good efficacy when injected i.m., and a study in mice has characterized conventional and monocyte-derived DCs in the muscle with a strong migratory capacity into the draining lymph node (LN) (Langlet et al., 2012). In contrast to DCs, little is known about the activation of iNKT cells by this route. However, when OVA peptide and αGC/CD1d-anti-HER2 fusion were co-injected i. m., a good expansion and activation of OVA-specific CD8 T cells and NK cells were observed, indicating that their iNKT-mediated transactivation had occurred. We hypothesize that the strong vascularization of the muscle allows an efficient systemic activation of iNKT cells and their cross talk with both muscle and splenic DCs. The subsequent treatments with the CD1d-fusion proteins are given i.v.

2.1.5 Combined vaccination with β2M-CD1d-anti-HER2 and OVA/CpG vaccine: Protocol

- Adoptive cell transfer of OT-1 and Vα14Jα18 iNKT cells

Usually 80–90% T cells of OT-1 mice are H-2Kb/OVA-specific in the spleen, and T414 mice have around 10% iNKT cells in the spleen as compared to 1–2% in wild type C57BL/6 mice (Griewank et al., 2007). Splenocytes from both transgenic mice are labeled with Vα2 and Vβ5 antibodies (OT-1) and CD3+ CD1d tetramer (type I NKT cells) to determine their exact respective frequencies. Based on the results, the equivalent of 10^5 OT-1 and 5×10^5 iNKT cells are transferred in CD45.1 mice grafted on the flank 6 days before with 10^5 B16-OVA-HER2 tumor cells (Corgnac et al., 2014). The transfer of iNKT cells from T414 mice is included to optimize the adjuvant effects of combined peptide/CpG and aGalCer.

- OVA peptide/CpG vaccine is administered intra-muscularly 1 day after the adoptive cell transfer, and the vaccination mix prepared just before use contains 10 μg OVA and 50 μg CpG per mouse, or 40 μg of CD1d-anti-HER2 protein in 100 μL final.
- CD1d-anti-HER2 treatment, immediately after the peptide/CpG vaccination, 40 μg of CD1d-anti-HER2 protein is injected i.v. (Fig. 3).

Groups of mice:

(1) No vaccine, no treatment
(2) No vaccine, i.v. treatment CD1d-anti-HER2
(3) Vaccine i.m. 10 μg OVA+ 50 μg CpG, no treatment
(4) Vaccine i.m. 10 μg OVA+ 50 μg CpG+ 40 μg CD1d-anti-HER2, i.v. treatment CD1d-anti-HER2

2.1.6 CD1d-antitumor fusion proteins: Results and concluding remarks

This approach greatly favors the maturation of pro-inflammatory DCs, as seen by the strong up-regulation of CD40, CD86 and CD70 on $CD8\alpha^+$ DCs, as well as by a high serum concentration of IL-12 (Fig. 4) from Corgnac et al. (2014). In addition to its adjuvant activity, the systemic treatment with the fusion protein allow redirecting the immune response to the tumor site as seen by increased frequency of tumor-specific TILs (Fig. 5) from Corgnac et al. (2014).

2.2 XCL1-SLP-fc fusion protein to target cross-presenting DCs

2.2.1 General considerations

The sequence of murine Xcl1 is fused to a peptide epitope, followed by the human IgG1 Fc part as illustrated below (Fig. 6).

The synthetic long peptides (SLP) contain both MHC I and MHC II epitopes, either contiguous, or overlapping each other, as described below:

Murine OVA SLP

K**ISQAVHAAHAEINEAGR**E**SIINFEKL**TEWT

MHC I H-$2K_b$: SIINFEKL

MHC II I-A_b: ISQAVHAAHAEINEAGR

Murine synthetic TRP1/TRP2 SLP

SGHNCGT**CRPGWRGAACNQKI**LTVR*GSGS***SVYDFFVWL**

MHC I H-$2K_b$: SVYDFFVWL

MHC II I-A_b: CRPGWRGAACNQKI

GSGS: small spacer placed between the two epitopes

Human melan A SLP

GHGHSYTTA**EEAAGIGILTVILGVL**

HLA-A2: EAAGIGILTV

HLA-DQ6: EEAAGIGIL

HLA-DR1: AAGIGILTVILGVL

2.2.2 Materials and reagents

pMP-PB vector expression vector (ExcellGene)
InFusion cloning kit (Clontech-Takara)
Synthetic DNA fragments (GeneArt-Thermofischer)
rProtein A-Sepharose (GE-Healthcare)
Elution buffer: 0.1 M glycine pH 3.0
Neutralization buffer: TRIS 1 M pH 8.0
PBS
Ethanol pure

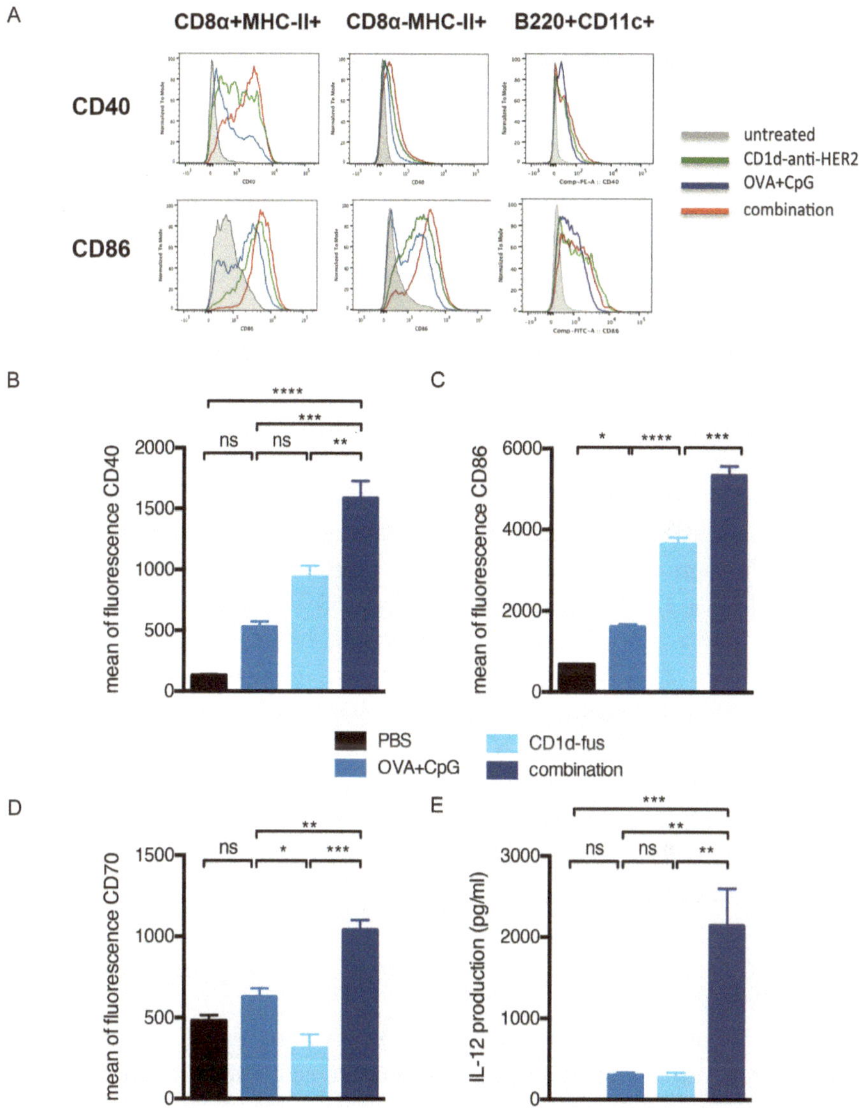

Fig. 4 Optimal maturation of CD8α^+ αDCs after CD1d-mediated activation of iNKT cells combined with CpG-ODN. Mice transferred with Vα14-Jα18 and OT-I splenocytes were immunized i.m. with OVA peptide alone or in combination with CpG or CD1d-fus/CpG followed or not by systemic treatment with the CD1d-antitumor protein. Mice were bled after 6 h and sacrificed 20 h after i.m. immunization. (A) 24 h post immunization, splenic DC subsets were discriminated according to their expression of B220, CD11c, CD11b, MHC II and CD8a, and each subset was analyzed for the expression of CD40 and CD86 maturation markers. Data are shown as histograms overlay of the different treatment groups as indicated. (B) Mean fluorescence intensity of CD40 on gated CD8α^+ DCs.

(Continued)

Murine Xcl1 (Hölzel Diagnostika Handels GmbH, Germany)
Amicon filter 0.2 μm
Filter Sartobind IDA 75 (Sartorius n°85030-517-46)

2.2.2.1 Mouse strains

C57BL/6J (B6) are obtained from Envigo laboratories (Gannat, France), and maintained in our animal house. TCR-transgenic OT-1 mice are bred in our animal house (Sakuishi et al., 2010). BATF3 KO mice were obtained from T. Rolink in Basel.

2.2.3 Protocol

The protocol is described in Botelho et al. (2019), and additional details are given below.

We opted for a human IgG1 Fc fusion protein to facilitate purification on Protein A and also to benefit from the dimerization and stability conferred by the Fc fragment. In order to avoid Fc receptor binding, we took advantage of the Asp to Ala mutation on position 265, which prevents Fc receptor binding (Clynes, Towers, Presta, & Ravetch, 2000).

The Xcl1-SLP-Fc fusion proteins were cloned in the expression vector pMP-PB (Excellgene). Similar to the CD1d-antitumor fusion proteins, Xcl1-SLP-Fc proteins were produced at the Laboratory of Cellular Biotechnology of EPFL, Lausanne, Switzerland, by transient transfection of Chinese Hamster Ovary cells. Transient transfection was preferred since higher yield were obtained as compared to stable clones established on puromycine resistance.

2.2.4 XCL1-SLP-fc fusion protein to target cross-presenting DCs: Protocol

1. Batches of 1 L supernatant from CHO cells were centrifuged and filtered on Amicon 0.2 μm filter bottle to remove cells and cell debris.

Fig. 4—Cont'd (C) Mean fluorescence intensity of CD86 on gated CD8α$^+$ DCs. (D) Mean fluorescence intensity of CD70 on gated CD8α$^+$ DCs. (E) Sera from mice were collected 6 h after the immunization and IL-12p70 was measured using a BD flex set assay. Bar graphs show mean of fluorescence or cytokine level in pg/mL as mean +/α SEM for 3 samples per group. Data are representative of 2 independent experiments. *$P<0.05$; **$P<0.01$; ***$P<0.001$; ****$P<0.0001$. *Reproduced from Corgnac, S., Perret, R., Zhang, L., et al. (2014). iNKT/CD1d-antitumor immunotherapy significantly increases the efficacy of therapeutic CpG/peptide-based cancer vaccine.* J. Immunotherapy Cancer *2, 39. https://doi.org/10.1186/s40425-014-0039-8.*

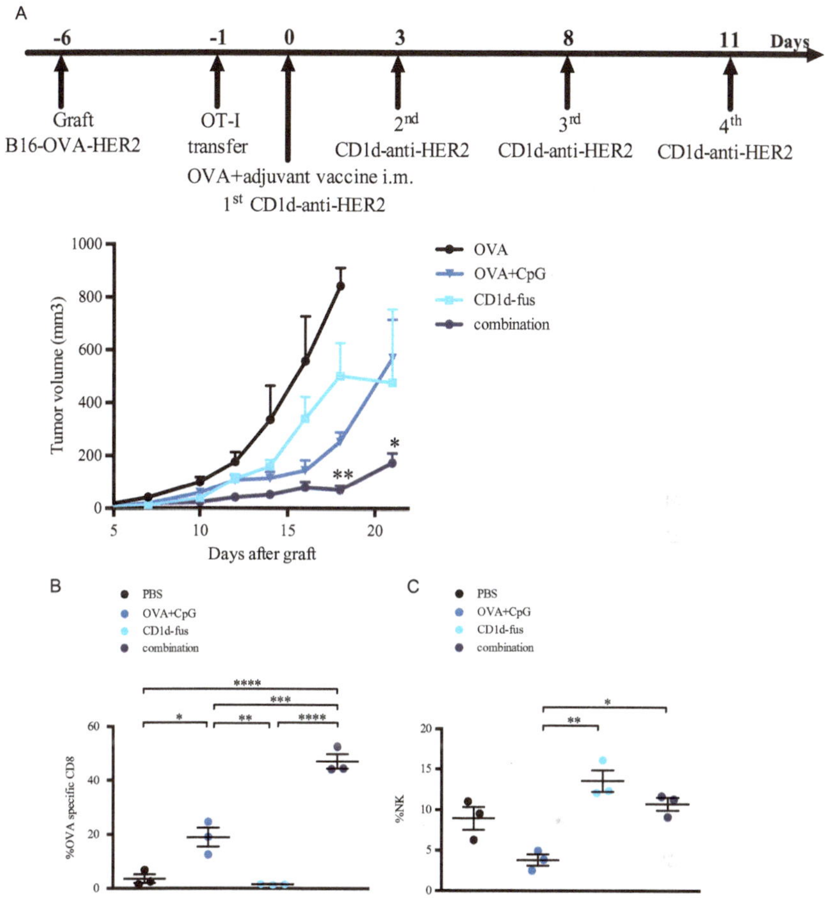

Fig. 5 Combined OVA/CpG vaccine and CD1d-antitumor fusion lead to enhanced anti-tumor effects correlating with accumulation of OVA-specific CTLs at the tumor site. Recipient CD45.1 mice were grafted s.c. with B16-OVA-HER2 tumor cells following Vα14-Jα18 and OT-I transfer and 5 days later, they were immunized and treated as described in Fig. 1 and in the methods. (A) Kinetic of tumor growth shown as mean tumor volume (mm^3) ± SEM of 5 mice/group. (B and C) Mice were sacrificed 7 days after the immunization and (B), frequency of H-2K^b/OVA-specific CD8 T cells among total tumor infiltrating CD8 T lymphocytes and (C), frequency of NK cells among total tumor lymphocytes were measured. Bar graphs show frequencies as mean ± SEM of groups of three mice. Data are representative of three independent experiments. *$P<0.05$; **$P<0.01$; ***$P<0.001$; ****$P<0.0001$. *Reproduced from Corgnac, S., Perret, R., Zhang, L., et al. (2014). iNKT/CD1d-antitumor immunotherapy significantly increases the efficacy of therapeutic CpG/peptide-based cancer vaccine.* Journal of Immunotherapy Cancer, 2, *39. https://doi.org/10.1186/s40425-014-0039-8.*

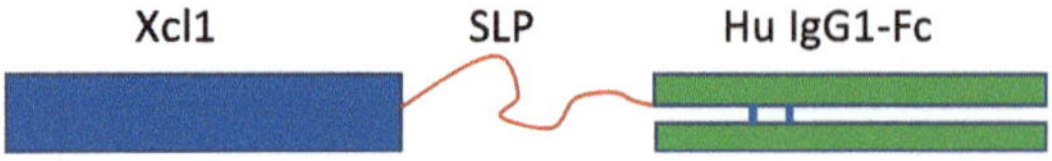

Fig. 6 Scheme of the fusion molecule made of murine Xcl1, a tumor-associated SLP and a human IgG1Fc part.

2. Filtered supernatant was purified using rProtein A-Sepharose Fast Flow from GE-Healthcare. Instead of batchwise purification, we opted for chromatographic purification on small disposable columns filled with 5 mL of Protein A-Sepharose. The supernatant is purified by recircularization using a peristaltic pump. Alternatively, Hi-Trap Protein A HP columns can be used (GE-Healthcare).
3. After 2–3 passages, elution is done with 10 mL of 0.1 M glycine pH 3.0. The eluate is immediately buffered by adding 1 mL of TRIS 1 M pH 8.0 in the elution tube.
4. The eluate is dialyzed O/N against PBS, proteins concentrations are measured.
5. Aliquots of flow through and dialyzed eluate are analyzed on SDS-PAGE to verify that all fusion protein was captured. In case, flow through is passed once again.
6. For further use, the column is washed sequentially with 0.1 M glycine and 20% Ethanol in PBS in which it can be kept at 4 °C.

- Yield

  In contrast to CD1d-antitumor fusion proteins, the production of Xcl1-SLP-Fc proteins was less efficient and ranged from 7 to 13 mg/L supernatant. Transient transfection of HEK293-EBNA cells or other expression systems such as bacteria and insect cells, did not further improve protein production.

  The purity and absence of degradation products is analyzed on a 1% SDS-PAGE.

  Aliquots can be kept several days at 4 °C or several months at −80 °C.

- Endotoxin contamination

For *in vivo* use, endotoxin removal is performed using anionic exchange on AcroPrep 96 filter plates based on Mustang Q anionic exchange from Pall. Endotoxin level is measured by the LAL bioassay (Adipogen), and is <0.5 EU/μg protein.

2.2.5 Xcl1-SLP-fc vaccination: Considerations

The Xcl1-SLP based vaccine is delivered intradermally in the footpad with the aim on one hand to target $CD103^+$ dermal DCs and on the other hand to redirect the fusion protein to the inguinal lymph node to target resident $CD8\alpha^+Xcr1^+$ DCs. CpG B ODN1826 is given to all mice in order to deliver a pro-inflammatory stress signal.

2.2.6 Xcl1-SLP-fc vaccination: Protocol and results

In order to evaluate the delivery of the SLP antigen to cross-presenting DCs, vaccination of Xcl1-SLP-Fc and Xcl1-Fc+SLP are compared. In that case, equimolar amounts of Xcl1-SLP-Fc and Xcl1-Fc+SLP are administered, based on a 20 μg dose of Xcl1-SLP-Fc protein. As seen in Fig. 7 from Botelho et al. (2019), *ex vivo* analysis of DC subpopulations demonstrated a preferential targeting of Alexa 488-labeled Xcl1-SLP-Fc protein to $CD8^{\alpha+}$ and $CD103^+$ cross-presenting DCs. This observation was further consolidated by a significantly decreased fluorescence in DCs from Batf3 KO mice, which are deprived in Xcr1+ DCs (Hildner et al., 2008).

Next, the Xcl1 fusion proteins were tested in therapeutic settings against the OVA-expressing EL-4 lymphoma model (EG7), and the more stringent B16-OVA melanoma model. On day 7, when tumors were established and measurable, mice received an intradermal vaccination with the XCL1 fusion proteins including or not the SLP or with free OVA SLP +/−XCL1. Except for the untreated group, all mice received 50 μg of CpG-ODN. As illustrated below in the B16-OVA model, even in the absence of a transfer of OT-1 precursors, around 20% of OVA-specific T cells were obtained on day 7 post immunization (Fig. 7B) from (Botelho et al., 2019). Although there was no difference in the frequencies of OVA-specific T cells between the Xcl1-OVA-SLP-Fc and the Xcl1-Fc+OVA SLP, the tumor growth delay of B16-OVA tumors was slightly improved with the fusion protein that includes the SLP.

2.2.7 Xcl1-SLP-fc vaccination: Concluding remarks

Using fluorescently-labeled Xcl1-SLP-Fc fusion proteins, we could demonstrate the preferential antigen targeting to cross-presenting DCs (Fig. 7) from Botelho et al. (2019), which results in a more potent CD8 T cell priming and tumor inhibition (Fig. 8) from Botelho et al. (2019). Extending this approach to additional tumor epitopes and newly defined neoantigens is highly promising.

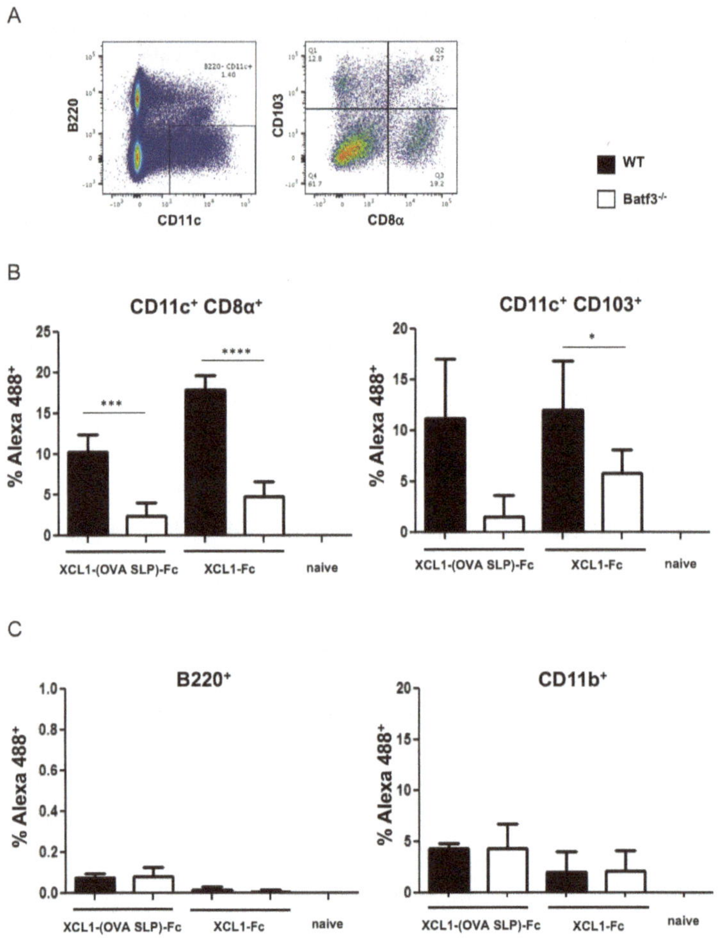

Fig. 7 *In vivo* uptake of Xcl1-(OVA SLP)-Fc in skin draining LN. WT and Batf3 KO mice were injected intradermally in the footpad with a mix of 50 μg of CpG and 6 μg of Alexa 488-labeled XCL1-(OVA SLP)-Fc or XCL1-Fc fusion proteins. Inguinal LNs were harvested 16 h post injection and the uptake of labeled fusion proteins was measured in different populations of APCs, (A) gating strategy for identifying CD103+ and CD8α subtypes in $CD11c^{+}B220^{neg}$DCs isolated from inguinal LNs. (B) Uptake of labeled XCL1-fusion proteins by CD8α DCs (left), CD103+ DCs (right), and (C) B220+ B cells (left) and CD11b macrophages (right). Data are shown as mean ± SEM ($n=3$–4 mice/group). Results are representative of two independent experiments. *Reproduced from Botelho, N. K., Tschumi, B. O., Hubbell, J. A., Swartz, M. A., Donda, A., & Romero, P. (2019). Combination of synthetic long peptides and XCL1 fusion proteins results in superior tumor control.* Frontiers in Immunology, 10, *294*.

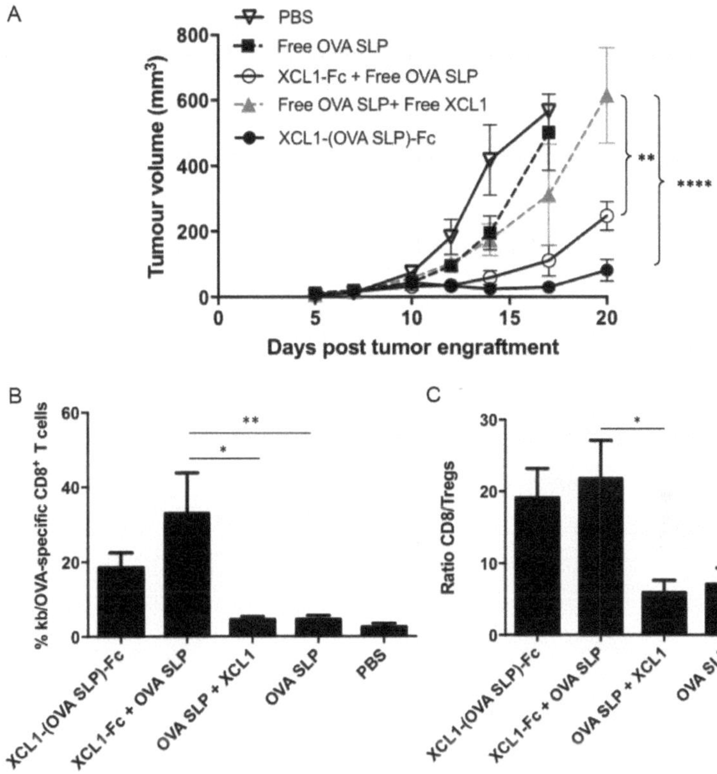

Fig. 8 Effective therapeutic vaccinations with XCL1-(OVA SLP) fusion proteins even in the absence of OT-1T cell transfer. (A) B16.OVA tumor growth of C57BL/6 mice engrafted subcutaneously on the left flank with of 2×10^5 B16.OVA cells followed by intradermal vaccination on day 3. Data is shown as mean ± SEM ($n = 5$–6 mice/group). Results are representative of three independent experiments. (B) Frequencies of SIINFEKL-specific $CD8^+$ T cells present in the B16.OVA tumors harvested 10 days post vaccination. (C) Ratio of SIINFEKL-specific $CD8^+$ T cells *versus* Tregs within B16.OVA tumors. Data is shown as mean ± SEM ($n = 5$–6 mice/group). Results are representative of two independent experiments. *Reproduced from Botelho, N. K., Tschumi, B. O., Hubbell, J. A., Swartz, M. A., Donda, A., & Romero, P. (2019). Combination of synthetic long peptides and XCL1 fusion proteins results in superior tumor control.* Frontiers in Immunology, 10, *294.*

3. Concluding remarks

The use of either CD1d-antitumor or Xcl1-SLP-Fc fusion proteins in therapeutic cancer vaccination allows to preferentially promote the maturation of pro-inflammatory DCs. More specifically, the first option is exploiting the adjuvant activity of iNKT cells on the priming of the adaptive response, while also redirecting to the tumor site the cytotoxic activity of activated iNKT and NK cells. The second approach involving the Xcl1-OVA SLP-Fc fusion protein is specifically targeting the antigen to cross-presenting DCs, which results in a potent expansion of tumor-specific effector CD8 T cells.

4. Notes

Note 1. Amino acids and nucleotide sequences of CD1d-anti-HER2 fusion protein.

Protein Sequence

M A R S V T L V F L V L V S L T G L Y A I Q K T P Q I Q V Y S R H P P E N G K P N I
L N C Y V T Q F H P P H I E I Q M L K N G K K I P K V E M S D M S F S K D W S F Y I
L A H T E F T P T E T D T Y A C R V K H A S M A E P K T V Y W D R D M G G G G S
G G S G S G G G A S Q Q K N Y T F R C L Q M S S F A N R S W S R T D S V V W L G
D L Q T H R W S N D S A T I S F T K P W S Q G K L S N Q Q W E K L Q H M F Q V Y
R V S F T R D I Q E L V K M M S P K E D Y P I E I Q L S A G C E M Y P G N A S E S F
L H V A F Q G K Y V V R F W G T S W Q T V P G A P S W L D L P I K V L N A D Q G
T S A T V Q M L L N D T C P L F V R G L L E A G K S D L E K Q E K P V A W L S S V
P S S A D G H R Q L V C H V S G F Y P K P V W V M C M R G D Q E Q Q G T H R G D
F L P N A D E T W Y L Q A T L D V E A G E E A G L A C R V K H S S L G G Q D I I L
Y W D A R Q A P V G G G G G S G G S G S G G G G A D Y K D I V M T Q S P S S L S
A S V G D R V T I T C R A S Q D V N T A V A W Y Q Q K P G K A P K L L I Y S A S F
L Y S G V P S R F S G S R S G T D F T L T I S S L Q P E D F A T Y Y C Q Q H Y T T P P
T F G Q G T K V E L K R A T P S H N S H Q V P S A G G P T A N S G E V K L V E S G
G G L V Q P G G S L R L S C A T S G F N I K D T Y I H W V R Q A P G K G L E W V A
R I Y P T N G Y T R Y A D S V K G R F T I S A D T S K N T A Y L Q M N S L R A E D
T A V Y Y C S R W G G D G F Y A M D Y W G Q G T T V T V S S G S G G H H H H H
H Stop

Blue: β2m; Red: Gly-Ser rich linker; Green: CD1d; Orange: 4D5 s cFv anti-HER2; yellow: His-Tag

Nucleotide sequence

ATGGCTCGCTCGGTGACCCTGGTCTTTCTGGTGCTTGTCTCACTGACCGGCCTGT
ATGCTATCCAGAAAACCCCTCAAATTCAAGTATACTCACGCCACCCACCGGAGA
ATGGGAAGCCGAACATACTGAACTGCTACGTAACACAGTTCCACCCGCCTCACA
TTGAAATCCAAATGCTGAAGAACGGGAAAAAAATTCCTAAAGTAGAGATGTCAG
ATATGTCCTTCAGCAAGGACTGGTCTTTCTATATCCTGGCTCACACTGAATTCACC
CCCACTGAGACTGATACATACGCCTGCAGAGTTAAGCATGCCAGTATGGCCGAG
CCCAAGACCGTCTACTGGGATCGAGACATGGGTGGAGGTGGATCAGGTGGATCC
GGTTCTGGAGGTGGA*GCTAGC*CAGCAAAAGAATTACACCTTCCGCTGCCTGCAGA
TGTCTTCCTTTGCAAACAGAAGCTGGTCCCGCACAGACAGCGTGGTCTGGCTGGG
GGATCTGCAGACTCACCGTTGGAGTAATGACTCAGCCACCATCAGCTTCACGAA
GCCATGGTCCCAGGGCAAGTTGAGTAACCAGCAGTGGGAGAAGTTGCAGCATAT
GTTTCAAGTCTATCGAGTCAGCTTTACCAGGGACATACAGGAATTAGTCAAAATG
ATGTCACCTAAAGAAGACTATCCCATTGAGATCCAGCTGTCTGCGGGCTGTGAAA
TGTACCCTGGGAATGCTTCGGAAAGCTTTTTACACGTAGCATTTCAAGGAAAATA
TGTCGTGAGATTCTGGGGAACGTCCTGGCAGACAGTCCCAGGGGCCCCATCTTGG
TTAGACTTGCCCATCAAAGTGCTCAACGCTGATCAAGGGACAAGTGCAACCGTG
CAGATGCTCCTGAATGACACCTGCCCCCTATTTGTCCGTGGTCTCCTAGAGGCAG
GGAAGTCAGACCTAGAGAAGCAAGAGAAGCCAGTGGCCTGGTTGTCCAGTGTCC
CCAGCTCTGCAGATGGCCATAGGCAGCTGGTGTGTCATGTCTCTGGCTTCTACCC
AAAACCTGTGTGGGTGATGTGCATGCGGGGTGACCAGGAGCAACAGGGTACTCA
CAGAGGTGATTTCCTGCCAAATGCTGATGAGACATGGTATCTTCAAGCAACCCTG
GATGTGGAGGCTGGAGAGGAAGCTGGCCTGGCCTGCAGGGTGAAGCACAGCAG
CCTAGGAGGACAGGATATCATCCTCTACTGGGATGCCAGGCAAGCACCCGTGGG
CGGTGGAGGTGGATCAGGTGGATCCGGTTCTGGAGGTGGA*GGCGCC*GACTACAA
AGATATCGTTATGACCCAGTCTCCGAGCTCCCTGTCCGCCTCTGTGGGCGATAGG
GTCACCATCACCTGCCGTGCCAGTCAGGATGTGAATACTGCTGTAGCCTGGTATC
AACAGAAACCAGGAAAAGCTCCGAAACTACTGATTTACTCGGCATCCTTCCTCTA
CTCTGGAGTCCCTTCTCGCTTCTCTGGATCCAGATCTGGGACGGATTTCACTCTGA
CCATCAGCAGTCTGCAGCCGGAAGACTTCGCAACTTATTACTGTCAGCAACATTA
TACTACTCCTCCCACGTTCGGACAGGGCACCAAAGTGGAGCTTAAGCGCGCTACC
CCGTCTCACAACTCCCACCAGGTTCCATCCGCAGGCGGTCCGACTGCTAACTCTG
GAGAAGTTAAACTGGTCGAGTCTGGTGGTGGTCTGGTACAGCCGGGTGGATCCC
TGCGTCTGTCTTGCGCTACCTCAGGCTTCAACATTAAAGACACCTATATACACTG
GGTGCGTCAGGCCCCGGGTAAGGGCCTGGAATGGGTTGCAAGGATTTATCCTAC
GAATGGTTATACTAGATATGCCGATAGCGTCAAGGGCCGTTTCACTATAAGCGCA
GACACATCCAAAAACACAGCCTACCTGCAGATGAACAGCCTGCGTGCTGAGGAC
ACTGCCGTCTATTATTGTTCTAGATGGGGAGGGGACGGCTTCTATGCTATGGACT
ACTGGGGTCAAGGTACCACCGTTACCGTTTCTTCTGGTTCTGGAGGACATCATCA
TCACCACCAT**TAA**

Same color code as for the protein sequence. Italic black: restriction sites used for cloning.

Note 2. Map of the PEAK8-β2M-CD1d-anti-HER2 vector.

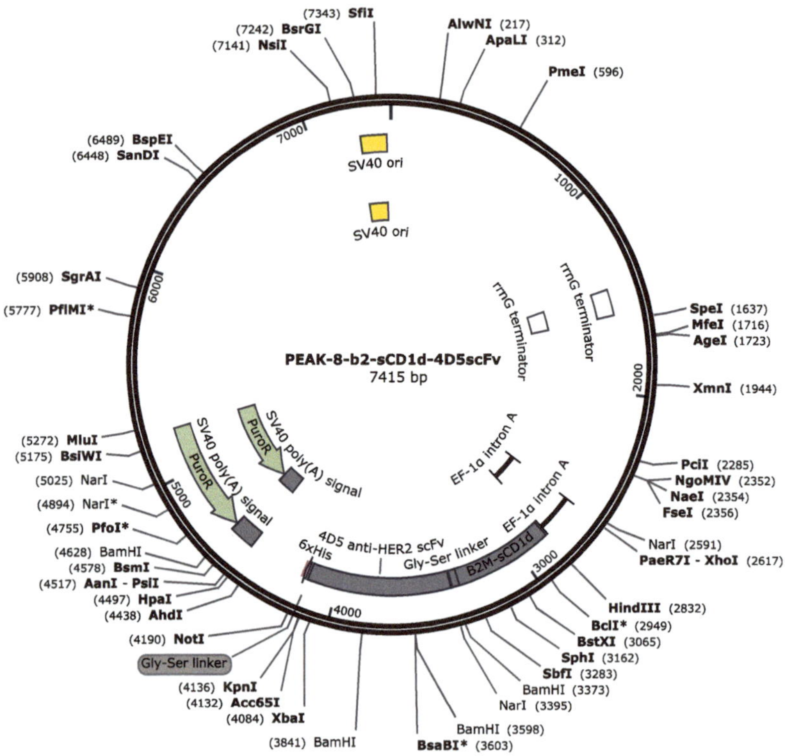

Note 3. Amino acids and nucleotide sequences of Xcl1-OVA SLP-Fc.

Protein sequence

MARLLLLTFLGVCCLTPWVVEGVGTEVLEESSCVNLQTQRLPVQKIKTYIIWEGAMR
AVIFVTKRGLKICADPEAKWVKAAIKTVDGRASTRKNMAETVPTGAQRSTSTAITLT
GGGGGSGGSGSGGG KISQAVHAAHAEINEAGRE **SIINFEKL** TEWT GSGG VPRDCGC
KPCICTVPEVSSVFIFPPKPKDVLMITLTPKVTCVVVAISKDDPEVQFSWFVDDVEVH
TAQTQPREEQFNSTFRSVSELPIMHQDWLNGKEFKCRVNSAAFPAPIEKTISKTKGRP
KAPQVYTIPPPKEQMAKDKVSLTCMITDFFPEDITVEWQWNGQPAENYKNTQPIMN
TNGSYFVYSKLNVQKSNW EAGNTFTCSVLHEGLHNHHTEKSLSHSPGK

Blue: murine Xcl1, Red: SLP, Green: hu IgG1 Fc, black: spacers.

Nucleotide sequence

ATGGCCAGACTTCTCCTCCTGACTTTCCTGGGAGTCTGCTGCCTCACCCCATGGG
TTGTGGAAGGTGTGGGGACTGAAGTCCTAGAAGAGAGTAGCTGTGTGAACTTAC
AAACCCAGCGGCTGCCAGTTCAAAAAATCAAGACCTATATCATCTGGGAGGGGG
CCATGAGAGCTGTAATTTTTGTCACCAAACGAGGACTAAAAATTTGTGCTGATCC
AGAAGCCAAATGGGTGAAAGCAGCGATCAAGACTGTGGATGGCAGGGCCAGTA
CCAGAAAGAACATGGCTGAAACTGTTCCCACAGGAGCCCAGAGGTCCACCAGCA
CAGCGATAACCCTGACT GGGGGTGGAGGTGGATCAGGTGGATCCGGTTCTGGAG
GTGGA AAGATCAGCCAGGCCGTGCACGCCGCCCACGCCGAGATCAACGAGGCCG
GCAGAGAGAGCATCATCAACTTCGAGAAGCTGACCGAGTGGACC GGATCAGGTG
GAGTGCCTCGGGACTGCGGCTGCAAGCCTTGCATCTGCACCGTGCCTGAGGTGTC
CTCCGTGTTCATCTTCCCTCCTAAGCCTAAGGACGTGCTGATGATCACCCTGACC
CCTAAAGTGACCTGCGTGGTGGTGGCCATCTCCAAGGACGACCCTGAGGTGCAG
TTCTCTTGGTTTGTGGACGACGTGGAGGTGCACACCGCTCAGACCCAGCCTCGGG
AGGAACAGTTCAACTCCACCTTCAGATCCGTGTCCGAGCTGCCTATCATGCACCA
GGACTGGCTGAACGGCAAAGAGTTCAAGTG CAGAGTCAACTCTGCCGCCTTCCC
AGCCCCTATCGAAAAGACCATCTCCAAGACCAAGGGCAGGCCTAAGGCCCCTCA
GGTGTACACCATCCCTCCTCCTAAAGAACAGATGGCCAAGGACAAGGTGTCCCT
GACCTGCATGATCACCGATTTCTTCCCTGAGGACATCACCGTGGAGTGGCAGTGG
AACGGCCAGCCTGCCGAGAACTACAAGAACACCCAGCCTATCATGAACACCAAC
GGCTCCTACTTCGTGTACTCCAAGCTGAACGTGCAGAAGTCCAACTGGGAGGCC
GGCAACACCTTCACCTGTTCTGTGCTGCACGAGGGCCTGCACAACCACCACACCG
AGAAGTCCCTGTCCCACTCCCCTGGCAAGTGA **TAA**

Same color code as for the protein sequence.

Note 4. Map of the pMP-PB-Xcl1-OVA SLP-Fc vector.

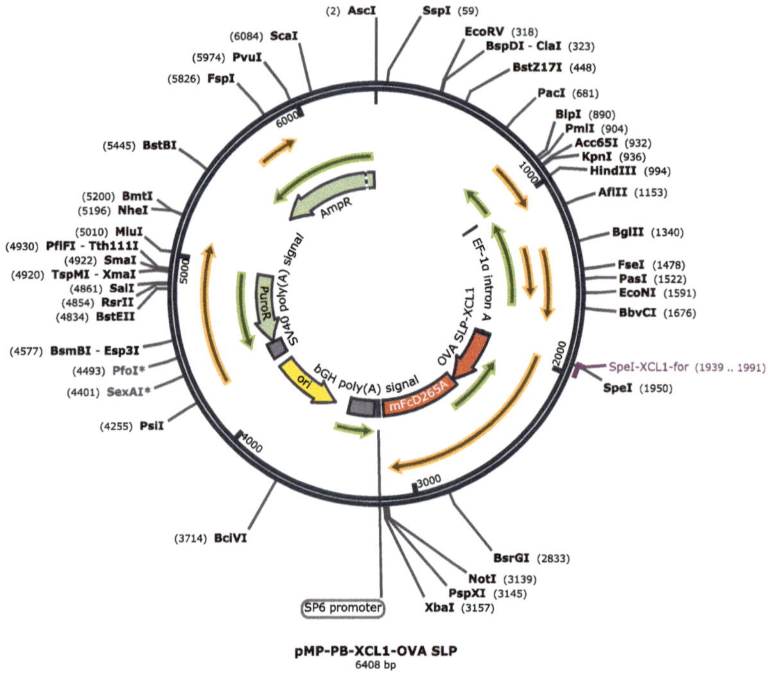

References

Baldi, L., Muller, N., Picasso, S., Jacquet, R., Girard, P., Thanh, H. P., et al. (2005). Transient gene expression in suspension HEK-293 cells: Application to large-scale protein production. *Biotechnology Progress*, *21*, 148–153.

Bendelac, A., Rivera, M. N., Park, S. H., & Roark, J. H. (1997). Mouse CD1-specific NK1 T cells: Development, specificity, and function. *Annual Review of Immunology*, *15*, 535–562.

Bendelac, A., Savage, P. B., & Teyton, L. (2007). The biology of NKT cells. *Annual Review of Immunology*, *25*, 297–336.

Bezu, L., Kepp, O., Cerrato, G., Pol, J., Fucikova, J., Spisek, R., et al. (2018). Trial watch: Peptide-based vaccines in anticancer therapy. *Oncoimmunology*, *7*, e1511506.

Botelho, N. K., Tschumi, B. O., Hubbell, J. A., Swartz, M. A., Donda, A., & Romero, P. (2019). Combination of synthetic long peptides and XCL1 fusion proteins results in superior tumor control. *Frontiers in Immunology*, *10*, 294.

Cerundolo, V., Silk, J. D., Masri, S. H., & Salio, M. (2009). Harnessing invariant NKT cells in vaccination strategies. *Nature Reviews Immunology*, *9*, 28–38.

Clynes, R. A., Towers, T. L., Presta, L. G., & Ravetch, J. V. (2000). Inhibitory Fc receptors modulate in vivo cytotoxicity against tumor targets. *Nature Medicine*, *6*, 443–446.

Corgnac, S., Perret, R., Derre, L., Zhang, L., Stirnemann, K., Zauderer, M., et al. (2013). CD1d-antibody fusion proteins target iNKT cells to the tumor and trigger long-term therapeutic responses. *Cancer Immunology, Immunotherapy*, *62*, 747–760.

Corgnac, S., Perret, R., Zhang, L., Mach, J. P., Romero, P., & Donda, A. (2014). iNKT/CD1d-antitumor immunotherapy significantly increases the efficacy of therapeutic CpG/peptide-based cancer vaccine. *Journal for Immunotherapy of Cancer*, *2*, 39.

Crowe, N. Y., Coquet, J. M., Berzins, S. P., Kyparissoudis, K., Keating, R., Pellicci, D. G., et al. (2005). Differential antitumor immunity mediated by NKT cell subsets in vivo. *The Journal of Experimental Medicine*, *202*, 1279–1288.

Dorner, B. G., Scheffold, A., Rolph, M. S., Huser, M. B., Kaufmann, S. H., Radbruch, A., et al. (2002). MIP-1alpha, MIP-1beta, RANTES, and ATAC/lymphotactin function together with IFN-gamma as type 1 cytokines. *Proceedings of the National Academy of Sciences of the United States of America*, *99*, 6181–6186.

Exley, M., Garcia, J., Balk, S. P., & Porcelli, S. (1997). Requirements for CD1d recognition by human invariant Valpha24+ CD4-CD8-T cells. *The Journal of Experimental Medicine*, *186*, 109–120.

Ghinnagow, R., De Meester, J., Cruz, L. J., Aspord, C., Corgnac, S., Macho-Fernandez, E., et al. (2017). Co-delivery of the NKT agonist alpha-galactosylceramide and tumor antigens to cross-priming dendritic cells breaks tolerance to self-antigens and promotes antitumor responses. *Oncoimmunology*, *6*, e1339855.

Griewank, K., Borowski, C., Rietdijk, S., Wang, N., Julien, A., Wei, D. G., et al. (2007). Homotypic interactions mediated by Slamf1 and Slamf6 receptors control NKT cell lineage development. *Immunity*, *27*, 751–762.

Hartung, E., Becker, M., Bachem, A., Reeg, N., Jakel, A., Hutloff, A., et al. (2015). Induction of potent CD8 T cell cytotoxicity by specific targeting of antigen to cross-presenting dendritic cells in vivo via murine or human XCR1. *Journal of Immunology*, *194*, 1069–1079.

Hildner, K., Edelson, B. T., Purtha, W. E., Diamond, M., Matsushita, H., Kohyama, M., et al. (2008). Batf3 deficiency reveals a critical role for CD8alpha+ dendritic cells in cytotoxic T cell immunity. *Science*, *322*, 1097–1100.

Klebanoff, C. A., Acquavella, N., Yu, Z., & Restifo, N. P. (2011). Therapeutic cancer vaccines: Are we there yet? *Immunological Reviews*, *239*, 27–44.

Kronenberg, M. (2005). Toward an understanding of NKT cell biology: Progress and paradoxes. *Annual Review of Immunology*, *23*, 877–900.

Kurts, C., Robinson, B. W., & Knolle, P. A. (2010). Cross-priming in health and disease. *Nature Reviews. Immunology*, *10*, 403–414.

Langlet, C., Tamoutounour, S., Henri, S., Luche, H., Ardouin, L., Gregoire, C., et al. (2012). CD64 expression distinguishes monocyte-derived and conventional dendritic cells and reveals their distinct role during intramuscular immunization. *Journal of Immunology*, *188*, 1751–1760.

Melief, C. J., & van der Burg, S. H. (2008). Immunotherapy of established (pre)malignant disease by synthetic long peptide vaccines. *Nature Reviews. Cancer*, *8*, 351–360.

Perret, R., Sierro, S. R., Botelho, N. K., Corgnac, S., Donda, A., & Romero, P. (2013). Adjuvants that improve the ratio of antigen-specific effector to regulatory T cells enhance tumor immunity. *Cancer Research*, *73*, 6597–6608.

Sakuishi, K., Apetoh, L., Sullivan, J. M., Blazar, B. R., Kuchroo, V. K., & Anderson, A. C. (2010). Targeting Tim-3 and PD-1 pathways to reverse T cell exhaustion and restore anti-tumor immunity. *The Journal of Experimental Medicine*, *207*, 2187–2194.

Shrimpton, R. E., Butler, M., Morel, A. S., Eren, E., Hue, S. S., & Ritter, M. A. (2009). CD205 (DEC-205): A recognition receptor for apoptotic and necrotic self. *Molecular Immunology*, *46*, 1229–1239.

Stirnemann, K., Romero, J. F., Baldi, L., Robert, B., Cesson, V., Besra, G. S., et al. (2008). Sustained activation and tumor targeting of NKT cells using a CD1d-anti-HER2-scFv fusion protein induce antitumor effects in mice. *The Journal of Clinical Investigation*, *118*, 994–1005.

Vasan, S., & Tsuji, M. (2010). A double-edged sword: The role of NKT cells in malaria and HIV infection and immunity. *Seminars in Immunology*, *22*, 87–96.

Zhang, L., & Donda, A. (2017). Alpha-Galactosylceramide/CD1d-antibody fusion proteins redirect invariant natural killer T cell immunity to solid tumors and promote prolonged therapeutic responses. *Frontiers in Immunology*, *8*, 1417.

FSC
www.fsc.org
MIX
Papier | Fördert gute Waldnutzung
FSC® C083411

Kolibri 360 GmbH
Ferdinand-Jühlke-Straße 7
99095 Erfurt, Deutschland
produktsicherheit@kolibri360.de